Green Development

The concept of sustainability lies at the core of the challenge of environment and development, and the way governments, business and environmental groups respond to it. *Green Development* provides a clear and coherent analysis of sustainable development in both theory and practice.

The third edition retains the structure of previous editions, but has been updated to reflect advances in ideas and changes in international policy. Greater attention has been given to political ecology, environmental risk and the environmental impacts of development.

This fully revised edition discusses:

- the origins of thinking about sustainability and sustainable development, and its evolution to the present day;
- the ideas that dominate mainstream sustainable development (Ecological Modernization, Market Environmentalism and Environmental Economics);
- the nature and diversity of alternative ideas about sustainability that challenge 'business as usual' thinking (for example eco-socialism, eco-feminism, Deep Ecology and political ecology);
- the dilemmas of sustainability in the context of dryland degradation, deforestation, biodiversity conservation, dam construction, and urban and industrial development
- the nature of policy choices about the environment and development strategies, and between reformist and radical responses to the contemporary global dilemmas.

Green Development offers clear insights into the challenges of environmental sustainability, and social and economic development. It is unique in offering a synthesis of theoretical ideas on sustainability and in its coverage of the extensive literature on environment and development around the world. The book has proved its value to generations of students as an authoritative, thought-provoking and readable guide to the field of sustainable development.

Professor Bill Adams has worked for over thirty years on the problems of conservation and development in Sub-Saharan Africa. He is Moran Professor of Conservation and Development in the Department of Geography at Cambridge. He was awarded the Busk Medal by the Royal Geographical Society with the Institute of British Geographers in 2004. His book *Against Extinction: the story of conservation* was published in 2004 (Earthscan, London).

Green Development
3rd edition

Environment and sustainability in a
developing world

W. M. Adams

Routledge
Taylor & Francis Group

LONDON AND NEW YORK

First edition published 1990

Second edition published 2001

Third edition published 2009
by Routledge
2 Park Square, Milton Park, Abingdon, Oxon

Simultaneously published in the USA and
Canada by Routledge
270 Madison Avenue, New York, NY 10016

Routledge is an imprint of the Taylor & Francis Group, an informa business

© 1990, 2001, 2009 W. M. Adams

Typeset in Galliard by
Bookcraft Ltd, Stroud, Gloucestershire
Printed and bound in Great Britain by
CPI Antony Rowe, Chippenham, Wiltshire

British Library Cataloguing in Publication Data
A catalogue record for this book is available from the British Library

Library of Congress Cataloging in Publication Data
Adams, W. M. (William Mark), 1955–
Green development : environment and sustainability in a developing world /
W.M. Adams. – 3rd ed.
   p. cm.
   Includes bibliographical references and index.
   1. Sustainable development—Developing countries. 2. Green movement—
   Developing countries. I. Title.
HC59.7.A714 2008
338.9'27091724–dc22   2008002208

ISBN13: 978-0-415-39508-3 (pbk)
ISBN13: 978-0-415-39507-6 (hbk)
ISBN13: 978-0-203-92971-1 (ebk)

ISBN10: 0-415-39508-9 (pbk)
ISBN10: 0-415-39507-0 (hbk)
ISBN10: 0-203-92971-3 (ebk)

In memory of my father,
H. C. Adams, 1917–1987

Contents

Plates

Figures

Tables

Preface

The first edition of this book was conceived in 1984 and written in the second half of the 1980s. Academically, as in so many other ways, that now seems a remote world, before email, the Internet or mobile phones. Since that first edition, the world's human population has grown by about 1.5 billion people, the axes of geopolitics have shifted, and the world economy has grown and globalized on a rising tide of consumption. Climate change has moved from a niche academic debate to a looming catastrophe.

In retrospect, the question of environment and development in the 1980s seems like a sleepy backwater, navigable reasonably simply within a single volume. Since then it has become a swollen and turbulent torrent with a vast academic and popular literature. Massive changes have taken place in the ways in which those active in development view the environment, and the ways in which environmentalists understand development. Academic writing about environment and development has boomed, with different social-science disciplines engaging theoretically and with a rich empirical literature drawing on field research. Internationally, the United Nations Conferences at Rio de Janeiro (1992) and Johannesburg (2002) transformed the terms within which debates about environment and development were held, and gave them significant political prominence.

Since the 1980s, there has been an astonishing growth in apparently 'green' ideas and statements of intent from development agencies, up to and including the World Bank, governments and businesses. In the early 1990s, what had been a minority concern about the course and costs of 'development' suddenly became conventional wisdom. Environmentalist radicalism became mainstream environmental planning. Environmentalists came in from the cold to talk at boardroom tables; indeed, the environment became a significant growth industry in itself, a vital part of the corporate portfolio of the eager executive in rapidly globalizing companies.

For an author, this explosion of activity is in a sense a delight – but it has to be admitted that it also presents problems. When *Green Development* was conceived, the debates it explored were not only marginal to development planning, but marginal academically. Three decades on, in a new century, nothing could be more different. In the 1980s it slowly became accepted by those working in development that the environment was important: through the 1990s, it became fashionable among social scientists as a whole, the centre of vibrant debates and a fertile sparring ground for new talent and ideas.

Academics are as susceptible to the allure of a bull market as any other group of entrepreneurs, and government policy concern has fed rapidly through into funding opportunities and in turn into theoretical development. Environmental economics, ecological economics, political ecology, eco-feminism and deep ecology have all developed extensive, demanding and exciting literatures, and debates about the social construction of nature have extended the once unfashionable area of 'environmental issues' by opening challenging links with wider developments in social science associated with the 'postmodern turn'. A sophisticated and vigorous theoretical debate about sustainability has grown to parallel policy debate about environment and development. The sheer weight of paper devoted to the subject of sustainability should give all those of us who have contributed to it pause for reflection.

My own understanding of both sustainability and development has continued to grow and evolve. I have been distressed by the way in which the radical potential of debates about poverty and environment has been dissipated, and the ease with which key words and phrases have been taken up and incorporated as a 'greenwash' over corporate, governmental and individual 'business as usual'. Truly, the path to sustainable development is paved with good intentions, but the rhetorical vagueness of that master phrase has made it too easy for hard questions to be ignored, stifled in a quilt of smoothly crafted and well-meaning platitudes.

My own ideas continue to be informed by the tension between a concern for human needs and dismay at the scale of destructive human demands on nature. These concerns relate to independent strands of moral argument. Both are vitally important: non-human nature has intrinsic value but also underpins economy and society by its capacity to provide resources to meet human needs. I do not warm to the argument that those who love nature must hate humankind, or to those whose concern for nature leads them to a conservation without a place for people. It seems to me that, if the 'sustainable development' debate is to have any value, it must address the challenge of relationships between people in their use of nature, and between humans and the rest of the biosphere.

This edition of *Green Development* is inevitably selective and less than encyclopaedic in its coverage of the literature. The diversity and sheer volume of writing on environment and development preclude the kind of panoptic treatment that I attempted in the first edition. Thanks to the enterprise of the many distant strangers who have put fingers to keyboard, the book remains a record of a steep learning curve. The first edition sought to be a book that made an argument – a book with a clear beginning, middle and end, rather than simply being a smorgasbord of ideas and examples from which busy teachers and students could pick a tasty plateful. I hope that this aspect of the first edition has survived. I hope, too, that, while this is obviously an academic's book, it is not an overly academic book, in the sense that I have been successful in making it readable.

A word is needed on the use of language. First, the phrase 'developing world', used in the book's subtitle, is a compromise. This is just one of the many euphemisms for poorer and less-industrialized countries (like the 'South', 'Third World' or 'underdeveloped' countries). None is entirely satisfactory, and all can be criticized. The term 'Third World' is less used today, the collapse of the Iron Curtain in Europe arguably making the label, once chosen by those non-aligned countries lying between capitalist 'First' and socialist 'Second' worlds, outdated. However, the term 'developing countries' is also problematic: many countries and regions within countries are not 'developing' very fast, and there is much in the conventional process of development that is destructive. Anyway, as will become clear, none of the words in the book's title is uncontested: the idea of development is shot through with contradictions, as is the concept of sustainability.

Second, I have chosen not to litter the text with '*sic*' where I have quoted authors from previous decades (or the names of current organizations) who choose to talk of 'man and nature' or 'man and the environment' rather than using less sexist language; times change – in this small way, at least, for the better – but past times used past language. I can but apologize if my usage offends.

Green Development tries to bridge two important gulfs. The first is between environmentalism and development, and the second between armchair theory and practice. To create the first bridge I have devoted most of the first seven chapters of the book to a discussion of the theory of sustainable development and conventional strategies for making it happen. These chapters argue that environmentalists still have much to learn from radical views of political economy, although I do not pretend that they offer a simple short cut to an adequate radical green theory. The second gulf that the book tries to stretch across is that between the ivory tower (in which, let it be clear, this has been written) and the real world of development decisions. To tackle this, later chapters explore aspects of sustainability and development thematically, focusing on environmental aspects of development projects and programmes, in drylands, conservation, forests and water resources and on urban and industrial hazards.

Behind the pragmatic (and within limits effective) 'mainstream' approach to sustainable development lie deeper and more subversive ideas about ways to respond to human impacts on non-human nature or to the limited success of the capitalist world economy in creating equitable and sustained outputs from environmental resource use.

The world is not a machine, to be run by privileged super-mechanics, however skilled in environmental housekeeping. Rather than simply contributing to the enhanced efficiency of bureaucratic, technocratic and economic management and society, 'green' development must also address the capacity of individuals and groups to plan and run their own lives, and control their own environments.

This book is not a handbook for making something called 'sustainable development' happen faster or in particular places. If there is a single conclusion to be drawn from what I have written, it is simply this: it is no good talking about how the environment is developed, or managed, unless this is seen as a political process. Any understanding of environment and development must come back to this issue. The radical potential of 'green' thinking about development, therefore, is to be sought not in its concern with ecology or environment *per se*, but in how it addresses questions of control, power and self-determination in the social engagement with nature.

Cambridge, December 2007

Acknowledgements

Many people have contributed to my education in the two decades over which this book has been written and revised. I have learned a great deal from colleagues with whom I have had the chance to work in the field, particularly Martin Adams, Alan Bird, Dan Brockington, David Hulme, Mark Infield, Max Graham, Jack Griffith, Ted Hollis, Jon Hutton, Mike Mortimore, James Murombedzi, Marshall Murphree, John Sutton, Kevin Kimmage and David Thomas. I would like to thank Andrew Warren and Tim O'Riordan for their support over many years, and especially Dick Grove, to whom I owe my engagement in research, and whose capacity for curiosity remains for me a standard by which academics should be judged.

I have learned a lot from the MacArthur Advancing Conservation in a Social Context planning group (Pete Brosius, Kent Redford, Tom McShane, Anne Kinzig, Michael Wright, Manuel Pulgar-Vidal, Doris Capistrano, Sheila O'Connor and Dan Miller), and (at different periods) from Patrick Dugan, Jean-Yves Pirot, Achim Steiner and Sally Jeanrenaud at IUCN.

In the Department of Geography in Cambridge I owe a great deal to many colleagues past and present, particularly members of the Political Ecology of Development group, Tim Bayliss Smith, Emma Mawdsley, Liz Watson, Wynet Smith and Bhaskar Vira, who are great to work with. I have gained greatly from the chance to teach the many enthusiastic and often frighteningly well-informed undergraduates reading Geography, and master's students on the M.Phil. in Society, Environment and Development. I have also been very lucky to work with many exceptional Ph.D. students, David Webb, Kevin Kimmage, Mike Mason, David Thomas, Uwem Ite, Polly Gillingham, Lisa Campbell, Liz Watson, John Murton, Lee Risby, Lucy Welford, Aaron Padilla, Malcolm Starkey, Caroline Cowan, Max Graham, Wynet Smith, Ivan Scales, David Webb, Sarah Milne and Richard Paley. Much of what I have learned they have taught me.

Chris Park invited me to write the first edition of *Green Development* long before sustainable development was fashionable. I am indebted to Paul Dickson for his intelligent and energetic help in finding research materials in the ever-expanding labyrinth that is the literature on sustainability. I would like to thank Ian Agnew for drawing the diagrams, Gareth Hughes for help with permissions and Emily Adams and Francine Hughes for help with proof-reading. Andrew

Mould at Routledge and Matthew Brown at Bookcraft, like their predecessors, have shown stoical patience in the face of missed deadlines. Hilary Walford was an outstanding copy-editor. I am also grateful to referees and readers for their advice. Taking it all would have required another book. Sins of omission (and commission) remain my own.

Above all, I would like to thank Franc, Emily and Tom, who have lived with this book through far too many evenings and weekends, and who continue to give my reflections on questions of sustainability and the relations between people and nature their rightful sharp and very personal edge.

Copyright acknowledgements

The author and publisher would like to thank the following for granting permission to reproduce material in this work:

Cambridge University Press, for Figure 1.1 from 'The urgent need for rapid transition to global sustainability', by R. J. Goodland, H. E. Daly and S. E. Serafy, in *Environmental Conservation*, 20: 297–309, 1993 (Figure 6.1).

Company School of Development Studies, University of East Anglia, for Figures 1.1 and 1.2 from 'Understanding environmental issues', by P. Blaikie, in *People and the Environment*, edited by S. Morse and M. Stocking, UCL Press, London, 1995 (Figures 8.2 and 8.3).

Dan Miller, for the cover photograph.

Edward Arnold (Publishers) Ltd, for Figure 3.3 from *Environmental and Social Impact Assessment: an introduction*, by C. J. Barrow, Arnold, 1997 (Figure 6.3).

Elsevier Science, for a figure from 'Tropical deforestation: balancing regional development demands and global environmental concerns', by William B. Wood, in *Global Environmental Change*, 1: 23–41, 1990 (Figure 9.2).

Flora & Fauna International, for Plates 9.1, 9.5, and 9.6.

Greenpeace for Plate 4.1.

IUCN Publications Service, for two text diagrams entitled 'The need to integrate conservation with development' and 'Depletion of living resources' from *The World Conservation Strategy*, by the International Union for Conservation of Nature and Natural Resources, United Nations Environment Programme, World Wildlife Fund, Geneva, 1980 (Figures 3.1 and 3.2).

Oxford University Press, for Figure 3, from 'Drought, agriculture and environment: a case study from the Gambia, West Africa', by K. M. Baker, in *African Affairs* 94: 67–86, 1995 (Figure 8.1).

SCOPE, for Figure 1.8 from SCOPE 36, 'Acidification and regional air pollution in the tropics', in *Acidification in Tropical Countries*, edited by H. Rodhe and R. Herrera, © SCOPE, J. Wiley and Sons, Chichester, 1988 (Figure 12.2).

Thomson Publishing Services, for Figure 1 from *Ecology and Equity: the use and abuse of nature in contemporary India*, by M. Gadgil and R. Guha, Routledge, London, 1995 (Figure 12.1).

Thomson Publishing Services, for Figures 3.2 and 3.3 from 'Towards a green political theory', by A. Carter, in *The Politics of Nature: explorations in green political theory*, edited by A. Dobson and P. Lucardie, Routledge, London, 1993 (Figures 7.2 and 7.3).

United Nations Environment Programme, Nairobi, for a figure based on the original map 'Desertification in the world's susceptible drylands', in *World Atlas of Desertification* (2nd edition), edited by T. Middleton and D. S. G. Thomas (Figure 8.4).

United Nations University, for 'Friberg and Hettne's "Green" project' from 'The greening of the world: towards a non-deterministic model of global processes', in *Development as Social Transformation: reflections on the global problématique*, by H. Addo *et al.*, Hodder and Stoughton, Sevenoaks, for the United Nations University, 1985 (Figure 7.1).

Wiley-Blackwell, for Figure 2 from 'Natural capital and sustainable development', by R. Costanza and H. E. Daly, in *Conservation Biology*, 6: 37–46, 1991 for (Figure 6.2).

World Conservation Monitoring Centre, Cambridge, for Figure 20.1 from *Global Biodiversity: status of the earth's living resources*, by B. Groombridge, Chapman & Hall, London, 1992 (Figure 9.1).

World Wildlife Fund, for two text diagrams entitled 'The need to integrate conservation with development' and 'Depletion of living resources' from *The World Conservation Strategy*, by the International Union for Conservation of Nature and Natural Resources, United Nations Environment Programme, World Wildlife Fund, Geneva, 1980 (Figures 3.1 and 3.2).

Every effort has been made to contact copyright holders for their permission to reprint material in this book. The publishers would be grateful to hear from any copyright holder who is not here acknowledged and will undertake to rectify any errors or omissions in future editions of this book.

Abbreviations and acronyms

ACIWLP	American Committee for International Wildlife Protection
AP6	Asia-Pacific Partnership on Clean Development and Climate
AWLF	African Wildlife Leadership Foundation
BCSD	Business Council for Sustainable Development
CAMPFIRE	Communal Areas Management Programme for Indigenous Resources (of Zimbabwe)
CBA	cost–benefit analysis
CBD	Convention on Biological Diversity
CBNRM	community-based natural resource management
CDC	Conservation for Development Centre (of IUCN)
CFC	chlorofluorocarbon
CGIAR	Consultative Group on International Agricultural Research
CIDA	Canadian International Development Agency
CILSS	Comité Inter-États de la Lutte contre la Sécheresse Sahélienne
CITES	Convention on International Trade in Endangered Species
COP	Conference of the Parties
CSA	Conseil Scientifique pour l'Afrique
CSD	United Nations Commission on Sustainable Development
CSR	corporate social responsibility
DFID	(UK) Department for International Development
DPCSD	(UN) Department for Policy Coordination and Sustainable Development
EA	environmental assessment
EC	European Community (now the European Union)
ECOSOC	United Nations Economic and Social Council
EDP	environmentally adjusted net domestic product
EIA	environmental impact assessment
EIC	environmental impact coefficient
EM	ecological modernization
EMAS	Eco-Management and Audit Scheme
ENSO	El Niño/Southern Oscillation
ESI	Environmental Sustainability Index
EU	European Union

FAD	food availability decline
FAO	Food and Agriculture Organization (of the United Nations)
FSC	Forest Stewardship Council
GATT	General Agreement on Tariffs and Trade
GDP	gross domestic product
GEC	global environmental change
GEF	Global Environment Facility
GEMS	Global Environment Monitoring System
GIS	geographical information science
GNP	gross national product
HCFC	hydrochlorofluorocarbon
HDI	Human Development Index (of the United Nations Development Programme)
IARC	International Agricultural Research Centre (e.g. International Institute for Tropical Agriculture or the International Rice Research Institute)
IBP	International Biological Programme
ICBP	International Committee for Bird Protection (later International Council for Bird Preservation; forerunner of BirdLife International)
ICDP	integrated conservation–development project or programme
ICID	International Commission in Irrigation and Drainage
ICOLD	International Commission on Large Dams
ICRAF	International Centre for Research on Agroforestry
ICRISAT	International Crops Research Institute for the Semi-Arid Tropics
ICSU	International Council of Scientific Unions
IFAD	International Fund for Agricultural Development
IIED	International Institute for Environment and Development
IITA	International Institute for Tropical Agriculture (Ibadan, Nigeria)
INCRA	National Institute for Colonization and Agrarian Reform (Brazil)
INEAC	Institut National pour l'Étude Agronomique du Congo Belge
IOPN	International Office for the Protection of Nature (forerunner of the International Union for the Conservation of Nature and Natural Resources)
IPAL	Integrated Project on Arid Lands
IPCC	Intergovernmental Panel on Climate Change
IPM	integrated pest management
IRRI	International Rice Research Institute
IRSAC	Institut pour la Recherche Scientifique en Afrique Centrale
ISEW	Index of Sustainable Economic Welfare
ITDG	Intermediate Technology Development Group
ITTO	International Tropical Timber Organization
IUBS	International Union of Biological Sciences
IUCN	International Union for the Conservation of Nature (World Conservation Union)

IUPN	International Union for the Protection of Nature (forerunner of the International Union for the Conservation of Nature and Natural Resources)
LIRDP	Lwangwa Integrated Resource Development Project
LPI	Living Planet Index (of the World Wide Fund for Nature)
MAB	Man and the Biosphere Programme
MEA	Millennium Ecosystem Assessment
MIPS	material input per unit of service
MSD	mainstream sustainable development
NEAP	National Environmental Action Plan
NEPA	(US) National Environmental Policy Act
NGO	non-governmental organization
NIE	new institutional economics
NOAA	(US) National Oceanic and Atmospheric Administration
NORAD	Norwegian Agency for Development Cooperation
ODA	(UK) Overseas Development Administration (now the Department for International Development)
OECD	Organization for Economic Cooperation and Development
OEHA	(World Bank) Office of Environmental and Health Affairs
OMS	(World Bank) Operational Manual Statement
OMVS	Organisation pour la Mise en Valeur du Fleuve Sénégal
OPEC	Organization of Petroleum Exporting Countries
ORSTOM	Office de la Recherche Scientifique et Technique d'Outre Mer
PA	protected area
PACD	Plan of Action to Combat Desertification
PCB	polychlorinated biphenyl
POP	persistent organic pollutant
RBDA	River Basin Development Authority
REDD	reduced emission from deforestation and degradation
SCOPE	Scientific Committee on Problems of the Environment
SD	sustainable development
SEA	strategic environmental assessment
SEAA	System of Integrated Environmental and Economic Accounting
SIA	social impact assessment
SNA	System of National Accounts
SPFE	Society for the Preservation of the Fauna of the Empire
SPWFE	Society for Preservation of the Wild Fauna of the Empire
SWIFT	Solomon Western Islands Fair Trade
TFAP	Tropical Forests Action Plan
TNC	transnational corporation
TRIPS	Trade Related Intellectual Property Rights (agreement)
TVA	Tennessee Valley Authority
UN	United Nations
UNCED	United Nations Conference on Environment and Development (Rio Conference or Earth Summit)

UNCOD	United Nations Conference on Desertification
UNCTAD	United Nations Conference on Trade and Development
UNDP	United Nations Development Programme
UNEP	United Nations Environment Programme
UNESCO	United Nations Educational, Scientific and Cultural Organization
UNFCCC	United Nations Framework Convention on Climate Change
UNGASS	United Nations General Assembly Special Session on Sustainable Development
UNSCCUR	United Nations Scientific Conference on the Conservation and Utilization of Resources (Lake Success Conference)
UNSO	United Nations Office to Combat Desertification and Drought (formerly the UN Sudano-Sahelian Office)
UNSTAT	United Nations Statistical Division
USAID	United States Agency for International Development
WBCSD	World Business Council for Sustainable Development
WCD	World Commission on Dams
WCED	World Commission on Environment and Development
WCS	*World Conservation Strategy*
WHO	World Health Organization
WID	Women in Development
WMO	World Meteorological Organization
WRI	World Resources Institute
WSSD	World Summit on Sustainable Development
WWF	World Wide Fund for Nature (formerly World Wildlife Fund)

1 The dilemma of sustainability

Telling me, a harried public official who must answer to 48 million restless, hungry and thirsty people, to 'Ensure development is sustainable and humane' is like warning me 'Operate, but don't inflict new wounds'. I know that. What I don't know is how to do it.

(Kader Asmal, Chair World Commission on Dams, 2000)

Sustainable development

In October 2007, the Nobel Peace Prize was awarded to the Intergovernmental Panel on Climate Change (IPCC) and the former US presidential candidate, Al Gore, for 'their efforts to build up and disseminate greater knowledge about man-made climate change, and to lay the foundations for the measures that are needed to counteract such change' (www.nobelprize.org, 12 October 2007). The award marked their work, in the IPCC's series of monumental reports and Gore's tireless lectures and successful documentary film *An Inconvenient Truth*, to identify and build awareness of the connection between human activities and climate change. Extensive climate changes were likely to alter and threaten the living conditions of much of humankind, placing particularly heavy burdens on the world's most vulnerable countries.

These connections between environment and human welfare are uncomfortable for world leaders. In April 2007, release of the second volume of the IPCC's Fourth Assessment Report *Impacts, Adaptation and Vulnerability* (Parry *et al.* 2007) was held up by last-minute political wrangling. As the *New Scientist* headline put it, 'as polluters quibble, the poor learn their fate' (Brahic 2007). The report showed the significance of climate change for the world's poor. Storms, drought, heatwaves, early flowering seasons, changes in insect migrations and dwindling water supplies from mountain regions were global problems the world's poorest countries and people were least well equipped to deal with. Greenhouse gas emissions in industrial and rapidly industrializing economies were directly linked to the day-to-day problems of the poor. The United Nations Development Programme (UNDP) regards climate change as 'a proven scientific fact', and comments 'we now know enough to recognize that there are large risks, potentially catastrophic ones' (UNDP 2007, p. 3). The connections between wealth and wealth creation, environmental change and poverty are laid bare by scientific understanding of planetary carbon metabolism. No wonder the diplomats squabbled over the small print.

In the first decade of the new century, the issue of human impacts on global climate change has mostly been framed within a broader debate about sustainability. The challenge of doing something about this and other global issues (such as biodiversity depletion and pollution), while simultaneously tackling global inequality and poverty and not letting the wheels come off the world economy, is labelled as sustainable development. In the aftermath of two global conferences, the United Nations Conference on Environment and Development (UNCED) in Rio de Janeiro in 1992 (the Rio Conference or the 'Earth Summit') and the World Summit on Sustainable Development in Johannesburg in 2002, these concepts have become staples in any debate about environment and development. The classic oxymoron 'sustainable development' (combining two seemingly contradictory concepts) had, as Michael Redclift (2005b) nicely puts it, 'come of age'. But where had it come from, and what did it mean?

The idea of development attracts new concepts at a ferocious rate. New terms are coined and adapted faster than old ones are discarded (Chambers 2005). This is an important process, for, as Robert Chambers observes, words change the way we think and what we do, modifying mindsets, legitimating actions and stimulating research and learning. The last twenty years of research in development studies, influenced by postmodernism and poststructuralism, leave no doubt of the enormous power of language and discourse to structure the way we think about – and therefore take action about – development (Crush 1995; Escobar 1995). Development action is driven forwards by texts ranging from humanitarian tracts to national development plans. The way these texts portray the world, often in a crisis of some kind, determines what knowledge (and whose knowledge) provides a frame for problems and solutions, constitutes the basis for action and determines who has the authority to act (Crush 1995).

The words we use to talk about development, and the way our arguments construct the world, are hugely significant. Cornwall (2007) comments that 'the language of development defines worlds-in-the-making, animating and justifying intervention in currently existing worlds with fulsome promises of the possible' (p. 471). Words like development and sustainability are 'buzzwords', unavoidable, powerful and floating free from concrete referents in a world of make-me-believe. Words matter – and the key question is whose words, and whose ideas, count most? There is a politics to the words we use about development: the words used by powerful global actors in central places such as Washington, New York, Paris, London or Beijing do most to shape development in the world periphery (Chambers 2005).

The concept of sustainability joined the lexicon of development that has been accumulating since formal development planning began following the Second World War, in the last decades of the twentieth century (Scoones 2007). Rapidly, the phrase 'sustainable development' had become ubiquitous in development discourse (Redclift 2005b – Sharachchandra Lélé (1991) correctly predicted that it would constitute 'the development paradigm of the 1990s' (p. 607). The capacity of the phrase to restructure development discourse and to reorganize development practice, a sure reflection of its power, will be discussed below.

Where did the new phrase come from? Its roots lie a long way back in the history of European and wider global thinking, but the concept began to be widely adopted following the United Nations Conference on the Human Environment in Stockholm in 1972 (see Chapter 3). The idea of environmental limits or constraints on development was explored by a number of authors around the start of the 1980s under the label of 'ecodevelopment', (e.g. Sachs 1979, 1980; Riddell 1981; Glaeser 1984), and it was a central concept in the *World Conservation Strategy* (*WCS*) published in 1980 (IUCN 1980). Most importantly, it was the foundation of the report of the World Commission on Environment and Development (WCED) seven years later (Brundtland 1987). At its launch in April 1988, it was claimed that this report, *Our Common Future*, set out a global agenda for change. This agenda soon began to command attention in the core of the development universe: in a major shift of culture and policy, the President of the World Bank spoke in May 1988 of the links between ecology and sound economics in a major statement of the Bank's policy on the environment (Hopper 1988). The idea that development thinking needed to be 'greened' was a challenging idea in the 1980s (Harrison 1987; Conroy and Litvinoff 1988). In the 1990s this argument became standard, the starting point for countless political speeches and student essays.

Sustainable development gained its salience largely as a result of the United Nations Conference on Environment and Development in Rio in 1992 (UNCED, or the Earth Summit). The 170 governments represented made public proclamations of support for the idea of environmentally sensitive economic development, egged on by a vast array of non-governmental organizations, meeting nearby in the parallel Global Forum (Holmberg *et al.* 1993; Chatterjee and Finger 1994). The media danced attendance, and the conference was promoted as a global event, although many a journalist pointed out the stark contrasts between the lifestyles and life chances of delegates and the poverty of people in Rio de Janeiro's *favelas*. The media had built up hopes that UNCED would bring about a new environmental world order, and, once the razzmatazz had died down, many commentators reported that the chance had been blown. A series of international agreements had been signed (see Chapter 4), but had anything really changed? Over the next decade, many commentators pointed out that the world economy was carrying on much as before, rich and poor, polluter and polluted. The words had changed, but it was said that deeds had not: it was 'business as usual' at Earth plc, despite the calls from its shareholders and the high-profile statements for chief executives.

Fifteen years later, we still look back on this event with some bemusement. Was this a critical point in the way the world thought about itself, or just another international talking shop? Did ideas of sustainability represent a real environmentalist critique of development, and if so of what kind? Was there anything really new in this sudden interest in environment and development? In terms of research insights, Rio added volume to existing debates but brought little that was novel. Concern about the environment in the developing world had been a feature of debate about development since the late 1970s, and awareness of the environmental dimensions of development, whether among scholars, practitioners or participants in development, was older than that. But, of course, Rio was not about academic ideas, and it certainly marked a change in the level of attention given to these issues. In the last decade of the twentieth century there was a step change in the scope and sophistication of critiques of the environmental dimensions of development in practice, and the higher profile being given to the environment in the context of social and economic change (McCormick 1992).

Credit for the insertion of environmental concerns into development discourse in the closing decades of the twentieth century lies in the first instance with environmentalists from Northern industrialized countries (Guha 2000). An urgent transition to sustainability was needed because 'global life support systems – the environment – have a time-limit' (Goodland *et al.* 1993, p. 297). The loss of species and natural habitat caused by development projects had been a potent focus for the extension of environmental pressure-group politics familiar in the industrialized world since the 1970s. 'Save the rainforest' campaigns followed logically enough from concerns about pollution, whales or the countryside. As globalization accelerated through the closing decades of the twentieth century, the media, the travel industry and improved telecommunications all brought the global South within the ambit of domestic environmental concern in the North. In the global village, the wildlife and landscapes of the developing world became the new countryside.

First World environmentalism, however, did more than simply broaden its field of concern (McCormick 1992; Guha 2000). There was a self-conscious effort to move beyond environmental protection and transform conservation thinking by appropriating ideas and concepts from the field of development. In extending their focus from hedgerows to rainforests, environmentalists found (or claimed to have found) much common ground with environmental groups in developing countries opposing development projects that threatened breakdown in indigenous and subsistence ways of life (Gadgil and Guha 1995; Guha 2000). In environmental opposition by environmental groups to investment in large projects such as dams, the threats they represent to the rights and interests of indigenous peoples are likely to be at least as prominently expressed as threats to biodiversity. The links between the two began to be drawn explicitly and prominently (e.g. Pearce 1992; Gadgil and Guha 1995).

The display of development agencies and environmental groups dancing to the same 'sustainable development' tune in the 1990s was remarkable, but not accidental. It reflected several factors. First, environmental concern expanded to address environmental problems at an explicitly transnational or 'global' scale, most notably the 'ozone hole' and the 'greenhouse effect' (McKibben 1990; O'Riordan and Jäger 1996). Second, environmentalists began to mount a successful critique of the performance of aid donors through the 1980s (Stein and Johnson 1979; Goodland 1984, 1990; Holden 1988). Third, the development planning responded to the more general 'greening' of politics in Western industrial countries in the 1980s, epitomized for UK observers by Margaret Thatcher's famous observation in 1988 that 'we have unwittingly begun a massive experiment with the system of this planet itself', and her political pitch in response: 'no generation has a freehold on this earth. All we have is a life tenancy – with a full repairing lease' (Thatcher 1998, 2008).

The nagging question remains, however, how deep this apparent revolution in development thinking goes. Has there really been a 'greening' of development? Has there, for example, been a revolution in ideology in any way analogous to Charles Reich's celebrated account of new thinking in the USA in the 1960s, *The Greening of America* (Reich 1970)? Commentators agree that the environmentalism of the 1960s and 1970s was a new social movement of profound significance (e.g. Cotgrove and Duff 1980; Hays 1987; Guha 2000), but to what extent did this embrace thinking about the developing world, let alone thinking *within* the developing world? Was the 'greening of development' evidence of a paradigm shift in development thought, or simply an exercise in relabelling? Do the terms 'sustainability' and 'sustainable development' relate to clearly defined concepts? How has the phrase sustainable development acquired the power to attract such a large and disparate following?

Sustainable development as panacea

The range of contexts in which the phrase 'sustainable development' is now employed is very wide. In research, it seems to offer the potential to unlock the doors separating academic disciplines, and to break down the barriers between academic knowledge and policy action. It does this because it seems to draw together ideas in ecology, ethics, economics, development studies, sociology and many other disciplines. Yet it looks forward to action and practical projects of social and environmental improvement. The term is beguilingly simple (O'Riordan 1988), yet at the same time capable of carrying a wide range of meanings. It can be used by political actors with divergent interests, a convenient rhetorical flag under which favoured projects can be launched. It has become a powerful term in the lexicon of development studies, but also a theoretical maze of remarkable complexity (Dixon and Fallon 1989; Daly 1990; Lélé 1991; Sneddon 2000; Robinson 2004; Kates *et al.* 2005).

It has been recognized for decades that sustainable development can be defined in many ways. Many definitions are rhetorical and vague (Lélé 1991). The most commonly quoted is that from the Brundtland Report, in *Our Common Future*: 'development that meets the needs of the present without compromising the ability of future generations to meet their own needs' (Brundtland 1987, p. 43). The longevity of this formulation stems from simultaneous appeal both to those concerned about poverty and development and to those concerned about the state of the environment, and the preservation of biodiversity (J. Robinson 2004). Moreover, it demands that attention be focused on both intragenerational equity (between rich and poor now) and intergenerational equity (between present and future generations). The appealing, moralistic but slightly vague form of words of the Brundtland Report allowed sustainable development to become, in Conroy's term, the 'new jargon phrase in the development business' (1988, p. xi). It also became a vital element in the discourse of researchers trying to explain the relations between economy, society and environment (e.g. Redclift 1984, 1987, 1996; Clark and Munn 1986; Redclift and Benton 1994; Kates *et al.* 2005).

Subsequent definitions have often been much more carefully phrased – the UK's Forum for the Future, for example, defines it as a 'dynamic process which enables all people to realize their potential and improve their quality of life in ways which simultaneously protect and enhance the Earth's life support systems' (www. forumforthefuture.org.uk, 20 October 2006). The holistic appeal of earlier definitions has proved perennially popular, and has endured manifold reformulations.

However, the Brundtland definition of sustainable development is a better slogan than it is a basis for theory. The phrase, whether in academic journals or the sound bites of politicians, very often proves to have no coherent theoretical core and no clear and consistent meaning (Redclift 1987). The very simplicity of the phrase allows users to make high-sounding statements that are vague in meaning. Its flexibility adds to its attraction.

Environmentalists speak of 'sustainable development' in trying to demonstrate the relevance to development planners of their ideas about proper management of natural ecosystems. The conviction behind works such as the *World Conservation Strategy* (IUCN 1980) was that sustainable development is a concept that could truly integrate environmental issues into development planning (Chapter 3). In using terminology of this sort since that time, environmentalists have attempted to capture some of the vision and to exert influence in development debates. Sadly, they often have no understanding of their context or complexity. Environmentalist prescriptions for development, shorn of any explicit treatment of political economy, can have a disturbing naivety.

Those working in development use the phrase 'sustainable development' to re-emphasize the importance of equity, the social outcomes of development or the desirability of participatory approaches to development planning. They have been less ready to heed environmentalist critiques of the model of development itself, or to address the structural causes of poverty or environmental degradation. Development bureaucrats and politicians have undoubtedly welcomed the opportunity to fasten on to a phrase that suggests radical reform without actually either specifying what needs to change or requiring specific action. As Luke (2005) points out, the phrase has increasingly been used to label lifestyles and modes of existence that are neither sustainable nor developmental.

Politicians and governments have been enthusiastic in their incorporation of the language of sustainable development. In the UK, for example, the 1997 UK government White Paper on international development made a specific commitment to the elimination of poverty in poorer countries through sustainable development; specific objectives include the promotion of 'sustainable livelihoods' (DFID 2000), a Sustainable Development Strategy was published in 2005. An independent Sustainable Development Commission was established in 2000, building on the work of the UK Round Table on Sustainable Development and the Panel on Sustainable Development (http://www.sd-commission.org.uk/index.php). In 2005 the UK government charged the Commission with the role of 'watchdog for sustainable development'. The UK has identified four priority areas for immediate action, shared across the UK: sustainable consumption and production ('working towards achieving less with more); natural-resource protection and environmental enhancement (protecting the natural resources on which we depend); 'from local to global' (building sustainable communities) and climate change and energy (www.sustainable-development.gov.uk, 18 June 2007).

There is no doubt of the sincerity of the attempt with which thinking about sustainability has been 'mainstreamed' into UK government policy in the UK, or other countries, since the Brundtland Report in 1987. Arguably the rushed application of green camouflage paint to existing policies that characterized the late 1980s has been replaced with more carefully constructed thinking and policies; at the very least, even the most hard-bitten cynic will admit that the quality of the paintwork has improved. Yet, beyond the simple interpretation of sustainable development as 'sustaining wealth' lie questions about the nature of consumption, the shape of the economy and the cultural definition of 'the good life'. These questions are hard, and most elected politicians duck them.

Business leaders have also found the rhetorical power of sustainable development, especially in emphasizing the power of the market to deliver social and environmental 'goods', and in setting out their own corporate 'green' credentials. The idea of 'green capitalism' was much discussed in the first decade of the twenty-first century (Hawken *et al.* 1999; Mason 2005; Porritt 2005). However, while it certainly became part of the language with which chief executives addressed their shareholders and critics, in their speeches its meaning often remained deliberately fluid. In business-speak, sustainable development tends to mean 'making our business sustainable', very often reflecting a determination to continue 'business as usual' in superficially greener times.

The discourse of development

The range of meaning attached to sustainable development reflects the contested question of what development itself means (Forsyth 2005). Debates about development threaten to lead into a semantic, political and indeed moral maze. At its most basic, development can be taken to mean the production of social change that allows people to achieve their human potential. Yet development remains an ambiguous and elusive concept, 'a Trojan Horse of a word' (L. Frank 1987, p. 231), meaning a term that can be filled by different users with their own meanings and intentions. As a 'buzzword' (Cornwall 2007) it is also what Howard (1978) described as a 'slippery value word' (p. 18) that can be used by 'noisy persuaders' (such as politicians or planners) to herd people in the direction they want them to go. Advocates for particular ends in development, or means to achieve those ends, make explicit use of the slipperiness of the word to promote their solutions.

There is no doubt of the concept's power (Crush 1995). Sachs (1992a) speaks of development as 'a perception which models reality, a myth which comforts societies, and a fantasy which unleashes passions' (p. 1). Such value-laden words easily become political battlegrounds of very real practical significance. Crush (1995) points out that the discourse of development promotes and justifies specific interventions in people's lives, and thus development discourse does not hang in some kind of academic abstract, but is inextricably linked to sets of material relationships, to specific policies and to the exercise of power. The idea of development is so powerful that some believe it has come to enclose debate. Escobar (1995) argues that reality has been colonized by the development discourse to such an extent that those who are dissatisfied with this state of affairs have 'to struggle for bits and pieces of freedom within it, in the hope that in the process a different reality could be constructed' (p. 5). The idea of development, and the idea of modernity that lies behind it, limit the extent to which alternative futures – of justice and a new international economic order – can be imagined (Escobar 2004).

The word 'development' has a complex pedigree. It is used both descriptively (to describe what happens in the world as societies, environments and economies change) and normatively (to set out what *should* happen (Goulet 1995)). The word itself came into the English language in the eighteenth century and soon acquired an association with 'organicism' and ideas of unfolding change and growth (Watts 1995). By the start of the nineteenth century, development had become a linear theory of progress, bound up with capitalism and Western cultural hegemony, something advanced through mercantilism and colonial imperialism (Cowen and Shenton 1995). The idea of 'improvement' was fundamental to European imperial expansion and the planting of colonies, whether in Ireland, the West Indies or further afield (Drayton 2000). Ideas of *under*development can be traced to nineteenth-century European thought (Cowen and Shenton 1995).

Despite the complex genealogy of development, there was a remarkable standardization of meanings in the second half of the twentieth century, following the end of the Second World War. The classic statement of this view was President Truman's inaugural address to the US Senate in January 1949 (Escobar 1995). He called for 'a programme of development based on the concepts of democratic fair dealing'; under the benign leadership of the USA, science and technical knowledge would be vigorously applied, would give rise to greater production and in turn to prosperity and peace for the whole globe, and particularly its 'undeveloped areas'. In the process, of course, the world's people would be freed from 'the deceit and mockery, poverty and tyranny' of the 'false philosophy of communism' (www.bartleby.com, 21 November 2007).

According to the standard 'developmentalist' worldview (Aseniero 1985), the modern West is re-created across the globe by the process of development: industrialized, urbanized, democratic and capitalist. Development, in this view, is a refiner's fire through which successful societies emerge singed but purified, both modern and affluent. In the classic framing of Walt Rostow, drummed into generations of students, development was presented as an iron-tight linear path of change. In *Stages of Economic Growth*, Rostow (1960) outlined five 'stages of economic growth', from traditional society, preconditions for take-off, take-off, maturity and the age of high mass consumption. Like Truman's, Rostow's idea of development was never far removed from his politics: his book was subtitled *A Non-Communist Manifesto*.

To Rostow and his many successors, progress down the flare path of change towards the all-important moment of 'lift-off' to a new world of mass consumption could be measured in terms of the growth of the economy, or some economic abstraction such as per capita gross domestic product. The word 'development' then came to mean the projects and policies, the infrastructure, flows of capital and transfers of technology that were supposed to make that imitation possible. Development thus involved the extension of the established world order on the newly independent periphery.

After the Second World War, it was assumed that rapid industrialization and improvement in the material conditions of life could quickly be achieved across the world by following the formula that had worked in reconstructing war-ravaged Europe (Goulet 1992). Orthodox development thinking sought to follow the success of the Marshall Plan by applying the same approach (injecting foreign aid for capital for investment in infrastructure) to the non-industrialized world, through 'aid', both bilateral (between governments) and multilateral (particularly through the new World Bank institutions, set up with the International Monetary Fund under the 1944 Bretton Woods agreement (Oman and Wignarajah 1991)).

The result was a one-size-fits-all conceptualization of development. Ivan Illich pointed out in 1973 that this formula had failed:

> There is a normal course for those who make development policies, whether they live in North or South America, in Russia or Israel. It is to define development and set its goals in ways with which they are familiar, which they are accustomed to use in order to satisfy their own needs, and which permit them to work through the institutions over which they have power or control.
>
> (p. 368)

With aid went hegemonic Western values, and an environmentally catastrophic idolization of consumption. The modernization paradigm was built on the conceptual separation of 'modern' and 'traditional' (or 'Western' and 'non-Western') societies. Such concepts, which welded seamlessly into ideas of development, came from the same roots in Western Enlightenment rationality, and built on profoundly encoded Western preconceptions about civilization and improvement versus barbarism (Slater 1993).

From the forging of the concept of development grew the exercise of economic and cultural power that has become development practice. Development discourse is built on a conceptual separation of the non-'developed' and non-Western 'other' as a fitting, needy and legitimate target for action. Edward Said's account (1979) of the power of orientalism applies strongly to the standard model of development (see also Crush 1995; Schech and Haggis 2002). The manner of that representation, and the material actions that flowed from it, have been highlighted and challenged in extensive writings about postcolonialism (Spivak 1990; Power 2003; Radcliffe 2005a).

Esteva (1992) noted that, with Truman's speech, two billion people became underdeveloped. 'Development', its meaning soon narrowed to economic growth, thenceforth was defined as the escape from that sorry condition. Of course, escape proved impossible for most countries and most people, even in these narrowly defined terms. Sachs (1992a) describes the project of development as 'a blunder of planetary proportions' (p. 3). For him, the concept of development is obsolete, standing 'like a ruin in the intellectual landscape' (p. 1).

By the time of the United Nations First Development Decade (1960–70), the certainties of developmentalism had begun to falter. Social and economic conditions for the majority of the population in many of the countries of the capitalist periphery steadily worsened in the immediate post-war years (Frobel *et al.* 1985). Commentators from a wide range of persuasions began to admit (and theorize about) the glaring gap between bland and simplistic expectation and reality. Debate about the nature and causes of the apparent failure to 'develop' has created the burgeoning disputes of development studies, and the proliferation of development theory.

The 1980s saw the rise to authority of a 'counter-revolution' in development theory and practice, opposed both to the established neo-Keynesian approach to planning, and to structuralist and Marxist theories of development (Toye 1993). The counter-revolution emphasized the benefits of free markets and the minimization of the activities of the state. The conversion of key Western governments (and hence of the World Bank) to the doctrine of economic liberalization (as in 'Reaganomics' and 'Thatcherism') for a while carried all before it. Thus the world financial institutions, spearheaded by the implacable economists of the World Bank and the International Monetary Fund, imposed structural adjustment to counteract 'longstanding weaknesses' in national economies and international markets, as revealed by the recession of the early 1980s, in pursuit of recovery and sustained and rapid growth 'of the kind the world enjoyed for twenty-five years after World War Two' (World Bank 1984b, p. 1). In particular, the counter-revolution demanded that governments slim down. As the *World Development Report* commented (in the related context of the transformation of the economies of Eastern Europe and the former Soviet Union), 'the state has to move from doing many things badly to doing its fewer core tasks well' (World Bank 1996, p. 110). Experience in the former Soviet Union, and elsewhere, demonstrated how hard the benefits of that simple prescription would prove to realize in practice.

Debate in development studies has reflected changing ideas about the meaning of development, and the policies necessary to achieve it. With the rise of conservative economic policy in the industrialized world, and the collapse of the Iron Curtain in Europe, old certainties broke down and old enemies wavered and became confused. In radical development theory, there was extensive but unresolved debate within and about Marxism and post-Marxism (Corbridge 1993), while the rise of postmodernism, cultural theory and postcolonialism undermined established certainties. It was widely seen that there was an impasse in development studies (Schuurman 1993). Academics, being enthusiastic arguers, mapped and remapped ways out of that impasse, proposing a renewed dependence on the redemptive powers of neo-populism, 'new social movements', a renewed and radically modernist post-Marxism and postcolonialism. Poststructural critique of development theory gave rise to 'post-development' theory (e.g. Sastchs 1992a; Escobar 1995), and this too was duly counter-critiqued and debated (e.g. Pieterse 1998; Escobar 2000; Matthews 2004). The crisis in development studies released a torrent of words. Meanwhile, the problem of global poverty persisted and deepened, the account of human misery growing almost unchecked.

The challenge of poverty

Whatever the state of development theory, there is no doubt of the ethical imperative of tackling human poverty (Corbridge 1993; Goulet 1995). Poverty remains 'a massive global outrage' (Goodland *et al.* 1993, p. 297). The perception of poverty as one of a series of dramatic and unsolvable problems in developing countries is common to politicians, aid agencies, academic analysts and the media. Indeed, such perceptions have long made crisis the commonplace motif of development writing (e.g. A. G. Frank 1981; Brandt 1983; Frobel *et al.* 1985). Africa, for example, is often stereotyped as locked, helpless, in a rictus of crisis (Watts 1989). Its apparent recurrent crises of war, drought, disease and famine are analysed as if separate from global political economy. As Julius Nyerere (1985), first President of Tanzania, commented 'African starvation is topical, but the relations between rich and poor countries which underlie Africa's vulnerability to natural disasters have been relegated to the sidelines of world discussion' (p. 491).

The dimensions of the 'crisis' of development, or the lack of it, are as broad as the brush of the analyst who paints it. Typically, the problems are held to include debt, low commodity prices, low per capita food production, lack of industrialization, growing poverty and growing inequalities between rich and poor both within and between developing countries. In the new millennium, the issue of poverty was brought to the forefront of international policy. The eradication of poverty became a formal international objective at the United Nations Millennium Summit in September 2000. Eight Millennium Development Goals were agreed, with 18 targets and 48 indicators as yardsticks for measuring improvements in people's lives (Sachs and McArthur 2005).

There has been much debate about the rigidity of the targets-led approach to poverty reduction, and the adequacy of the Millennium Development Goals themselves (e.g. Attaran 2005). Underlying this was a wider recognition of the complexity and multidimensional nature of poverty at the turn of the millennium. Thus as even the World Bank (http://web.worldbank.org, 11 July 2007) notes

> poverty is hunger. Poverty is lack of shelter. Poverty is being sick and not being able to see a doctor. Poverty is not being able to go to school and not

knowing how to read. Poverty is not having a job, is fear for the future, living one day at a time. Poverty is losing a child to illness brought about by unclean water. Poverty is powerlessness, lack of representation and freedom.

This approach reflects the widening of debates about development – for example Amartya Sen's vision (1999) of development as the enhancement of the individual freedoms (political, economic and social).

The *World Development Report 2000/1: attacking poverty* (World Bank 2001) set out the Bank's new pro-poor focus (or renewed that focus: there had been previous *World Development Reports* on poverty in 1980 and 1990). It drew on extensive research, including a study, *Voices of the Poor*, that was novel for the Bank in scale and method, drawing on interviews with poor people in sixty developing countries. Like all such reports, it was upbeat, suggesting that it was possible to make substantial reductions in poverty in the twentieth century. But, for all its optimism, the scale of the problem its data revealed was huge, with 2.8 billion people living on less than $2 a day, and lacking access to education and healthcare, lacking political power and voice, and vulnerable to illness, economic dislocation, personal violence and natural disasters. Globally, 1.2 billion people lack access to safe water and 2.6 billion lack access to sanitation (UNDP 2006).

The idea that decades of formal 'development' effort have created a world where all countries are experiencing economic growth and gains in quality of life (let alone all people in those countries) is an illusion. There has been substantial progress in poverty reduction: the proportion of people subsisting on less than a dollar a day globally halved between 1981 and 2001, and the absolute number of people living at this level stopped growing and started to decline. But that still left 21.3 per cent of the world's population living in extreme poverty, some 1.1 billion people (Wolfensohn and Bourguignon 2004).

Moreover, these gains have been concentrated in particular regions of the world, especially in Asia, and especially in China. Indeed, if China is excluded, the number of people living on less than a dollar a day has actually increased, growing from 836 million to 841 million between 1981 and 2004 (Chen and Ravallion 2007). This increase in the number of the poor was most marked in Africa, where the number living at this level rose from 164 million to 314 million between 1981 and 2001, a brutal 46 per cent of the population (Wolfensohn and Bourguignon 2004).

Success in reducing poverty suggests that it can be overcome for some individuals, families and communities. The 2015 Millennium Development Goals aim to reduce by 50 per cent the proportion of people in absolute poverty. Thanks largely to growth in China and India, this headline goal may well be met, if the wider environmental costs of rampant industrialization and consumer-led growth are discounted. However, even if the 2015 Goals are met in full, there will still be approximately 900 million people, mostly in sub-Saharan Africa and South Asia, whose poverty is intractable (Chronic Poverty Research Centre 2005). Many of these chronically poor people have been poor for years – often since birth: poverty is not only multidimensional, but often passed on from generation to generation.

Plate 1.1 Boy learning to write in Arabic, northern Nigeria. In sub-Saharan Africa, about 70 per cent of adult men and 50 per cent of women are literate. Access to primary education varies within countries by income, gender and urban/ rural location. For every 100 boys there are only 83 girls enrolled in primary education. The disparities between urban and rural areas are two to three times greater than between boys and girls. The World Bank believes that radically improved education is a prerequisite for achieving sustained economic growth at a level sufficiently high to reduce poverty, and help build more inclusive, democratic and equitable societies (www.worldbank.org).

The world economy is not the airport that Walt Rostow imagined, where economies lift off on schedule for a new life in the skies. Some economies move quickly and some slowly, but it is not very helpful to imagine them lined up on a runway, waiting until their engines build up power for an inevitable take-off. Some regional and national economies shrink even as others grow. Less than 10 per cent of the world's gross national product (GNP) of $28,862.2 trillion stems from low-income countries, this figure falling to less than 2 per cent if India and China are excluded (World Bank 2000). A vast number of countries are parked on the grass, well adrift from any rush to growth, or an ideal of equitable world development. Average annual income is less than $300 per head in Burundi, Cambodia, Chad, the Democratic Republic of Congo, the Central African Republic, Eritrea, Ethiopia, Malawi, Mali, Nepal, Niger, Nigeria, Rwanda, Sierra Leone, Tanzania and the Yemen Republic. Many, although not all, of these poorest countries are in Africa, and many are also suffering the destruction brought by civil or international war (for example, Sudan, Democratic Republic of Congo and Somalia). The share of global wealth enjoyed by the world's poorest countries, and by the world's poorest people in all countries, is low and falling.

With income poverty goes a whole host of other forms of deprivation. Amartya Sen urges that we concentrate not simply on income poverty, but on a more inclusive idea of capability deprivation (Sen 1999). As the 1996 *Human Development Report* noted, 'human development is the end – economic growth is a means' (UNDP 1996, p. 1). Since 1990 UNDP has calculated a Human Development Index (HDI) on the basis of longevity (life expectancy at birth) and educational attainment (adult literacy and primary, secondary and tertiary education enrolment ratios, and standard of living (measured as gross domestic product (GDP) per capita). The global index in 2006 was 0.74. The highest ranking country (Norway) scored 0.97, and the USA (eighth) scored 0.95. Developing countries overall scored 0.68, and the least developed countries scored 0.46. The lowest ranked country (Niger, 177th) scored 0.31 (UNDP 2006). In terms of just one measure, these numbers translate into life expectancies at birth of 67.3 years globally, 79.6 in Norway and 77.5 in the USA, compared to 65.2 in developing countries and 52.4 in least developed countries. In Niger life expectancy at birth is just under 45 years (UNDP 2006).

The overall picture of the human dimensions of the challenge in the developing world is clear. Globally, decades of development investment have not driven the problem away. Indeed, the disparities between the world's rich and poor have increased; the gap between the economic power of industrialized countries and the levels of consumption of the majority of their people, and the economic weakness and grinding poverty of the least industrialized countries, has grown. Furthermore, the globalization of economies and the inter-visibility provided by technology make these inequalities more and more glaring. The magnitude of the continuing problem of poverty is the chief evidence for the failure of the practical project of 'development' (that is, the failure of development bureaucracies to solve obvious problems). Behind this failure lie the limitations of both conventional economic thought and its new-right and reconstructed-left critiques, and the discomforting impasse of development studies. This gloomy scene has provided fertile ground for ideas about sustainable development to flourish in the 1980s and 1990s. Perhaps sustainable development could provide an alternative paradigm, certainly a new start. Out with the old (Keynesianism, Marxism, dependency theory, even harsher versions of the new 'market' orthodoxy), in with the new: by the middle of the 1990s the concept of development had deepened and broadened to include dimensions of empowerment, cooperation, equity, sustainability and security (UNDP 1996). By 2000 poverty was the central theme, with the World Bank adopting the striking headline definition of 'deprivation of wellbeing' (Chambers 2001).

The idea of sustainable development was welcomed by development thinkers and practitioners, because it seemed to provide a way out of the impasse and away from past failure, a means of rerouting the lumbering juggernaut of development practice without endangering belief in the rightness and feasibility of its continued forward movement. That mainstream sustainable development is analysed in Chapters 5 and 6.

The challenge of environmental change

Ideas about sustainable development draw on critiques of the development process, for example, from populist writings (including failures of distribution and the plight of the poorest), from radical ideas (such as dependency theory), and from more pragmatic critiques of development project appraisal and implementation. However, in sustainable development these have been wedded to rather different concerns about the *environmental* impacts of development, both the costs in terms of lost ecosystems and species, and (latterly) the impacts of development action on natural resources for human use. Above all, environmentalist critiques of development have presented a picture of the Third World environmental crisis. Indeed, the notion of *global* crisis was an important element of environmentalism in the 1960s and 1970s, and became a central element in debates about sustainable development.

The problems of the environmental impacts of development will be explored in detail in later chapters. I will simply note here that the literature on the global environment has portrayed a second global crisis, of environmental degradation, paralleling that of poverty. From the 1980s, academics and journalists identified many heads to this particular monster, most notably desertification (Grainger 1982), fuelwood shortage (Munslow *et al.* 1988) and the logging of tropical rainforest (Caufield 1982; N. Myers 1984). From 1972 the United Nations Environment Programme (UNEP) published annual 'state-of-the-environment' reports on particular issues, and carried out a review of 'world environmental trends' ranging from atmospheric carbon dioxide and desertification to the quality of drinking water (e.g. Holdgate *et al.* 1982).

Many accounts of the 'state of the world's environment' have been completed since, all of them relentlessly negative in their account of rapid declines in forest cover, rapidly rising global levels of energy use and carbon dioxide production (particularly in Asia), overexploitation of fisheries, depletion of soil resources and shortages of food not met by international trade and aid flows (e.g. Groombridge 1992; Holdgate 1996; WRI *et al.* 1996; UNEP 2000; Jenkins *et al.* 2003; Millennium Ecosystem Assessment 2005; Kennedy *et al.* 2006). Statistics on global change suffer from the same problems of quality and completeness as data on development or poverty. Despite more than three decades of satellite remote sensing data, and the application of increasingly sophisticated computers to the analysis, storage and retrieval of data, information on global or regional environmental change is still patchy and in some instances (in spite of the work of international organizations such as the World Resources Institute (WRI) in the USA and the UNEP–World Conservation Monitoring Centre in the UK) of limited reliability (Groombridge 1992; Jenkins *et al.* 2003). The urgency and melodramatic style of environmental groups desperate for enough media attention to win a hearing from politicians and government decision-makers have sometimes further muddied the waters.

Debates about global environmental crisis have often descended into slanging matches between environmentalist Cassandras crying disaster and conservative (often corporate) sceptics claiming that they exaggerate (e.g. Lomborg 2001). The debate is dogged not only by lack of data, but also by the lack of a clear 'headline' statistic – for example, something that might parallel the debates about poverty in the development field (although that is by no means easy to define, as the vast literature debating its measurement shows). One attempt to derive such a 'headline' is the World Wide Fund for Nature's Living Planet Index (LPI) (Loh *et al.* 1999; Hails *et al.* 2006). This was first calculated in 1998. It tracks global populations of 1,313 species. All these are vertebrates (fish, amphibians, reptiles, birds, mammals), and, although vertebrates represent only a fraction of known species, it is assumed that trends in their populations are typical of biodiversity overall. Separate indices are produced for terrestrial, marine and freshwater species, and then averaged to create an aggregated index. Between 1970 and 2003, the index fell by about 30 per cent. The terrestrial index (695 species) fell by 31 per cent, the marine index (274 species) by 27 per cent and the freshwater index (344 species) by 29 per cent (Hails *et al.* 2006).

Whatever the shortcomings of exercises of this sort and of the data on which they are based, there can be no doubt that the human impact on the biosphere is very extensive, and that it accelerated rapidly over the twentieth century. Global mapping suggests that three-quarters of the habitable surface of the earth has been disturbed by human activity (Hannah *et al.* 1994). The UNEP *Global Biodiversity Assessment* (UNEP 1995) suggested that between 5 per cent and 20 per cent of the perhaps fourteen million plant and animal species on earth are threatened with extinction. Rates of species extinction are hard to estimate with any accuracy, but Edward Wilson suggests that human activities have increased previous 'background' extinction rates by at least between 100 and 10,000 times. He comments, 'we are in the midst of one of the great extinction spasms of geological history' (E. O. Wilson 1992, p. 268). The rapid loss of species owing to human action has become an incontrovertible fact (Prance 1991; Smith *et al.* 1993; Hails *et al.* 2006; Secretary of the Convention on Biological Diversity 2006).

The fundamental dynamic of environmentalist concern about development in its broad sense – the expansion of industrial capacity, and the urbanization and sociocultural changes that accompany it – is the scale of human demands on the biosphere. The scale of human annexation of biological processes has slowly become apparent. Vitousek *et al.* (1986) calculated that 40 per cent of potential terrestrial net primary production was used directly by human activities, co-opted or forgone as a result of those activities. This consumption embraces food and other products both directly consumed (for example, crops, fish, wood, and so on), and consumed by livestock, as well as production consumed less directly (for example, in fires or human-induced soil erosion). Subsequent analyses using global survey data suggest similar levels (Rojstaczer *et al.* 2001 suggest 31 per cent), and marked regional patterns in the footprint of human consumption and its associated environmental impacts (Imhoff *et al.* 2004). Fertilizer made from industrially produced ammonia sustains roughly 40 per cent of the human population and comprises 40–60 per cent of the nitrogen in the human body (Fryzuk 2004). Vitousek *et al.* (1986) are surely right in suggesting that 'an equivalent concentration of resources into one species and its satellites has probably not occurred since land plants first diversified' (p. 372). Humans are not simply annexing the earth's productivity, and driving the reduction of living diversity, but are also starting to affect evolutionary trajectories in the species and ecosystems that can persist alongside their seemingly insatiable demands (Palumbi 2001).

At different times there have been various species selected as canaries to show that all is not well in the global coalmine. In the twentieth century, for example, concern grew about human impacts on the oceans. The fact that diversity and productivity were concentrated in particular places (tropical and temperate coral reefs, for example, or deep ocean seamounts) began to be recognized, as were the vulnerability of productive shallow seas and the damage of industrial fish-harvesting techniques such as trawling. The ecology of the oceans has been transformed by human action, to an extent that even marine scientists have only recently begun to appreciate (Pauley *et al.* 2003; C. M. Roberts 2003, 2007). Research shows that, globally, the population of large predatory fish is now less than 10 per cent of preindustrial levels (Myers and Worm 2003). As Vitousek *et al.* (1997) noted, 'the rates, scales, kinds and combinations of changes occurring now are fundamentally different from those at any other time in history; we are changing the earth more rapidly than we are understanding it'.

Since the 1990s, the issue of climate change has come to dominate environmentalist discourse about human impacts on the biosphere. The World Climate Conference in Geneva in 1978 called attention to the problem of greenhouse gases and anthropogenic climate change, and the work of the IPCC from 1988 established a strong global scientific consensus that human action was indeed affecting global climatic patterns. Human impacts on climate are superimposed on natural variation, and the global (and even more the regional) ocean–atmosphere system is notoriously hard to model satisfactorily. However, the IPCC consensus has held. The *Second Assessment Report* in 1995 concluded that the global mean temperature of the twentieth century was at least as warm as any since 1400, and discussed what was rather blandly described as 'an enhanced global mean hydrological cycle', meaning more droughts and floods and storms (Houghton *et al.* 1995). The *Third Assessment Report* (2001) found a 100-year trend in temperature (1901–2000) was +0.6 °C, while the *Fourth Report* in 2007 noted that the period 1906–2005 had been hotter, at +0.74 °C (IPCC 2007a). Temperature rises were experienced widely across the globe, but were greatest at northern latitudes and on land masses. Eleven of the twelve years 1995–2006 were among the twelve warmest years in the instrumental record, which began in 1850 (IPCC 2007a). The impacts of anthropogenic climate change, globally and regionally, are acknowledged to be highly complex (Parry *et al.* 2007). The implications for environment and society (particularly in the tropics) are, however, also recognized to be huge.

Anthropogenic climate change has become more than some environmentalist bogey; it is now accepted scientific fact. The Framework Convention on Climate Change, signed at Rio, reflects that acceptance, although views about who should take what action and when vary a great deal (see Chapter 4). To environmentalists,

Plate 1.2 Mount Kenya National Park, Kenya. Mount Kenya lies across the equator, 140 kilometres north of Nairobi in Kenya. The mountain is 5200 metres high, with slopes of forest, bamboo, scrub and moorland rising to permanent snow and ice. It is an important water catchment area. The Mount Kenya National Park covers 715 square kilometres, and includes all land above 3,200 metres. It is surrounded by the Mount Kenya National Reserve (2,095 square kilometres). It is a UNESCO World Heritage Site and a Biosphere Reserve. Mount Kenya's snow and ice, its alpine vegetation and animals, and its capacity to supply ecosystem services in the form of water supply are all threatened by anthropogenic climate change in the twenty-first century.

the evidence for human impacts on climate has offered clear evidence of the unacceptably large scale of human demands on the biosphere. It represents a significant challenge to developmentalism and its conventional strategies of industrialization and economic expansion.

Clearly, just as the idea and practices of development are an unprecedented human enterprise of the past century or two, there is a novelty to that enterprise's demands on the natural systems of the earth. As environmentalists, from the 1960s onwards, have said repeatedly, humans have not been here before: there are no road maps for the future. How is life (human and non-human) to be sustained? It was to answer such questions that the discourse of sustainable development was created.

Environment and development: one problem, two cultures

The threat of multidimensional global crisis has, therefore, been a key theme within debates about sustainability: a crisis of development, of environmental quality and of threats to the material benefits supported by natural biogeochemical processes and sinks. The 1992 *World Development Report* opened with the assertion that 'the achievement of sustained and equitable development remains the greatest challenge facing the human race' (World Bank 1992, p. 1). By the 1990s the point that there are close links between the problems of development and environment had sunk in, although, despite significant advances (Redclift 1984, 1987; Blaikie 1985; Blaikie and Brookfield 1987; Redclift and Benton 1994; J. A. Elliott 1999), theoretical understanding of the links has continued to lag behind practical and rhetorical recognition of the problem (see Chapters 4, 5 and 6).

It is recognized that tight and complex links exist between development, environment and poverty (e.g. Broad 1994; Blaikie 1995; Reardon and Vosti 1995, Bass *et al.* 2005). The poor often endure degraded environments, and in some instances contribute to their further degradation. The idea that poverty and the environment are linked was fundamental to the work of the Millennium Ecosystem Assessment (MEA) (2005), and to the work of initiatives such as the Equator Initiative (Timmer and Juma 2005). The MEA noted that progress achieved in addressing the goals of poverty and hunger eradication, improved health, and environmental protection was unlikely to be sustained if the ecosystem services on which humanity relies continue to be degraded (Millennium Ecosystem Assessment 2005).

However, in the three decades of sustainable development that followed publication of the Brundtland Report in 1987, the fields of developmental and environmental studies have been far from unified. The one language of sustainability has hidden the separation of two cultures, which have often remained remote from each other both conceptually and practically. Despite the rise of careers in 'environment and development', and of massive sources of funding such as the Global Environmental Facility that fuel them, debates about the politics and economics and sociology of development and the science of environmental change have remained separate fields. Development and environment work are still the fruit of distinct cultures. Although the two overlap a great deal, and indeed make confident inroads onto each other's territory with scant regard for the exact meaning or purpose of terminology, there is rarely if ever any integration. Sociologically, different disciplines still have their own separate cadres and culture, their own self-contained arenas of education and theory formation, their own technical language and research agendas, and – above all – their own literature.

The need for effective interdisciplinarity to make sense of the problems of environment and development is blindingly obvious. As Piers Blaikie (1995) comments, 'environmental issues are by definition also social ones, and therefore our understanding must rest on a broader interdisciplinary perspective that transcends institutional and professional barriers' (p. 1). In practice, however, both academics and practitioners are reluctant to cross disciplinary boundaries. Our individual 'disciplinary bias' is deeply coded by our training, and is a severe constraint on innovative thinking (Chambers 1983). This problem is not confined to the developing world. Thus it is recognized that research on global environmental change must be pursued through collaboration between the natural and social sciences; however, such work is by no means easy to achieve successfully (Miller 1994). Unrealistic expectation, problems of data and measurement, and problems with the ways in which research questions are framed all represent challenges to interdisciplinarity (L. M. Campbell 2005). The differing perspectives of ecologists and economists have long provided difficult terrain for effective engagement on issues of sustainability (Tisdell 1988). While the development of new fields such as conservation biology attempt to transcend existing disciplines (Meine *et al.* 2006), problems of communication persist (Agrawal and Ostrom 2006; Fox *et al.* 2006).

There is a critical ideological component in understanding environmental resource use and environmental change and degradation. This was an important element of the developing field of political ecology (Forsyth 2003; Neumann 2004c; Peet and Watts 2004). Narratives or discourses about the environment have enormous power to condition and constrain even apparently 'impartial' scientific research on the environment (Hajer 1995; Leach and Mearns 1996). Environmental scientists, and environmentalists, persistently fail to recognize the ideological burden of ideas and policies, yet the way environmental problems are formulated and understood has considerable significance for the possibility of dealing with them equitably and effectively (Adams *et al.* 2003).

Development crises and environmental crises exist side by side in the literature, and together on the ground, yet explanations often fail to intersect. Environmentalists and social scientists speak different languages. Very often theirs is a dialogue of the deaf, carried on at cross purposes and frequently at high volume (Agrawal and Ostrom 2006). The complex and multidisciplinary nature of the links between development, poverty and environment makes them difficult to identify and define. They often go unnoticed, fall down the cracks between disciplines, or get ignored because they fit so awkwardly into the structures of academic analysis or discourse. Nonetheless, in the real world these links are real enough. They explain why development policy often causes rather than cures environmental problems. Development and environmental degradation often form a deadly trap for the poor.

Chambers argued in his book *Rural Development: putting the last first* (1983) that it is the plight of the poor that should set the agenda for development action. The idea of sustainable livelihoods (Conroy and Litvinof 1988; Carney 1998) remains a powerful and aspirational concept (Chambers 2005). There are both moral and practical imperatives for making sustainable livelihood security, defined as the secure access to sufficient stocks and flows of food and cash to meet basic needs, the focus for development action. However, there is something comfortably technical about this phrase. It perhaps permits a sense that the challenges of sustainable development are in essence an engineering problem, something that 'we' have a duty to organize for 'them'. But sustainable development is a concept with two sharp ends, which must challenge established wealth and power, and highlight the moral responsibility of the wealthy and powerful in the universal search for wellbeing (Chambers 2005).

The touchstone for debate about environment and development is the human needs of the poor, in terms of both the environment and the economic social and political means that can support a good life. Debate about the environment, like that about development, is inherently political; but, as Michael Redclift (1984) argued, it is an illusion to believe that environmental objectives are 'other than political, or other than distributive' (p. 130).

Outline of the book

This book is not another attempt to find the winning formula, the mix of sticks and carrots, rhetoric, capital flows and environmental knowledge that will achieve 'real' development. Rather, it is about the peculiar difficulty of talking sensibly about the environmental dimension to development in the Third World. Its aims are, first, to discuss the nature and extent of the 'greening' of development theory. It does this by examining the key concept of sustainable development. It looks at the origins and evolution of these ideas, and offers a critique of their articulation in the *World Conservation Strategy* (*WCS*), the Brundtland Report and the documents of the Rio and Johannesburg conferences. It is argued that the ideology of sustainable development is eclectic and often confused. Sustainable development is essentially reformist, calling for a modification of development practice, and owes little to radical ideas, whether claiming a green or a Marxist heritage. Second, the book attempts to draw a link between theory and practice by discussing the nature of the environmental degradation and the impacts of development. In doing so it attempts to address the question of the limitations of reformist approaches. It argues that, ultimately, 'green' development has to be about political economy, about the distribution of power, and not about environmental quality.

The first part of the book is largely concerned with ideas and theories about environment and development, and focuses in particular on the global scale. Chapter 2 discusses the origins and growth of sustainable development ideas, looking in particular at their roots in nature preservation, colonial science and the internationalization of scientific concerns in the 1960s and 1970s. This is an account of institutions and organizations as well as ideas, and takes the form of a historical account. Attention is focused on the 1970s and the Conference on the Human Environment in Stockholm in 1972, at which sustainable development became a specific and identified area of concern. The *WCS* was a direct development of the thinking at that time.

Chapters 3 and 4 discuss the evolution of 'mainstream sustainable development', arguing that a coherent set of ideas has persisted through the 1980s and 1990s, in the *WCS* and the Brundtland Report, in *Caring for the Earth*, and (in Chapter 4) in the work of the United Nations Conferences at Rio and Johannesburg. Mainstream sustainable development draws on both technocentrist and ecocentrist worldviews, the first being rationalist and technocratic, and leading to approaches to the environment involving management, regulation and 'rational utilization', and the latter being romantic and transcendentalist, embracing ideas of bioethics and the intrinsic values of non-human nature (O'Riordan 1981; O'Riordan and Turner 1983; Worster 1985; Turner 1988a).

This history of ideas goes some way to explaining why current visions of sustainable development are rather messy: enthusiastic and committed without, in general, being overtly political.

The theoretical dimensions of sustainable development are explored in Chapters 5, 6 and 7. Mainstream sustainable development is essentially reformist, a broadly neo-populist vision of the world being allied with a call for more technically sophisticated environmental management. Chapter 5 discusses market environmentalism, ecological modernization, environmental populism and the continuing power of neo-Malthusian ideas, and the role of private corporations in delivering sustainable development. Chapter 6 discusses environmental and ecological economics and the measurement of sustainability at different scales from the national economy to the individual project. More radical ideas about environment and development are discussed in Chapter 7, including eco-socialism, eco-anarchism, deep ecology and eco-feminism, in terms of their relations both to the conventional reformism of mainstream sustainable development thinking and to wider radical thought.

The second half of the book provides a commentary on the theoretical ideas in the first half by discussing environment and development in practice. Chapter 8 explores issues of sustainability and environmental degradation, looking particularly at the power, persistence and adequacy of ideas about desertification and overgrazing, and the way they relate to scientific research. Chapter 9 discusses sustainability in the context of tropical forests, reviewing narratives of forest loss and the political ecology of deforestation and the timber trade. Chapter 10 considers the politics of biodiversity conservation, including the social and economic impacts of protected areas, and attempts to combine conservation and development. In Chapter 11, the environmental costs of imposed development are described through the specific lens of water resources and the impacts of dams. Chapter 12 analyses industrial and urban hazard, and the politics of environmental risk. The final chapter of the book (Chapter 13) discusses the tensions between various 'reformist' approaches to sustainability and more radical strategies. It suggests that there are significance unanswered questions to be faced by those who wish to promote 'sustainable' development.

This book is not an encyclopaedia. It does not offer a compete synthesis of sustainable development thinking, nor attempt to set out a blueprint that will make 'sustainable development' work. Its aim is not to celebrate the achievements of sustainable development but to tease out the tensions in the heart of the calculated reformism of its mainstream. Important elements in green critiques of development are radical and not reformist, and they are both awkward and inconvenient. Mainstream sustainable development is bureaucratically and politically acceptable, because it seeks to reprogramme the juggernaut of development through reformist thinking, involving better measurement of social and environmental impacts, better assessment of costs and benefits, better 'clean' technologies and efficient planning procedures. Alternative countercurrents within sustainable development offer politically far more risky waters, for they challenge the global status quo and raise painful and radical questions. Between these two broad views there is unavoidable tension. The ethics of sustainability demand rather more than merely reform of the development process. The 'greening' of development demands a more radical analysis, and a more transformative response.

Charles Reich (1970) wrote with passion and hope in *The Greening of America* of the new consciousness abroad in the USA, 'arising from the wasteland of the corporate state like flowers pushing up through the concrete pavement' (p. 328). Almost four decades later, a generation for many and a lifetime for many in the South, the confident exuberance of that time is gone. However, many of the seeds sown in the 1960s have taken root, and environmentalism is one of them. Despite their flaws, the growth of sustainable development ideologies has had an impact on the consciousness that informs development thought and action. There is a new environmental awareness in development that is perhaps evidence of a 'greening' of a kind. To date it has been largely superficial, a thin green layer painted onto existing policies and programmes. The challenge of more profound change still lies ahead.

Summary

- Sustainable development has become a central concept in development studies, building on environmental, social and political critiques of development theory and practice.
- There is no simple single meaning of 'sustainable development': a wide range of different meanings are attached to the term. Far from making the phrase useless, it is precisely because of its ability to host divergent ideas that sustainable development has proved so useful, and has become so dominant.
- One reason for the complexity of concepts of sustainable development is the confused and contested meaning of development itself. The idea of sustainable development gained currency in the 1990s, at a time when development thought was widely held to have reached an impasse.

• The use of the term 'sustainable development' reflects in particular the prominence at the end of the twentieth century and the beginning of the twenty-first of the problem of acute global poverty and global environmental degradation. Although it is now acknowledged that these crises are linked, problems of environment and development are often addressed independently. They have to be tackled in an integrated way; the challenge of doing so is inevitably political. There are choices to be made between reformist and radical ideas about sustainability and development.

Further reading

Bass, S., Reid, H., Satterthwaite, D. and Steele, P. (2005) *Reducing Poverty and Sustaining the Environment: the politics of local engagement*, Earthscan, London.

Brown, L. R. (2006) *Plan B. 2.0: rescuing a planet under stress and a civilization in trouble*, W. W. Norton, New York, for the Earth Policy Institute.

Chambers, R. (2005) *Ideas for Development*, Earthscan, London.

Elliott, J. A. (2005) *An Introduction to Sustainable Development*, Routledge, London (3rd edn).

Forsyth, T, (2005) *Encyclopedia of International Development*, Routledge, London.

Guha, R. (2000) *Environmentalism: a global history*, Oxford University Press, Delhi.

Kennedy, D. and the Editors of *Science* (2006) (eds) *Science Magazine's State of the Planet 2006–7*, Island Press, for the American Association for the Advancement of Science, Washington.

Middleton, N. (2003) *The Global Casino: an introduction to environmental issues*, Arnold, London (3rd edn).

Millennium Ecosystem Assessment (2005) *Ecosystems and Human Wellbeing: synthesis*, Island Press, Washington.

Redclift, M. R. (1996) *Wasted: counting the cost of global consumption*, Earthscan, London.

Redclift, M. R. (2000) *Sustainability: life chances and livelihoods*, Routledge, London.

Redclift, M. R. (2005) *Sustainability: critical concepts in the social sciences*, Routledge Major Works (four volumes), Taylor and Francis, London.

Sachs, W. (ed.) (1992) *The Development Dictionary: a guide to knowledge as power*, Witwatersrand University Press, Johannesburg, and Zed Books, London.

Schech, S. and Haggis, J. (2002) (eds) *Development: a cultural studies reader*, Blackwell, Oxford.

Sen, A. (1999) *Development as Freedom*, Oxford University Press, Oxford.

Willis, K. (2004) *Theories and Practices of Development*, Routledge, London.

Web sources

http://www.ipcc.ch/ Intergovernmental Panel on Climate Change

http://www.fao.org/waicent/faoinfo/sustdev/ Natural Resources Management and Environment Department of the Food and Agriculture Organization (FAO) of the United Nations, with divisions on environment, climate change and bioenergy, land and water and research and extension.

http://www.wri.org/ The World Resources Institute: environment, resources and sustainable development, including biodiversity, forests, oceans and coasts, water and health.

http://www.unep-wcmc.org/ The UNEP–World Conservation Monitoring Centre: data and maps on conservation and sustainable use of the world's living resources, including the status of species, freshwaters, forests and marine environments.

http://www.footprintnetwork.org/ The Global Footprint Network: data on global and national footprints.

http://www.worldbank.org/poverty/ The World Bank's 'PovertyNet': poverty and its alleviation.

http://devdata.worldbank.org/wdi2006/contents/cover.htm World Bank Group World Development Indicators.

http://www.un.org/esa/sustdev/ United Nations Department of Economic and Social Affairs Division for Sustainable Development (including information on the United Nations Commission on Sustainable Development, on individual countries, and the particular problems of 'Small Island Developing States').

http://www.undp.org/ The United Nations Development Programme: UNDP Human Development Reports, and the 'End Poverty 2015' Millennium Campaign.

2 The origins of sustainable development

We are now in the middle of a long process of transition in the nature of the image which man has of himself and his environment.

(Kenneth E. Boulding, 'The economics of the coming spaceship earth', 1966)

Environmentalism and the emergence of sustainable development

The phrase 'sustainable development' has become the focus of debate about environment and development. It is not only the best-known and most commonly cited idea linking environment and development; it is also the best documented, in a series of publications, beginning with the *World Conservation Strategy* (*WCS*) (IUCN 1980) and the Brundtland Report, *Our Common Future* (1987), and leading to the documents arising out of the Rio Conference in 1992 and the Johannesburg Summit in 2002. These mainstream documents are the subject of the next two chapters, which discuss their arguments and assess the nature of the ideology that shapes their ideas about development. However, the concept of sustainable development cannot be understood in a historical vacuum. It has many antecedents, and over time has taken on board many accretions and influences. These are the subjects of this chapter.

The history of thinking about sustainable development is closely linked to the history of environmental concern and of the conservation of nature in Western Europe and North America. An understanding of the evolution of sustainable development thinking must embrace the way essentially metropolitan ideas about nature and its conservation were expressed on the periphery in the twentieth century, initially on the colonial periphery and latterly within the countries of the independent developing world. This focuses attention in particular on the rise of international environmentalism in the second half of the twentieth century (Boardman 1981; McCormick 1989; Holdgate 1999; Guha 2000). The phenomenon of that emergence is well described elsewhere (e.g. O'Riordan [1976] 1981; Cotgrove 1982; Hays 1987; McCormick 1989; Rawcliffe 1998; Guha 2000). Its intellectual roots lie a great deal further back and are beyond the scope of this book to unravel. They have been explored, for example, by Merchant (1980),

K. Thomas (1983), Pepper (1984), R. H. Grove (1990a, 1995) and Grove *et al.* (1998).

As McIntosh (1985) points out in his history of ecological science, *The Background of Ecology*, stories of the development of ideas divorced from social and intellectual context are of little value. It is unhelpful to look for clear and simple 'roots' to ideas that in fact relate to each other through time in a complex and fluid way, and that at any given time are held and articulated in diverse ways by different people. Ideas about non-human nature, and particularly about ways people or society should treat or manage nature, are both subtle and intractable. They reflect changing ideas about society itself, and are slippery, and hard to trace through space and time. An account of the evolution of different strands of thought about what we have come to call 'sustainable development' is therefore not entirely straightforward. As Guha (2000) points out, many environmental thinkers (like the social ecologist Patrick Geddes, or his followers Lewis Mumford and Radhakanal Mukherjee) defy attempts to pigeonhole them in particular debates.

However, if the nature of the thinking about sustainable development that emerged in the 1960s is to be understood, it is necessary to tease out some of the strands in the complex fabric of its past. This chapter does this by selecting nine themes, overlapping in time. The first is the rise of environmentalism as a global concern. The second is the development of nature preservation, and later of conservation, both in industrialized countries and their global empires. The third theme is the institutionalization of global environmental concern through international organizations. The fourth is the importance of ecological ideas about the 'balance of nature', the fifth the place of ecology in tropical (especially colonial) development and the sixth the growth of ecological managerialism. The seventh theme is the growth of concern about the ecological impacts of development, and the eighth is the rise of perceptions of global environmental crisis, and particularly the perceived threat of human population growth. The final theme is the increasingly international organization of scientific concern about the environment, particularly in the form of the Man and the Biosphere Programme of the United Nations Educational, Scientific and Cultural Organization (UNESCO).

Global environmentalism

The attempt to write about the history of sustainable development is made difficult by the abundance of, and the Eurocentric and Americocentric focus of, the literature on environmentalism (Guha and Martinez-Alier 1997; Guha 2000). Accounts of the 'global environmental movement' have tended to portray its history as an almost exclusively northern-hemisphere phenomenon. This ethnocentrism should make us wary of international comparisons that are in fact based on European or North American experience (Redclift 1984; Guha 2000). Southern environmental non-governmental organizations (NGOs) began to appear from the 1970s onwards, and their number and capacity have grown rapidly (Guha and Martinez-Alier 1997; Dalton *et al.* 2003; Dwivedi 2005). By the end of the twentieth century an environmental movement was even developing in China (Ho 2001). However, the size and influence of environmental NGOs based in industrialized countries are such that they remain the dominant force internationally (Princen and Finger 1994; Guha 2000). The effectiveness of developing world grass-roots organizations is often precisely in their ability to transcend locality and connect to international arenas, and this is often done through better-connected metropolitan partners (Dwivedi 2005).

Concern about human relationships with the environment ran deep through the medieval and modern periods in Europe (K. Thomas 1983). Mediterranean classical writing provided numerous forerunners of Western thinking (Glacken 1967). They influenced thinking about the destructive power of human activities in North America and Europe, particularly in the second half of the nineteenth century, most famously in George Perkins Marsh's *Man and Nature* (1864). Marsh ([1864] 1965) observed: 'man is everywhere a disturbing agent. Wherever he plants his foot, the harmonies of nature are turned to discords. The proportions and accommodations which ensured the stability of existing arrangements are overthrown' (p. 36). Marsh's classical education, his boyhood in Vermont and his sojourns in Europe created a truly modernist critique of industrialization and the environmental demands of economic growth. Nature, he explained and demonstrated, 'avenges herself upon the intruder, by letting loose on her defaced provinces destructive energies hitherto kept in check by organic forces destined to be his best auxiliaries, but which he has unwisely dispersed and driven from the field of action' (p. 42). Similar perceptions, if less scientifically expressed, had surfaced elsewhere in the industrializing world, most significantly perhaps in the Romantic movement and the ideas of people like the British poet William Wordsworth or John Ruskin (Bate 1991; Veldman 1994). Concern for the conservation of nature in the USA and in Europe tapped these concerns in a direct way (Adams 1996, 2004).

Twentieth-century environmentalism, with its rather belated concern for the unindustrialized, tropical and colonial parts of the world, therefore, has immediate precursors in the industrializing world in the nineteenth century. However, this is far from a complete picture (R. H. Grove 1990a, 1995, 1998b; Barton 2002; Grove and Damodaran 2006). Global environmental concern was neither simply a local response to the conditions of Western industrialization, nor something derived exclusively from Northern attitudes. From the fifteenth century onwards, global trade and travel transformed European ideas of nature, and these changes developed further with European colonial conquest. New ideas fed back into Europe from imperial possessions and informed changing philosophical ideas about nature, and about the significance of human claims on the earth. The newly described tropical world was appropriated spiritually, becoming associated with ideas of paradise or Eden. The tropics (and especially tropical islands) became the symbolic location for the 'idealised landscapes and aspirations of the Western imagination' (R. H. Grove 1990a, p. 11). Indeed, in the seventeenth century, in the heady mix of religious belief and infant rational science, scholars speculated that the Garden of Eden might be a real place, and explorers set out to find it (Withers 1999).

One element in this process was the rise of European science, and its transnational, indeed increasingly global, reach. The work of Swedish taxonomist Carl Linnaeus in the eighteenth century provided a system of classifying organisms that drew directly upon biological collections from explorers and colonists. Indeed, Mary Pratt suggests that this North European industry of naming and classifying the bizarre species revealed by exploration 'created a new kind of Eurocentred planetary consciousness' (Pratt 1992, p. 39). The development of science, and the Enlightenment separation of natural and human that lay behind it, went hand in hand with the growing power of European imperialism from the sixteenth century onwards, a process of tightening 'government' of nature (Drayton 2000).

In the mid-seventeenth century, it began to be realized that human activities on tropical and subtropical islands such as the Canary Islands, Madeira or Mauritius threatened to destroy both their natural beauty and their bounty as a source of food and timber for passing ships trading across far-flung European empires. From these isolated (and increasingly devastated) fragments of paradise grew an awareness of the ecological impacts of emergent capitalism and colonial rule (R. H. Grove 1990a, 1995, 1998a; Grove and Damodaran 2006). In due course this led to an awareness of environmental limits and the need for conservation, as experience of ecological change on islands was translated into a more general awareness of the possibility of environmental destruction on a global scale (Grove 1990b, 1995).

Scientists employed by imperial trading companies as surgeons and botanists (the Dutch and English East India Companies and the French *Compagnie des Indes*) developed and disseminated ideas about desiccation, drought and famine, and environmental limits to human action (Grove and Damodaran 2006). The Indian Forest Service, in particular, was a critical arena within which environmentalist ideas were developed (Grove 1998b; Grove *et al.* 1998; Rajan 1998). These drew, in particular, on German traditions of scientific forestry, which spread widely – for example, to Australia and the USA (Guha 2000; Barton 2002). Ideas of environmental crisis, environmental limits and environmental management, developed in the colonial periphery, where full-throated capitalist and imperialist expansion met tropical societies and ecosystems for the first time, became the familiar basis for twentieth-century environmentalist concern, and an important source of ideas about sustainable development.

Nature preservation and sustainable development

In many ways, wildlife or nature conservation has been the most deep-seated root of sustainable development thinking. Indeed, sustainable development was put forward as a concept partly as a means of promoting nature preservation and conservation. The history of nature conservation in countries of the industrial metropole – for example in Britain (D. E. Allen 1976; Sheail 1976; D. Evans 1992) or the USA (Nash 1973; Worster 1985) – is well established. Although the intellectual roots of a concern for nature (either for its own sake, or for fear of repercussions of its misuse for people) lay deeper and further back (K. Thomas 1983; Pepper 1984; R. H. Grove 1995), the foundation of formal organizations to carry out and promote conservation began in the nineteenth century.

Thus in Britain the second half of that century saw legislation for the protection of seabirds and the establishment of a number of conservation organizations such as the Commons Open Spaces and Footpaths Preservation Society (1865), the Royal Society for the Protection of Birds (founded initially in 1893), the National Trust for Places of Historic Interest and Natural Beauty (1894), and the Society for the Promotion of Nature Reserves (1912). The early years of the twentieth century saw the foundation in 1909 of the Swiss League for the Protection of Nature (primarily to raise funds for a national park, achieved in 1914), and of the Swedish Society for the Protection of Nature. There were parallel developments in Germany (Conwentz 1914). In the USA the Yellowstone National Park was established in 1872, the Boone and Crocket Club formed in 1887, and the Sierra Club in 1892. National parks were established in Australia in 1879, Canada in 1887 and New Zealand in 1894. In Britain debate about the need for national parks ran through the early decades of the twentieth century, before they and government nature reserves were eventually made possible by Act of Parliament in 1949 (Sheail 1976, 1984, 1996).

These developments were primarily aimed at promoting the protection of nature within the industrialized nations themselves. However, from an early date there was also concern about conservation on a wider geographical scale, in imperial or colonial possessions (McCormick 1989; Grove 1995). In Africa, for example, concern about the depletion of forests in the Cape Colony developed in the early nineteenth century. This, with pressure for government money for the botanic garden (*inter alia* from the Royal Botanic Gardens at Kew, itself taken over by the government in 1820), led to the appointment of a Colonial Botanist in 1858. Legislation to preserve open areas close to Cape Town was passed in 1846, and further Acts for the preservation of forests (1859) and game (1886) followed (Grove 1987). Similar institutions were created in other imperial territories: in India, for example, 30 per cent of non-agricultural land in some provinces had been brought under the control of the Forest Department (R. H. Grove 1990b).

Conservation in the British Empire had several strands. The concern about environmental degradation and climatic change discussed above, and the utilitarian and holistic conservation ideas to which they gave rise—for example, in the African Cape in the 1880s—brought them into conflict with settlers, and 'the driving interests of local European capital' (Grove 1987, p. 36). On the other hand, more narrowly focused concerns for the protection of species could be more easily accommodated to settler demands (Adams 2004). In particular, the establishment of protected areas could be fitted into a spatial jigsaw of land apportionment between settler, native and nature (Neumann 2004a, b). Moreover, the preservation of species reflected persistent idealist (and orientalist (cf. Said 1979)) ideas of tropical 'nature' as 'Eden' and the need to protect it from rash humanity (R. H. Grove 1990b).

The spectre of extinction because of over-hunting was a potent element in the establishment of wildlife conservation institutions and organizations at the end of the nineteenth century. European hunting was important in colonial Africa commercially (especially for ivory), as a way in which African rulers first began 'riding the tiger of European advance' (MacKenzie 1988, p. 43) and as a subsidy for that advance, a source of meat for railway construction workers, or a means of feeding and financing trade and missionary activity. Most importantly, however, hunting was important as a defining social practice of the colonial elite, with its Victorian and Edwardian obsession with trophies, sportsmanship and the ideals of British boys' education (MacKenzie 1988). As rifles improved and the number of hunters rose, astonishing numbers of game animals were killed. Alistair Graham (1973) picturesquely commented, 'the swirling torrents of bloodlust that were gratified are beyond our powers of measurement' (p. 54). By the last decades of the nineteenth century, substantial areas of southern Africa were more or less emptied of game, certainly near white settlements, railways and wagon trails.

In London, the Society for Preservation of the Wild Fauna of the Empire (SPWFE) was founded in 1903 by a distinguished and powerful roll-call of politicians and aristocracy, colonial administrators, businessmen, hunters, scientists and naturalists (Neumann 1996; Prendergast and Adams 2003). These elite British conservationist-hunters ('penitent butchers' (Fitter and Scott 1978)) had experience and interests of various kinds in Africa, and between 1905 and 1909 the SPWFE took three delegations to the Secretary of State for the Colonies to argue for stronger hunting regulations and for game reserves. In 1906 a delegation from the society told the Secretary of State for the Colonies that it was 'the duty and the interest of Great Britain' to follow the US example and establish protected areas in East Africa (Adams 2004). The government game reserve became the mainstay of colonial conservation in the British Empire, a resort for gentleman hunters, whether traveller or colonial servant (MacKenzie 1988; Neumann 1996, 2004a; Adams 2004).

The most obvious aspect of the conservation based on this hunting ethos was the complete denial of hunting to Africans. White men hunted; Africans poached. This denial was achieved through controls on firearms, and latterly by the establishment of game reserves. The Cape Act for the Preservation of Game of 1886 was extended to the British South African Territories in 1891. In 1892 the Sabie Game Reserve was established (to become the Kruger National Park in 1926), and in 1899 the Ukamba Game Reserve was created in Kenya, including land in what became the Amboseli National Park. In 1900 the Kenyan Game Ordinance was passed, effectively banning all hunting except by licence (Graham 1973; MacKenzie 1988; Adams 2004).

Hunting in Africa also led to significant institutional developments. A conference of African colonial powers (Britain, Germany, France, Portugal, Spain, Italy and the Belgian Congo) met in London in 1900 and signed a Convention for the Preservation of Animals, Birds and Fish in Africa (Fitter and Scott 1978; McCormick 1989). The 1900 Convention was never set in operation, but in the early decades of the century there were a number of further developments. For example, in the mid-1920s a French Permanent Committee for the Protection of Colonial Fauna was established, and in 1925 King Albert created the gorilla sanctuary that became the Parc National Albert (now the Virunga National Park), the first African national park (Fitter and Scott 1978; Boardman 1981).

After the Second World War, interest in international conservation, particularly in Africa, spread. Bernard Grzimek led work by the Frankfurt Zoological Society in Kenya and Tanzania, and Frank Fraser Darling was appointed to lead the Conservation Foundation (set up in 1948 by the New York Zoological Society under Fairfield Osborn). National parks were declared in a series of colonial territories as the prospect of decolonization loomed – for example Nairobi National Park in Kenya in 1946 and Tsavo two years later; Wankie in Southern Rhodesia, Serengeti in Tanganyika in 1951, and Murchison Falls and Queen Elizabeth National Parks in Uganda in 1952 (Fitter and Scott 1978).

A special conference on African conservation problems at Bukavu in the Belgian Congo in 1953 made what McCormick (1989) describes as 'a tangential departure from previous thinking' (p. 43) towards the idea of sustainable development. It proposed broadening the concerns of conservation from simply the preservation of fauna and flora to the wider human environment, and suggested a new convention to address the whole natural environment and focus on the needs of Africans (McCormick 1989). However, the work of game departments in Africa remained much more narrowly focused, dominated by the 'poaching problem' (Graham 1973; Steinhart 1989; Beinart and Coates 1995; Adams 2004). The classic argument about the destructiveness of hunting by Africans (as opposed to Europeans) was well expressed by Richard Hingston, who was sent by the Society for the Preservation of the Fauna of the Empire (SPFE; the word 'wild' was dropped from their title after the First World War) to East Africa to investigate (and promote) national parks in the 1930s. He said:

> It is commonly thought that the visiting sportsman is responsible for the decline of the African fauna. That is not so. The sportsman does not obliterate wild life. True, he kills. But seldom is the killing wholesale or indiscriminate. What the sportsman wants is a good trophy, almost invariably a male trophy, and the getting of that usually satisfies him … The position is not the same with the native hunter. He cares nothing about species or trophies or sex, nor does he hunt for the fun of the thing. What the native wants is as many animals as possible for the purpose either of meat or barter.
>
> (Hingston 1931, p. 404)

Such views were typical of hunter-conservationists in the early twentieth century. They remained an important element in thinking about wildlife in developing countries (Adams 2004). The importance of such ideas in the establishment of reserves for wildlife conservation is considerable (see Chapter 10).

Plate 2.1 Elephant skull, Zimbabwe. Hunting for ivory in the second half of the
nineteenth century took a staggering toll of elephant populations. John
MacKenzie reports that 40,000lb of ivory was traded on the Zambezi in
1876, implying the killing of 850 elephants; one hunter, Henry Hartley, shot
1,000–1,200 elephants in his career (MacKenzie 1987). Elephant populations
declined drastically, contributing to the wider perception by 'penitent butchers'
of the need for conservation. Through the twentieth century, elephants
continued to hold a prominent if controversial place in debates about African
conservation, from the IUCN African Special Project of the 1960s through
to the burning of stockpiled ivory in Kenya in the 1990s, and the ongoing
argument about the moral rights and wrongs of safari hunting, and the merits
of banning or legalizing international trade in ivory. Photo: W. M. Adams.

International environmental organization

An important factor in the development of the idea of sustainable development was the creation of global organizations, not simply links between colonial
metropole and colonies. The idea of an international organization to promote
conservation was mooted at an International Congress for the Preservation of
Nature held in Paris in 1909, and in fact an 'Act of Foundation of a Consultative
Commission for the International Protection of Nature' was signed at Berne in
1913 by delegates from seventeen European countries. However, war prevented
any substantive action, and, despite the recommendation of another congress at
Paris in 1923, it was not until 1928 that progress was made. In that year, the
assembly of the International Union of the Biological Sciences established the
Office International de Documentation et de Corrélation pour la Protection de la
Nature (Boardman 1981). In 1934 this was consolidated into the International
Office for the Protection of Nature (IOPN) (Holdgate 1999). By that time (in

1922), a second organization, the International Committee for Bird Protection (ICBP, later International Council for Bird Preservation and now BirdLife International) had been created to promote international nature protection.

Internationally, concern for nature between the two world wars was strongly focused on Africa. Hingston's paper (1931) to the Royal Geographical Society on the need for national parks was plain and persuasive: 'It is as certain as night follows day that unless vigorous and adequate precautions be taken several of the largest mammals of Africa will within the next two or three decades become totally extinct' (p. 402). In the 1930s the IOPN published a series of reports on African colonial territories. In the wake of hunting and 'scientific' safaris by prominent Americans such as former president Theodore Roosevelt and the cinematographer George Eastman, American concern about Africa's fauna began slowly to grow (Boardman 1981; Jeffers 2003; Adams 2004). In 1930 the Boone and Crockett Club set up the American Committee for International Wildlife Protection (ACIWLP) to promote nature protection and to carry out research. In Europe there was an International Congress for the Protection of Nature in Paris in 1931, and an intergovernmental conference proposing national parks in London in 1933 (Boardman 1981). The American Committee assisted the IOPN financially, particularly during the Second World War, when its work (based in Amsterdam) was severely disrupted.

After the war, efforts to strengthen international nature protection were renewed, led by the IOPN, the ICBP and the Swiss League for the Protection of Nature. A new organization was mooted at a conference in Basle in 1946, and taken up by the newly established UNESCO under the leadership of its first director, Julian Huxley (1977). Huxley had been chairman of the government committee that in 1947 had recommended government involvement in nature conservation in Britain and the establishment of a Biological Service (later substantially implemented in the creation of the Nature Conservancy (Sheail 1976, 1984; Adams 1996)). Huxley also had long-standing interests in Africa – for example, as a member of the African Survey Research Committee formed in Britain in the early 1930s (Huxley 1930; Worthington 1983).

Given this experience, it is perhaps not surprising that the UNESCO General Conferences in 1946 and 1947 were persuaded to take nature conservation on board. A 'Provisional Union for the Protection of Nature' was set up at a further meeting the same year and a constitution of the International Union for the Protection of Nature (IUPN) was adopted at a conference at Fontainebleau in 1948 (Holdgate 1999). It comprised an unusual blend of governmental and non-governmental organizations, and its purposes were to promote the preservation of wildlife and the natural environment, public knowledge of the issues, education, research and legislation (Holdgate 1999). UNESCO granted financial support a month later.

In 1949 IUPN ran an 'International Technical Conference on the Protection of Nature' concurrently with the United Nations Scientific Conference on the Conservation and Utilization of Resources (UNSCCUR) at Lake Success in New York State. McCormick (1989) described UNSCCUR as 'the first major landmark in the rise of the international environmental movement' (p. 37). The parallel IUPN meeting (attended by representatives from thirty-two countries and seven international organizations) proposed that IUPN should, with development agencies, consider carrying out surveys of the ecological impact of development projects (McCormick 1989). This notion, which lay unimplemented for another twenty years, demonstrates the stirring of a specifically conservationist concern for environmental aspects of development.

In 1956 the IUPN changed its name to the International Union for the Conservation of Nature and Natural Resources (IUCN), reflecting the more utilitarian approach of conservation of the USA. IUCN's focus was firmly placed wider than the industrialized countries of the West. Data collection on endangered species began, and was strengthened by extra funding after 1955 to put a biologist into the field in Africa, South Asia and the Middle East to report on threatened mammals. Red Data Books began to be published from 1966, listing endangered species, many in developing countries. IUCN established a Commission on Ecology in 1954, and Project MAR (with an ecosystem focus on wetlands) was launched in 1961 (McCormick 1989). The Convention on Wetlands of International Importance, Especially as Waterfowl Habitat, was signed in 1971 at Ramsar in Iran (Holdgate 1999).

By 1960 Africa had become 'the central problem overshadowing all else' for IUCN (Boardman 1981, p. 148). African countries were becoming independent, and political control was shifting away from the metropole; the poachers were turning gamekeepers, and might not follow the same policies. In 1961 IUCN therefore joined the Food and Agriculture Organization (FAO) to launch the 'African Special Project' to influence African leaders and promote conservation policies (Holdgate 1999). At a Pan African Symposium on the Conservation of Nature and Natural Resources in Modern African States, held in Arusha in Tanzania in 1961, the Tanganyikan leader Julius Nyerere voiced a strong personal commitment to wildlife conservation, but placed it firmly in the context of the human needs of African people. He stated:

> The survival of our wildlife is a matter of grave concern to all of us in Africa. These wild creatures amid the wild places they inhabit are not only important as a source of wonder and inspiration but are an integral part of our natural resources and of our future livelihood and well-being.
>
> (Worthington 1983, p. 154)

However, although Africa was important, it was not unique in its problems. Conferences to try to reproduce the spirit of Arusha were also held around the world, in Bangkok in 1965 and at San Carlos de Bariloche in Argentina in 1968 (Fitter and Scott 1978).

A clear pattern of international environmental action now emerged. The new international organizations were strongly based in industrialized countries, reaching out to the developing world. As Boardman (1981) comments: 'Western Europe and North America were the habitat of the conservationist' (p. 114).

This was particularly true of the World Wildlife Fund (WWF; now the World Wide Fund for Nature), founded in London in 1961 (Holdgate 1999). Although originally thought of as a means to raise funds for IUCN, WWF rapidly developed its own programme: between 1962 and 1967, less than 13 per cent of WWF expenditure went to IUCN (McCormick 1989). By 1976 twenty African countries were represented by government agencies or non-governmental organizations as members of IUCN, but they were swamped by the number of members from industrialized countries. This relative domination by developed-country organizations led to periodic complaints (for example, by the Kenyan delegation at the IUCN General Assembly in New Delhi in 1969 (Boardman 1981)).

The other important development of the post-war period was the growing involvement of the USA in international conservation. In the late 1950s the New York Zoological Society began a series of wildlife surveys and projects in Asia (Burma and Malaya) and in Africa (Kenya, Tanganyika, Uganda, Ethiopia and Sudan). In 1957 the Fulbright Commission began a programme to send scholars to look at biological problems in parks in Uganda. In 1961 the Washington Safari Club's conservation committee formed the African Wildlife Leadership Foundation (AWLF) to build a new cadre of conservation leaders among Africans. In 1962 it established the College of African Wildlife Management at Mweka in Tanganyika and a year later it sent four African students to study wildlife management in America. In 1965 the American Conservation Association sent an emergency grant of $10,000 to the Congo to support the staffing of the national parks. By the mid-1960s such substantial investments in African conservation dwarfed the efforts of the post-war Fauna Preservation Society and other ex-colonial institutions.

By the 1960s there was growing concern within IUCN to make conservation broaden its concerns beyond simply the preservation of wildlife. There was, for example, discussion of population and resource issues at the 1963 IUCN General Assembly in Nairobi. Conservationists needed to consider the long-term management of the environment and species, and therefore they needed to address the wider picture of resource exploitation (Holdgate 1999). This reflected, in particular, the influence from the USA, where conservation as the 'wise use' of resources had been part of public life since the progressive conservation movement associated with Gifford Pinchot at the start of the twentieth century (Hays 1959). This change in focus had already been expressed at the Bukavu Conference in the Belgian Congo in 1953, and by 1956 was by the change of title from IUPN to IUCN (in response to US pressure) to include the words 'conservation' and 'natural resources' (McCormick 1989). This change 'symbolised the conviction reached over the previous eight years that "nature", the fauna and flora of the living world, is essentially part of the living resources of the planet; it also implied that social and economic considerations must enter into the problem of conservation' (Munro 1978, p. 14)

In 1969 the 10th IUCN General Assembly in New Delhi adopted a new mandate, defined as 'the perpetuation and enhancement of the living world – man's natural environment – and the natural resources on which all living things depend' (Holdgate 1999, p. 108). Conservation was defined as 'the management ... of air, water, soil, minerals, and living species including man, so as to achieve the highest possible quality of life' (McCormick 1989, p. 46). In his history of IUCN, Martin Holdgate points out that, in adopting the principle of sustainability two decades before Rio, it was explicitly seeking to move with the tide of environmentalist concern in the developed world, while carrying the support of developing countries.

The African Convention on the Conservation of Nature and Natural Resources, adopted by the Organization of African Unity in Addis Ababa in 1968 (Boardman 1981) also took a broad definition of conservation. The convention embraced not only the established preservationist concerns of fauna and flora, but also the conservation of the more immediately obviously 'economic' resources of soil and water. It suggested that these were all resources to be managed 'in accordance with scientific principles' and 'with due regard for the best interests of the people' (McCormick 1989, p. 46).

Within IUCN there was increasing recognition of the need to make conservation more 'relevant' to the needs of the emerging Third World in the 1960s, and it progressively repackaged its message to embrace development. IUCN took part in a conference at Washington University in Virginia in 1968 on ecology and international development (subsequently published (Farvar and Milton 1973)), and went on to cooperate with the Conservation Foundation in producing a 'guidebook' for development planners (McCormick 1989, p. 155), *Ecological Principles for Economic Development* (Dasmann *et al.* 1973). This is discussed later in this chapter.

This thinking was a direct forerunner of the idea of 'sustainable development' contained in the *World Conservation Strategy* in 1980. The emergence of this document is discussed in the next chapter. Before that, it is necessary to backtrack slightly to examine other themes that fed into concepts of sustainable development. The African Convention referred to 'scientific principles'; where were these to come from, and what was the role of science and scientific thinking in views of tropical environments?

Ecology and the balance of nature

The science of ecology developed at the end of the nineteenth century in Europe and the USA (Lowe 1976; McIntosh 1985; Worster 1985; Sheail 1987). There were close links from an early date between the new science and the preservation or conservation movement, particularly in the UK, especially through the work of plant ecologist Arthur Tansley (Salisbury 1964; Sheail 1976, 1987). Perhaps it was not inappropriate that the word 'ecology' was used in popular discussion of the rise of environmentalism from the 1960s, even where ideas owed little or nothing to scientific ideas or method (Enzensberger 1974, 1996), although it was true that many prominent early figures in the environmental movement were trained in ecology (Chisholm 1972).

Ecology has contributed to thinking about sustainability in a series of related ways. First, ecological theory has underpinned much broader thinking about the environment and human impacts upon it. This relationship, and particularly the idea that there is some kind of 'balance of nature', is reviewed in this section. Second, ecology has been particularly important in thinking about tropical environments and therefore the development of colonial territories and developing countries. This is discussed in the next section. Third, there is a close resonance between the acquisition of scientific understanding and the application of that knowledge to both environmental management and development. In some ways ecology was a 'science of empire', and there were strategic links between science and politics, ecology and empire (Robin 1997). This ecological 'managerialism', which was particularly attractive in places such as Africa at the end of the Second World War, formed an important strand in the growth of sustainable development thinking.

Ecology's most obvious contribution to sustainable development has been its scientific description and analysis of the living environment. Within ecology, a whole series of concepts had been developed to describe patterns of change in natural systems, and these came to provide a powerful conceptual basis for sustainable development. Chief among them was the concept of the ecosystem and the idea of balance between predator and prey species. These concepts underpin the close links between the science of ecology and the development of conservation discussed earlier in this chapter.

Ecology's first priority was the description of variety and order in vegetation, of plant communities. In the UK, A. G. Tansley's *Types of British Vegetation* (Tansley 1911) provided a classificatory framework for the first lists of nature reserves in the UK (Sheail 1976; Adams 1996). In *Research Methods in Ecology*, the American ecologist F. E. Clements provided a scientific basis for the identification of vegetation 'types' (McIntosh 1985). He saw plant succession in terms of progressive change towards a 'climatic climax', with the vegetation formation like a complex organism 'developing' through time in a way deliberately analogous to the growth of individual organisms. This view was challenged both by another American ecologist, H. A. Gleason, and also by Tansley (McIntosh 1985). In a famous paper in 1935, Tansley proposed a more complex pattern of succession, with soils, physiography and human action all driving change under different conditions. To do so, he framed the new concept of the ecosystem (Tansley 1935, 1939; Sheail 1987). This enabled ecology to offer a frame that could encompass both natural and human-induced change.

The development of animal ecology was also important to the development of thinking about sustainability. In 1927 Charles Elton's book *Animal Ecology* had presented the concepts of food chain, pyramid of numbers and niche. He drew on quantitative data on animal population dynamics, especially fur-trapping records from the Canadian Arctic, and he emphasized the dynamics of populations in space and time. The ambition of animal population ecology expanded empirically in the 1930s and 1940s (for example, in early studies of the African lakes (Worthington 1932)) but also theoretically. Work on animal ecology was strongly quantitative, while studies of plant population dynamics lagged and did not focus on the dynamics of populations until the 1970s (McIntosh 1985).

Plate 2.2 Wicken Fen, Cambridgeshire. Wicken Fen is one of Britain's oldest nature
reserves. Its 100th anniversary was celebrated in 1999; the first 2-acre strip
was purchased on 1 May 1899 and donated to the National Trust. The reserve
represents a tiny fragment of the thousands of square kilometres of fenland that
existed before the great drainage projects of the seventeenth century. Yet over
7,000 species have so far been recorded, including more than 120 species on
the Red List of invertebrates. The reserve is managed by continuing traditional
sedge cutting and attempting to maintain water levels, as surrounding land
levels fall owing to shrinkage of drained peat for agriculture. The reserve is now
the centre of a major habitat restoration project (Colston 2003; Hughes *et al.*
2005; see www.wicken.org.). Photo: W. M. Adams

The foundation of theoretical population ecology was the development of the
logistic curve in the 1920s. The potential for geometric growth of populations
had been noted by Thomas Malthus in the eighteenth century, and the logistic
curve was formulated by Velhust in 1828 (McIntosh 1985). It was rediscov-
ered by Raymond Pearl in the 1920s, who suggested (controversially) that there
was a 'law of population growth', with specific reference to human population
(Pearl 1927). Two mathematicians, Lotka and Volterra, independently developed mathematical equations to describe the fluctuations of two interacting
species, and from these, and extensive laboratory experimentation with highly
simplified 'ecosystems' (with flour or wheat beetles for example) by Chapman,
Pearl and Gause, the discipline of population ecology grew (McIntosh 1985).
Within it developed the notion that animal population existed in balance in
nature, sustained by density-dependent competition. This idea, developed by

the Australian A. J. Nicholson in the 1930s, persisted as the centre of research and disputation between ecologists (McIntosh 1985). It also sank into the wider consciousness of ecologists and environmentalists formulating ideas about sustainable development.

These ideas drew also on the experience of marine biology and the scientific management of fisheries. Fisheries science had two roots, one in Victorian marine biology (for example, the voyage of the *Challenger* in the 1870s), and the other in the systematic decline of fish catches as capture became industrialized with the advent of steam-driven boats and other innovations (Cushing 1988). In 1883 the British biologist Thomas Huxley, august President of the Royal Society, blithely argued that sea fisheries – for example, for herring, mackerel, pilchard and cod – were effectively inexhaustible, and any attempt to regulate them was therefore pointless (Cushing 1988; C. M. Roberts 2007). He could not have been more wrong. Already new technologies and burgeoning demand from urban consumers were driving fishing further and further away from port as stocks were fished out. The reduction of flatfish stocks off the UK was clear by the 1880s. The purse seine net, invented in the 1850s, was widely used by the 1870s, and the steam trawler expanded the range and capacity of fishing fleets. Catch per unit effort in the English North Sea trawl fishery declined by 50 per cent in the decade 1888–98. Declining stocks generated UK government inquiries in the second half of the nineteenth century, and in North America the fishery for pelagic menhaden and mackerel off the east coast, and later for halibut off the west, also led to stock depletion (C. M. Roberts 2007).

In 1899 the International Conference for the Exploration of the Sea proposed scientific enquiries to promote 'rational exploitation of the seas' (Cushing 1988, p. 194). Fisheries biology eventually came to provide some of the leading ideas about population ecology, notably in W. F. Thompson's work on the Pacific halibut in the 1920s. By the 1930s the idea (and the mathematics) of a maximum sustainable yield had been worked out. Furthermore, international institutions had been established to try to use this emerging science to regulate fishing, from the International Council for the Exploration of the Sea established in 1902 in Copenhagen, through a series of other regional institutions, the Overfishing Convention agreed in London in 1946, and the International Whaling Commission. Neither these nor their successors achieved the sustained exploitation of any significant open-water stock either of fish or of marine mammals. The boom–crash cycle of sealing in the nineteenth century (Cushing 1988), the collapse of commercial whale harvests (Harrop 2003) and the commercial extinction of the northern cod of the Newfoundland Grand Banks in the twentieth century (Kurlansky 1999) are all testament to the failure either of science to model stocks and harvests effectively, or to turn scientific research into effective management.

However, the idea of scientifically based management was established. The principle of using science to define what fishing harvest was sustainable provided a powerful model for application outside fisheries, in other areas of resource development. The language within which analysis of fish populations was expressed suggests a considerable influence of economics: stocks were renewed or depleted,

Plate 2.3 Salt cod, Barcelona. Cod has long been a highly prized fish in Europe, particularly in Spain. The transatlantic fishery is centuries old, with European boats, from the Basque country in particular, crossing the Atlantic to Newfoundland to catch and salt cod for the European market. Traditionally, most fishing was small-scale, and confined to the coast in summer, using traps, gill nets and hook and line. From the 1960s industrial trawling expanded in scale and intensity, and the fishery collapsed and did not recover. European markets are dependent on residual stocks in overfished but not yet extinct fisheries in the north-east Atlantic– for example, off the Lofoten Islands in Norway, or Iceland (Kurlansky 1999; see the Marine Conservation Society FISHONLINE website www.fishonline.org/). Photo: W. M. Adams

and calculations included estimates of catch per unit effort. In turn, economics reflected evolving understanding of the dynamics of resources, particularly in the distinction between renewable (flow) and non-renewable (stock) resources (Ciria-cy-Wantrup 1952). In this form, too, ideas about the dynamics of natural systems and their response to human management fed into thinking about sustainable development.

The application of fisheries science to resource management offered a similar conceptual frame to that of scientific forestry, developed first in Prussia in the eighteenth century and subsequently spread across the world (Scott 1998; Demerrit 2001; Barton 2002). Science enabled forests to be seen in the abstract, allowing the volume of standing timber to be calculated, and patterns of cutting and regeneration to be planned. Here, too, the idea of sustained yield (and sustained income) were central, a science-based prediction of how far the harvest could be optimized for human benefit. This 'ecological managerialism' is discussed further below.

In the first decades of the twentieth century, 'nature' was mostly understood to be essentially static, an array of habitat fragments as natural objects set in a landscape of change. This drew on a powerful organic metaphor of nature, a view of nature balanced and integrated and threatened by change from 'outside', from human action (Botkin 1990; Livingstone 1995). However, as theoretical and experimental approaches to ecology developed, the 'organismic' approach gave way to an essentially mechanistic framework of analysis, involving the concept of the ecosystem ecological energetics and systems analysis (Tansley 1935; Lindemann 1942; McIntosh 1985; Botkin 1990). However, even in the 'systems ecology' associated particularly with the work of E. P. Odum, the fundamental notion that ecosystems tended towards equilibrium endured (McIntosh 1985). Donald Worster (1993) comments that both Odum's view of nature as 'an automated factory' and its predecessor the 'Clementsian super-organism' implied that nature 'tended toward order' (p. 160).

Ideas of equilibrium and stability, the 'classical paradigm' or 'equilibrium paradigm' (Steward *et al.* 1992), dominated ecology until the 1970s. It suggested that ecological systems were closed, and that ecosystems were self-regulating, so that if disturbed they would tend to return towards an equilibrium state. This paradigm in turn fed ideas in the wider environmental movement that there was a 'balance of nature', easily upset by inappropriate human action. Both systems ecology and population ecology emphasized equilibrium and stability. Nature was portrayed as a homeostatic machine (Pahl-Wostl 1995), and ecosystems were analysed as if they were 'nineteenth-century machines, full of gears and wheels, for which our managerial goal, like that of any traditional engineer, is steady-state operation' (Botkin 1990, p. 12). Nature was a system whose state was maintained by processes of internal feedback, but it was also susceptible to external control. Human action could upset the machine, but fortunately the ecologist could predict how and where this upset might occur, and diagnose how to put the balance right. Ecological science could therefore be used to generate technocratic recipes for managing nature. Words and concepts drawn from thermodynamics and engineering (such as energetics, equilibrium and control) were used blithely by ecologists to describe nature. Conservationists, schooled in ecology, presented themselves as 'ideal scientific "managers" of the environment, the engineers of nature' (Livingstone 1995, p. 368). In the decades that saw the rise of formal 'development', between the end of the Second World War and the rising wave of the environmental revolution in the 1960s, ecology provided a powerful script for the emerging dialogue between environmental protection and economic development.

Ecology and tropical development

The tropics were an important hearth of innovation in biological science, and particularly ecology, in the twentieth century (as in previous eras – see discussion earlier in this chapter). The science of ecology had its immediate roots in the temperate habitats of northern Europe (Denmark, Germany and the UK) and the USA, and the challenge of describing and analysing vegetation (McIntosh 1985). However the attention of ecologists was soon extended to the study of the diversity of nature overseas. This engagement simply carried forward the close and reciprocal links that had existed between science and exploration in the tropics through the eighteenth and nineteenth centuries (Crosby 1986; Stoddart 1986; Pratt 1992).

The British Ecological Society was founded in 1914 (Salisbury 1964; Allen 1976), and the Ecological Society of America a year later. The first volume of the *Journal of Ecology* in 1914 (edited by Tansley) already showed a strong international flavour, with reviews of publications on the forests of British Guiana (C. W. Anderson 1912) and studies of vegetation in Natal and the eastern Himalayas. The second volume reviewed works on the vegetation of Aden and the highlands of north Borneo and Sikkim, and the following year added work on the dipterocarp forests of the Philippines and montane rainforest in Jamaica.

Interest in plant ecology and vegetation outside Britain was maintained in subsequent years, and in volume 12 (1924) a specific section was instituted, quaintly entitled 'Notices of publications on foreign vegetation'. The majority of publications reviewed concerned the USA, Australia and South Africa, but there were others – for example, Cuba (volume 15, 1927). The journal also carried research papers on the vegetation of South Africa (Bews 1916), the forests of south Brazil (McLean 1919) and the Garhwal Himalaya (Osmaston 1922). US studies on the vegetation and soils of Africa were begun in 1918 at the instigation of the American Commission to Negotiate Peace, involving a journey from the Cape to Cairo from 1919 to 1920 and a monograph on African vegetation, soils and land classification (Shantz and Marbut 1923).

Such ecological science was readily harnessed in the service of colonial development. The Imperial Botanical Conference in London in 1924 stressed the need for 'a complete botanical survey of the different parts of the Empire' (F. T. Brooks 1925, p. 156). This conference followed International Botanical Congresses in Paris (1900), Vienna (1905) and Brussels (1910), and was deferred until the British Empire Exhibition took place. The argument being made was that science should serve imperial commerce: 'in such a Survey the Imperial Government has good reason to take a prominent part, as it depends so much on the overseas portions of the Empire for the supply of raw materials for manufacture, and of foodstuffs' (Davy 1925, p. 215). Throughout, the tone of the conference was strongly economic and utilitarian:

> I submit that it is our duty as botanists to enlighten the world of commerce, as far as may lie in our power, with regard to plants in their relation to man and their relation to conditions of soil and climate.
>
> (Hill 1925, p. 198)

The Imperial Botanical Conference established the British Empire Vegetation

Committee, under the chairmanship of A. G. Tansley (already a leading scientific figure in conservation in the UK). The committee published *Aims and Methods in the Study of Vegetation* in 1926, although a further twelve years passed before the first in a planned series of regional monographs appeared, *The Vegetation of South Africa* (Adamson 1938).

Ecology had more to offer than simply the scientific task of describing vegetation. Ecologists discovered that their science had considerable relevance to economic problems, and found a ready audience among officers of the colonial state. In reviewing research on the East African Great Lakes many years later, E. B. Worthington (1983) commented, 'advice on the existing and potential fisheries was a primary objective, but to provide this it was necessary to elucidate the ecology of each lake' (p. 13). Development needed careful and extensive scientific research. Vegetation analysis and classification could serve 'a most practical purpose', allowing the definition of natural regions 'for use in the development of both agriculture and forestry' (Shantz and Marbut 1923, p. 4). In 1931 Phillips advocated a 'progressive scheme' of ecological investigation in Africa to help develop resources. He believed that ecology could contribute to agriculture ('the rational use of biotic communities'), grazing (the 'wise utilisation of natural grazing'), forestry (the development of 'progressive forestry policies'), soil conservation (the 'prevention of soil erosion and its concomitant evils'), catchment water conservation and research into tsetse fly (Phillips 1931, p. 474). Phillips was practical enough to include detailed suggestions on staffing, a research programme and administrative arrangements.

Practical application of botanical survey to development began in what was then Northern Rhodesia (contemporary Zambia) in the 1930s, in the vegetation and soils surveys of three scientists, Colin Trapnell, Neil Clothier and William Allan (Trapnell and Clothier 1937; Trapnell 1943; cf. Moore and Vaughan 1994). These concepts were significant early expressions of ideas that came to prominence half a century later, when they formed one line of thinking about sustainable use of drylands (see Chapter 8). Ecological research was also proving itself important in the study of disease in Africa, notably on the tsetse fly (J. L. Huxley 1930; Buxton 1935; Ford 1971).

The argument about the usefulness of ecology as an input to development in Africa was set out in the work of the African Survey (Hailey 1938), carried out under the aegis of an African Research Survey Committee, on which the two biologists were Julian Huxley (later Director of UNESCO) and John Orr (later Director of the FAO). In 1935 Sir Malcolm Hailey (later Lord Hailey) drove from Cape Town to the Mediterranean. With him for part of the way went Barton Worthington, who wrote a contributory volume *Science in Africa* (Worthington 1938). This took the utilitarian line that science (particularly ecology) was useful in development (or 'bonification', as Worthington called 'the promotion of human welfare'):

A key problem was how *Homo sapiens* could himself benefit from this vast ecological complex which was Africa, how he could live and multiply on the income of the natural resources without destroying their capital ... and how he could conserve the values of Africa for future generations, not only the economic values but also the scientific and ethical values.

(Worthington 1983, p. 46)

Science in Africa not only placed science in the service of a loosely defined 'development', it also used an explicitly ecological structure in its presentation. The report moved from geology and meteorology through botany and zoology to agriculture, fisheries, disease, population and anthropology, and stressed the fact that interrelations between the sciences had 'important practical applications' (Worthington 1938, p. 3).

Following the end of the Second World War, institutions were put in place to secure the capacity for science in Africa. The Conseil Scientifique pour l'Afrique (CSA), established following an Empire Scientific Conference held by the Royal Society in London in 1946 and a further conference in Johannesburg in 1949, developed similar arguments about the importance of science to development (Worthington 1958). Increased interest in scientific research in African territories at this time saw the establishment of the British Colonial Research Council (under Lord Hailey), the French Office de la Recherche Scientifique et Technique d'Outre Mer (ORSTOM) and the Belgian Institut pour la Recherche Scientifique en Afrique Centrale (IRSAC) which joined the existing agricultural body, the Institut National pour l'Étude Agronomique du Congo Belge (INEAC) (Worthington 1983). Post-war colonial development drew heavily on a scientific approach to the environment and its management.

Ecological managerialism

As the idea of development planning began to take root in post-war Africa, ecology was seen to provide not only valuable environmental data for planning, but *a model for the practice of development itself*. Early post-war development plans were crude affairs. In 1948 the journal *Nature* complained of one plan that was 'an agglomeration of departmental suggestions masquerading as a development plan' (*Nature* 1948). In a remarkable polemic in support of a managerial role for environmental science, Culwick (1943) suggested that the task of development was 'primarily a scientific problem in which all branches of science, physical and social, have an essential part to play' (p. 1). The idea of sustaining benefits through correct environmental management was central to ecology's contribution to development. Culwick saw development in Tanganyika essentially as a problem of 'controlling the environment and exploiting it in such a way that not only this but succeeding generations may be able to attain their full potentialities, both physical and mental' (p. 1).

Furthermore, development not only had to involve scientists, but arguably needed to be conceived of within the paradigms of ecology. Culwick called for reform of the institutions of government after the war, so that life could be 'planned as an oecological whole'. In the past, scientists had been marginal to the essentially political business of government, regarded as 'appendages to the main administrative body, called in every now and again for consultation when a scientific question cropped up'. However, the notion of development (at this time, of course, still only loosely conceived – see Chapter 1) demanded a new approach. Government should be regarded 'primarily as a scientific affair', and administration should therefore become 'merely the mechanism for putting the big scientific plan into action' (Culwick 1943, p. 5). Science, and particularly ecology and environmental science, could not only inform development, but direct it.

Development in the post-Second World War period was conceived of as an organized and coherent attempt to overcome constraints on economic growth, and often explicitly aimed at overcoming environmental constraints on that growth. Croll and Parkin (1992) suggest that development may be conceived of 'as a form of self-conscious or planned construction, mapping and charting both landscapes and mindscapes' (p. 31). That planning and action were increasingly driven by science, technology, a capitalist market economy and formal organiza-tion. In his book *Rationality and Nature*, Raymond Murphy (1994) describes these as dimensions of 'rationalization', following the work of Max Weber ([1922] 1978). Rationalization is the process by which human reason frames and allows ordering and control of both nature and society. The attraction of the science of ecology in development planning was its power to rationalize both comprehen-sion of development 'problems' and action. Science provided knowledge that could be used to control environment, economy and society in such a way that 'developmental' change could be directed in desired directions.

Murphy identifies a number of dimensions of rationalization. The first is the development of science and technology: 'the calculated, systematic expansion of the means to understand and manipulate nature', and the scientific worldview, 'belief in the mastery of nature and of humans through increased scientific and technical knowledge' (Murphy 1994, p. 28). The second dimension of ration-alization is the expansion of the capitalist economy (with its rationally organ-ized and in turn organizing market); the third dimension is formal hierarchical organization (the creation of executive government, translating social action into rationally organized action). The fourth is the elaboration of a formal legal system (to manage social conflict and promote the predictability and calculability of the consequences of social action). All these things were features of the work of the colonial state, in Africa as elsewhere. Ecology provided ample contributions to the first of these dimensions, as a science able to produce rational stories in the face of novel environmental complexity (for example, the 'useful purpose' served by surveys that 'properly analysed and classified vegetation' (Shantz and Marbut 1923, p. 4)).

Ecology was an element within the wider rationalizing and ordering power of science for planning and structuring action in colonial territories. As discussed above, the advent of scientific forestry in eighteenth-century Prussia had allowed the calculation and measurement of productivity and efficient physical management. It did the same in the nineteenth century, in the USA (Demerrit 2001), in France, in British colonial possessions, notably in India (Barton 2002), and in Australia (Guha 2000). The governance of colonial states drew on the idea that nature could be understood, manipulated and controlled for social benefit through the development of schematic (and increasingly scientific) knowledge (Scott 1998). Government bureaucracies were established to organize relations between state, society and nature (Hays 1959; Mackenzie 2000; Demerrit 2001). Issues like soil erosion stimulated complex attempts at social and environmental control (D. M. Anderson 1984; Mackenzie 2000; Carswell 2003; see also below, Chapter 8).

In the colonial world of the tropics, the hegemony of ecology in development thinking was at best only partial. Even in Africa, where colonial states were slender and recently rooted, and the environmental unknowns were glaring, the contribution of ecology was limited. The colonial administration in Tanganyika, for example, never reduced itself to Culwick's 'mere mechanism' for applying science. However, what Low and Lonsdale (1976) have called a 'second colonial occupation' began in East Africa after the Second World War. Scientists acquired new and important roles in some fields, most notably agriculture. The engagement with the development problems of rural Africans inevitably demanded an engagement with the ecology of their production systems and the wider landscapes of which they were part. The importance of the biology of agriculture, grazing, forestry and disease, and the physical geography of erosion and water supply, were recognized, and scientific expertise in each field won a place within what had been predominantly legal, political and fiscal institutions of colonial government.

This attention to indigenous production systems sometimes generated positive official understandings, as in West Africa (e.g. Faulkner and Mackie 1933; Stamp 1938), or in Northern Rhodesia, where William Allan's work (published long afterwards as a book, *The African Husbandman*, 1965) demonstrated the logic of indigenous agricultural systems. More generally, the 'second colonial occupation' featured a conventional scientific wisdom that human use of resources led to degradation, as, for example, in the work of Pole-Evans and Phillips on range management in South Africa or that of Stebbing in Nigeria (Stebbing 1935; Scoones 1996; Grove and Damodaran 2006; see also below, Chapter 8).

Whether favourable or hostile in its portrayal of indigenous resource users, ecological managerialism penetrated the development planning process in the colonial world following the end of the Second World War. Thus in Uganda, the first development plan was explicitly ecological: 'fundamentally, the problem of development was one of human ecology, the diverse people reacting with their varied environments' (Worthington 1983, p. 97). Similar approaches to the issue of human development were proposed elsewhere, a famous example being Frank Fraser Darling's *West Highland Survey* (1955). By the middle of the twentieth century, ecological science had acquired a strong influence on thinking about the development process, reflecting the wider significance of science in government: the British scientific novelist and bureaucrat C. P. Snow wrote in 1959 that scientists 'had the future in their bones' (Snow [1959] 1998, p. 10).

The ecological impacts of development

By the 1960s and early 1970s, there was growing understanding of the adverse ecological impacts of development, particularly on the part of the International Union for Conservation of Nature (IUCN), the International Biological Programme (IBP) and the Man and the Biosphere Programme (MAB). This duly led to attempts to identify specific formulae for avoiding or minimizing these impacts, in an essentially technocentrist search for 'environmentally benign' development.

A conference on 'the ecological aspects of international development' was held late in 1968 at Airlie House in Virginia, organized by the Conservation Foundation and Washington University of St Louis. Its proceedings were published as *The Careless Technology: ecology and international development* (Farvar and Milton 1973). These papers presented a readable and authoritatively researched catalogue of environmental problems associated with or caused by economic development, from the downstream impacts of dams to pollution. It set the style for many subsequent accounts of environment and development, and many of the issues raised remain important (and unresolved).

A series of meetings followed this between international conservation, environment and development organizations, including IUCN, the Conservation Foundation, UNDP, UNESCO and FAO. In 1970 there was a meeting at the FAO headquarters in Rome of interested parties, including now the United States Agency for International Development (USAID) and the Canadian International Development Agency (CIDA) to consider further the ecological impacts of development. It was decided that IUCN and the Conservation Foundation should publish guidelines for development planners. These appeared as *Ecological Principles for Economic Development* (Dasmann *et al.* 1973). Initially the book was to include discussion of 'interrelationships between economic development, conservation and ecology', but in the event it was restricted to an exploration of ecological concepts 'useful in the context of development activities'. Particular emphasis was placed on tropical rainforests and semi-arid grazing lands, the effects of tourism (particularly in fragile environments such as high mountains and coasts) and the development of agriculture and river basins (Dasmann *et al.* 1973, p. vi).

Ecological Principles for Economic Development encapsulated evolving thinking by conservationists and ecologists since the end of the Second World War on the subject of development. It was an important precursor of the idea of sustainable development. Its central premiss was the need to apply concepts and insights from the science of ecology to development activities to decide what could, or should, be done with respect to the environment. Ecology would help planners 'to make sure of success' (Dasmann *et al.* 1973, p. 21). However, if the 'lessons' of ecology were ignored, 'entirely unexpected consequences can often result from what are intended to be straightforwardly beneficial activities'. Thus, for example, the replacement of tropical rainforest with a palm-oil plantation would set in motion 'complex ecological forces', which might involve the 'loss of equilibrium' and pest outbreaks (Dasmann *et al.* 1973, p. 44). Adverse environmental impacts of development had to be understood and dealt with during the process of planning development if problems were to be avoided.

Dasmann *et al.* pictured ecology as an integrative science, an interdisciplinary way of thinking that could be instilled in the minds of 'forester, agricultural specialist, range manager ... development economist or engineer'. The use of ecology in development planning has the aim of both 'enhancing the goals of development' and 'anticipating the effects of development activities on the natural resources and processes of the larger environment'. Ecology had in the past been used to assess the potential productivity of a resource; now it needed to be used to assess the adverse impacts of development and technology on local and global environments, so that these impacts could be anticipated and decisions about developments made 'in full knowledge of possible consequences'. A decision to develop despite environmental impacts might be justified by counterbalancing benefits, but 'it should never be taken blindly' (Dasmann *et al.* 1973, pp. 21–2).

This pragmatic and explicitly ecological approach to development was duly applied in a number of fields, notably with respect to the environmental impacts of dams. By the 1960s large dams were sufficiently common globally for it to be worth the construction industry compiling a global register, and for environmentalists and researchers to be starting to record extensive and complex and largely unforeseen environmental effects, which were themselves noted by engineers (D. J. Turner 1971). By the second half of the 1970s such reviews were commonplace in journals and magazines read by engineers and hydrologists (e.g. Biswas and Biswas 1976, writing in *Water Power and Dam Construction*), and it was possible to synthesize more than a decade of ecological research (Baxter 1977). The MAB Project 10 focused on the effects of major engineering works on humans and their environment (UNESCO 1976). In 1977 the UN Water Conference received a report called *Large Dams and the Environment: recommendations for development planning* (Freeman 1977). The International Commission on Large Dams (ICOLD) appointed a committee on dams and the environment in 1972, which in 1981 published *Dam Projects and Environmental Success*, 'intended to illustrate the concern and knowledge of dam engineers related to environmental matters' (ICOLD 1981, p. 8). This publication, with its review of impacts and its stress on the fact that environmental effects could be beneficial as well as adverse, neatly combined professional training and public-image creation. It also demonstrates the success, at least on a rhetorical level, of attempts to imbue development with ecological awareness.

Such attempts were further developed through the formulation of principles of environmental impact assessment (EIA) in the late 1960s. EIA became an important part of public policy in the industrialized world with the US National Environmental Policy Act (NEPA) of 1969. A number of industrialized countries followed the USA to a greater or lesser degree down the path of institutionalizing the assessment of environmental impacts (Barrow 1997). One of the international applications of EIA procedures was again by ICOLD, which designed and tested a version of the Leopold Matrix (Leopold *et al.* 1971) to take account of the impacts of large dam construction (ICOLD 1980). The procedure of EIA was taken up in the 1970s by the Scientific Committee on Problems of the Environment (SCOPE), which in 1975 published an authoritative review, *Environmental Impact Assessment: principles and procedures* (Munn 1979).

Environmental limits, population and global crisis

The rise of environmentalism in the industrialized world in the 1960s had enormous significance for debates about the role of ecology and conservation in development, and was also influenced by those debates. In 1982 Myers and Myers argued that perception of environmental issues rarely transcended national boundaries or national interests. This may be true, as regards the extent to which environmentalism reflected self-interest as opposed to some more egalitarian global consciousness. However, the perception that there *were* environmental issues of global significance was a distinctive and novel feature of the 'new environmentalism' (Cotgrove 1982) that arose in North America and Western Europe in the 1960s and 1970s. The image of the earth as a blue ball spinning in the darkness of space ('Spaceship Earth', a term popularized by Barbara Ward (1966) and Kenneth Boulding (1966)) became environmentalism's icon. The threats to its perfection were portrayed in the global 'doomsday syndrome' about which Maddox (1972) complained.

The global vision of environmentalism in the 1960s and 1970s is nicely captured in the title of the book written by Max Nicholson in 1970, *The Environmental Revolution: a guide for the new masters of the world*, and that published by Barbara Ward and René Dubos for the 1972 UN Conference on the Human Environment in Stockholm, *Only One Earth* (Ward and Dubos 1972). Partly through the Stockholm Conference, awareness of environmental problems (particularly pollution (Dahlberg *et al.* 1985)) on a global scale became a key theme in the environmental revolution of Europe and North America in the 1970s. Parallel with it grew an apocalyptic vision of neo-Malthusian crisis. It is within this context that ideas about sustainable development emerged.

Such globalism in environmental concern was not new. Fairfield Osborn (1954) wrote, 'man is becoming aware of the limits of his earth. The isolation of a nation, or even a tribe, is a condition of an age gone by' (p. 11). One fruit of this realization was the rise of concern about global population. Ecologists such as Raymond Pearl (1927) discussed the phenomenon of human population growth in the 1920s, and the spectre of population growth was raised by commentators such as Carr-Saunders (1922, 1936). In the late 1940s Osborn (1948) commented: 'the tide of the earth's population is rising, the reservoir of the earth's living resources is falling' (p. 68).

Under the British scientist Julian Huxley, UNESCO became involved in the 1950s in the application of scientific evidence in the debate about population and development. For example, UNESCO ran the World Population Conference with the FAO in Rome in 1954, thereby linking development issues to one of the central concerns of environmentalism. The 'population problem', as it became widely known, was commented on extensively in the 1950s – for example, by Boyd-Orr (1953), Stamp (1953) and Russell (1954), and in the famous conference volume *Man's Role in Changing the Face of the Earth* (W. L. Thomas 1956). Neo-Malthusian arguments were a prominent feature of environmentalism in the 1960s and 1970s, most notably in the work of Ehrlich and Ehrlich (1970), and in the apocalyptically titled *The Population Bomb* (Ehrlich 1972). Garret Hardin's paper in the journal *Science* in 1968, 'The tragedy of the commons', also reflected this neo-Malthusian concern about the exhaustion of living resources. Hardin (1968) observed that 'a finite world can support only a finite population; therefore population growth must eventually equal zero' (p. 1243). The issue of common-pool resource management, which this highlighted, subsequently became the focus of the discipline of new institutional economics (e.g. North 1990; Ostrom 1990). However, its starting point was an environmentalist fear of global population growth.

Neo-Malthusian thinking about the global environment has been remarkably persistent. In 1970 the world population was 3.5 billion, and 'Spaceship Earth' was said to be 'filled to capacity and beyond and is running out of food' (Ehrlich and Ehrlich 1970, p. 3). In the early 1980s Barton Worthington (1982) wrote: 'whichever way one looks at the population problem – whether as a biologist, sociologist, theologian, medical doctor, industrialist, administrator or politician – it is obvious that it presents the greatest menace to the future of the biosphere' (p. 98). To some environmentalists, population growth offered a dire and global threat: 'the remedy is left to Nature's ways of shortage and deprivation, famine or pestilence, and to Mankind's own way of increasing violence and slaughter' (Polunin 1984, p. 296). (It is worth noting that with a global population of over 6.5 billion and rising, views about this issue continue to range from the sanguine to the panic-stricken. Neo-Malthusianism is discussed further in Chapter 5.)

'Catastrophist' environmentalist thinking about pollution and population growth in the 1960s and 1970s was matched by a parallel and closely related debate about economic growth. Over the 1960s and 1970s, economic growth in industrialized countries gave way to recession and inflation. Even then, most analysts – for example, Kahn and Wiener (1967) – offered a fairly optimistic range of scenarios of the future. Others were less sanguine. Mishan ([1967] 1969) looked at the costs of economic growth (coining the delightful term 'growthmania'), and concluded rather cautiously that 'the continued pursuit of economic growth by Western societies is more likely on balance to reduce rather than increase social welfare'.

This still cautious view was overtaken by the results of attempts to produce global computer models of the 'world system', notably Forrester's *World Dynamics* (1971). This approach was developed by a team from the Massachusetts Institute of Technology for the 'Club of Rome', an international group set up with the backing of European multinational companies (Golub and Townsend 1977). The work was published as *The Limits to Growth* (Meadows *et al.* 1972), and with the British utopian polemic *Blueprint for Survival* published in the same year (Goldsmith *et al.* 1972), it forms one of the two most commonly quoted (although perhaps less commonly read) treatises of 1970s environmentalism.

The idea of zero growth and a steady-state economy has attracted sober support (e.g. Daly 1973, 1977; Mishan 1977), but has also generated a body of fairly vituperative criticism. *The Limits to Growth* has been the more marked target. Beckerman (1974) suggested that the Club of Rome report was 'guilty of various kinds of flagrant errors of fact, logic and scientific method' (p. 242), while Simon (1981) dismisses it as 'a fascinating example of how scientific work can be outrageously bad and yet be very influential' (p. 286). Maddox (1972) said simply: 'the doomsday cause would be more telling if it were more securely grounded in facts' (p. 2).

The issue of sustainability and growth will be discussed in more detail in Chapter 6. The important point to note here is the way the notion of 'global crisis' in the 1960s and 1970s contributed to the internationalization of both ecology and environmentalism. Although the 'ecological prescriptions for managing the human use of the earth' that were on offer were indeed extremely limited (Stoddart 1970, p. 2), the grandiose claims and global fears of environmentalism were far from ineffective.

Global science and sustainable development

The final thread in the story of the roots of sustainable development is the establishment of specific international scientific organizations with the state of the global environment as a fundamental party of their remit. Theoretical and practical links between ecological science and development were fostered in 1964 by the establishment of the International Biological Programme (IBP), launched by the International Union of Biological Sciences (IUBS) under the International Council of Scientific Unions (ICSU). The model of international scientific cooperation came from the International Geophysical Year 1957–8 (Worthington 1975, 1983). Planning began in 1959, with the vision of studying 'the biological basis of man's welfare' (Worthington 1975, p. 5). However, it was only at a meeting at the IUCN headquarters in Morges in 1962 that the IBP planning committee adopted the sevenfold grouping of research that came to form the structure of the programme. Perhaps unsurprisingly given the venue, one group was dedicated to terrestrial conservation (E. M. Nicholson 1975).

In the event, extensive work was done by the terrestrial conservation section, in particular the gathering of global data on areas of 'scientific' (that is, conservation) importance, and an attempt to establish a global network of research stations in different biomes. However, Max Nicholson argues that the biological science community 'never fully endorsed in practise the inclusion of conservation' (1975, p. 14). The terrestrial conservation section faced particular problems. It had to be Janus-headed and address two audiences: research biologists, to encourage them to apply their ideas, and natural resource managers, to encourage the application of ecological theory (Worthington 1975, p. 86). There were also substantial problems of integration, with limited cooperation between industrial world scientists and their developing world counterparts (Boffey 1976). The results were mixed.

The IBP, however, was only one dimension of the expansion of research on the global environment, and its problems. A series of meetings were called in the

post-war period to 'bring together key actors of the globe' to address environmental issues on a global scale (Dahlberg *et al.* 1985). One was a symposium held
by the Wenner-Gren Foundation at Princeton, New Jersey, in 1955, *Man's Role in
Changing the Face of the Earth* (W. L. Thomas 1956), which was chaired by Carl
Sauer, Marston Bates and Lewis Mumford. This symposium claimed roots in the
work of George Perkins Marsh ([1864] 1965) on the influence of man on nature,
and similar thinkers from the late nineteenth century onwards. Man, 'the ecological dominant on this planet', needed the insights of scholars to understand his
'impress' on the earth. The symposium was 'a first attempt to provide an integrated
basis for such an insight and to demonstrate the capacity of a great number of fields
of knowledge to add to our understanding' (W. L. Thomas 1956, p. xxxvii). This
theme was one that fitted easily into the new internationalism of science in the IBP,
and with the rising globalism of environmentalist thinking in the 1960s.

The global nature of environmental problems was debated at the Third
General Assembly of the IBP in Bulgaria in April 1968, and in 1969 the Scientific
Committee for Problems of the Environment (SCOPE) was established under the
ICSU (Worthington 1983). SCOPE focused on ways of understanding specific
environmental problems, particularly at a global scale (Table 2.1). It reported
in 1971 and 1973 on global environmental monitoring, in 1975 and 1979 on
global geochemical cycles, and in 1979 on Saharan dust. The proceedings of
the SCOPE/UNEP symposium on 'environmental sciences in developing countries' in Nairobi were published in 1974. SCOPE's work also collated material
on specific environmental issues – for example environmental impact assessment
(discussed in the previous section (Munn 1979)).

To an extent, the work of SCOPE was overtaken (although not replaced) by
the Man and the Biosphere progamme (MAB), which grew out of the 'Biosphere
Conference' held in Paris in 1968 (the Intergovernmental Conference of Experts
on a Scientific Basis for Rational Use and Conservation of the Biosphere). This
made explicit the growing engagement by conservationists and environmentalists with the development process (Caldwell 1984). It was a further step in
the incorporation of developing countries into the world of international environmental concern. The initiative stemmed primarily from a realization of the
continuing failure to create a truly international environmentalism. Boardman

Table 2.1 The mandate of the Scientific Committee for Problems of the Environment
(SCOPE)

- To assemble, review and assess the information available on man-made environmental
  changes and the effects of these changes on man
- To assess and evaluate the methodologies of measurement of environmental parameters
- To provide an intelligence service on current research
- By the recruitment of the best available scientific information and constructive
  thinking to establish itself as a corpus of informed advice for the benefit of centres
  of fundamental research and of organizations and agencies operationally engaged in
  studies of the environment

Source: Munn (1979).

(1981) comments that, 'in the last analysis, the one major political cleavage that the issue of nature conservation has failed to bridge adequately is that between industrialised and developing countries' (p. 19).

Through the 1960s it had been increasingly clear to conservationists within UNESCO that they could not influence decisions about the use of natural resources in the developing world unless they were prepared at least to talk in the new language of development. UNESCO had adopted key resolutions that explicitly linked conservation and development at its 1962 General Conference. It sponsored the symposium Man's Place in the Island Ecosystem at the Tenth Pacific Science Congress in Honolulu in 1961, which developed the idea of human actions as a functioning part of an ecosystem (Fosberg 1963). UNESCO also joined other UN agencies (the FAO and UNDP), IUCN, the Conservation Foundation and the World Bank in the discussions about ecology and development that followed this conference (Dasmann *et al.* 1973), and had been involved in a review of natural resources in Africa at the request of the Economic Commission for Africa in 1959 (UNESCO 1963).

The Biosphere Conference in 1968 had a complicated history. The idea of an international conference on endangered species was initially suggested at the IUCN General Assembly in Nairobi. The broader 'biosphere' approach came from the United Nations Economic and Social Council (ECOSOC) and UNESCO (Boardman 1981). The Biosphere Conference called for the establishment of an interdisciplinary and international programme of research on the rational use of natural resources and 'to deal with global environmental problems' (Gilbert and Christy 1981). The practical emphasis of this proposal was heavily influenced by those in national delegations with experience of the IBP (Worthington 1983, p. 175). However, the MAB programme that was launched in 1971 had a strong scientific base, and was in many ways a direct successor of the IBP (Gilbert and Christy 1981; Worthington 1983; Holdgate 1999). There was considerable passage of scientific information between the two programmes (Worthington 1975, 1983).

MAB's function was 'to develop the basis within the natural and social sciences for the rational use and conservation of the resources of the biosphere and for the improvement of the global relationships between man and the environment' (Gilbert and Christy 1981). Consequently, MAB was to be both useful and down-to-earth: 'ivory tower research' was of little use to 'those who have to make management decisions in a world of increasing complexity'. MAB would be different, breaking down 'obsolete barriers' between natural and social scientists and decision-makers, and offering instead 'an interdisciplinary, problem-oriented approach to the management of natural and man-modified ecosystems' (UNESCO, n.d.). MAB was given an exhaustive and astonishingly open-ended range of specific objectives. These included the study of the 'structure, functioning and dynamics of natural, modified and managed ecosystems' and of the relations between 'natural' ecosystems and 'socio-economic processes', and the identification and assessment of human impacts on the biosphere. There were also aims to promote 'global coherence of environmental research', environmental education and specialist training, and 'global awareness of environmental problems' (Gilbert and Christy 1981, pp. 704–5).

These aims were obviously not all attainable in full, so a series of fourteen specific fields or 'projects' was defined. These projects focused either on particular environments of concern to environmentalists (for example, rainforests or semi-arid zones) or on particular impacts of development (for example, major engineering works, urban systems and energy, or pollution). The main shortcomings of the programme were due quite simply to its ambition. Government response in the developing world was favourable, but overall international action was slow to get off the ground and was under-funded (Batisse 1975). Some projects did become operational. For example, under the MAB Project on the ecological effects of increasing human activities on tropical and subtropical forest ecosystems, the San Carlos de Rio Negro project was begun in 1976 through a newly established International Centre for Tropical Ecology in Caracas. Studies of the structure, composition and production of undisturbed rainforest soils were followed by research on nutrient cycling and the effects of disturbance (Gilbert and Christy 1981). Under the MAB Project on the impact of human activities and land-use practices on grazing lands, the Integrated Project on Arid Lands (IPAL) was begun in 1976 with UNESCO and UNEP funding, and produced a stream of integrated research studies on semi-arid vegetation and the ecology of pastoralism. Under the MAB Project on island ecosystems, a pilot project was organized in Fiji from 1974 to 1976, supported by the UN Fund for Population Activities (Bayliss-Smith *et al.* 1988), and a second in selected islands of the eastern Caribbean in 1979 (di Castri 1986).

These projects, and others, have a good claim to be the forerunners of 'sustainable development' thinking, linking natural ecosystems and human use in an innovative or wholly research-based structure. However, within MAB, traditional 'nature' conservation also remained important, even if dressed in the new clothes of human ecology. The best example of this is the notion of 'biosphere reserves' (Project 8). These were to be zoned nature reserves, whose aim was to conserve 'natural areas and the genetic information they contain' in core zones, while allowing suitable human activities to continue in outer zones. Existing nature reserves could be reclassified (and sometimes extended) to fit the MAB framework as biosphere reserves. This initiative was an important linkage between 'pure' wildlife conservation and the much broader aims of MAB, and between nature preservation and the idea of conservation of natural resources for human use. Thus Batisse (1982) wrote: 'The greatest merit of the "Biosphere Conference" was perhaps the assertion, for the first time in an intergovernmental context, that the conservation of environmental resources could and should be achieved alongside that of their utilisation for human benefit' (p. 101).

Although the MAB programme as a whole failed to live up to the hopes it raised at its inception, it undoubtedly contributed through the 1970s to the growing belief that there was an ecologically sound approach to development that would be 'sustainable' and acceptable, and that this could be discovered for specific environments and circumstances through research done in new, open and interdisciplinary ways.

The biosphere reserves represent this vision at its rosiest tint. Batisse (1982) wrote that

> experience already shows that when the populations are fully informed of the objectives of the biosphere reserve, and understand that it is in their own and their children's interests to care for its functioning, the problem of protection becomes largely solved. In this manner, the biosphere reserve becomes fully integrated – not only into the surrounding land-use system, but also into its social, economic, and cultural, reality.
>
> (p. 107)

This vision – of conservation integrated with and serving some rather vaguely defined human (and hence economic) purpose – was central to the MAB programme, and through the 1970s it was fostered by it. The identical notion resurfaces in 1980 in ideas about sustainable development in the *WCS*. This, and the other documents of what I term the 'sustainable development mainstream', are the subject of the next two chapters.

Summary

- Ideas about sustainable development that emerged in the 1980s had deep roots. Important themes include the development of environmentalism, concern for nature preservation, the development of international environmental organizations, the development of the science of ecology (and ideas about the balance of nature and the need for science-based ecological management), concern about global population growth and the development of global scientific networks.
- The tropics, which were the focus of much of the environmental concern and development action in the late twentieth century, were also important to the development of environmentalism at much earlier periods, particularly in the seventeenth and eighteenth centuries.
- Nature preservation is important to sustainable development both as a source of the impulse to balance human need and human claims on nature, and also because of the role of international conservation organizations (especially the International Union for the Conservation of Nature (IUCN), the World Conservation Union) in generating the thinking that stimulated the formulation of the concept of sustainable development, and organizing the meetings where it was first set out.
- The science of ecology contributed to development, and development planning, in various ways, particularly in the period following the Second World War.
- Ecological ideas such as 'the balance of nature', the concept of the ecosystem and maximum sustainable yield provide an essential underpinning of concepts of sustainable development.

- Concern about limits to growth and global population growth were fundamental to the environmental revolution of the 1970s, and provide the background to the emergence of formal statements of the idea of sustainable development. Global scientific collaboration, notably in the International Biological Programme, provided an authoritative, apolitical and effective arena within which ideas of sustainable development could develop and be discussed in the 1970s.

Further reading

Adams, W. M. (2004) *Against Extinction: the story of conservation*, Earthscan, London.

Barton, G. A. (2002) *Empire Forestry and the Origins of Environmentalism*, Cambridge University Press, Cambridge.

Boardman, R. (1981) *International Organizations and the Conservation of Nature*, Indiana University Press, Bloomington.

Drayton, R. (2000) *Nature's Government: science, imperial Britain and the 'improvement' of the world*, Yale University Press, New Haven.

Dresner, S. (2002) *The Principles of Sustainability*, Earthscan, London.

Glacken, C. J. (1967) *Traces on the Rhodian Shore: nature and culture in Western thought from ancient times to the end of the eighteenth century*, University of California Press, Berkeley and Los Angeles.

Griffiths, T. and Robin, L. (1997) (eds) *Ecology and Empire: environmental history of settler societies*, Keele University Press, Keele.

Grove, R. H. (1995) *Green Imperialism: colonial expansion, tropical island Edens and the origins of environmentalism, 1600–1800*, Cambridge University Press, Cambridge.

Grove, R. H., Damodaran, V. and Sangwan, S. (1998) (eds) *Nature and the Orient: the environmental history of South and South East Asia*, Oxford University Press, Delhi.

Guha, R. (2000) *Environmentalism: a global history*, Oxford University Press, Delhi.

Holdgate, M. (1999) *The Green Web: a union for world conservation*, Earthscan, London.

Kurlansky, M. (1999) *Cod: a biography of the fish that changed the world*, Vintage, London.

McCormick, J. S. (1989) *The Global Environmental Movement: reclaiming paradise*, Belhaven, London.

MacKenzie, J. M. (1988) *The Empire of Nature: hunting, conservation and British imperialism*, Manchester University Press, Manchester.

Rawcliffe, P. (1998) *Environmental Pressure Groups in Transition*, Manchester University Press, Manchester.

Worster, D. (1985) *Nature's Economy: a history of ecological ideas*, Cambridge University Press, Cambridge.

Worthington, E. B. (1983) *The Ecological Century: a personal appraisal*, Cambridge University Press, Cambridge.

Web sources

http://www.birdlife.org BirdLife International (founded in 1926 as the International Council for Bird Preservation).

http://www.fauna-flora.org/ Fauna & Flora International, founded in 1903 as the Society for the Preservation of the Wild Fauna of the Empire.

http://www.unesco.org/mab/ The UNESCO Man and the Biosphere Programme: information on the world network of biosphere reserves.

http://www.iucn.org/ The World Conservation Union (IUCN).

http://www.sanparks.org/parks/kruger/ Kruger National Park in South Africa, established 1926: History of the park; Eden in cyberspace, including webcams.

http://www.fao.org/fi/default.asp The Food and Agriculture Organization's Fisheries programme: the continuing struggle to make fisheries management sustainable.

http://www.populationconnection.org/ Population Connection (until 2002 Zero Population Growth).

3 The development of sustainable development

In our time, man's capability to transform his surroundings, if used wisely, can bring to all peoples the benefits of development and the opportunity to enhance the quality of life. Wrongly or heedlessly applied, the same power can do incalculable harm to human beings and the human environment.

(Declaration of the United Nations Conference on the Human Environment, 1972)

Before the mainstream: the Stockholm Conference

Sustainable development was codified for the first time in *The World Conservation Strategy* (*WCS*), a document prepared over a period of several years in the later 1970s by IUCN with finance provided by UNEP and the World Wildlife Fund (IUCN 1980). It was then further developed through the report of the World Commission on Environment and Development, *Our Common Future* (Brundtland 1987), and the follow-up to the *WCS*, *Caring for the Earth* (IUCN 1991), before its appearance in *Agenda 21* at the Rio Conference in 1992. By 2002, when the World Summit on Sustainable Development (the Earth Summit) was held in Johannesburg, the idea of sustainable development was central to debates about the environment and development.

Although there were changes in the way sustainable development has been presented, these documents were built around a remarkably consistent core of 'mainstream' ideas. The stock from which all these mainstream documents descended, and the forum at which ideas of sustainable development were first brought onto the international agenda, was the United Nations Conference on the Human Environment held in Stockholm in June 1972.

McCormick (1989) argues that in many ways the Stockholm Conference simply developed ideas already raised at the Biosphere Conference in Paris in 1968 (Chapter 2). However, the Stockholm meeting is usually identified as the key event in the emergence of sustainable development. It was only partly, and belatedly, concerned with the environmental and developmental problems of the emerging developing world. The primary motivation behind the UN's decision to hold such a conference came from industrialized countries, and was the product of the classic concerns of First World environmentalism, particularly pollution associated with industrialization such as acid rain (McCormick 1986). Sweden itself was motivated by a desire for an alternative to the latest UN proposal for a conference on atomic energy (Ivanova 2007). The Swedish ambassador to the United Nations submitted a proposal for a conference on the human environment to ECOSOC in July 1968, and the resolution was approved in December of that year (Ivanova 2007). Conference planning began in 1968, and a twenty-seven-nation 'Preparatory Committee' began meeting under the chairmanship of Maurice Strong in 1970 (Holdgate 1999).

However, the proposed conference did not command support from all countries. Developing countries in particular believed that environmental problems and development problems had become separated, and the sense of integration and of shared problems between developing and industrialized countries was lost (Russell 1975). Developing-country governments mistrusted neo-Malthusian ideas, whether of zero growth or lifeboat ethics:

> some 'developing' countries felt that the concept of global resources management was an attempt to take away from them the national control of resources. Furthermore, as industrialised countries used the lion's share of resources and contributed to most of the resulting pollution, the Third World countries did not see much reason to find and pay for the solutions.
>
> (Biswas and Biswas 1984, p. 36)

Faced with urgent short-term problems of poverty, hunger and disease, longer-term environmental problems associated with industrialization seemed not only remote, but a possible means by which industrialized economies might wriggle off the hook of responsibility for supporting a rapid drive for development.

It seemed possible that controversy over the relative priorities to be accorded environment and development in the Third World might cause the Stockholm Conference to fail. A meeting of a Preparatory Committee of twenty-seven experts at Founex in Switzerland in June 1971 sought to soothe the concerns of Third World countries, allaying fears about the economic effects of environmental protection policies. It proffered assurance that environmental protection would not go against their interests and would not affect their position in international trade (e.g. by anti-pollution barriers), and that rapid industrialization could still be pursued, but in such a way that its most adverse impacts were avoided (McCormick 1989).

The Founex meeting was certainly a political success inasmuch as the developing countries duly came to Stockholm. It was becoming clear that the position of non-industrialized countries was gaining recognition, just as (through their voting power in the UN) they were gaining power. The scope of the conference was expanded in December 1971 to include issues such as soil erosion, desertification, water supply and human settlement. Founex made the case that the environment was relevant to less industrialized countries, indeed that environmental issues were a central issue in successful development (McCormick 1989). The argument was made that the apparent dichotomy of 'environment versus development' was false, and should not be recognized, let alone fostered (Biswas and Biswas 1984).

At the same time, the Founex meeting did not break new conceptual ground. Like Maurice Strong in his meetings with Third World governments, it simply made a statement of faith that development and environment *could* be combined in some way that would optimize ecological and economic systems, without explaining how. It promised that the Stockholm meeting would 'point the way towards the achievement of industrialization without side-effects', but it did not say *how* this desirable trajectory of change was to be achieved (Clarke and Timberlake 1982, p. 7).

It was the same at the Stockholm Conference itself. The Indian Prime Minister, Indira Gandhi (the only head of state apart from Sweden's Olaf Palme to attend), made a conference-defining speech in which she argued 'are not poverty and need the greatest polluters? How can we speak to those who live in villages and in slums about keeping the oceans, the rivers and the air clean when their own lives are contaminated at the source? The environment cannot be improved in conditions of poverty. Nor can poverty be eradicated without the use of science and technology' (Ivanova 2007; quotation from www.centerforunreform.org, 16 November 2007).

However, there was little discussion of the links between poverty and environmental degradation. Few of the Stockholm Conference's Declaration of Principles did more than recognize the nature of the particular problems of the Third World. Many Third World countries remained sceptical, and a common theme in their leaders' speeches was that 'environmental factors should not be allowed to curb economic growth' (McCormick 1989, p. 99). Even the humane and encyclopaedic popular book written for the conference by Barbara Ward and René Dubos, *Only One Earth* (1972), offered relatively little to the 'developing regions'. It recognized the hard inheritances of colonialism and exploitative trade, and discussed the problems of population, possible policies for growth in agriculture and industry, and the question of urban environments. The synthesis, however, was global, and there was little beyond general exhortation in this volume – which became one of the classics of 1970s environmentalism – about how environment and development could be integrated in the Third World. It was clear that this *should* happen, but less clear how it could be done.

The conference itself in June 1972 was attended by the governments of 113 countries. The German Democratic Republic was not invited, and in protest the USSR and most East European countries did not attend. Five hundred non-governmental organzations participated in a parallel 'fringe' meeting, the Environmental Forum. There were fierce debates – for example about colonialism, Vietnam, whaling and nuclear weapons testing. Eventually, 26 Principles and 109 Recommendations for action were agreed (Clarke and Timberlake 1982). The Principles were wide-ranging, from human rights (Principle 1) and nuclear disarmament (Principle 26) through to the need for environmental education and research (Principles 19 and 20). There were general exhortations about pollution (Principles 6 and 7), the need to 'safeguard' wildlife and natural resources, the

Plate 3.1 Bicycle repairman, Nigeria. The bicycle remains the perfect example of
intermediate technology, a concept drawn from the work of the radical
economist Fritz Schumacher, best known for his book *Small is Beautiful*. He
founded the Intermediate Technology Development Group in 1966; this still
exists as Practical Action (www.itdg.org). Photo: W. M. Adams

need to 'share' non-renewable resources (Principle 5), and the need to cooperate over international issues. There was stress on the right of individual nations to determine population and resource policies (Principles 16 and 21).

Most importantly, in the light of the influence of the 'spirit of Stockholm', there were some deliberate attempts to address the problems of the Third World. The fundamental point was that development need not be impaired by environmental protection (Principle 11). This was to be achieved by integrated development planning (Principle 13) and rational planning to resolve conflicts between environment and development (Principle 14). Furthermore, development was needed to improve the environment (Principle 8), and this would require assistance (Principle 9), particularly money to pay for environmental safeguards (Principle 12), and reasonable prices for exports (Principle 10).

Like Founex, the Stockholm meeting itself stated the need to resolve conflicts between environment and development without demonstrating how. The suggested solutions, 'rational planning' or 'integrated development', were words without any detailed substance. They owed much to the technocratic element in ecological and environmentalist thinking developing at that time. In very mild form they reflected the authoritarian idea that environmental harmony should be sought through central control (see Pepper 1984): they were based on the premiss that planning was neutral and perfectible, that conflicts could be planned away.

The need to see environment and development as an integrated whole was well argued at Stockholm. However, few of its Recommendations addressed the issue: only 8 of the 109 Recommendations for Action referred to development and environment, and they were 'extraordinarily negative' (Clarke and Timberlake 1982, p. 12), concerned chiefly with minimizing possible costs of environmental protection. In the ensuing decade there was little progress in this field, the only specific report by the United Nations Conference on Trade and Development (UNCTAD) failing to identify the effects of economic policies on trade, and by 1982 the debate was said to be 'largely dead' (ibid., p. 23). Stockholm focused most of its energy on industrial country concerns (Holdgate 1999).

The most conspicuous result of the Stockholm Conference was the creation of the United Nations Environment Programme (UNEP). This was established by resolution of the General Assembly of the United Nations in December 1972 to act as a governing council for environmental programmes; a secretariat would focus environmental action within the whole UN system and an environment fund would finance environment programmes (Ivanova 2007). UNEP was located in Nairobi in Kenya, the first UN body outside the developed world, a symbolic (and politically astute) decision. The conference secretariat had proposed a wholly new intergovernmental body within the UN to deal with environmental problems, but the suggestion was strongly opposed by existing UN organizations (Ferau 1985). The outgoing UN Secretary-General in 1971 favoured a 'switchboard' linking separate sectoral organizations (McCormick 1989, p. 93), and this is essentially what was created. UNEP therefore is not a UN agency like UNESCO or FAO, and these remain responsible for the environmental aspects of their own activities. UNEP seeks to act as a catalyst and think tank, the 'conscience of the UN system' (Clarke and Timberlake 1982, p. 49).

Sadly, UNEP's small size, poverty, relative weakness within the UN system and peripheral position in Nairobi have limited its effectiveness (Ivanova 2007). It is officially a unit of the UN Secretariat and gets administrative funds from that source, but money for projects comes from the Environment Fund. Contributions to that are voluntary, and have fallen short of needed targets. UNEP's influence on the UN agencies has been relatively small, and they have gone about their business much as before. As Myers and Myers (1982) commented, 'we wanted an Environmental Programme of the United Nations. Instead we got a United Nations Environment Programme' (p. 201).

The Stockholm Conference saddled UNEP with an impossibly broad remit and a vague list of priorities. It went on to develop several notable activities such as the Global Environment Monitoring System (GEMS), begun in 1975 (Gwynne 1982), and the Regional Seas Programme. This was launched in 1974, and had roots directly in the Stockholm Declaration on marine pollution. UNEP acts as a coordinator of intergovernmental action based on an agreed Action Plan, an approach tackling a global problem through regional action. By 1982 the programme covered 10 regions and 120 coastal states (Bliss-Guest and Keckes 1982). UNEP also organized the United Nations Conference on Desertification (UNCOD) in Nairobi in 1977 (United Nations 1977). This was brought about by a resolution of the UN General Assembly in 1974, which had discussed the Sahelian drought of 1972–4. The UN General Assembly endorsed the conference's Plan of Action to Combat Desertification (PACD), and, although other UN bodies had significant experience in the field (such as UNESCO in arid zone research), coordination of the PACD was entrusted to UNEP (Karrar 1984). In the event, international, national and local 'anti-desertification' action fell far short of expectations, and the movement spearheaded by UNEP achieved very little. Indeed, by the 1980s and 1990s the whole subject had become beset by controversy (not lessened by the Convention on Desertification that eventually followed the Rio Conference two decades later (Swift 1996)). Desertification is discussed in detail in Chapter 8.

Following Stockholm, international debate about sustainable development began to be more extensively influenced by concerns about poverty in the developing world. An international meeting at Geneva in April 1974 flew in the face of the environmentalist rhetoric of the time by eschewing the notion of limits to growth, and emphasizing that the key problem was the distribution of natural resources and the benefits that flowed from them, not their global scarcity (McCormick 1989). The advent of the Organization of Petroleum Exporting Countries (OPEC) on the international scene had demonstrated the political power of resource-rich states, and in May 1974 the UN General Assembly adopted a declaration calling for a New International Economic Order. This was to 'correct inequalities and redress existing injustices, making it possible to eliminate the widening gap between the developed and developing countries and ensure steadily accelerating economic development' (Lummis 1992, p. 44).

The needs of developing countries, which had been an issue at Stockholm, were now receiving much stronger emphasis. A meeting of experts held at Cocoyoc in Mexico in October 1974, which looked at environmental problems from the perspective of the developing world, and particularly its poor, was attended by Maurice Strong, Secretary-General of the Stockholm Conference, and chaired by Barbara Ward. The resulting 'Cocoyoc Declaration' pointed to the problem of the maldistribution of resources and to the 'inner limits' of human rights as well as the 'outer limits' of global resource depletion. It stressed the priority of basic human needs, and called for a redefinition of development goals and global lifestyles. It called for global resource management, international regimes for the management of common resources, and development policies aimed at the poor (McCormick 1989).

These ideas drew extensively from the debates in development which had emerged through the 'First Development Decade' of the 1960s, and were to emerge again in the global interdependence arguments of the Brandt Report (1980). They represented the productive fusion of those debates with those of Western environmentalism. By the time *The World Conservation Strategy* emerged six years later, the two had, superficially at least, merged.

The World Conservation Strategy

In the 1960s, thinking within IUCN began to embrace greater concern for economic development (McCormick 1986). The idea of a strategic approach to conservation was considered at the IUCN General Assembly in New Delhi in 1969, and conservation and development were the theme of the 1972 General Assembly at Banff, Canada. Work on a strategy for nature conservation began in 1975, when IUCN joined UNEP, UNESCO and the FAO to form the 'Ecosystem Conservation Group', and gradually the notion of a 'world conservation strategy' took shape (Boardman 1981; McCormick 1986, 1989; Holdgate 1999).

In 1977 UNEP commissioned IUCN to draft a document to provide 'a global perspective on the myriad conservation problems that beset the world and a means of identifying the most effective solutions to the priority problems' (Munro 1978). Preliminary drafts were discussed at the IUCN General Assembly in Ashkhabad, in the USSR, in 1978. At that stage the strategy was effectively an extended textbook of wildlife conservation, about the conservation of species and special areas rather than the integration of conservation and of development (Munro 1978). The focus was subsequently changed substantially to include questions of population, resources and development (Boardman 1981).

Thinking about the wider issues of people and environment in the 1970s, and the attempt to imbue development with environmental ideas and principles, was often expressed in terms of 'ecodevelopment' (I. Sachs 1979, 1980). This term was coined by Maurice Strong, Secretary-General of the Stockholm Conference (Boardman 1981). It was subsequently developed and promoted by UNEP (1978), and widely discussed internationally – for example, at meetings in Belo Horizonte in Brazil in 1978 (the International Workshop in Ecodevelopment and Appropriate Technology), in Berlin in 1979 (the Conference on Ecofarming and Ecodevelopment (Glaeser and Vyasulu 1984)) and in Ottawa in 1986 (the

IUCN Conference on Conservation and Development (Svedin 1987)). Behind the notion of ecodevelopment lay the awareness of the intrinsic complexity and dynamic properties of ecosystems and the ways they respond to human intervention, and the need to ensure the 'environmental soundness' of development projects (*Ambio* 1979, p. 115). The challenge was that of 'improving the economic wellbeing of people without impairment of the ecological systems on which they must depend for the foreseeable future' (Dasmann 1980, p. 1331). These ideas about ecodevelopment are fundamental to *The World Conservation Strategy* (*WCS*).

The *WCS* was eventually published in 1980 in the name of IUCN, UNEP and the World Wildlife Fund (WWF) (IUCN 1980). It had already been presented to the FAO and UNESCO, and publication had been delayed to include their amendments (McCormick 1986).

The *WCS* was intended to be an outgoing, even evangelistic document, seeking to show the relevance of conservation to the development objectives of others, in governments, industry and commerce, organized labour and the professions (Allen 1980). In the words of the chairman of the WWF, Sir Peter Scott, it suggested for the first time that development should be seen as 'a major means of achieving conservation, rather than an obstruction to it' (Allen 1980, p. 7). It was aimed at government policy-makers, conservationists and development practitioners, and was intended 'to stimulate a more focused approach to the management of living resources and to provide policy guidance on how this can be carried out' (IUCN 1980, p. vi). The *WCS* was divided into three parts. First, it described objectives for conservation, their relevance for human survival, and priority requirements for achieving them; second it set out a strategy for action at national and subnational levels, and identified obstacles and possible ways to deal with them; third, it outlined the international action required to stimulate and support action at smaller scales.

The *WCS* identified three objectives for conservation (see Table 3.1), broken down into priority requirements (see Table 3.2). They ranged from the sublime

Table 3.1 Objectives of conservation

---

1 To maintain essential ecological processes and life-support systems (such as soil regeneration and protection, the recycling of nutrients and the cleansing of waters) on which human survival and development depend
2 To preserve genetic diversity (the range of genetic material found in the world's organisms), on which depend the functioning of many of the above processes and life-support systems, the breeding programmes necessary for the protection and improvement of cultivated plants, domestic animals and micro-organisms, as well as much scientific and medical advance, technical innovation, and the security of the many industries that use living resources
3 To ensure the sustainable utilization of species and ecosystems (notably fish and other wildlife, forests and grazing lands), which support millions of rural communities as well as major industries.

Source: *The World Conservation Strategy* (IUCN 1980).

intention to 'prevent the extinction of species' to the detailed requirements of site protection for species conservation (see Table 3.2).

The first of the three objectives was the maintenance of 'essential ecological processes and life-support systems'. These processes were essential for food production,

Table 3.2 Priority requirements of the *World Conservation Strategy*

A Priority requirements: ecological processes and life-support systems (Section 5)

1 To reserve good cropland for crops
2 To manage cropland to high, ecologically sound standards
3 To ensure that the principal management goal for watershed forests and pastures is protection of the watershed
4 To ensure that the principal management goal for coastal wetlands is the maintenance of the processes on which the fisheries depend
5 To control the discharge of pollutants

B Priority requirements: genetic diversity (Section 6)

1 To prevent the extinction of species
2 To preserve as many kinds as possible of crop plants, forage plants, timber trees, livestock, animals for aquaculture, microbes and other domestic organisms and their wild relatives
3 To ensure on-site preservation programmes protect:

- the wild relatives of economically valuable and other useful plants and animals and their habitats
- the habitats of threatened and unique species
- unique ecosystems
- representative samples of ecosystem types

4 To determine the size, distribution and management of protected areas on the basis of the needs of the ecosystems and the plant and animal communities they are intended to protect
5 To coordinate national and international protected area programmes

C Priority requirements: sustainable utilization (Section 7)

1 To determine the productive capacities of exploited species and ecosystems and to ensure that utilization does not exceed those capacities
2 To adopt conservation management objectives for the utilization of species and ecosystems
3 To ensure that access to a resource does not exceed the resource's capacity to sustain exploitation
4 To reduce excessive yields to sustainable levels
5 To reduce incidental take as much as possible
6 To equip subsistence communities to utilize resources sustainably
7 To maintain the habitats of resource species
8 To regulate international trade in wild animals and plants
9 To allocate timber concessions with care and to manage them to high standards
10 To limit firewood consumption to sustainable levels
11 To regulate the stocking of grazing lands to maintain the long-term productivity of plants and animals
12 To utilize indigenous wild herbivores, alone or with livestock, where domestic stock alone would degrade the environment

Source: IUCN (1980).

health and 'other aspects of human survival and sustainable development' (para. 2.1). Their maintenance demanded maintenance of the ecosystems that govern, support or moderate them, including agricultural land and soil, forests and coastal and freshwater ecosystems. Threats to these systems included soil erosion, pesticide resistance in insect pests, deforestation and associated sedimentation, and aquatic and littoral pollution. Their conservation demanded rational planning and allocation of land use, so that crops were given priority on the best land, and areas such as watersheds and littoral zones were set aside and used appropriately.

The second objective of conservation set out in the *WCS* was the preservation of genetic diversity, in terms of both the genetic material in different varieties of locally adapted crop plants and livestock and that in wild species (see Table 3.2). Genetic diversity was both an 'insurance' (for example, against crop diseases), and an investment for the future (for example, for crop breeding or pharmaceuticals). The conservation of genetic diversity demanded site-based protection of ecosystems (essentially the familiar nature conservation strategies for the protection of the habitats of rare and unique species and typical ecosystems in protected areas) and the timely creation of banks of genetic material.

The third objective of conservation in the *WCS* was to ensure 'the sustainable utilization of species and ecosystems', particularly fisheries, wild species that are cropped, forests and timber resources, and grazing land (see Table 3.2). No fewer than twelve priority tasks were associated with this objective, linking research and action to determine and achieve resource utilization at sustainable levels. These included some that even at the time of publication were well established – for example, the regulation of international trade in wildlife products, and others such as the ingenuous suggestion to 'limit firewood consumption to sustainable levels' (see Table 3.2), which must surely defy policy implementation in the real world. The implications of some of this thinking are discussed below.

The *WCS* proposed that governments or NGOs should prepare separate national strategies to review development objectives in the light of the conservation objectives. These should establish priority requirements, identify obstacles and propose cost-effective ways of overcoming them, determine priority ecosystems and species for conservation and establish a practical plan of action. 'Strategic principles' suggested the integration of conservation and development by doing away with narrow sectoral approaches, managing ecosystems so as to retain future options on use (this reflecting the poor state of knowledge about tropical ecosystems in particular), mixing cure and prevention, and tackling causes as well as symptoms.

Two key problems were highlighted: first, the relative weakness of conservation institutions in the context of national policy-making, combined with the sectoral nature of such planning, and, second, the fact that environmental planning rarely allocated land uses rationally. It was proposed that these should be tackled by 'anticipatory and cross-sectoral' environmental policies and better planning (more and better evaluation of the capacity of ecosystems to meet human demands, improved prediction of the environmental effects of development, and better procedures for matching capacities and uses of land and water resources). For

Plate 3.2 Local varieties of sorghum and rice, northern Nigeria. Industrial agriculture depends not only on a small number of crops, but on a very restricted number of commercial varieties. Locally grown crops in the developing world are far more diverse, and represent a significant source of genetic diversity. The need to conserve crop biodiversity has been a theme of sustainable development from the *World Conservation Strategy* to the Convention on Biological Diversity. Many public collections of seed from around the world have been taken over by private-sector organizations, raising questions about the ownership of such genetic resources. Photo: W. M. Adams

development policy to be 'ecologically as well as economically and socially sound', conservation objectives needed to be considered at the beginning of the planning process, not factored in at the end once impacts had been caused (see Figure 3.1). Other problems identified were the inadequacy of legislation and weak and over-lapping natural resource management agencies, lack of ecological data, shortcomings in training and education of conservation personnel and in research, and lack of support for conservation policies.

Thus far, this is a fairly familiar analysis of the inadequate scope of ecological and environmental planning. However, the *WCS* also highlighted a 'lack of awareness of the benefits of conservation and of its relevance to everyday concerns' (para. 13.2). To remedy this, planning must not only be better technically, but must also involve more public participation and community involvement, and public education. Finally, the *WCS* also highlighted the lack of conservation-based rural development, 'rural development that combines short term measures to ensure human survival with long term measures to safeguard the resource base and improve the quality of life' (para. 14.5).

In conclusion, the *WCS* turned to international actions needed to promote conservation, recognizing that many living resources lay partly or wholly outside national boundaries. It addressed the limitations of institutions for

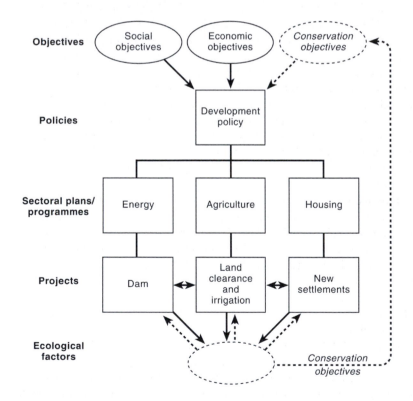

Figure 3.1 The need to integrate conservation and development (after IUCN 1980)

global environmental management in international law and conventions and the responsibilities of bilateral and multilateral aid donors. It called for the cooperative management of the global commons of the open ocean, the atmosphere and the Antarctic, international river basins and seas, and for support of the Plan of Action on Desertification. More interesting, perhaps, in the light of subsequent conservation concern, is a rather unfocused call for 'international action' to conserve tropical forests, and for a 'global programme for the protection of genetic resource areas'. This programme would involve the site-based protection of concentrations of economic or useful varieties (for example, wild relatives of cultivars), concentrations of threatened species, ecosystems of 'exceptional diversity' and ecosystems that were poorly represented in existing protected areas. This would require financing internationally, because 'many countries particularly rich in genetic resources are developing ones that can ill-afford to bear alone the burden of their on-site protection' (para. 17.11), or possibly through commercial participation or sponsorship.

Integrating conservation and development

What impact did the *WCS* have? The promotion of sustainable development formed one of IUCN's seven Programme Areas for the period 1985–7. IUCN established a Conservation for Development Centre (CDC) in 1979, marking a move into field programmes with partner organizations (Holdgate 1999). The CDC undertook projects to advise donor agencies how to build conservation into their work, projects to help Southern countries achieve conservation, and projects to promote international conventions in Southern countries. A series of projects supported the creation of national conservation strategies. The original plan to revise the *WCS* every three years gave way to progressive adaptation, as national conservation strategies were produced under IUCN guidance (IUCN 1984; McCormick 1986). A number of countries in both the developed and developing worlds duly produced national strategies (Nelson 1987; Bass 1988; Holdgate 1999).

At this nominal level, the *WCS* might therefore be judged a success. Caldwell (1984) describes it as 'the nearest approach yet to a comprehensive action-oriented programme for political change' (p. 306). It was successful too in aiding the spread of the phrase 'sustainable development' as a part of development terminology. On the other hand, Michael Redclift (1984) argued that, despite its diagnostic value, the *World Conservation Strategy* 'does not even begin to examine the social and political changes that would be necessary to meet conservation goals' (p. 50). This reflects the way that the *WCS* was the child of 1970s environmentalism. This can be seen in several ways.

First, the *WCS* (IUCN 1980) was neo-Malthusian in its approach. It is argued that every country should have a 'conscious population policy' to achieve 'a balance between numbers and environment' (para. 20.2). New approaches to resource management were needed because of the impact of population growth and rising demand, and it was 'the escalating needs of soaring numbers' that

have led to 'short-sighted' approaches to natural resources' (p. i). The *WCS*
graphically illustrated the challenge of growth in population and consumption
by juxtaposing predicted rates of degradation of arable land (a stalk of wheat),
reduction in unlogged productive tropical forest (a shrinking tree) and the global
population (a human giant) (see Figure 3.2). This neo-Malthusianism is evident
in the essentially determinist vision of the *WCS*. It offered a 'conservation or
disaster' scenario not far removed from the classic polemics of the 'ecodoomsters'
(P. Hill 1986). It identified ecological and environmental limits for human action,
and applied ideas drawn from wildlife management (particularly the notion of
'carrying capacity') directly to people without discussion of the political, social,
cultural or economic dimensions of resource use. The *WCS* argued not only that
ecology should determine human action in the way development is attempted,
but also that it set limits on the scope of human action.

The second way in which the *WCS* (IUCN 1980) reflected the thinking of
1970s environmentalism was the way it expressed the environmental challenge
to development as a global problem, emphasizing the 'global interrelatedness
of actions, with its corollary of global responsibility' (p. i). Global responsibility
demanded global strategy for development and conservation, which the *WCS* set
out to provide. Both IUCN and UNEP were global agencies with profiles and
ambitions far greater than their limited budgets. The *WCS* tried to tie a global
problem to national responses, with these organizations playing a vital coordinating role.

A third way in which the *WCS* reflected 1970s environmentalism was in its
ethics. It welded together scientific utilitarianism and romantic 'holist' or 'vitalist'
thinking into a form of 'bioethics' (O'Riordan [1976] 1981; Pepper 1984;
Worster 1985). It was argued that wild species should be conserved for two quite
different reasons: first, because they were useful to human society and economy,
and, second, because it was morally right to conserve them. The utilitarian argument for conservation, which adhered to the *WCS*'s conservative principle of

Why a conservation strategy is needed: depletion of living resources

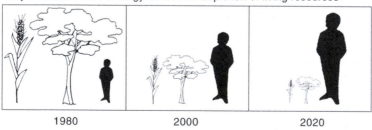

1980   2000   2020

Figure 3.2 Depletion of living resources, as portrayed by the *World Conservation Strategy*
(after IUCN 1980). Relative rates of degradation of arable land are shown by
the stalk of wheat symbol, reduction in unlogged productive tropical forest is
shown by the tree symbol, and the global population growth is shown by the
human figure.

keeping options open, was simple: 'we cannot predict what species may become useful to us' (para. 3.2). This principle underlay the relatively numerous and well-thought-out proposals concerning the sustainable utilization of living resources (fisheries management, game farming, regulation of wildlife trade (see Table 3.2)). The *WCS* also noted the non-monetary value of nature, pointing out the symbolic, ritual and cultural importance of wildlife (para. 4.11). However, the *WCS* did not put all its eggs in the basket of utilitarianism. It did not shrink from a moral argument for conservation. Because of the power of humans to transform the biosphere (and influence the process of evolution), conservation is a matter of moral principle: 'we are morally obliged – to our descendants and to other creatures – to act prudently' (para. 3.3).

This attempt to argue for conservation along two parallel tracks is by no means accidental. It reflects old divisions between technological and ecological envi-ronmentalism (Pepper 1984), or between techno-centrism and eco-centrism (O'Riordan [1976] 1981), which create such schizophrenic confusion within the conservation movement and for individual conservationists. In practical terms this dualism is extremely useful (Norton 1991). On the one hand, the utilitarian argument allows conservation to be packaged so as to be attractive to the anthropocentric materialism that underlies thinking about development. The *WCS* makes the self-evident case that ecosystems and species must be used sustainably to ensure their continued availability 'almost indefinitely' (para. 4.1). On the other hand, moral arguments can be employed where they are more effective – for example, among conservationists in industrialized countries. Here economies have substantially been freed from primary dependence on renew-able resources, a freedom that developing countries would like to emulate. The rational argument for sustainable utilization of ecosystems seems less critical to development strategies, but the moral argument can be rapidly marshalled to fill the gap. Although greater economic diversity and flexibility might reduce the need to utilize certain resources sustainably, the *WCS* argues that there is less excuse not to do so (para. 4.1). The *WCS*'s double-barrelled justification of sustainable development is both versatile and robust.

The *WCS* can thus be interpreted as conservationist environmentalism refo-cused for a new decade, and attempting to engage with issues of development. Its attempts to demonstrate the indivisibility of conservation and development lie in the way they are defined. The key concept here is, of course, sustainability. Indeed, the whole message of the *WCS* turns on the question of what 'sustainability' means. The opaque nature and flexibility of these terms have been discussed in Chapter 1. Within the *WCS*, development and conservation are defined in such a way that their compatibility becomes inevitable. Development is presented as 'the modification of the biosphere and the application of human, financial, and living and non-living resources to satisfy human needs and improve the quality of human life' (para. 1.4). Meanwhile, conservation is 'the management of human use of the biosphere so that it may yield the greatest sustainable benefit to present generations while maintaining its potential to meet the needs and aspirations of future generations' (para. 1.4).

If it is taken for granted that development ought to be 'sustainable' (meaning that it must be capable of being extended indefinitely for the benefit of future generations), conservation and development are of course 'mutually dependent' and not incompatible (para. 1.10). Conservation is essential to every sector (health, energy, industry), for conservation is 'that aspect of management which ensures that utilisation is sustainable' (para. 1.6). Whereas, in the past, development practitioners might have seen conservation as irrelevant to the task of development, or even in opposition to it, this was because they failed to understand about 'real' conservation. Properly understood (the *WCS* argued), 'real' conservation would have helped avoid ecological damage and the failure of development, a failure that demonstrates that much development is not real development at all.

The *WCS* accepted that environmental modification was a natural and necessary part of development, but argued that not all such modification would achieve the social and economic objectives of development. In suggesting that development planning should not only be socially and economically sound but also fit conservation objectives (see Figure 3.1), the *WCS* established the basic triptych of mainstream sustainable development thinking in the 1990s, of economic, social and environmental sustainability (see Chapter 5). Conflict between conservation and development could be avoided if conservation and development were integrated at every stage of planning (see Figure 3.1). Meanwhile, conservation needed to address the causes and not just the symptoms of environmental change, and avoid coming across as 'anti-development (hence anti-people)' (para. 8.6).

This is the core of the *WCS*. It represented a significant repackaging of conservation. It is less clear if it really offered the new principle and purpose that is claimed. The *WCS* was Janus-headed, addressing the very different worlds of conservation and development at the same time. It tried to make a case for conservation in a new way, without losing established priorities. It argued that sustainable utilization of ecosystems for human benefit can coexist with the preservation of nature. To conservationists, therefore, the *WCS* seemed to make a logical and effective inroad into the hitherto unfamiliar and inexplicably destructive world of economic development while reaffirming the moral basis for conservation, and while finding ways to justify the protection of land in reserves and parks. The influence of its 'conservation with development' ideas on subsequent conservation policy is explored in Chapter 10.

However, at the same time, the *WCS* also sought to make that advance into the development field effective, by making its arguments plausible within development. This it tried to do by emphasizing that species and ecosystems are resources for human subsistence and development, defending even the establishment of parks and reserves on the grounds of indirect ecological benefits (such as runoff from forested watersheds, now termed 'ecosystems services' (e.g. Millennium Ecosystem Assessment 2005)) or direct economic returns from 'ecotourism' (Campbell 2002a). On this second front the *WCS* can be judged less successful.

The presentation of the *WCS*, although well rounded and plausible, failed to recognize the essentially political nature of the development process. This matters on two levels. First, conservation – like science – was presented as somehow beyond ideology. The *WCS* shows no real understanding of the way in which nature and culture interact, such that views of nature are created by society. There is no apparent awareness of arguments about the social production of nature (N. Smith 1984). Second, the *WCS* suggested that conservation could in some way bypass structures and inequalities in society. It seems to assume that 'people' can exist in some kind of vacuum, outside the influence of inequality, class or the structures of power. The goal of the *WCS* was stated as 'the integration of conservation and development to ensure that modifications to the planet do indeed secure the survival and wellbeing of all people' (para. 1.12). This is pious, liberal and benign, but also naive.

The *WCS* avoided the explicit use of the language and ideas of political economy. It does not engage in the inescapable politics of the development process, the impacts of economic change and the constraints of capital, technical knowledge and manpower. This hugely limits its attempt to place essentially environmentalist ideas within a development matrix, and what it has to say about development is not particularly convincing. Subsequent mainstream sustainable development thinking largely shared these limitations. The reasons for this, and some critiques and alternatives, are examined in later chapters.

The Brundtland Report

The *WCS* was also simplistic in what it has to say about economic development. It barely began to address issues of national economic management (questions, for example, of the relative weight given to different sectors, or the pros and cons of economic globalization), let alone questions of international economic management. It said little beyond bland generalities about the gulf in wealth between North and South, or the dependence of the one upon the other, and, as we shall see later, nothing at all about the various radical theories on the global economy. The global scope of the *WCS* embraced neither the real world and the practical politics of international development, nor the theoretical ideas being discussed in development studies. Robert Prescott-Allen, who wrote the *WCS*, commented of IUCN in 1998 that 'the problem was that it wanted to sell conservation to the development constituency, but it didn't understand what the development constituency was like. The conservationists didn't see that development was the driving force in human affairs' (quoted in Holdgate 1999, p. 123).

However, as the 1980s wore on, this was to change. Sustainable development was brought into the established political arena of international development through the work of the World Commission on Environment and Development (WCED) chaired by the Prime Minister of Norway, Gro Harlem Brundtland. The Commission was established by the UN General Assembly in December 1983. Its report, *Our Common Future* (Brundtland 1987), was presented to the General Assembly in 1987.

Our Common Future claimed a very specific heritage. The chair wrote: 'After Brandt's *Programme for Survival* and *Common Crisis*, and after Palme's *Common Security*, would come *Common Future*' (p. x). The issues of environment and poverty were here being treated as a global threat, as war had been. Like its forerunners, the WCED had as its target the promotion of multilateralism and the interdependence of nations: 'the challenge of finding sustainable development paths ought to provide the impetus – indeed the imperative – for a renewed search for multilateral solutions and a restructured international economic system of cooperation' (p. x). The Brundtland Report reflected what Chatterjee and Finger (1994) call 'same boat ideology' (p. 80), proposing that global crisis could be staved off by dialogue between enlightened individuals, global environmental awareness and planetary stewardship. Indeed, in places, Brundtland's vision verged on the sugary, as in the aim of its sustainable development strategy: 'to promote harmony among human beings and between humanity and nature' (p. 65).

The Brundtland Commission was important for several reasons. First, it obviously and rather self-consciously attempted to recapture the 'spirit of Stockholm 1972', which had been so celebrated by environmentalists in the early 1970s and whose demise as recession bit was so lamented. The *WCS* tried to do exactly the same thing, of course, but Brundtland achieved this resuscitation far more expertly and effectively, largely because of its origins in the UN General Assembly and not out in the wildwoods of UNEP and IUCN.

Second, *Our Common Future* placed elements of the sustainable development debate within the economic and political context of international development. Its starting point was deliberately broad, and a move to limit its concern simply to the 'environment' was firmly resisted:

> This would have been a grave mistake. The environment does not exist as a sphere separate from human actions, ambitions, and needs, and attempts to defend it in isolation from human concerns have given the very word 'environment' a connotation of naivety in some political circles.
>
> (Brundtland 1987, p. xi)

Third, Brundtland placed environmental issues firmly on the formal international political agenda. Arguably, of course, it simply reflected the de facto situation created by its predecessors, but nonetheless it achieved something that Stockholm, UNCOD and the *WCS* had failed to do, and got the UN General Assembly to discuss environment and development as one single problem.

The Brundtland Report therefore started from the premiss that development and environmental issues could not be separated, recognizing that it was futile to try to tackle environmental problems without considering broader issues of 'the factors underlying world poverty and international inequality' (p. 3). This view was at the same time less abstract and less simplistic than that of the *WCS*. *Our Common Future*'s argument was not presented in the rather general terms of linkages between sustainable use of ecosystems and human wealth and welfare, nor was the spectre of 'ecodisaster' prominent. Instead, the essentially reciprocal links between development and environment were drawn more explicitly. *Our Common Future* recognized that development could 'erode the environmental resources on which they must be based', and hence that environmental degradation could undermine economic development. Furthermore, the links, again reciprocal, between poverty and environment were also recognized, poverty being seen 'as a major cause and effect of global environmental problems' (Brundtland 1987, p. 3).

Our Common Future's familiar definition of sustainable development (see above, Chapter 1) is based on two concepts. The first is the approach to basic needs and the corollary of the primacy of development action for the poor. The second involves the idea of environmental limits. These limits are not, however, those set by the environment itself, but those set by technology and social organization. This involves a subtle but extremely important transformation of the ecologically based concept of sustainable development, leading beyond concepts of physical sustainability to the socio-economic context of development. Physical sustainability could be pursued in 'a rigid social and political setting', but it cannot be secured without policies that actively consider issues such as access to resources and the distribution of costs and benefits. In other words, the sustainable development of *Our Common Future* was defined by the achievement of certain social and economic objectives, and not by some notional measurement of the 'health' of the environment. Whereas the *WCS* started from the premiss of the need to conserve ecosystems and sought to demonstrate why this made good economic sense (and – although the point was underplayed – could promote equity), *Our Common Future* starts with people and goes on to discuss what kind of environmental policies are required to achieve certain socio-economic goals. The answers (perhaps unsurprisingly) are remarkably similar. The premisses, however, do differ, and, of the two, *Our Common Future* was by far the more effective document in its ability to address and engage government policy-makers.

The elements of the sustainable development ideas in *Our Common Future* are listed in Table 3.3. They represent an interesting blend of environmental and developmental concerns. Among the former is the need to achieve a sustainable level of population. This is a softened version of the neo-Malthusian message that recognizes the greater demands on resources made by a First World child, but it still argues that population should 'be stabilised at a level consistent with the productive capacity of the ecosystem' (Brundtland 1987, p. 56). Other rather familiar concerns are conserving (and enhancing) the resource base and reorienting technology, particularly with regard to risk. Prominent among the latter is the fundamental concern with meeting basic needs. Prominent too is the need to 'merge' environment and economics in decision-making. Most prominent of

Table 3.3 Critical objectives for environment and development policies that follow from the concept of sustainable development

1	Reviving growth
2	Changing the quality of growth
3	Meeting essential needs for jobs, food, energy, water and sanitation
4	Ensuring a sustainable level of population
5	Conserving and enhancing the resource base
6	Reorienting technology and managing risk
7	Merging environment and economics in decision-making

Source: Brundtland (1987, p. 49).

all, however, is the focus on growth: economic growth is seen as the only way to tackle poverty, and hence to achieve environment–development objectives. It must, however, be a new form of growth: sustainable, environmentally aware, egalitarian, integrating economic and social development. Above all, the Brundtland Report's vision of sustainable development was predicated on the need to maintain and revitalize the world economy. This means 'more rapid economic growth in both industrial and developing countries, freer market access for the products of developing countries, lower interest rates, greater technology transfer, and significantly larger capital flows, both concessional and commercial' (p. 89).

However, this prescription was based on an economic and not an environmentalist vision. It used some of the language of 1970s environmentalism, but it did not question growth or technology, and it avoided arguments about eco-disaster. In the classification of Cotgrove (1982) it was firmly cornucopian rather than catastrophist. *Our Common Future* argued that it is poverty that puts pressure on the environment in the Third World, and it is economic growth that will remove that pressure. Furthermore, that growth cannot be conceived of in a geopolitical vacuum; it is only the ending of dependence that will enable these countries to 'outpace' their environmental problems.

The Brundtland Report was, therefore, built on the need to promote economic growth. But what of the pressures of that growth itself? What about demands for energy and raw materials, or pollution? *Our Common Future* hoped to have its cake and eat it: 'The Commission's overall assessment is that the international economy must speed up world growth while respecting environmental constraints' (p. 89). However, it did not say how this balancing trick was to be achieved.

What does this form of sustainable development require? Quite simply a restructuring of national politics, economics, bureaucracy, social systems, systems of production and technologies, and a new system of international trade and finance (Table 3.4). This is no small agenda. Sustainable development must be global in scope and internationalist in formulation. This demands first that 'the sustainability of ecosystems on which the global economy depends must be guaranteed', and, second, equitable exchange between nations. It is this latter requirement that lifts the Brundtland Report out of the mould of previous eco-development writing.

Table 3.4 Requirements of a strategy for sustainable development

1	A political system that secures effective citizen participation in decision-making
2	An economic system that is able to generate surpluses and technical knowledge on a self-reliant and self-sustained basis
3	A social system that provides for solutions for the tensions arising from disharmonious development
4	A production system that respects the obligation to preserve the ecological basis for development
5	A technological system that can search continuously for new solutions
6	An international system that fosters sustainable patterns of trade and finance
7	An administrative system that is flexible and has the capacity for self-correction

Source: Brundtland (1987, p. 65).

Environmentalists might have seen it as a logical extension of the *WCS*, and welcomed its apparently new and authoritative tone and the credibility with which it handled development terminology. In fact it was only rather secondarily the fruit of environmentalism. It is better understood as an extension of the thinking of the Brandt Reports *North–South* and *Common Crisis* (Brandt 1980, 1983). Indeed, while the *WCS* represented the attempt by conservationists to capture the rhetoric of development to repackage old ideas, *Our Common Future* is the result of the reverse process. The existence of a global environmental crisis is cited as evidence for the need for a multilateralist solution:

> At first sight, the introduction of an environmental dimension further complicates the search for cooperation and dialogue. But it also injects an additional element of mutual self-interest, since a failure to address the interaction between resource depletion and rising poverty will accelerate global deterioration.
>
> (Brundtland 1987, p. 90)

In *Our Common Future*, the establishment of the sustainable use of resources was shown to demand more than simply dealing with the micro-scale questions of industrial methodology or the issues of refining or reforming project planning procedures. It needed to embrace international trade and international capital flows. Patterns of international trade (for example, in hardwood timber products) that impoverished developing countries also promoted unsustainable resource use. Lack of external capital (as aid and rescheduling of debts) limited the improvement of living standards in developing countries: 'Without reasonable flows, the prospect for any improvements in living standards is bleak. As a result the poor will be forced to over-use the environment to ensure their own survival' (Brundtland 1987, p. 68).

Debt, poverty and population growth restrict the capacity of developing countries to adopt environmentally sound policies. The proposed solutions involve, first, increased capital flow to the developing world, but redirected to promote 'sustainable' projects; second, new deals on commodity trade (particularly attention to hidden pollution costs of industrial processes in developing countries); and, third, ending protectionism and reforming transnational investment to ensure 'responsibility' (for example, through technology transfer and environmentally sound technologies). This analysis of the potential for productive reform of world trade and finance stems from a very particular vision of the working of the world economy and the pattern of economic and political forces, based on ideas of multilateralism and notions of global cooperation and dialogue.

The Brandt Reports set out much of the ground on which Brundtland built, both in terms of general principles and in the approach to (if not the priority given to) environmental problems. *North–South* discussed measures 'which together would offer new horizons for international relations, the world economy, and for developing countries' (Brandt 1980, p. 64). Those 'new horizons' include the environment, both globally ('the biosphere is our common heritage and must be preserved by cooperation' (Brundtland 1987, p. 73)) and in the countries of the South themselves. The package argued for by the Brandt Commission involved growth (in both North and South), 'massive transfers' of capital, expansion of world trade, the end of protectionism, an orderly monetary system and a move towards international equality and peace. Like Brundtland, Brandt argued that only the abolition of poverty would bring an end to population growth, and that this was a global problem and not one confined in its impacts to the poor or the Third World alone. Poverty required multilateral action, not just because of the obvious moral imperative to eradicate it, but because of mutual self-interest. Other imperatives 'rooted in the hard self-interest of all countries and people' reinforce the claim of human solidarity (Brundtland 1987, p. 77).

The Brandt Reports were themselves part of a longer evolution of thinking about economic interdependence (cf. Brookfield 1975). The Bretton Woods system of international financial management (the World Bank family, initially the International Monetary Fund and the International Bank for Reconstruction and Development) in 1944 was based on an essentially Keynesian vision, to create a stable, growing and interdependent world economy, 'an environment for liberal trade and to promote economic cooperation' (World Bank 1981). In the 1950s and 1960s the world economy, and international trade, indeed grew. However, fundamental problems were emerging by the 1960s, and in 1971 fixed exchange rates were abandoned, leading to volatility in currency markets, and eventually the destabilization of oil prices and the oil crisis of 1973 (Strange 1986).

The 1970s oil crisis coincided with the sudden flowering of concern about 'limits to growth'. This too stressed global interdependence, but an interdependence of crisis not stability, from which environmentalists drew an altogether different message about the desirability and possibility of continued growth. The attitude of 'global' environmentalism to economic growth at the time of the Stockholm Conference in 1972 was discussed above. The reconciliation of these difficulties created formal statements of sustainable development. Those ideas were based on the resurgence of Keynesian thinking represented by the Brandt Reports, a return to principles of an organized, managed and growing world economy.

The approach taken by Brandt (and Brundtland) to the world economy had many contemporary critics (e.g. Frank 1980; Seers 1980; Corbridge 1982). *North–South* argued that the protectionist trade policies of industrialized countries were the root cause of global economic problems and, in particular, persistent slow growth in the South. Tariff barriers and quotas stifled Southern economies and caused stagnation of Northern economies as Southern markets shrank. The solution was to open up the world economy, to pump capital and technical aid into the South to encourage trade, and to accept economic restructuring in the North (Brandt 1980, p. 186).

However, this analysis failed to demonstrate adequately the alleged dependence of the North on Southern markets. Furthermore, it was already clear in the 1980s that, on the scale of individual countries, the effect of trade liberalization is at best difficult to predict and at worst deleterious (e.g. Corbridge 1982).

North–South also adopted a caricature of development in the South and the constraints upon it. It assumed that actions on the international scale would actually reach and benefit the poor, thus ignoring the problem of political and economic structures within developing countries (see Frank 1980; Seers 1980). Similarly, *North–South* assumed that political and economic interests in the North were uniform, explicitly articulated and susceptible to rational debate. Northern governments were expected to recognize the mutual benefits from global economic cooperation and change their policies to achieve it. However, their capacity to do this was limited by the international political economy, and the first Brandt initiative foundered in the early 1980s amid growing protectionism. *North–South* adopted an unrealistic picture of the power and logic of capitalism, offering only a 'tepid programme of political action' (Corbridge 1982, p. 263) to achieve ambitious ends. The same epitaph could be justly applied to *Our Common Future*.

Brandt argued that development could not be achieved by tinkering with world trade, but only by altering the relations of production within developing countries and globally. As a result his commission's concept of mutuality was too loose to provide an effective basis for international political action. The world proceeded to become ever more interconnected economically and financially, a veritable 'casino of capitalism' with accelerating flows of money and rising uncertainty (Strange 1986). This hectic globalization of investment, banking and production was very different from Brandt's calm, managed, interdependent world economy (Lash and Urry 1994).

The simple premiss of 'mutuality' of both Brandt and Brundtland was naive. *Our Common Future* shows greater awareness of the real world of economics and politics than its predecessor, the *WCS*, but it comes no closer to explaining how that system works. Michael Redclift (1987), writing before the publication of *Our Common Future*, suggested that the Brundtland Commission might produce a radical departure from previous writing on sustainable development. This proved not to be the case. Certainly, *Our Common Future* set ideas about sustainable development more credibly within the overall matrix of development thinking, but its untheorized mutualism placed it firmly within the camp of a rather comfortable Keynesian reformism. *Our Common Future* did not change the intellectual landscape of development thinking; it placed sustainable development within it, but in a far from commanding position.

Caring for the Earth

In the year before *Our Common Future* was published, IUCN held an international conference in Ottawa to discuss progress since publication of the *WCS*. This proposed a revision of the *WCS*, among other things to give it a 'human face', and indeed, through IUCN partners, a 'southern face' (Holdgate 1999, p. 181). Although it was planned that this would be ready by 1988, in fact it was not published until 1991, again by IUCN with WWF and UNEP (*Caring for the Earth: a strategy for sustainable living* (IUCN 1991)).

Caring for the Earth was the result of a much more participatory process than its predecessor, and was the fruit of extensive consultation around the IUCN regions. Martin Holdgate (who had a major hand in writing it, with David Munro and Robert Prescott-Allen) described it as 'unashamedly a social and political document' (Holdgate 1999, p. 209). *Caring for the Earth*'s aim was 'to help improve the condition of the world's people' (IUCN 1991, p. 3). It argued that this required two things: first, a commitment to a new 'ethic for sustainable living'; and, second, the integration of conservation and development, conservation to keep human actions within the earth's carrying capacity, and development to 'enable people everywhere to enjoy long, healthy and fulfilling lives' (p. 3). It argued that poor care for the earth had raised the risk that the needs of the present generation, and the needs of their descendants, would not be met. That risk could be eliminated 'by ensuring that the benefits of development are distributed equitably, and by learning to care for the Earth and live sustainably' (p. 4).

Caring for the Earth picked up themes from the *WCS*. It presented nine 'principles for sustainable development', and took these as its structure (Table 3.5). It opened with a chapter setting out 'principles to guide the way towards sustainable societies', and these nine principles provided a structure for the rest of the report. They blended the ethical ('respect and care for the community of life'), the humanitarian ('improve the quality of human life'), the classically environmentalist ('keep within the Earth's carrying capacity' and 'minimise the depletion of non-renewable resources'), the conservationist ('conserve the Earth's vitality and diversity') and the pragmatic ('provide a national framework for integrating development and conservation').

The central argument of *Caring for the Earth* was much the same as its predecessor's, although more carefully and fully expressed:

> we need development that is both people-centred, concentrating on improving the human condition, and conservation-based, maintaining the variety and productivity of nature. We have to stop talking about conservation and

Table 3.5 Principles of sustainable development in *Caring for the Earth*

1	To respect and care for the community of life
2	To improve the quality of human life
3	To conserve the earth's vitality and diversity
4	To minimize the depletion of non-renewable resources
5	To keep within the earth's carrying capacity
6	To change personal attitudes and practices
7	To enable communities to care for their own environments
8	To provide a national framework for integrating development and conservation
9	To forge a global alliance

Source: *Caring for the Earth: a strategy for sustainable living* (IUCN 1991).

development as if they were in opposition, and recognise that they are essential parts of one indispensable process.

(IUCN 1991, p. 8)

The report represented a much more sophisticated presentation of traditional conservationist and environmentalist ideas than its predecessor. Gone is the awkward neo-Malthusianism of the *WCS*, replaced with a homely reminder of the problems of meeting human needs faced with rapid population growth. Human needs have a high profile, as does the ethical basis for conservation ('respect and care for the community of life').

Caring for the Earth shared with the *WCS* a focus on environmental management, itemizing principles for human action in farmland, rangeland, forests, fresh and salt waters, and also in settlements and in the care for the environment required of business, industry and commerce. It shared with the Brundtland Report its emphasis on mutuality and good global management, although it had little to say about the large-scale drivers of the global economy or environmental change. It recognized the importance of tackling poverty, meeting basic needs of food, shelter and health, and addressing issues of quality of life such as illiteracy and unemployment. It recognized the importance of debt (calling for official debt to be written off and commercial debt to be reduced) and the need for reform of South–North financial flows, the possible benefits of trade liberalization and removal of non-environmental trade barriers, and the need for new and better-targeted aid (Actions 9.5, 9.6, 9.7). General population growth and the notion of global limits to the earth's pollution-absorption capacity were set in the context of gross disparities in levels of resource consumption in North and South. It stated clearly that a 'concerted effort is needed to reduce energy and resource consumption by upper income countries' (IUCN 1991, p. 44). Its vision of a 'sustainable community' was quite radical, almost eco-socialist:

> a sustainable community cares for its own environment and does not damage those of others. It uses resources frugally and sustainably, recycles materials, minimises wastes and disposes of them safely. It conserves life support systems and the diversity of local ecosystems. It meets its own needs so far as it can, but recognises the need to work in partnership with other communities.

(p. 571)

Caring for the Earth was a cleverly drafted and integrated package, and represented a significant maturing of understanding about development on the part of IUCN. Holdgate (1999) argued that what made it different from other 'green manifestos' was the way it linked principles and suggested action: it listed 132 actions and 113 specific and dated targets, although these still did not engage with the structure of global or local political economy.

The publication of *Caring for the Earth* was a major event: almost 47,000 copies were printed in three languages, and it was launched in sixty-five countries. Its immediate impact was muted by the proximity of the Rio Conference (just six months away), by the babble of green rhetoric current in the early 1990s and by the lack of shock or novelty in its proposals (a direct corollary of their careful practicality). In the longer term, progress in meeting *Caring for the Earth*'s targets proved in many cases disappointing (Holdgate 1996). However, the impact of *Caring for the Earth* was considerable. Critically, its ideas and approach matched thinking in the various meetings of the preparatory committee (PrepComs) leading up to the Rio Conference, and its ideas fed very directly into the statements about sustainable development agreed at UNCED. These are the subject of the next chapter.

Summary

- There is a dominant 'mainstream' to ideas about sustainable development. This development was formulated and elaborated in a series of documents drafted in the 1980s, the *World Conservation Strategy* (*WCS*) (IUCN 1980), *Our Common Future* (Brundtland 1987) and *Caring for the Earth* (IUCN 1991). This chapter discusses the creation of these documents and assesses their significance.
- Sustainable development was first explicitly discussed in the context of the UN Conference on the Human Environment at Stockholm in 1972. This was dominated by the environmental concerns of industrialized countries, but in an attempt to meet the fears of Southern countries, the idea was put forward that concern for the environment need not adversely affect development.
- The *WCS* was the culmination of more than two decades of work by conservationists, especially through IUCN, to get conservation taken seriously in development. It argued for the maintenance of essential ecological processes and life-support systems, the preservation of genetic diversity and the sustainable use of species and ecosystems. It suggested that development and conservation could be made compatible through better and more timely planning.
- The Brundtland Report, *Our Common Future*, attempted to locate the debate about the environment within the economic and political context of international development. It was a successor to the Brandt Commission reports, *North–South* and *Common Crisis* (1980, 1983), and argued that poverty drove environmental degradation and required multilateral (global) action. International self-interest should drive a more equitable world economy, and with appropriate economic growth was necessary to achieve proper environmental management.
- In 1986 IUCN decided to update the *WCS*, eventually producing *Caring for the Earth* in 1991. This synthesized many of the ideas in the *WCS* and *Our Common Future*, setting them within the context of a new 'ethic for sustainable living'. It offered an analysis of how change could be made to happen at local, national and global scales, and set out targets. *Caring for the Earth*'s impact was absorbed into the preparations for the Rio Conference in 1992.

Further reading

Brundtland, H. (1987) *Our Common Future*, Oxford University Press, Oxford, for the World Commission on Environment and Development.

Dresner, S. (2002) *The Principles of Sustainability*, Earthscan, London.

Guha, R. (2000) *Environmentalism: a global history*, Oxford University Press, Delhi.

Holdgate, M. (1996) *From Care to Action: making a sustainable world*, Earthscan, London.

Holdgate, M. (1999) *The Green Web: a union for world conservation*, Earthscan, London.

McCormick, J. S. (1989) *The Global Environmental Movement: reclaiming paradise*, Belhaven, London.

Redclift, M. R. (2005) 2005 *Sustainability: critical concepts in the social sciences*, Routledge Major Works (four volumes), Taylor and Francis, London.

Swart, L. and Perry, E. (2007) (eds) *Global Environmental Governance: perspectives on the current debate*, Center for UN Reform, New York.

Web sources

http://www.iucn.org/ World Conservation Union (IUCN): full-text versions of the Resolutions and Recommendations adopted by IUCN at its General Assemblies from 1948 to 1996.

http://www.wwf.org/ World Wide Fund for Nature: information on the global network of national organizations and programme offices.

http://www.unep.org The United Nations Environment Programme (UNEP): information on programmes, including information on the 'state of the global environment'.

http://www.fao.org The United Nations Food and Agriculture Organization (FAO): details of programmes on agriculture, forestry, fisheries, etc.

http://www.peopleandplanet.net People and Planet: a 'global review and internet gateway' into the issues of population, poverty, health, consumption and the environment.

4   Sustainable development: making the mainstream

> Unless the penguin and the poor evoke from us an equal concern, conservation will be a lost cause. There can be no common future for humankind without a better common present. Development which is not equitable is not sustainable in the long term.
>
> (Monkombu Swaminathan, Opening Address to IUCN General Assembly, Perth, 1991, cited from Holdgate 1999, p. 206)

The Rio Conference

The last decade of the twentieth century and the start of the twenty-first saw the United Nations organize two environmental mega-conferences, the United Nations Conference on Environment and Development (UNCED, or more commonly simply 'the Rio Conference'), and the World Summit on Sustainable Development (WSSD) at Johannesburg (Seyfang and Jordan 2002). These defined and consolidated the international agenda of sustainable development. They followed directly from the work of the Brundtland Commission. This reported to the United Nations General Assembly, and in December 1989 the UN resolved to convene a conference on environment and development to take place five years after the Brundtland Report, and consider what progress had been made. This was the Rio Conference, held at Rio de Janeiro in Brazil in June 1992.

Expectations of this meeting were immense, although the auspices were not good. The Secretary-General of UNCED was Maurice Strong, who had filled a similar role at Stockholm two decades earlier. He billed it as a meeting at which decisions would be made 'that will literally decide the fate of the earth' (F. Pearce 1991, p. 20). However, the Preparatory Commission meetings (styled 'Prep-Coms') revealed bitter conflicts of interest between industrialized and non-industrialized countries. As at Stockholm, the Rio Conference unleashed a debate about disparities of wealth and poverty that was barely contained by the process, and the emollient strategies of international organizations. In 1991 UNCED looked like 'a crunch meeting between management and shop stewards at a company facing bankruptcy' (F. Pearce 1991, p. 21).

UNCED was a massive undertaking, a major outing for the international diplomatic circus. It was attended by 172 states and 116 heads of state or government.

There were 8,000 delegates and 9,000 representatives of the press; over 3,000 representatives of NGOs were accredited (N. Robinson 1993). The cost of the whole process, and the contrast between that cost and the urban poverty of Rio de Janeiro, were much commented upon by attending journalists. Press appetite for such criticism grew as the exaggerated hopes for the conference broke up under the realpolitik of international vested interest. It was not helpful that Rio fell in an election year in the USA: the government of George H. W. Bush was a reluctant participant in several critical arenas.

The PrepCom met five times, twice in New York, twice in Geneva and once in Nairobi. It was vast, including all the member states of the UN. After a series of marathon debates, the PrepCom brought an agreed text of twenty-seven principles to the conference (subsequently adopted as the Rio Principles), but the other documents considered by the conference contained 350 sections of bracketed or disputed text. At the conference itself, no undisputed text was reopened for discussion, but even so, consensus was hard to achieve. The final session of the Main Committee ran from 9.00 p.m. on 10 June to 6.00 a.m. the following morning. By its end all but two disputes had been resolved. Consensus was finally achieved through negotiations at ministerial level on 12 June, and *Agenda 21* and the Statement of Principles on Forest Management were adopted on 14 June (Koh 1993). Despite the fraught nature of discussion, a considerable amount was achieved (see Table 4.1).

The formal procedures of the PrepCom meetings, and the conference itself, were only part of a wider circuit of discussion and negotiation that ran through the five years before the conference. Beneath the formal proceedings lay a vast iceberg of international conferences and meetings (such as the Dublin Conference on Water and the Environment in January 1992), national reports on environment and development (172 of which had been received by June 1992) and meetings to coordinate responses by particular groups, such as the World Industry Conference on Environmental Management and the Business Council for Sustainable Development (Grubb *et al.* 1993).

UNCED was also a major focus of action for NGOs. They were deliberately brought into the UNCED process by Maurice Strong from the first PrepCom in Nairobi in 1991. The Centre for Our Common Future (established in 1988 to carry forward the work of the Brundtland Commission) also set up an 'International Facilitating Committee' (Chatterjee and Finger 1994). Despite these initiatives, extensive funding of NGOs at Rio and a specific NGO Conference in Paris in 1991, many NGOs were disappointed by their lack of influence on the UNCED process.

At Rio itself, NGOs were represented at a Global Forum, and had an opportunity for networking and debate at an 'Earth Parliament'. However, all this was physically and psychologically distant from the main conference. Some chose to express their views in distinct and forthright ways – for example, in Greenpeace's banner above the city of Rio de Janeiro (Plate 4.1). Although some NGOs remained close to government delegations through the conference (to which 1,400 lobbyists were accredited), NGOs were excluded from the official negotiating sessions (Holmberg *et al.* 1993).

Table 4.1. The achievements of the Rio Conference

The United Nations Conference on Environment and Development in 1992 (UNCED, or the Rio Conference) saw a series of documents and agreements that developed thinking on sustainable development from the *World Conservation Strategy, Our Common Future* and *Caring for the Earth.*

* *The Rio Declaration:* a consensus document, listing 27 'principles' for sustainable development.
* *Agenda 21:* a 600-page document drafted through intense negotiation between government diplomats, and as a result a balancing act between different interests. *Agenda 21* describes many vital actions to promote 'sustainability, but it makes none of them mandatory.
* In place of the anticipated global forest convention, a simple and limited set of *Forest Principles* was agreed. Countries with substantial areas of rainforest (mostly poor with limited industrialization, and determined to derive maximum economic benefit from forestry) could not agree on the need to halt deforestation, with industrialized countries driven by preservationist domestic environmental lobbies (but often practising unsustainable forestry at home, and with a long history of forest conversion to other land uses).
* The Convention on Biological Diversity was signed at Rio, to conserve biological diversity, promote the sustainable use of species and ecosystems and the equitable sharing of the benefits of genetic resources. It came into force in 1993.
* The Framework Convention on Climate Change agreed at Rio was the fruit of growing recognition of the problem of human-induced climate change through the 1980s, and particularly the scientific consensus achieved in the first report of the IPCC in 1990. The Convention came into force in 1994, but it laid no binding commitments to reduce greenhouse-gas emissions on individual countries. This continued to be debated and negotiated at meetings of the Conference of Parties, eventually being agreed at Kyoto in 1997, and coming into force in 2004.

Nonetheless, their presence may have influenced the way some issues were approached in *Agenda 21* – for example, in the emphasis on 'empowerment'; chapter 27 of *Agenda 21* explicitly discusses the importance of their role in achieving sustainable development, although it must be said that it does so through a stream of bland and empty statements. While Rio may have broadened the recognition that grassroots groups, particularly women's groups, were key elements in debates about environment and development (Ekins 1992), the conference also began to emphasize the distance between the powerful, wealthy and influential NGOs of industrialized countries and the 'grass roots' in the sense of groups formed among the poor of the urban and rural developing world. Chatterjee and Finger (1994) conclude that only the largest and most globally organized NGOs (almost all of which were North American) had much influence on the Rio documents. Most NGOs failed to make effective use of the US-style lobbying process and ended up confused, frustrated and divided.

Debate at Rio de Janeiro in 1992 built very directly onto the evolving mainstream of ideas dominating public debate about environment and development of the 1980s. The same themes appear in both the Rio Declaration and the much larger text of *Agenda 21.* However, the creation of these texts was far from

Plate 4.1 Greenpeace protest at the Rio Conference. A large number of NGOs attended the United Nations Conference on Environment and Development at Rio de Janeiro in 1992. They met at a parallel 'Global Forum' that was physically separated from the conference itself. Many NGOs felt excluded and that their views were poorly represented. Southern and grassroots NGOs resented the lobbying power and corporate muscle of big North-based environmental NGOs. Popular hopes that Rio would usher in a new environmentally conscious world order were widely disappointed. This sense of an opportunity missed was well captured by the Greenpeace protest, hanging a vast banner above the city of Rio. Photo: Greenpeace.

straightforward and harmonious. The 'Rio process' was, in practice, a mutual bludgeoning between teams of diplomats to produce texts that gave least away to perceived national interests. In particular, the distinction between the views of countries in the industrialized North and the underdeveloped South became steadily more glaring in the run-up to and during the conference. There was difference over the key problems (for the industrialized countries, global atmospheric change and tropical deforestation; for unindustrialized countries, poverty and the problems that flow from it), and responsibility for finding solutions. As at Stockholm in 1972, there was fear on the part of Third World countries that their attempts to industrialize would be stifled by restrictive international agreements on atmospheric emissions. They also feared that their freedom to use natural resources within their boundaries would be constrained by agreement imposed by industrialized countries that had themselves become wealthy precisely by squeezing their environments – for example, by clearing the vast majority of their forest cover and latterly by allowing industries to develop and operate with limited regard for environmental externalities such as pollution.

Sustainable development at Rio

The outputs from the Rio Conference (Table 4.1), agreed after long nights of diplomatic negotiation and horse-trading, included a slightly rambling Rio Declaration, a much watered-down set of principles for forest management, and the vast compendium of good intentions in *Agenda 21*. This contained more than 600 pages of text in 40 separate chapters. These were divided into four sections, covering socio-economic and environmental aspects of sustainable development, the actors who could make it happen, and the means of implementation (N. Robinson 1993). *Agenda 21* covered a great deal of ground, and included much rhetoric about good environmental management and poverty alleviation.

The tensions between Northern and Southern governments are clear in the texts of documents agreed at UNCED. The Rio Declaration (Table 4.2) was not the strong and sharp 'Earth Charter' originally conceived by the conference chairman, Maurice Strong. Its 27 principles constituted 'a bland declaration that provides something for everybody' (Holmberg *et al.* 1993, p. 7). It opened with the statement that 'human beings are at the centre of concerns for sustainable development. They are entitled to a healthy and productive harmony with nature' (see Table 4.2). Many of the principles were uncontentious (for example, Principle 4, on the need to integrate conservation and development, or Principle 5, on the eradication of poverty). Others were more closely fought over at Rio, particularly those that addressed the central issue of the conference: international action and international responsibility. Thus Principle 2 noted the sovereign right of countries to develop, while Principle 7 established the notion of 'common but differentiated responsibilities' for the global environment. Hidden behind a bland comment that 'states shall cooperate in a spirit of global partnership to conserve, protect and restore the health and integrity of the world's ecosystem' was the question of responsibility for the burden of action of developed countries (Holmberg *et al.* 1993). The US delegation released an 'interpretative statement' that effectively dissociated it from a number of the principles agreed. These included the notion of a right to development in Principle 3 (they argued that 'development is not a right ... on the contrary development is a goal we all hold' (Holmberg *et al.* 1993, p. 30), and also rejecting any interpretation of Principle 7 that suggested any form of international liability.

The main output of the Rio Conference was the encyclopaedic compendium of ideas and principles of *Agenda 21* (United Nations 1993). This ran to more than 600 pages, and its text was subject to detailed diplomatic debate. It was claimed that it reflected 'a global consensus and political commitment at the highest level on development and environmental cooperation' (Holmberg *et al.* 1993).

The name 'Agenda 21' came from the first PrepCom meeting in Nairobi, when Maurice Strong proposed a document to set out how to make the planet sustainable by the start of the twenty-first century. By the time the conference itself began, the document had become bloated, 'quite indigestible and impossible to implement'

Table 4.2 The Rio Declaration on Environment and Development

1 Human beings are at the centre of concerns for sustainable development. They are entitled to a healthy and productive life in harmony with nature.

2 States have, in accordance with the Charter of the United Nations and the principles of international law, the sovereign right to exploit their own resources pursuant to their own environmental and developmental policies, and the responsibility to ensure that activities within their jurisdiction or control do not cause damage to the environment of other states or of areas beyond the limits of national jurisdiction.

3 The right to development must be fulfilled so as equitably to meet developmental and environmental needs of present and future generations.

4 In order to achieve sustainable development, environmental protection shall constitute an integral part of the development process and cannot be considered in isolation from it.

5 All states and all people shall cooperate in the essential task of eradicating poverty as an indispensable requirement for sustainable development, in order to decrease the disparities in standards of living and better meet the needs of the majority of the people in the world.

6 The special situation and needs of developing countries, particularly the least developed, and those most environmentally vulnerable, shall be given special priority. International actions in the field of environment and development should also address the interests and needs of all countries.

7 States shall cooperate in a spirit of global partnership to conserve, protect and restore the health and integrity of the earth's ecosystem. In view of the different contributions to global environmental degradation, states have common but differentiated responsibilities. The developed countries acknowledge the responsibility that they bear in the international pursuit of sustainable development in view of the pressures their societies place on the global environment and of the technologies and financial resources they command.

8 To achieve sustainable development and a higher quality of life for all people, states should reduce and eliminate unsustainable patterns of production and consumption and promote appropriate demographic policies.

9 States should cooperate to strengthen endogenous capacity-building for sustainable development by improving scientific understanding through exchanges of scientific and technical knowledge, and by enhancing the development, adaptation, diffusion and transfer of technologies, including new and innovative technologies.

10 Environmental issues are best handled with the participation of all concerned citizens, at the relevant level. At the national level, each individual shall have appropriate access to information concerning the environment that is held by public authorities, including information on hazardous materials and activities in their communities, and the opportunity to participate in decision-making processes. States shall facilitate and encourage public awareness and participation by making information widely available. Effective access to judicial and administrative proceedings, including redress and remedy, shall be provided.

11 States shall enact effective environmental legislation. Environmental standards, management objectives and priorities should reflect the environmental and developmental context to which they apply. Standards applied by some countries may be inappropriate and of unwarranted economic and social cost to other countries, in particular developing countries.

12 States should cooperate to promote a supportive and open international economic system that would lead to economic growth and sustainable development in all countries, better to address the problems of environmental degradation. Trade policy measures for environmental purposes should not constitute a means of arbitrary or unjustifiable discrimination or a disguised restriction on international trade.

Continues

Table 4.2 Continued

Unilateral actions to deal with environmental challenges outside the jurisdiction of the importing country should be avoided. Environmental measures addressing transboundary or global environmental problems should, as far as possible, be based on an international consensus.

13   States shall develop national law regarding liability and compensation for the victims of pollution and other environmental damage. States shall also cooperate in an expeditious and more determined manner to develop further international law regarding liability and compensation for adverse effects of environmental damage caused by activities within their jurisdiction or control to areas beyond their jurisdiction.

14   States should effectively cooperate to discourage or prevent the relocation and transfer to other states of any activities and substances that cause severe environmental degradation or are found to be harmful to human health.

15   In order to protect the environment, the precautionary approach shall be widely applied by states according to their capabilities. Where there are threats of serious or irreversible damage, lack of full scientific certainty shall not be used as a reason for postponing cost-effective measures to prevent environmental degradation.

16   National authorities should endeavour to promote the internationalization of environmental costs and the use of economic instruments, taking into account the approach that the polluter should, in principle, bear the cost of pollution, with due regard to the public interest and without distorting international trade and investment.

17   Environmental impact assessment, as a national instrument, shall be undertaken for proposed activities that are likely to have a significant adverse impact on the environment and are subject to a decision of a competent national authority.

18   States shall immediately notify other states of any natural disasters or other emergencies that are likely to produce sudden harmful effects on the environment of those states. Every effort shall be made by the international community to help states so afflicted.

19   States shall provide prior and timely notification and relevant information to potentially affected states on activities that may have a significant adverse transboundary environmental effect and shall consult with those states at an early stage and in good faith.

20   Women have a vital role in environmental management and development. Their full participation is therefore essential to achieve sustainable development.

21   The creativity, ideals and courage of the youth of the world should be mobilized to forge a global partnership in order to achieve sustainable development and ensure a better future for all.

22   Indigenous people and their communities, and other local communities, have a vital role in environmental management and development because of their knowledge and traditional practices. States should recognize and duly support their identity, culture and interests and enable their effective participation in the achievement of sustainable development.

23   The environment and natural resources of people under oppression, domination and occupation shall be protected.

24   Warfare is inherently destructive of sustainable development. States shall therefore respect international law providing protection for the environment in times of armed conflict and cooperate in its further development, as necessary.

25   Peace, development and environmental protection are independent and indivisible.

26   States shall resolve all their environmental disputes peacefully and by appropriate means in accordance with the Charter of the United Nations.

27   States and people shall cooperate in good faith and in a spirit of partnership in the fulfilment of the principles embodied in this Declaration and in the further development of international law in the field of sustainable development.

Source: www.unep.org, 15 April 2008.

(Chatterjee and Finger 1994, p. 54). *Agenda 21* is a monument to the problems of making the rhetoric of international cooperation about environment and development concrete. It has become an icon of sustainable development, held up for symbolic veneration for its encapsulation of all possible arguments, a scripture dipped into for proof-texts to legitimate particular points of view but not subjected to detailed analysis.

The scope of *Agenda 21* was enormous, covering issues from water quality and biodiversity to the role of women, children and organized labour in delivering sustainable development. The chapters were divided into four sections (see Table 4.3); first, 'Social and Economic Dimensions' (i.e. development, chapters 2–8); second, 'Conservation and Management of Resources for Development' (chapters 9–22); third, 'Strengthening the Role of Major Groups' (chapters 23–32); and, fourth, 'Means of Implementation' (chapters 33–40). Chapters varied greatly in length, with those addressing environmental management (section 2) being the longest, and comprising almost half the total volume.

Commentaries on *Agenda 21* demand a large measure of creativity, for the document itself is so convoluted as to defy straightforward précis. Each chapter sought to set out the basis for action, the objectives of action, a set of activities and the means to be used to implement them (Grubb *et al.* 1993). In this, each part of *Agenda 21* was in a sense a microcosm of the whole, with a particular emphasis on the means of implementation.

There were a number of key themes in *Agenda 21*. The first was the idea of 'growth with sustainability'. As at Stockholm, debate about sustainable development at

Table 4.3 The structure of *Agenda 21*

The Structure of Agenda 21

Section 1   Social and Economic Dimensions
Eight chapters, covering international cooperation, combating poverty, consumption patterns, population, health, settlements and integrated environment and development decision-making.

Section 2   Conservation and Management of Resources for Development
Fourteen chapters on the environment. These cover the atmosphere, oceans, freshwaters and water resources, land-resource management, deforestation, desertification, mountain environments, sustainable agriculture and rural development. They also cover the conservation of biological diversity and biotechnology, toxic, hazardous, solid and radioactive wastes.

Section 3   Strengthening the Role of Major Groups
Ten chapters discussing the role of women, young people and indigenous people in sustainable development; the role of non-governmental organizations, local authorities, trade unions, business and scientists and farmers.

Section 4   Means of Implementation
Eight chapters, exploring how to pay for sustainable development, the need to transfer environmentally sound technology and science; the role of education, international capacity-building; international legal instruments and information flows.

Source: N. Robinson (1993).

Rio did not question the importance of continued economic growth, for either rich or poor countries. Sustainable development was about tuning the economic machine, not redesigning it. The second theme was 'sustainable living', under which come issues such as poverty, health and population growth. The third theme addressed the problems of urbanization (water supplies, wastes, pollution and health). These concerns had been underplayed in previous 'mainstream' documents, but here they properly came to the fore: the problems of sustainable management of rural resources and the rural poor were perhaps more readily recognized, but the deprivation of the urban poor was the more intractable (and more rapidly growing) problem. The fourth theme was 'efficient resource use', under which heading was included everything from combating deforestation and desertification through to conservation of biological diversity. Just as growth is the foundation stone of mainstream sustainable development, efficient resource use is the mechanism for achieving it: Chatterjee and Finger (1994) comment, 'in the name of environmental protection ... *Agenda 21* extends the economic rationality to the most remote corners of the earth' (p. 56). The fifth theme concerned global and regional resources (atmosphere and oceans), the sixth the management of chemicals and wastes. The seventh and final theme was 'people's participation and responsibility' (United Nations 1993).

Agenda 21 bears the strong inheritance of its predecessors. This is evident in various ways. The first is in the centrality it gave to growth. This is the familiar Brundtland agenda re-expressed: in the mainstream interpretation of sustainable development, everything is predicated on economic growth, both globally and nationally.

Second, *Agenda 21* showed a familiar dominance (in volume and position) of straightforward issues of environmental management. In the second section of *Agenda 21* all the familiar environmental issues from the *World Conservation Strategy* appeared, developed but unmistakable.

Third, *Agenda 21* was techno-centric. The first six key themes make this quite clear: growth will power and technology will direct the evolution of policy towards more efficient use of the environment and hence towards a more sustainable world economy. The 'essential means' to achieve sustainability also reflect this techno-centrism, building on information, science and environmentally sound technology (United Nations 1993).

Fourth, *Agenda 21* inherited the multilateralism of the Brundtland Report. The dominant mechanism for making any of its provisions happen was seen to be the common interest of industrialized and non-industrialized countries, of present generations in both caring about the future. International flows of financial resources and technology would reflect this mutual interest, international agencies would direct and promote these flows and their effectiveness, and international legal instruments would structure and regulate their product.

Fifth, like its predecessors, *Agenda 21* called for sustainable development through participation. As in *Caring for the Earth*, women, children, young people, indigenous people, trade unionists, businesses, industry, farmers, local authorities and scientists were all summoned to play a role, a rainbow coalition to put flesh on the endless skeleton of the text of *Agenda 21*. Here the text had all the emotive power that motherhood and apple-pie statements could render. Chapter 25, 'Children and youth in sustainable development', for example, suggests that 'it is imperative that youth from all parts of the world participate actively in all relevant levels of decision-making processes because it affects their lives today and has implications for their futures' (para. 25.2). Participation was a vital watchword of *Agenda 21*, but, like its predecessors, it was much stronger on hopeful sentiments about involvement than political analysis of power.

Alongside *Agenda 21*, the Rio Conference saw agreement on a set of *Forest Principles*. The Rio Conference should have seen a Convention on Forests signed. This did not happen, and, instead, a much shorter and lesser document of forest management was agreed. Its title reveals its character, and the problems that beset the convention for which it substituted: a 'non-legally binding authoritative statement of principles for a global consensus on the management, conservation and sustainable development of all types of forests' (F. Sullivan 1993). Pressure for specific action on forests came primarily from Northern environmental organizations concerned at the rate of clearance of tropical moist forests (rainforests). It followed a series of international initiatives during the 1980s such as the Tropical Forests Action Plan (TFAP, under the UNDP, FAO, World Bank and the World Resources Institute (WRI)) and the International Tropical Timber Organisation (under UNCTAD).

The idea of a global forest convention was made in a review of the TFAP in 1990, and the proposal made at a meeting of the G7 group of industrialized countries in Houston later that year (F. Sullivan 1993). However, the North–South divide became very clear, and debate bitter and intransigent. Southern countries (the 'G77', led by Malaysia and India) opposed a global forest convention. (The G77 countries are a group of 128 less developed and less industrialized countries set up as a counter-lobby to the developed 'G7' countries.) They argued that industrialized countries had cleared their own forests during their industrialization and that non-industrialized countries had a sovereign right to do the same. They could point to both an established history of non-sustainable forestry (notably in the USA) and continuing unsustainable practices in the harvesting of old-growth forests – for example, in the Pacific Northwest of the USA and in the western and boreal forests of Canada (Maser 1990; see also Plate 4.2). Furthermore, if tropical forests served a global benefit (whether through their biodiversity or by locking up CO_2), Southern countries argued that the costs of maintaining them should be borne globally. If there was to be a global forest convention, it should have a mechanism for compensating Southern countries for revenue forgone in setting aside forest reserves (Holmberg *et al.* 1993).

This debate rapidly over spilled the tight confines of the PrepCom meetings, and by PrepCom 4 it was clear that a legally binding agreement on forests could not be achieved at Rio. Energies were focused instead on capturing the high ground and trying to establish some kind of global consensus on forest management. The resulting 'Forest Principles' constituted a political document and not an operational tool (Holmberg *et al.* 1993). The principles closely reflected chapter 11 of *Agenda 21*, on 'Combating deforestation', and explicitly addressed

Plate 4.2 Clear-felled old-growth forest, on Vancouver Island, Canada. At the Rio
    Conference, Southern countries with extensive forests opposed Northern
    proposals for a global forest convention, arguing that industrialized
    countries had cleared their own forests during their industrialization and that
    non-industrialized countries had a sovereign right to do the same. They point
    to both an established history of non-sustainable forestry and continuing
    unsustainable practices in the harvesting of old-growth forests – for example,
    in the Pacific north-west of the USA and in the western and boreal forests of
    Canada, Finland and other countries. This photograph shows industrial felling
    of previously un-cut moist forests in western Canada. Photo: W. M. Adams.

all forests – that is, temperate and boreal as well as tropical forests. They avoided
specific commitments. They repeated the familiar arguments about the social,
environmental and economic importance of forests, and the need for them to
be managed sustainably. They mentioned the need for international cooperation
and the need for funds from industrialized countries to meet management needs
and broadly support free trade in timber and forest products (against calls, for
example, for environmentally defined trade bans in the North). They called for
scientific assessment and management of environmental impacts of forestry, and
they discussed the need for local participation in forest management decisions.

All these principles were widely recognized as desirable, and, while they
presented a challenge to dominant forest management practices in almost every
country (in developed as well as developing countries), they also reflected existing
ideas within the forestry industry about 'best practice'. Most critically, the *Forest
Principles* emphasized national sovereignty for forests within national borders (N.
Robinson 1993; Sullivan 1993). They did not provide a basis for Northern intervention in Southern forest management on environmental grounds. If anything,

their even-handedness reflected the unwillingness of any government (including those in North America and Scandinavia) to constrain logging industries midway through the liquidation of the assets in old-growth forests. The hopes of Northern NGOs that Rio might generate significant constraints on rates of tropical defor-estation fell foul of international politics, and the awkward fact of the unsustain-ability of logging practices in parts of the North (which, of course, Northern NGOs such as Greenpeace and Earth First! also vigorously opposed).

The Convention on Biological Diversity

The two conventions signed at Rio, the Convention on Biological Diversity (CBD) and the Framework Convention on Climate Change, were negotiated not through the PrepCom, but through international negotiating committees (Chatterjee and Finger 1994). Both reflect fairly closely the relevant chapters of *Agenda 21*, which considered 'environmentally sound management of biotech-nology' and the 'conservation of biological diversity' (chapters 16 and 15), and 'protection of the atmosphere' (chapter 9).

The CBD was one of the elements of Rio with the longest pedigree. A draft convention was prepared in the mid-1980s by IUCN, in conjunction with other international organizations (including the WWF, UNEP, WRI and the World Bank). This initiative was the fruit of the continuation of the conventional conservation agenda that had inspired *The World Conservation Strategy* (*WCS*) a decade before. The idea of a global conservation convention had been mooted at the Second World Congress on National Parks in Bali (organized by IUCN's Commission on National Parks and Protected Areas), and between 1988 and 1992 there was sustained pressure both for the convention itself, and for effective measures to preserve global biodiversity (and hence in large measure Southern biodiversity). All the major international bodies with an interest in environment and development (the WRI, IUCN, UNEP, WWF, the World Bank, the FAO and UNESCO) contributed to a series of meetings and reports that culminated in the Global Biodiversity Strategy in 1992 (WRI *et al.* 1992). Completion and adoption of the convention was the priority requirement of this strategy (Holdgate 1999).

Negotiations over a convention were initiated by UNEP in 1990, reflecting essentially Northern concern about rainforest loss. However, at the second Geneva PrepCom meeting, the G77 countries demanded inclusion of the issue of bioprospecting and biotechnology, and the sharing of wealth generated by the exploitation of biodiversity in the South by Northern biotech companies. In this odd hybrid form, the convention was agreed at Rio and signed by 156 countries (Chatterjee and Finger 1994). The USA refused to sign at that time, although it did so subsequently.

The aim of the CBD was to conserve biological diversity and to promote the sustainable use of species and ecosystems, and the equitable sharing of the economic benefits of genetic resources. It is this last element that set this conven-tion apart from all previous international conservation agreements. Signatory nations committed themselves to the development of strategies for conserving biological diversity, and for making its use sustainable. Biodiversity conservation can be achieved *in situ* (that is, through conventional methods such as the designation of systems of protected areas) or *ex situ* (for example, through captive breeding), but the convention also requires cross-cutting measures (for example, affecting forestry or fishing). While all these elements of the convention are qualified by a get-out clause (all is to be done 'as far as possible and appropriate'), the CBD's provisions were a logical development of the traditional conservationist concern for sustainable ecosystems use, and drew directly on the thinking in the *WCS* and *Caring for the Earth*.

What was novel (and controversial) in the convention was its provisions for the exploitation of genetic resources through biotechnology. In principle, this is no different from the provisions of the earlier mainstream documents that species and ecosystems should provide resources for human benefit, if used in such a way that their availability was sustained. However, by 1992 the rapid development in genetic science had opened up vast new areas of potential exploitation at the sub-specific and molecular level, including the creation of novel organisms (which might perhaps be patented by the organization that created them) and products (drugs, for example) derived from wild species. It was perceived that this technology had the potential to generate vast wealth; however, the biotechnological capacity was almost entirely held by industrialized countries (because of the high costs of research laboratories, research infrastructure and training), and moreover was increasingly held by private corporations within those countries and not by states themselves. Third World countries feared stripping of their genetic resources by bioprospectors, and loss of access to economic benefits derived by First World corporations (Shiva 1997; Hayden 2003). Developed countries (particularly the USA, which dominated in this area of science) feared restriction of economic opportunity if trade in biotechnology were restricted by a benefit-sharing agreement. The convention reflected the balance of these opposite fears, and does contain provision for sharing benefits from commercial exploitation of genetic resources (Article 15). This debate was cross-cut by the desire of those pushing traditional conservation arguments to achieve their conventional goals. Few of those negotiating the treaty could have had direct knowledge of the present (let alone the future) potential of the scientific revolution in biotechnology that took place in the 1990s.

The Convention on Biological Diversity came into force on 29 December 1993, and by 1997 it had been ratified by 162 countries. This rapid entry into force reflected the level of concern about continued biodiversity loss, but also the rapid development of biotechnology in the 1990s, and the potential commercial value of genetic material both in its raw (wild) state and as patentable 'improved' forms. Widespread ratification has not by any means ended controversy, and interest in the Conferences of the Parties has been considerable: more than 130 governments sent over 700 delegates to the first Conference of the Parties (COP) in Nassau in the Bahamas in 1994. Debate over the issues embraced by the convention has continued to be fierce, addressing among other things integration with other biodiversity conventions (Ramsar, on wetlands, and CITES, on trade in endangered species, for example) and relations with the World Trade Organization, and its agreement on Trade Related Intellectual Property Rights (TRIPS) (Bragdon 1996; Pimbert 1997).

International conservation organizations have worked to make the CBD the conservation convention they originally wanted it to be. Critical to this process was the decision of the sixth COP in April 2002 on a strategic plan, which contained a commitment 'to achieve by 2010 a significant reduction of the current rate of biodiversity loss at the global, regional and national level as a contribution to poverty alleviation and to the benefit of all life on earth' (CBD Decision VI/26, CBD). Making good on this commitment presented conservationists with a considerable challenge. Their response was to propose indicators for biodiversity and ecosystem services that were 'rigorous, repeatable, widely accepted and easily understood' (Balmford *et al.* 2005, p. 212). The seventh COP, in Kuala Lumpur in 2004, duly established goals and sub-targets for the protection of biodiversity (CBD Decision VII/30; see Table 4.4).

The Framework Convention on Climate Change

The idea of a climate change convention also pre-dates the immediate preparations for Rio. Scientific studies of global warming were stimulated by the International Geophysical Year (1957–8), which also stimulated international biological science and, with the International Biological Programme and related developments, played a role in the stimulation of late-twentieth-century environmentalism by fostering awareness of the environment as a global issue (see Chapter 2). The notion that countries should take responsibility for transnational pollution was established at the Stockholm Conference in 1972 (see Chapter 3). Scientific evidence from observations and early computer models was sufficiently strong to lead UNEP, the World Meteorological Organization (WMO) and the International Council of Scientific Unions (ICSU) to convene a World Climate Conference in Geneva in 1979 (Jäger and O'Riordan 1996). Scientific research continued to develop rapidly through the 1980s. The impact of carbon-dioxide concentrations in the 'greenhouse effect' were recognized, as was the potential of other greenhouse gases such as water vapour, methane, nitrous oxide and ozone.

The third of a series of scientific meetings at Villach in Austria in 1985 discussed scenarios for future emissions of all greenhouse gases and began to establish a clear scientific consensus about anthropogenic climate change (Jäger and O'Riordan 1996). This brought the issue clearly onto the political agenda; indeed, it was itself in large measure responsible for the political importance of the environment in that decade. The conference in Toronto in 1988 entitled 'The Changing Atmosphere' produced a statement calling on all developed countries to reduce their CO_2 emissions by 20 per cent from 1987 levels by 2005. The Intergovernmental Panel on Climate Change (IPCC) was established in 1988 (by WMO and UNEP), with three working groups, on scientific evidence, environmental

Table 4.4 The Convention on Biological Diversity, 2010 global biodiversity target

Focal area	Goal		Target	
To protect the components of biodiversity	1	To promote the conservation of the biological diversity of ecosystems, habitats and biomes	1.1	To conserve at least 10% of each of the world's ecological regions effectively
			1.2	To protect areas of particular importance to biodiversity
	2	To promote the conservation of species diversity	2.1	To restore, maintain or reduce the decline of populations of species of selected taxonomic groups
			2.2	To improve status of threatened species
	3	To promote the conservation of genetic diversity	3.1	To conserve genetic diversity of crops, livestock, and harvested species of trees, fish and wildlife and other valuable species, and to maintain associated indigenous and local knowledge
To promote sustainable use	4	To promote sustainable use and consumption	4.1	To derive biodiversity-based products from sources that are sustainably managed, and to manage production areas consistently with the conservation of biodiversity.
			4.2	To reduce unsustainable consumption of biological resources, or consumption that impacts upon biodiversity
			4.3	To ensure no species of wild flora or fauna are endangered by international trade
To address threats to biodiversity	5	To reduce pressures from habitat loss, land-use change and degradation, and unsustainable water use	5.1	To decrease rate of loss and degradation of natural habitats
	6	To control threats from invasive alien species	6.1	To control pathways for major potential alien invasive species
			6.2	To put in place management plans for major alien species that threaten ecosystems, habitats or species
	7	To address challenges to biodiversity from climate change, and pollution	7.1	To maintain and enhance resilience of the components of biodiversity to adapt to climate change
			7.2	To reduce pollution and its impacts on biodiversity

Focal area	Goal		Target
To maintain goods and services from biodiversity to support human well-being	8	To maintain capacity of ecosystems to deliver goods and services and support livelihoods	8.1 To maintain capacity of ecosystems to deliver goods and services
			8.2 To maintain biological resources that support sustainable livelihoods, local food security and health care, especially of poor people
To protect traditional knowledge, innovations and practices	9	To maintain socio-cultural diversity of indigenous and local communities	9.1 To protect traditional knowledge, innovations and practices
			9.2 To protect the rights of indigenous and local communities over their traditional knowledge, innovations and practices, including their rights to benefit sharing
To ensure the fair and equitable sharing of benefits arising out of the use of genetic resources	10	To ensure the fair and equitable sharing of benefits arising out of the use of genetic resources	10.1 To ensure all transfers of genetic resources are in line with the Convention on Biological Diversity, the International Treaty on Plant Genetic Resources for Food and Agriculture and other applicable agreements
			10.2 To share benefits arising from the commercial and other utilization of genetic resources with the countries providing such resources
To ensure provision of adequate resources	11	Parties have improved financial, human, scientific, technical and technological capacity to implement the Convention	11.1 To transfer new and additional financial resources to developing country Parties, to allow for the effective implementation of their commitments under the Convention, in accordance with Article 20
			11.2 To ensure technology is transferred to developing country Parties, to allow for the effective implementation of their commitments under the Convention, in accordance with its Article 20, paragraph 4

Source: www.biodiv.org, 15 November 2005.

Note
Goals and sub-targets for each of the identified focal areas to clarify the 2010 global biodiversity target (adopted by the Conference of the Parties of the Convention on Biological Diversity at its seventh meeting, held in Kuala Lumpur in February 2004).

and socio-economic impacts, and response strategies respectively. These reported for the first time in 1990.

The Science Assessment Working Group set out to establish a global consensus on the complex science of climate change. In the words of its chair, the IPCC reports could 'be considered as authoritative statements of the contemporary views of the international scientific community' (Houghton [1994] 1997, p. 159). Working Group I reported scientific certainty that human action was affecting atmospheric concentrations of greenhouse gases, and estimated that this was responsible for over half the enhanced greenhouse effect, both past and future (Jäger and O'Riordan 1996).

Debate about the United Nations Framework Convention on Climate Change (UNFCCC) reflected the divergent reactions to the IPCC First Assessment Report in 1990. The IPCC's consensus view of the importance of fossil fuel consumption and carbon dioxide (CO_2) output cut directly at the heart of the interests of the industrialized Northern countries, while also having significant implications for rapidly industrializing countries in the South such as India and China. The International Negotiating Committee on Climate Change began work in 1990, following the Second World Climate Conference in Geneva, with the aim of preparing a convention for signature at Rio in 1992. It rapidly fell foul of fundamental differences between different parties. There was a broad divergence between industrialized and non-industrialized countries, with the North urging the priority of environmental protection and that any measures agreed should be cost effective, while the South pushed the need for development and industrialization, and the principle of historical responsibility (Rowbotham 1996). Industrialized countries were unwilling to countenance a significant reduction in CO_2 output. Oil-producing states were also opposed to this, while small island states vulnerable to sea-level rise wanted urgent action on precisely this. The EU favoured agreement on targets and a timetable for implementation, the USA was reluctant (the latter even refusing, in the run-up to Rio, to agree to cut back emissions in the year 2000 to 1990 levels). In April 1992, at the last Intergovernmental Negotiating Committee meeting, the compromise was agreed on a non-binding call for an attempt to return to 1990 emissions of CO_2 and other greenhouses gases not controlled by the Montreal Protocol.

The convention was a delicate balance between divergent political and economic interests, and rather full of rather pious intentions (Rowbotham 1996). Like *Agenda 21*, it stressed (in Article 3, 'Principles') the significance of the protection of the climate system for both present and future generations, and stated that there must be equity between industrialized and non-industrialized countries in taking action. This equity must reflect historic responsibility, state of development and capacity to respond (Holmberg *et al.* 1993). The diversity of interest was such that the text was ambiguous, left open to subsequent interpretation at the meetings of COP. The Framework Convention was weak (arguably 'toothless' (Chatterjee and Finger 1994, p. 45)) in that it contained no legally binding commitments for the stabilization (let alone reduction) of CO_2 emissions. Again, this was left until later. However, the convention's negotiation, in a mere sixteen months, was a remarkable testament to the urgency with which global climate change was viewed in the early 1990s, and is a major achievement of the Rio process.

The Framework Convention on Climate Change was signed by over 150 states and the European Community (now the European Union) at Rio, and came into force in March 1994 (having received its fiftieth signatory in December the previous year). COP1 took place in Berlin in 1995. The issue of binding targets and timetables for reducing greenhouse gas emissions (proposed by the Association of Small Island States) remained controversial, as did the notion of 'joint implementation', under which one country can aid another to implement a project or change a policy that will result in reduced mutual greenhouse gas emissions (Rowbotham 1996; Bush and Harvey 1997). The IPCC has continued to assess the evidence for and predict the impacts of anthropogenic climate change. Its Second Assessment Report in 1995 drew on new data and analysis and confirmed the conclusion of the 1990 report: that the balance of the evidence suggested a discernible human influence on climate. This made a critical contribution to the negotiations that led to adoption of the Kyoto Protocol of the UNFCCC in December 1997. The Protocol was open for signature from 16 March 1998 to 15 March 1999 at United Nations Headquarters, New York.

Kyoto set out a binding obligation on developed countries ('Annexe 1' countries) to reduce emissions for six greenhouse gases by 5.2 per cent below 1990 levels by 2008–12. Different countries agreed to different reductions (8 per cent for the European Union and 7 per cent for the USA, for example, while Australia was allowed to *increase* emissions by 8 per cent and Iceland by 10 per cent). Changes in land use and forestry since 1990 that could be held to have a positive impact on net emissions could be counted against national targets.

Critics from various positions have claimed that the Kyoto Protocol is flawed, both economically inefficient and politically impractical. Others are more optimistic about its solid basis in economic principles and the way it moved forwards on burden-sharing (e.g. Bohringer 2003). Flexibility has been a principle of the approach to the regulation of greenhouse gas emissions. In 1998 an Action Plan was agreed by the COP at Buenos Aires, focusing on flexible strategies for implementing targets. These include Joint Implementation and emissions trading between Annex 1 countries, and arrangements between Annex 1 and developing countries under the Clean Development Mechanism (Grubb 1998; Vira 2002).

However, progress towards the emissions targets agreed at Kyoto has remained problematic (Najam *et al.* 2003). Kyoto failed to offer a decisive breakthrough in climate policy. There was a major setback at COP6 in The Hague in November 2000, where, despite two weeks of intensive negotiations, signatories failed to agree on how to operationalize the Kyoto Protocol. There was profound disagreement about measures to cut emissions on the part of a handful of wealthy industrialized countries, led by the USA and Australia. In March 2001 the US administration declared that it saw the Kyoto Protocol as 'fatally flawed'. At a summit in Bonn in July 2001 international agreement was reached to rescue the Kyoto process, albeit without the USA. A few months later, negotiations resumed at COP7 in Marrakesh (2001), and successively at COP8 in New Delhi (2002) and COP9 in Milan (2003). The so-called Umbrella Group of non-EU developed countries (including Canada, Australia, Japan, Russia and New Zealand) remained reluctant to agree binding targets, but eventually the Russian Federation ratified the Kyoto Protocol in November 2004, and it came into force on 16 February 2005 (Hovi *et al.* 2003). China ratified in January 2003, the Russian Federation in December 2004 and Australia in December 2007. The USA has not ratified. By June 2007, 174 countries and the EEC had deposited instruments of ratifications, accessions, approvals or acceptances (http://unfccc. int/2860.php).

In December 2005 the United Nations Climate Change Conference was held in Montreal, along with a Conference of the Parties and the first ever 'Meeting of the Parties'. This was a major event, and brought the USA back into discussions. New ideas included the idea of carbon credit for reducing deforestation, and clarification of the Clean Development Mechanism.

However, it has remained difficult to keep negotiations about greenhouse gas emissions and responses to anthropogenic climate change in a single chamber. In the most significant departure the Asia-Pacific Partnership on Clean Development and Climate (AP6) was established in January 2006 between Australia, China, India, Japan, Republic of Korea and the United States. This sought to establish a different 'pro-growth' approach to climate change without binding agreements, but drawing on collaborations in new technology – for example, for clearer burning fossil fuels or renewable energy (www.dfat.gov.au/environment/ climate/ap6/).

Meanwhile, the IPCC has continued to issue periodic Assessment Reports, the third in 2001 and the fourth in 2007. The IPCC has become more interdisciplinary and broader in its scope (Shackley 1997). Its reports have maintained pressure on the international community by tending to confirm the scientific consensus on the significance of increases in greenhouse gas concentrations in the atmosphere as a direct result of human action since 1750. The Fourth Assessment Report in 2007 concluded that concentrations now far exceed preindustrial values, as determined from ice cores spanning hundreds of thousands of years. Increases in carbon dioxide are attributed primarily to fossil-fuel use and land-use change, increases in methane and nitrous oxide that are primarily due to agriculture (IPCC 2007b). A series of natural phenomena in the middle of the first decade of the new century (notably Hurricane Katrina, which devastated New Orleans in 2005), together with effective advocacy and campaigning (notably Davis Guggenheim's film featuring US Senator Al Gore, *An Inconvenient Truth*, and the accompanying book, Gore 2006) made climate change the dominant environmental issue of the opening years of the twenty-first century.

This continuing development in the international regime for the regulation of greenhouse gas emissions reflects the technical and reductionist nature of the global discourse of climate change. It is a classic example of instrumental rationality, applying technical knowledge to an agreed purpose by the most efficient means possible (Cohen *et al.* 1998), an example of the approach of ecological modernization discussed in Chapter 5 of this book. Climate change is arguably too complex for Kyoto's approach to emissions control, 'treating tonnes of carbon dioxide like stockpiles of nuclear weapons to be reduced via mutually identifiable targets and timetables' (Prins and Rayner 2007, p. 973).

Although most world leaders agreed with the view expressed by UK Prime Minister Tony Blair at the G8 Summit in Gleneagles in 2005 that climate change was 'long-term the single most important issue we face as a global community' (http://www.g8.gov.uk), the issue is far from solved. There remain profound difficulties in bridging the protective self-interest of developed countries that have already industrialized yet are alarmed at the prospect of drastic climatic change, and those of developing countries for whom these fears are felt though the immediate crises of poverty (Vira 2002; Najam *et al.* 2003).

From Rio to Johannesburg

UNCED left a legacy of solid achievements – in international law, in new and partially refocused international institutions, in the actions of national and local governments and (above all) in the composition of the political and policy rhetoric that is taken to represent international consensus about environment and development. However, commentators are doubtful concerning the significance of these achievements. Many of the problems on which *Agenda 21* focused became worse following Rio (K. Brown 1997). Poverty deepened and became more entrenched, and the gap between rich and poor countries grew. The enhanced resource flows needed to implement *Agenda 21* have not been forthcoming, and indeed countries such as the UK have reduced their aid budget. The activities of transnational companies have not been significantly influenced. Moreover, many countries have failed to respond to the Framework Convention on Climate Change and curb CO_2 emissions (Parikh *et al.* 1997), and issues of sovereignty over genetic resources have remained intractable (Pimbert 1997).

Overall, Rio did little to promote sustainable development as such; however, it did open the debate about choices in development, about the ways in which the biosphere is restructured in pursuit of profit, of the costs of technology, of the inequalities in wealth, technologies, environmental hazard and life chances (Brown 1997). In many ways, the Rio Conference should not be interpreted as a single event that can be assessed in terms of success or failure. Roddick (1997) suggests that it set out to create not a single regime, but 'an entire framework for the management of environmental problems' (p. 147). This framework included 'soft law': voluntary reporting and action by communities and other non-state actors at local, national and international levels. In practice, however, the established practice of forging formal international agreements has continued to dominate, despite the chronic deficit on enforcement.

Probably the chief failure of the Rio Conference was that it did not stimulate the scale of financial support necessary to implement *Agenda 21*. Before the Fourth UNCED PrepCom started in New York, Maurice Strong estimated the cost of implementing the Rio agreements at $125 billion in new finance every year between 1992 and 2000 (Chatterjee and Finger 1994). At the conference itself, the secretariat estimated that implementation would cost something like $600 billion per year, of which $125 billion per year would have to be in the form of gifts and concessional loans. This was more than twice the current total disbursement of official development assistance to developing countries, and close to the official UN target of 0.7 per cent of GNP from industrialized countries (Grubb *et al.* 1993). In the event, the money available has been a tiny fraction of that. Of the $125 billion per year needed to take necessary actions, only about $2.5 billion was pledged (Holmberg *et al.* 1993).

The financing mechanism chosen for funding actions arising from Rio was the Global Environment Facility (GEF). The GEF already existed. It was set up in 1990 by the World Bank, UNDP and UNEP. The World Bank administers the GEF and acts as repository of the trust fund. UNEP provides the secretariat for the Scientific and Technical Advisory Panel and supplies environmental expertise on specific projects; UNDP is responsible for technical assistance and project preparation. The idea of channelling funds to implement the Rio agreements through the GEF was strongly opposed by developing-country governments (which feared the dismal scientists of the World Bank, and the iron grip of the North on its policies) and by environmentalists (who had maintained a barrage of criticism of the environmental impacts of World Bank lending since the 1980s; see Chapter 6).

The GEF is widely held to be too small to be effective. It held $1.3 billion in the first phase; by the Fourth GEF Meeting (in December 1992), seventy projects had been approved, amounting to $584 million (Chatterjee and Finger 1994). By September 1998 (seven years after its establishment) the GEF had allocated $2 billion (World Bank 2000). While the GEF has brought much-needed funds to some sectors (national parks, for example), the overall influence of the GEF's funds, in the context of debt and total resource flows between the First and Third Worlds, has been very limited (Brown *et al.* 1993).

At the same time as it adopted *Agenda 21* (in December 1992), the General Assembly of the UN requested ECOSOC to establish a Commission on Sustainable Development (CSD) (N. Robinson 1993). This had been suggested in chapter 38 of *Agenda 21*, and was established in 1992. It was seen in many quarters as an opportunity to deal with the 'unfinished business' of Rio, to monitor progress with implementation of *Agenda 21* in member countries, and to oversee (and promote) the other recommendations of the conference. The CSD Bureau has representatives from fifty-three countries, elected from geographical regions and serving for three years each. The CSD has held a series of formal sessions, from 1994 onwards (Bigg 1995), culminating in the Special Session of the UN General Assembly, the so-called 'Earth Summit II' in June 1997, which extended the CSD's work for a further five years (Jordan and Voisey 1998; see discussion below). However, the powers of the CSD are limited; it can report to the General Assembly of the UN only through ECOSOC, and cannot peer-review national *Agenda 21* statements. Overall the CSD has been a disappointment.

One piece of unfinished business at Rio was a Convention on Desertification. This was proposed as a part of the Rio process, but negotiation fell behind in the run-up to the conference. In the event, a formal commitment was made at Rio to negotiate a convention on desertification after the conference was over. This was duly done, and the Convention to Combat Desertification was open for signature by June 1994, coming into force in December 1996. It addresses implementation through four regional annexes that allow for different approaches in different areas (Africa, Asia, Latin America and the Caribbean, and the northern Mediterranean). This convention will be discussed further in Chapter 8 (see also www.unccd.ch).

The UN asked its various organizations (including UNEP, UNDP, UNCTAD and UNSO (the United Nations Sudano-Sahelian Office, now the United Nations Office to Combat Desertification and Drought) for specific proposals as to how they would implement the provisions of *Agenda 21* (N. Robinson 1993). An Inter-Agency Committee on Sustainable Development (CSD) was established to draw together representatives of the nine main UN organizations. In 1993 the UN created the Department for Policy Coordination and Sustainable Development, based in New York, to support the CSD and coordinate UN response to the various conferences that followed Rio (Flanders 1997). However, many observers feel that there was a lack of strategic leadership within the UN system in the aftermath of Rio, and in particular that UNEP failed to define an effective leadership role. Those commentators, remembering their hopes for a brave new greener world in 1992, see in the many tendrils that have grown from Rio little evidence of change; rather, business as usual. Others, perhaps with more appetite for the brushfires of bureaucratic battles and committee rooms, saw evidence of real hope for change. Murphy and Bendell (1997), writing about the response of business to the sustainable development agenda, described sustainable development as an 'emerging, positive myth which has the potential to bring together diverse and often competing causes' (p. 35).

Partly to counter disenchantment at Rio's limited achievements, Maurice Strong argued that UNCED should be seen not as the end of international discussion about the challenge of sustainable development, but as the start of a process of adjustment by national governments (Chatterjee and Finger 1994). This 'Rio process' was furthered in a succession of follow-up conferences after 1992, including the Conference on Sustainability in Small Island States (1993), the International Conference on Population and Development in Cairo (1994), the World Summit on Social Development in Copenhagen (1995), the World Conference on Women in Beijing (1995) and the World Food Summit in Rome (1996) (Fresco 1997; Murphy and Bendell 1997; Jordan and Voisey 1998). Each of these can be regarded as forming part of a connected process of international engagement with different aspects of the agenda debated at Rio (www.earthsummit2002.org).

The most significant follow-up to Rio was a special session of the UN General Assembly in 1997 to review progress since UNCED. This meeting, often referred to as 'Earth Summit II' (or, less memorably, 'UNGASS'), was held in the summer of 1997 in New York. In retrospect it is clear that preparation was somewhat hurried, and there was too little preparatory work, and its agenda was too broad and lacked focus (Osborn and Bigg 1998). Documents were prepared for the Earth Summit II at the preceding CSD meeting, but agreement was reached only on a short six-paragraph 'statement of commitment' and a 'Programme of Action for the Further Implementation of Agenda 21'.

Earth Summit II lacked the scale of Rio, and did not attract the same media attention. Debate was fierce about climate change (the meeting preceded Kyoto), forests (with pressure for the UK, Canada and the EU to establish an international negotiating committee for a forest convention, and Brazil, Malaysia, the USA and many G77 countries opposing the notion on the grounds that forests are a matter for national jurisdiction) and trade. Nonetheless, industrialized countries such as the UK were disappointed that no new initiatives were launched on forests, oceans and freshwater (Jordan and Voisey 1998).

More seriously, developing countries were disappointed at the lack of progress on the finance and support to implement the actions agreed at Rio. The trust underpinning the Rio 'deal' had begun to evaporate in the years following the conference (Jordan and Voisey 1998). Under the 'deal' that had emerged at Rio, developing countries agreed to environmental commitments sought by developed countries in return for assurances about 'new and additional' financial resources and technical assistance. However, aid disbursements in the 1990s were falling rather than rising, and with the exception of those from Scandinavian countries were far below the headline figure of 0.7 per cent of donor-country GDP. As discussed above, financial flows were far less than the sums calculated as necessary at Rio to implement *Agenda 21*.

The World Summit on Sustainable Development

Through the 1990s, debate between developers and environmentalists, between those who wished to exploit natural resources and those who wished to conserve them, between poor and rich countries, grew in scale and intensity. For all its success in setting an agenda for sustainable development, the Rio Conference had failed to reconcile the different demands of industrialized and developing countries, and failed to bring about the kinds of changes demanded since the Brundtland Report fifteen years before. Following Rio, 'sustainable development' had become one of the most hard-used linguistic tropes of the international system, freighted with a number of different agendas. But Rio offered no decisive breakthrough in business as usual, for governments, business or citizen. To a large extent momentum was lost (Annan 2002), and the policy evolution of the 'Rio process' through the 1990s was grindingly slow. By 2000 the loose movement promoting sustainable development was 'more muted, more fractured, and perhaps a little more realistic' (Scoones 2007, p. 594).

In the new millennium, a new environmental 'mega-conference' (Seyfang and Jordan 2002) was proposed ('Rio Plus 10'). This meeting was charged with achieving what the 1997 meeting (Earth Summit II) had failed to do: 'rekindle[d] the fire' of the Rio accord (www.earthsummit 2002.org) and promote implementation of existing commitments (La Viña et al. 2003). The WSSD was held in Johannesburg, South Africa, in 2002, ten years after Rio. There was no shortage of encouragement for bold initiatives (e.g. W. Sachs 2002).

By 1992 poverty had returned to the top of the international agenda (Mabogunje 2002), a move marked by the United Nations Millennium Summit, which took place in New York in September 2002. The United Nations Millennium Declaration made pledges on peace, security and disarmament, and also on development and poverty eradication, saying:

> We will spare no effort to free our fellow men, women and children from the abject and dehumanizing conditions of extreme poverty, to which more than a billion of them are currently subjected. We are committed to making the right to development a reality for everyone and to freeing the entire human race from want.
>
> (Resolution adopted by the General Assembly, 55/2.
> United Nations Millennium Declaration)

Among the fundamental values 'essential to international relations in the twenty-first century' was 'respect for nature'. This stated:

> Prudence must be shown in the management of all living species and natural resources, in accordance with the precepts of sustainable development. Only in this way can the immeasurable riches provided to us by nature be preserved and passed on to our descendants. The current unsustainable patterns of production and consumption must be changed in the interest of our future welfare and that of our descendants.
>
> (United Nations Millennium Declaration, 8 September 2000,
> http://www.un.org/millennium/declaration/
> ares552e.htm)

The Millennium Summit reached agreement on eight Millennium Development Goals (see Table 4.5). One goal addressed sustainability. Eighteen targets and forty-eight indicators were defined, intended to be yardsticks for measuring improvements in people's lives.

The goal to 'ensure environmental sustainability' involved three targets, the integration of the principles of sustainable development into country policies and programmes and the reversal of the loss of environmental resources, the halving of the proportion of people without sustainable access to safe drinking water by 2015 and the achievement of a significant improvement in the lives of at least 100 million slum-dwellers by 2020.

The implications of degraded environments for the poor, and the implications of poverty for environmental degradation, had been recognized by the Brundtland Report, and at Rio. They were central to debate at Johannesburg at the start of the new century (Najam *et al.* 2002). The summit focused on five areas, water, energy, health, agriculture and biodiversity, in an attempt to move from the numerous broad commitments of Rio to action. Kofi Annan, Secretary-General of the UN, spoke of the conference as the chance to 'catch up' with the changes needed. He wrote:

Table 4.5 The Millennium Development Goals

Goal	MDG	Examples of links to the environment
Goal 1	To eradicate extreme poverty and hunger	Livelihood strategies of the poor often depend directly on the functioning ecosystem and the diversity of goods and services they provide.
		Insecure rights of the poor to environmental resources limit their capacity to protect the environment and to improve their livelihoods and well-being.
Goal 2	To achieve universal primary education	Children's labour (e.g. collecting fuelwood or water) restricts learning.
Goal 3	To promote gender equality and empower women	Time spent by women collecting fuelwood or water reduces opportunity for revenue generation.
		Unequal rights and insecure access to land and other resources for women limit livelihoods and well-being.
Goal 4	To reduce child mortality	Diseases of poor water and sanitation and respiratory infections are leading causes of mortality in children under 5.
Goal 5	To improve maternal health	Indoor air pollution and heavy load-carrying in pregnancy put women's health at risk.
Goal 6	To combat HIV/AIDS, malaria, and other diseases	Up to one-fifth of disease in developing countries is linked to environmental factors.
Goal 7	To ensure environmental sustainability	
Goal 8	To develop a global partnership for development	Industrialized countries consume more resource and produce more waste than developing countries: many global environmental problems (climate change, loss of species diversity, management of oceans) can be tackled only through global partnership

Source: www.developmentgoals.org and UN Millennium Project (2005).

the issue is not environment versus development, or ecology versus economy. Contrary to popular belief, we can integrate the two. Nor is the issue one of rich versus poor. Both have a clear interest in protecting the environment and promoting sustainable development.

(Annan 2002, p. 14)

This framing of the possibility of win–win outcomes in sustainable development is familiar from Johannesburg's predecessors. So, too, is the existence of a discrepancy in the importance placed on particular issues in developed and developing countries (Najam *et al.* 2002; see also Table 4.6). Poverty alleviation topped concerns in both regions, although almost two-thirds of those in developing

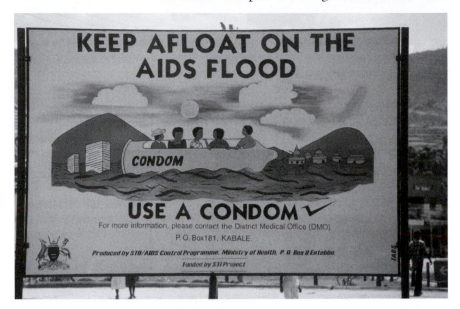

Plate 4.3 AIDS awareness notice, Uganda. AIDS is the single largest cause of mortality
in sub-Saharan Africa. More than two-thirds of adults (68 per cent) and nearly
90 per cent of children infected with HIV live in Africa, although it holds only
10 per cent of the world's population. More than three-quarters of global
deaths due to an AIDS-related illnesses in 2007 occurred in sub-Saharan Africa.
Education and affordable and accessible anti-retroviral treatment are important.
Goal six of the Millennium Development Goals is to halt and begin to reverse
the spread of HIV/AIDS, malaria and other major diseases (www.unaids.org).
Photo: W. M. Adams.

countries listed it in their top five concerns, compared to only 45 per cent of those
in developed countries. However, while the issue of atmosphere/climate change
came second in developed countries with almost the same score, it ranked sixth for
respondents in developing countries, with less than a third rating it in their top 5
concerns (see Table 4.6). Interestingly, biodiversity (by 2002 the preferred term
for wildlife conservation, which had been a primary concern of those drafting the
World Conservation Strategy in 1980) was ranked ninth by both sets of respondents (Najam *et al.* 2002).

The WSSD, like Rio before it, commanded worldwide attention from governments. It was attended by over 22,000 people, including 82 Heads of State.
Conspicuously, the US President (George W. Bush) did not attend. The meeting
generated more hype and attracted more participants than Rio, but produced
less, both in terms of agreed outputs and in terms of progress on the challenges
left from Rio. The USA fought successfully against tough targets and timetables:
the US representative, the Secretary of State Colin Powell, was heckled. NGOs
denounced the lack of commitment (Greenpeace called the Plan of Action 'not
much of a plan and it contains almost no action' (Speth 2003)).

Table 4.6 Issues important for sustainable development in developed and developing countries

Rank	Developed country priorities	Developing country priorities
1	Poverty alleviation (45%)	Poverty alleviation (62%)
2	Atmosphere (including climate change) (44%)	Education and awareness (42%)
3	Consumption and production (33%)	Financing sustainable development (41%)
4	Financing sustainable development (29%)	Civil society participation (34%)
5	*Business participation (29%)*	Capacity building (33%)
6	Civil society participation (28%)	Atmosphere (including climate change) (30%)
7	*Freshwater (28%)*	*Integrated decision making (26%)*
8	Education and awareness (26%)	Consumption and production (25%)
9	Biodiversity (24%)	Biodiversity (24%)
10	Trade and environment (23%)	Trade and environment (16%)
11	Capacity building (22%)	*Debt relief (16%)*
12	*Population (22%)*	Agriculture (16%)
13	Agriculture (20%)	*Technology transfer (15%)*
14	*Energy (18%)*	

Source: Najam *et al.* 2002.

Note
Percentage of respondents who identified each issue as one of their five most important priorities for future sustainable development (each issue was identified by at least 15% of the group). Italicized entries signify issues not on the other group's list.

Many of the commitments (and much of the rhetoric) at Johannesburg were recycled. Governments committed themselves to achieving poverty-related targets and goals, including those in *Agenda 21* and the United Nations Millennium Declaration. The meeting issued the Johannesburg Plan of Implementation, which started with the issue of the eradication of poverty, before moving on across issues such as unsustainable patterns of consumption and production, the protection and management of the natural resource base of economic and social development, globalization and health (see Table 4.7).

A number of successes are claimed for Johannesburg. There was, for example, agreement on a new target of halving the proportion of people unable to reach or afford safe drinking water or obtain basic sanitation by 2015. There were also moves on sustainable production and consumption (basically seeking to divorce economic growth from environmental degradation), and a reaffirmation of Rio's commitments on freedom of environmental information, public participation and access to justice. At the same time, commentators identify a series of key failures, including inadequate progress on specific and time-bound targets, energy, climate change and the problems posed by globalization, for example, for environmental regulation (La Viña *et al.* 2003).

One group disappointed by Johannesburg was the wildlife conservationists. The emphasis on poverty led some conservationists to express concern that their

Table 4.7 Johannesburg Plan of Implementation

I	Introduction
II	Poverty eradication
III	Changing unsustainable patterns of consumption and production
IV	Protection and managing the natural resource base of economic and social development
V	Sustainable development in a globalizing world
VI	Health and sustainable development
VII	Sustainable development of small island developing states
VIII	Sustainable development for Africa
IX	Other regional initiatives

    A   Sustainable development in Latin America and the Caribbean
    B   Sustainable development in Asia and the Pacific
    C   Sustainable development in the West Asia region
    D   Sustainable development in the Economic Commission for Europe region

  X Means of implementation
 XI Institutional framework for sustainable development

    A   Objectives
    B   Strengthening the institutional framework for sustainable development at the international level
    C   Role of the General Assembly
    D   Role of the Economic and Social Council
    E   Role and function of the Commission on Sustainable Development
    F   Role of international institutions
    G   Strengthening institutional arrangement for sustainable development at the regional level
    H   Strengthening institutional frameworks for sustainable development at the national level
    I   Participation of major groups

Source: www.un.org, 15 April 2008.

grip on the development agenda was slipping, that 'poverty alleviation has largely subsumed or supplanted biodiversity conservation' (Sanderson and Redford, 2003b, p. 390; cf. Adams *et al.* 2004). The main policy focus for those concerned at the loss of global biodiversity remained the CBD, and the '2010 Biodiversity Target' (see above), which the Johannesburg Conference endorsed.

The evolution of thinking about sustainable development in the last three decades of the twentieth century and the first decade of the twenty-first was to a large extent shaped by the three mega-conferences at Stockholm (1972), Rio (1992) and Johannesburg (2002). All three suffered from a mismatch between expectations and achievements, and whether they are seen as a disappointment or a limited success depends partly on the frame of reference used (Seyfang and Jordan 2002). A realistic view of what such extravaganzas might hope to achieve suggests that they all made some progress in six critical areas: setting global agendas, facilitating 'joined-up' thinking on environment and development, endorsing common principles, providing global leadership for national and local governments, building institutional capacity and legitimizing global governance by making the process more inclusive (Seyfang and Jordan 2002). Johannesburg, for example, drew in an unprecedented range of stakeholders, including (controversially for some) business interests. They certainly clarified and consolidated thinking about sustainable development, moving it to centre stage in terms of global policy-making and codifying it (Seyfang and Jordan 2002). The 'mainstream' sustainable development that they defined is the subject of Chapter 5.

Summary

- The United Nations Conference on Environment and Development in 1992 (UNCED, or the Rio Conference) was preceded by protracted international negotiation at meetings of the Preparatory Commission (PrepCom). One hundred and seventy-two governments were represented at Rio, which fell in a US election year. There was a large unofficial NGO contingent.
- A series of documents were agreed at Rio, carrying forward the ideas of the mainstream of sustainable development from *The World Conservation Strategy* (*WCS*), *Our Common Future* and *Caring for the Earth*.
- The Rio Declaration was a bland and somewhat rambling consensus document, consisting of twenty-seven 'principles' for sustainable development.
- *Agenda 21* was the main output of Rio, a bloated document over 600 pages long, a compendium of environmental thinking and a balancing act between different interests. *Agenda 21* was subject to intense negotiation between national teams of diplomats. It provides the basis for many kinds of action in pursuit of 'sustainability', but it makes none of them mandatory.
- An agreed set of Forest Principles fell far short of proposals for a global forest convention. Agreement was not reached on deforestation between rainforest countries (determined to derive maximum economic benefit from forestry) and industrialized countries driven by preservationist domestic environmental lobbies (but often practising unsustainable forestry at home, and with a long history of forest conversion to other land uses).
- A Convention on Biodiversity was signed at Rio, to conserve biological diversity and promote the sustainable use of species and ecosystems and the equitable sharing of the benefits of genetic resources. It came into force in 1993.
- The UN Framework Convention on Climate Change (UNFCCC) agreed at Rio was the fruit of growing recognition of the problem of human-induced climate change through the 1980s, and particularly the scientific consensus achieved in the first report of the Intergovernmental Panel on Climate Change (IPCC) in 1990. The convention came into force in 1994, but it laid no binding commitments to reduce greenhouse gas emissions on individual countries. This continued to be debated and negotiated at meetings of the Conference of the Parties (COP), eventually being agreed at Kyoto in 1997.
- The Rio Conference established mainstream sustainable development thinking, but it failed of itself to achieve binding or timetable commitment to systematic change in national or international policy. Debate about what sustainable development means, and what action should be taken to achieve

it, has continued, through the work of international organizations, and the COPs of the treaties.

- The Johannesburg Conference confirmed and clarified the outcomes of Rio, but also failed to generate the political commitment by developed countries to provide the resources necessary to implement the new world order to which Rio had pointed.

Further reading

Bigg, T. (2004) (ed.). *Survival for a Small Planet: the sustainable development agenda*, Earthscan, London.

Chatterjee, P. and Finger, M. (1994) *The Earth Brokers: power, politics and world development*, Routledge, London.

Dresner, S. (2002) *The Principles of Sustainability*, Earthscan, London.

Holdgate, M. (1996) *From Care to Action: making a sustainable world*, Earthscan, London.

Middleton, N. (2003) *The Global Casino: an introduction to environmental issues*, Arnold, London (3rd edition).

Redclift, M. (1996) *Wasted: counting the costs of global consumption*, Earthscan, London.

Redclift, M. R. (2005) *Sustainability: critical concepts in the social sciences*, Routledge Major Works (four vols), Taylor and Francis, London

Robinson, N. (1993) (ed.) *Agenda 21: earth's action plan*, IUCN Environmental Policy and Law Paper 27, Oceana Publications, New York.

UNEP (2000) *Global Environment Outlook 2000*, Earthscan, London, for the United Nations Environment Programme.

United Nations (1993) *The Global Partnership for Environment and Development: a guide to Agenda 21*, post Rio edition, United Nations, New York.

Web sources

http://www.biodiv.org/ The Convention on Biological Diversity (the full text is on http://www.unep.ch/bio/conv-e.html).

http://www.igc.apc.org/habitat/agenda21/ Agenda 21 and other Rio agreements; also links to related conferences and conventions, and to the meetings of the Commission on Sustainable Development.

http://www.un.org/esa/sustdev/csd.htm The United Nations Commission on Sustainable Development.

http://unfccc.int// The website of the Framework Convention on Climate Change.

http://unfccc.int/kyoto_protocol/items/2830.php The text of Kyoto Protocol. Coverage of the Thirteenth United Nations Conference on Climate Change in Bali in December 2007 is at http://unfccc.int/meetings/cop_13/items/4049.php.

http://www.earthsummit2002.org The United Nations Environment and Development Forum website.

http://www.climatenetwork.org/ The Climate Action Network (CAN), a cooperative network of over 280 non-governmental organizations working on climate change; organized through a series of area networks including networks in Africa, South Asia, South-East Asia and Latin America.

5 Mainstream sustainable development

> Can we envisage a more ecologically benign modernity, or is modernity ecologically irredeemable?
>
> (J. S. Dryzek, 'Toward an ecological modernity', 1995)

Market environmentalism

Chapters 3 and 4 described a continuum of thinking that stretched from the UN Conference at Stockholm in 1972 through Rio in 1992 and Johannesburg in 2002, a mainstream in the diversity of ideas about sustainable development. The definition of that 'mainstream' is obviously in a sense artificial, because there is a great diversity of thought about environment and development. This diversity (and the radicalism and divergent thinking it represents) will be explored in Chapter 7. Of interest in this chapter and Chapter 6 is the broad consensus of the sustainable development mainstream itself. This has not been very radical, diverging in critical but often small and technical particulars from established thinking about development. The mainstream has been essentially reformist, and convergent in its propositions. It has sought to refocus existing development initiatives and policy action rather than transform their principles or practice. It has joined two positive-sounding concepts ('sustainability' and 'development'). In doing so, it has sought 'to resolve at a stroke the conflict between an economy based on everlasting growth and a planetary environment of permanent high quality' (Low and Gleeson 1998, p. 12). Sustainable development has been built on the premise that these could be reconciled if the economy were organized around productive activities that do not harm the environment.

Mainstream sustainable development (MSD) became progressively more strongly defined through the 1990s. Following Low and Gleeson (1998), this chapter identifies three important groupings of thought within it: *market environmentalism* (including the role of business corporations), *ecological modernization* and *environmental populism*. It also considers the importance of neo-Malthusian concerns about limits to growth.

The most important feature of MSD is that it shares the dominant industrialism and 'developmentalism' of the modern world system, of which the constituent processes are modernization, economic growth and nation-state building (Aseniero 1985). The capacity of the modern state to measure, regulate and order nature and society is fundamental to the twentieth-century process of development (Scott 1998), and these capacities are also central to MSD.

As discussed in Chapter 1, developmentalism both underpins ideas about the proper direction of economic change in industrialized countries, and (because 'development' is seen to be a path along which all economies travel, albeit at different speeds) acts as a model for 'developing' countries. Behind the strategies for economic change lies the rich ideological web of the modern industrial world: 'a common corporate industrial culture based on the values of competitive individualism, rationality, growth, efficiency, specialization, centralization and big scale' (Friberg and Hettne 1985, p. 231). MSD does not challenge the dominant capitalist industrializing model. Instead it opens debate about its methods and outcomes. Thus *Our Common Future* focused on better planning techniques, more careful use of state capital and more careful use of economic appraisal to reduce development that causes ecological disruption. *Agenda 21* addressed the question of retooling the wealth-producing industrial plant of the world economy and changing the priorities of its management team. It did not suggest that its fundamental business, its methods or its products needed to be radically reimagined.

Part and parcel of the mainstream's acceptance of developmentalism lie in its emphasis on the market as an institution for achieving sustainability. The sets of ideas that Low and Gleeson label 'market environmentalism' are utilitarian, individualistic and anthropocentric. These ideas present the market as the most important mechanism for mediating between people, and regulating their interaction with the environment. They involve a political agenda of 'rolling back' the state, deregulating markets and extending market relations into society and its relations with the environment. This is a familiar recipe to anyone who lived through the 1980s and 1990s, either in industrialized countries or in those enduring the enforced rigours of the economic doctrines of the International Monetary Fund and the World Bank, and their obsession with 'structural adjustment' (e.g. Hanlon 1996). Welfare-utilitarian economists such as Wilfred Beckerman (1974, 1994) argue that the market is the only efficient way to regulate human use of the environment. Huffman (2004) notes:

> Markets are more flexible than the political process. They respond more quickly to change, and usually generate better information because participants have direct incentives to obtain good information about the effects of their actions. Markets also create incentives that guide decision-making that has an impact on the environment.
>
> (p. 65)

Market prices rise as resources become scarce, and logically people will innovate to find cheaper sources or ways of using resources more efficiently. As natural resources become scarce, their relative prices change, and a chain of market responses will follow that tend to discourage its use and the development of substitutes (Beckerman 1995a). As Low and Gleeson (1998) describe the future under market environmentalism:

> The 'green economy' will be a capitalist economy. And just as the economy theoretically reaches a level of equilibrium in which social needs are met, so the green economy will theoretically reach a level of 'sustainable development' in which the capacity of the planet to provide raw materials and absorb wastes is not overstretched.
>
> (p. 81)

Free-market environmentalists argue that attempts by the state to make rules about resource use are therefore arguably inherently inefficient, and bound to fail both to allow economic welfare to be maximized and to maintain resources at desirable levels. In this analysis, environmental problems are seen to follow from the misallocation of resources. Open-access resources are liable to over-exploitation, while private resources are managed efficiently and conserved. Market-oriented environmental policy in the Third World has therefore tended to focus on the privatization of resources – for example, the privatization of communally held land and other resources (Bassett 1993; Woodhouse *et al.* 2000; Lesorogol 2005).

Market environmentalism argues that the further market exchange penetrates into the environment, the greater the efficiency of environmental management. Policy proposals therefore involve the commodification of nature, and the setting of prices for environmental 'goods' and 'services'. Economists have long been interested in the use and conservation of resources as a source of raw materials, although the discipline of environmental economics dates only from the 1950s – for example, in the work of Ciriacy Wantrup (Spash 1999). Interestly economists in the concept of sustainable development dates to the 1980s (e.g. Goodland and Ledec 1984; Pearce *et al.* 1989; Turner 1988b). The possibility of applying economic calculations to the management of nature and to issues of sustainability spread rapidly in the wake of the work by David Pearce and his colleagues in *Blueprint for a Green Economy* (Pearce *et al.* 1989). The field of ecological economics was established in the 1980s, seeking to make a decisive bridge between environmentalist concern about development and the world of policy and government decisions (Spash 1999; Åkerman 2003). The range of environmental and ecological economics is now very wide, and its scope increasingly sophisticated (e.g. Costanza 1991; Costanza and Daly 1992; Barbier *et al.* 1994 List and de Zeeuw 2002; Markandya *et al.* 2002).

Economic analysis provided a metric that allowed the costs and benefits of using environmental resources into the conventional calculus of economic decision-making (Munasinghe 1993a). It became central to policy debate about environment and development. Its power to build and transform debate is well demonstrated by the *Stern Review* on the economics of climate change (N. Stern 2007). There has also been more than a decade's experience of market-based instruments for environmental management, based on ideas of competition to achieve efficiency, notably in the area of gas emissions to the atmosphere (Bailey 2007; Prins and Rayner 2007). Following the Rio Conference, the leadership of the World Bank and its army of economists led the development of policy and the greening of governments and development bureaucracies, in some ways a remarkable case of the poacher turned gamekeeper. There was reform of the way economies are understood and measured (for example, in national accounts), and reform of the economic methods that support development decision-making (for example, cost–benefit analysis (CBA)). These are described further in Chapter 6.

Market environmentalism is predicated on continued capitalist growth. It 'defends the status quo, the globalising institution of the market, and resists the notion that any fundamental change is needed' (Low and Gleeson 1998, p. 163). Thus Hamilton and Johnson (2004) argue that faster growth is the key to meeting Millennium Development Goal targets, but that its benefits must be widely spread, and it must be 'responsible' in environmental and social terms. There are clearly important questions about the energy, carbon and pollution intensities of any future economies that meet these criteria.

MSD clearly does not endorse the pessimistic ideas of 'zero growth' or 'limits to growth' that were prominent in the 1970s (Mishan [1967] 1969, 1977; Daly 1973, 1977; see also Chapter 2). The place of ideas about limits to growth within the mainstream is explored later in this chapter.

Conventional economists argue that the task for governments and their economic advisers has not been fundamentally changed by the sustainability debate: to seek the economic growth rate that maximizes welfare over time – that is, the optimal growth rate. In his combative book *Small is Stupid*, for example, Wilfred Beckerman (1995b) argued that 'in developing countries there is no conflict between growth and the "quality of life"' (p. 35). The World Bank (1992) argued that 'the world is not running out of marketed non-renewable energy and raw materials, but the unmarketed side effects associated with their extraction and consumption have become serious concerns' (p. 37). It is to remedy this that market-based approaches to environmental problems (such as tradable pollution permits) have been developed in industrialized countries, and in international environmental regimes such as those for greenhouse gases (D. Pearce 1995; Lewis 1992; Norregaard and Reppelin-Hill 2000; Vira 2002; Najam *et al.* 2003). Economic tools to address wider environmental aspects of development have also been developed (see Chapter 6).

In practice, development and the environment have a 'two-way relationship' (World Bank 1992, p. 1), with economic growth having both good and bad effects on the environment. There is a vast literature analysing empirical evidence of the environmental consequences of economic growth. Some environmental indicators do tend to show improvement as incomes rise (access to clear water and sanitation). Some indicators show unambiguous deterioration (volumes of municipal waste and production of CO_2). Some indicators show early deterioration then improvement and then deterioration, such as air quality (suspended particles and SO_2), sewage pollution or deforestation (Neumayer [1999] 2003). This third pattern is referred to as the Environmental Kuznets Curve, after Simon Kuznets, who in 1955 hypothesized that, under economic growth, inequalities in income would grow at first, but later decline.

Most developing countries are still on the rising limb of the graph, with environmental degradation rising with incomes and economic growth. However, there has been considerable interest in the possibility of countries 'tunnelling through' using technology and cleverly designed policies to maximize growth while minimizing environmental harm (Munasinge 1999). The statistical basis for the Environmental Kuznets Curve is weak (D. I. Stern 2004). In slow-growing economies new technologies can reduce levels of certain pollutants (for example, airborne particulates), but in faster-growing economies this effect is swamped by rising incomes. Furthermore, economic analysis does not make it clear just what might actually cause declines in environmental quality to reverse under economic growth. The process does not seem to be automatic. The way policy-makers respond to environmental degradation is important, and there is evidence that environmental outcomes are better where democracy and political freedoms are stronger (Neumayer [1999] 2003). It is important to note that arguments based on pollution declines in now-wealthy countries are not reliable, because the surrounding circumstances of today's industrialization are quite different from those of the past. It is also relevant that the 'cleaned-up' environments of many industrialized countries are in part possible because the pollution involved in the manufacture of commodities and products has been shifted offshore, to developing countries where pollution problems abound as a result of the pursuit of growth. This issue is discussed further in Chapter 12.

Growth creates the possibility of 'win–win' opportunities through poverty reduction and improved environmental stewardship in both high- and low-income countries. *Our Common Future* and its successors expressed the benign possibilities of global capitalism, with sustainable development made possible by economic growth yielding a broader distribution of economic goods while avoiding environmental damage. In *Agenda 21* the drive for economic growth was recognized as a prime contributor to unsustainability.

However, growth also brings risks of serious side effects if markets fail to capture environmental values and deal with externalities such as pollution, whether local, regional or global, and if the public regime of regulation at scales from local to international are inadequate. Thus the structure of the global economy undermined many of the principles in the Rio agreements (K. Brown 1997). The effect of the vigorous expansion of free trade under the General Agreement on Tariffs and Trade (GATT) and (from 1995) the World Trade Organization on sustainability is debated, with regard to both unequal benefits to richer and poorer partners, and the environment. Most observers conclude that free trade has made it harder for governments to regulate effectively to protect the environment: certainly there is no evidence of the reverse process (Mol 2001; Winter 2003; Neumayer 2004; Thomas 2004).

Corporations and sustainability

A key element in the market environmentalism has been the growing phenomenon of corporate 'greening', particularly by major international companies. Corporate environmentalism, actions taken by firms that have a substantive or symbolic commitment to ecological protection (Mason 2005), has had an important influence on the way MSD is imagined and delivered. At one level, it can be argued that corporate environmentalism developed as a strategy to meet shareholder calls for environmental and social responsibility (often as part of a more general engagement with Corporate Social Responsibility). Such calls followed directly from decades of public campaigning by environmental organizations in industrialized countries from the 1960s. Major industrial pollution disasters, such as the Bhopal disaster in India in 1984 (see Chapter 12) and the *Exxon Valdes* oil spill in Alaska in 1989 have also been important (Mason 2005). At another level, it represents a rethinking of corporate strategy that reflects competition for markets, investment and 'brand' values.

The conventional relationship between environmentalists and corporations has been openly antagonistic, using direct action to attract media coverage and highlight the corporate performance they opposed. Iconic examples include the action by Friends of the Earth against the decision in 1971 by the soft-drink company Schweppes to use non-returnable glass bottles (dumping 1,500 bottles outside their headquarters) or Greenpeace's controversial and successful occupation of Shell's redundant *Brent Spar* oil platform as it was dragged from the North Sea towards the Atlantic to be dumped at sea (Rose 1998): Greenpeace believed it should be recycled, and eventually it was.

A critical element of this approach to campaigning was to use media coverage of actions to highlight a campaign that called for change. By the 1980s an increasingly important dimension was to influence consumers, asking them to boycott particular products from targeted companies. The logic was that actual and feared reductions in sales would force companies to change production methods. Thus the Friends of the Earth Rainforest Campaign, launched in 1985, involved a boycott on high-street outlets using tropical timber obtained through clear-felling or non-replacement selective logging. As such campaigns grew in scope and sophistication, environmental campaigners also bought shares in companies, and spoke out at shareholders' meetings about environmental and human-rights performance. Such publicity increased pressure on chief executives and board members to avoid arguments, and to reassure corporate investors worried about both their impact on the reputation and the future profitability of the company. It might be said, too, that some investors and corporate executives also acknowledged the need to take sustainability seriously, especially as the generation exposed to environmentalism through school and university began to enter positions of authority.

The 1990s saw a change in the crude oppositionism of environmental groups to corporations. Greenpeace, for example, advocated 'solutions-led' campaigning. This required research and development of alternative technologies and then a campaign directed towards making businesses adopt them. One example of such an approach was the development of propane refrigeration technology as an alternative to the use of CFCs or HCFCs (Rose 1993; Rawcliffe 1998).

Once environmental performance is something that differentiates companies in terms of both consumer choice and investor confidence, it can be argued that normal processes of competition between businesses becomes a driver of corporate change towards more sustainable forms of production and operation. Business response to the challenge of sustainable development can be thought of as having passed through three phases: pollution prevention (around 1970), self-regulation (1980s) and sustainability (in the 1990s) (Murphy and Bendell 1997). The simplest strategy for industry of externalizing costs (such as pollution from a factory) became increasingly constrained by government regulation, backed by environmentalist pressure. A more sophisticated strategy of offloading liability for the environment to the state and coping with only those externalities such diffuse pollution as necessary to sustain profit also became untenable, particularly for those companies large enough to have a global reputation to defend. In some instances, the nature and scale of environmental transformation (for example, climate change) might be so great that the earning power of corporations themselves is compromised (Enzensberger 1996). More prosaically, the costs of meeting regulations (for example, by retro-fitting pollution-control technology to a factory) can be high, and their timing difficult to predict. Such costs can have significant effects on corporate performance. Self-regulation by industrial sectors (setting a level and affordable playing field) can therefore be attractive. Profits can be protected by direct engagement in moves towards environmental regulation. Self-regulation by industrial sectors helps manage public and state demands for greater environmental accountability and regulation.

Moreover, competitive advantage can be won by influencing the nature and speed of regulation: if a move to more sustainable production is inevitable, then it is in the interest of individual corporations to make that move at a convenient time (for example, when a new plant is commissioned), especially if by doing so they can obtain a competitive advantage over rivals (for example, by producing a 'greener' product, or making investments in a timely and predictable way), and also as a new field for competition with rival enterprises. In a succession of sectors, individual companies in the 1990s and 2000s took a deliberate step away from rivals to establish a 'green' brand and to claim emerging environmentally conscious markets. Examples include BP in the oil and gas sector (using its slogan 'beyond petroleum' in a brand relaunch following merger with Amoco), or the car manufacturer Toyota with its 'hybrid' (petrol/electric) car Prius. In the British high street, retailers competed to attract the new green consumer, notably by stocking certified timber (see Chapter 12). Thus in 1995 the managing director of the UK hardware chain B&Q wrote: 'we recognised that our stakeholders, that is our customers, staff, local communities, young people, government and shareholders, believed that the environment did matter' (B&Q 1995, p. 3).

The Rio Conference in 1992 was an important landmark in the development of corporate engagement in sustainability issues. Corporations funded a fifth of the

Plate 5.1 Corporate advertising, Nigeria. The tobacco industry is widely vilified in Europe and North America because of the impacts of smoking on health. Sales in the developing world are central to future profitability. Player's Gold Leaf is a brand of British American Tobacco, which has sought to establish a 'green' image for its products, and which has engaged extensively in issues of environmental management and human rights (www.bat.com). Photo: W. M. Adams.

costs of the conference secretariat and parts of the programme (Chatterjee and Finger 1994). In 1990 the Swiss businessman Stephen Schmidheiny founded the World Business Council for Sustainable Development (WBCSD), following an initiative by Maurice Strong, Secretary-General of the Rio Conference (Timberlake 2006). In June 1992 their report on environment and development, *Changing Course*, was published in time to feed into the Rio process (Schmidheiny 1992). The conference and the WBCSD provided a platform for international business to present its view of environment and development, and indeed to present itself as a central part of the solution. The phenomenon of 'green capitalism' was an important feature of the 1990s (e.g. Welford and Starkey 1996; Utting 2002), and a significant contribution to wider thinking within mainstream sustainable development.

There has been widespread corporate adoption of environmental management systems (notably the International Standards organizations (ISO 14001) and the European Union's Eco-Management and Audit Scheme (EMAS)), as corporations have sought to reassure critics and investors that they have moved beyond minimalist compliance with environmental regulations (Mason 2005). There has been increasing engagement in corporate social responsibility (CSR) as a duty of the responsible corporation with regard to human rights, social development and environment, although such commitments can be much easier to draft than to implement, as Shell's experiences in the Niger Delta in Nigeria attest (Ite 2004; see also Chapter 12).

Agenda 21 called for partnerships between environmental groups and businesses, and the development of such relationships has been an important element in the post-Rio agenda. There have also been many experiments with partnerships between businesses and NGOs (Murphy and Bendell 2005). Such alliances fit the wider context of market environmentalism, arguably delivering improved environmental policy while bypassing the deadening hand of state bureaucratic regulation (Arts 2002).

Clearly, there have been company chief executives and boards who have been persuaded of the moral imperative to take environmental aspects of their business seriously, and rather more who have seen marketing and growth opportunities in realigning their business to meet the new demands of consumers and the new opportunities in 'green' products. Equally, there has been much hype, and claims of 'greenwash'. The specific commitments of individual manufacturers or traders have sometimes been shown up by catastrophic performance, as were BP's claims to be green by the disastrous fire at its Texas City Refinery in March 2005 and a large oil spill on Alaska's North Slope in 1986. Corporate protestations of environmental claims have spread to become ubiquitous. Yet the vertical integration of commodity chains has made effective oversight of the environmental and social conditions of production by consumer or regulator effectively impossible (Lebel 2005). Nonetheless, since 1992, corporations have claimed a significant role in the delivery of MSD. The WBCSD has about 200 members from more than 35 countries and 20 major industrial sectors, involving over 1,000 business leaders. It remains a significant global actor, considerably less conservative than some governments in discussing, and calling for change in, fields such as climate change (WBCSD 2005).

The market environmentalism that underpins MSD holds that the world can literally grow out of global environment and development problems, and consumption can be the engine through which sustainable environments and livelihoods are to be achieved. To business advocates of market environmentalism, it is the corporation and the consumer that can deliver on the mainstream agenda of sustainable development. Capitalist markets can clean up and 'green' the planet (Low and Gleeson 1998, p. 81). Thus Martyn Lewis (1992) concludes that, 'regardless of extremist fantasies, we can expect that once capitalist energies begin to be harnessed to environmental protection, a virtuous spiral will begin to develop' (p. 183). A sustainable environment is to be achieved through self-regulation (not heavy-handed regulation by the state), through a revolution in corporate thinking that 'greens' industry from within, combined with a hard-headed (and increasingly global) pursuit of the 'green' consumer dollar. Environmentalists, on the other hand, express severe doubts about the possibility of a society 'buying its way to sustainability', and there is increasingly sharp debate about the validity and utility of the concept of 'sustainable consumption' (T. Jackson 2006).

MSD is firmly anchored within the existing economic paradigms of the industrialized North. It is constructed on a platform of continued capitalist growth, 'green growth'. It focuses on the potential for extension of the market to organize social interaction and human engagement with nature. To Lewis, opposition to a market approach to environment and development suggests that some 'ecoradicals' may have 'more hostility toward capitalism than concern for nature' (Lewis 1992, p. 182). The nature and strength of such radicalism are assessed in Chapter 7. Here I turn to consider the wider context of market environmentalism, in ecological modernization.

Ecological modernization

Market environmentalism is the dominant force in MSD. However, while there are advocates of completely free markets, most thinking about sustainable development (even by the corporate sector) proposes markets that are carefully regulated to deliver environmentally and socially optimal outcomes. Even Beckerman (1994) notes that 'economists have been well aware of the fact that, left to itself, the environment will not be managed in a socially optimal manner. There are too many market imperfections' (p. 205). In MSD, the state, and its capacity to regulate the market, are of critical importance.

Market environmentalism does not admit to the existence of an intractable environmental crisis, and it does not envisage major restructuring of political economy, or the relations between people and non-human nature, to deal with it. Instead, it proposes to make markets work better, *for* rather than *against* society and environment. During the 1990s and 2000s there have been substantial changes in public environmental policy, both nationally (particularly within industrialized countries) and to a lesser extent internationally, and also in the operational practices of global corporations (Gibbs 2000). It has become commonly argued, especially in Northern Europe, that these constitute a new and systematic phenomenon, the fruit of a set of technical changes in systems of production and exchange required to avert environmental disaster. It is also suggested that capitalism can and should be steered to make these changes by an enabling state (Low and Gleeson 1998). This view has been labelled 'ecological modernization' (Spaargaren and Mol 1992; Hajer 1995; Christoff 1996; Mol 1996, 2001; York and Rosa 2003), and it constitutes the second main strand within MSD.

Ecological modernization is a reformist and regulatory approach, which recognizes the ecological dangers posed by unfettered markets, but believes in 'the self-corrective potential of capitalist modernisation' (Low and Gleeson 1998, p. 165). Innovation, greater competitiveness and technological change are key elements of ecological modernization, as they are of other attempts to modernize developed economies (Gibbs 2000). It holds that modernization is fully compatible with ecological sustainability (York and Rosa 2003).

Here is the vision of Brundtland, with economic growth in a capitalist economy working within the constraints of ecological sustainability. The definition of ecological modernization began in the work of social scientists in the early 1980s in Western Europe, especially in the Netherlands, West Germany and the UK (Mol 2001). In policy terms the approach is exemplified by the German government's response to the impacts of acidification and forest die-back through acidification (*Waldsterben*) in the 1980s, or the Japanese response to air pollution in the 1970s (Hajer 1995, 1996). Following Rio, and *Agenda 21*, MSD became broadly ecologically modernist in its approach (Hajer 1996). Academically, Mol (2001) argues that ecological modernization theory is distinct in three ways: first, it conceives of environmental deterioration as a challenge to technical and economic reform rather than as an inevitable consequence of the current institutional structure; second, it emphasizes the transformation of modern institutions (of science, technology, nation state and global politics) to achieve environmental reform; third, it takes a position that is distinct from Marxist and postmodernist analysis.

Ecological modernization, therefore, represents a break with radical environmental critiques of political economy. At its core lies a conviction that the ecological crisis can be overcome by 'technical and procedural innovation' (Hajer 1996, p. 249). Future industrial development, far from continuing to degrade the environment, 'offers the best option for escaping from the global ecological challenge' (York and Rosa 2003, p. 273). In 1981 Tim O'Riordan suggested that environmentalism could be thought of as containing two dimensions, one ecocentrist (romantic, transcendental and concerned, for example, with the rights of other species) and one techno-centrist (rationalist and technocratic, and concerned with the better ordering and regulation of the environment). In these terms, ecological modernization is techno-centrist in its pursuit of rational, technical solutions to environmental problems and more efficient institutions for environmental management and control. It suggests that 'the only way out of the environmental crisis is by going further into industrialisation, toward hyper- or superindustrialisation' (Spaargaren and Mol 1992, p. 336).

Techno-centrism's idea of science applied to 'solve' human problems has been important within mainstream sustainable development since *The World Conservation Strategy* (*WCS*) (IUCN 1980), whose agenda was not only expressed but also conceived in terms of ecology. Ecological modernization involves working towards improved and more 'rational' planning, management, regulation and utilization of human use of the environment. It seeks to 'make environmental issues calculable and facilitate rational social choice' (Hajer 1996, p. 252). Such approaches would include ideas about the role of the environment in development planning (for example, the importance of such procedures as environmental impact assessment), and the ways in which economic development can be achieved without undue environmental costs.

Whether consciously or not, ecological modernization draws on many of the principles of the utilitarian 'wise use' philosophies of conservation in the USA in the first decades of the twentieth century (Hays 1959, ch. 2). Rationality is a key concept in ecological modernization. The ordering of non-human nature to suit human ends was a key element in the twentieth-century project of rationalization (R. Murphy 1994). In *Conservation and the Gospel of Efficiency* (1959), Samuel Hays says of the US conservation movement:

its essence was rational planning to promote efficient development and use of natural resources ... Conservationists envisaged, even if they did not realise their aims, a political system guided by the ideal of efficiency and dominated by the technicians who could best determine how to achieve it.

(p. 2)

The idea of sustainability is a major means by which society has sought to rationalize its interaction with nature: Raymond Murphy (1994) defines instrumental rationality as 'conscious reasoning in which action is viewed as a means to achieve particular ends and is oriented to anticipated and calculable consequences' (p. 30). Murphy points out that by failing to take ecological factors into account, rationalization has been 'inadequate in terms of durability' (p. 43). In this sense, the environmental degradation associated with industrialization is a 'design fault' (Spaargaren and Mol 1992, p. 329), a failure of rational environmental management. Ecological modernization might be referred to as 'the intensification of ecological rationalisation under capitalism'.

Spaargaren and Mol (1992) argue that ecological modernization 'indicates the possibility of overcoming the environmental crisis without leaving the path of modernisation' (p. 334). Capitalist economic growth may be reconciled to the requirements of ecological sustainability by a series of strategies. These involve the injection of improved techniques and technologies into production, the refinement and regulation of markets to tune them to ecological constraints and the 'greening' of corporate ethics and objectives (Low and Gleeson 1998). The new approaches demand a move from reactive 'end-of-pipe' solutions to environmental problems (which tend to be late, unpopular with business and ineffective) to integrated and predictive and holistic frameworks for environmental regulation and management. These frameworks demand new partnerships between state and private enterprise, including market-based incentives and cooperative, voluntary and self-regulatory approaches by industry itself (Christoff 1996).

Clearly there is more to ecological modernization than simply 'greenwash' and politicians' puff. But how much more? In operationalizing sustainable development through ecological modernization, policy-makers are left with great freedom: 'they can either make a few aesthetic alterations but basically continue with business as usual, or they can use sustainable development as a crowbar to break with previous commitments' (Hajer 1996, p. 262). Christoff (1996) points out that ecological modernization is 'a strategy of political accommodation of the radical environmentalist critique of the 1970s' as much as a technical response to environmental degradation. It offers a strategy for governments and corporations seeking to manage ecological dissent.

Ecological modernization has real potential, but offers no magic solution to the self-interest of corporate power and sclerotic government. Hajer (1996) points to the risks inherent in the rationalization of ecology through the amendment of existing bureaucratic structures and the invention of new ones. There are risks in the strategy of intensifying modernity in the hope of finding the seeds of rejuvenation and ecological sustainability (York and Rosa 2003).

There are important questions about the significance of ecological modernization outside the comfortable circle of industrialized countries. Without doubt, the post-Rio system of international environmental regulation reflects the same trends as those labelled ecological modernization in Europe. Market reform and efficient enabling governance have become key elements in the post-Rio 'green' development aid agenda. However, it is not clear how useful ecological modernization is outside the industrial core where it was imagined. Ecological modernization is predicated on the notion of the nation state, and therefore offers little insight into the significance of globalization, international trade or the limited power of non-industrialized countries to enforce environmental regulations (Christoff 1996; Speth and Haas 2006). It is also not clear how far weaknesses in governance, and in the institutions of civil society, in developing countries may prevent the full flowering of ecological modernization as envisaged by theorists in Europe (Christoff 1996).

More fundamentally, ecological modernization is offered as a strategy for making minor adjustments to the conventional growth model. In that sense it is best understood as an extension of (or part of) market environmentalism. Developmentalism remains hegemonic as the inevitable and inescapable process of change and improvement. Ecological modernization is 'the next necessary or even triumphant stage of an evolutionary process of industrial transformation' (Christoff 1996, p. 487). Countries of the global periphery, lacking industrial might and technology, must first transform themselves into facsimiles of the industrialized North. At best, this offers the unhelpful prospect of a perhaps prolonged period of environmental degradation associated with aggressive industrialization (powered by global capital essentially unregulated by any national jurisdiction) before sufficient affluence is generated to make the 'green' turn to ecological modernization.

Of course, if critiques of the conventional development paradigm are valid, developing countries may never escape their poverty, and may add to it the appalling risks of poorly regulated high-technology production. The issues of sustainability and 'Risk Society' will be discussed later in this book, in Chapter 10. Meanwhile, pursuit of sustainability through ecological modernization in wealthy industrialized countries in the global core could have the effect of displacing excessive resource depletion, production of wastes and pollution to the countries of the poorer periphery (Low and Gleeson 1998).

Environmental populism

Politically, MSD draws heavily on ideas about the power of civil society, deliberative democracy and political modernization (Mol 2006). Ecological modernization places a premium on partnerships between state and citizen, and citizen and business: an informed civil society, business that takes corporate social responsibility seriously, a green economy and slim efficient governance combining to create a sustainable economy and environment (Aarts 2002; Ite 2004; Mason 2005). The shift in governmental and business practice implied by ecological modernization is associated with the shift in public values, a wider 'greening' of society, and this has taken place in certain industrialized countries, particularly perhaps Germany and the Netherlands (Capra and Spretnak 1984). This is related to the rise of 'green consumerism' and 'green capitalism' discussed above (Hawken *et al.* 1999; Utting 2002; Mason 2005).

Ecological modernization is built on the principle that institutions can change, and that actors within them can learn. Hajer (1996) attributes the new consensus on ecological modernization to 'a process of maturation of the environmental movement' (p. 251). In the 1980s and 1990s, environmental groups in countries such as the UK grew and restructured (Rawcliffe 1998). Radical issues became normalized and incorporated into the ideology of dominant political and economic institutions. Some radical pressure groups rethought their strategies, notably perhaps in Greenpeace's focus on 'solutions' campaigning described above (Rose 1993).

MSD emphasizes the capacity for citizens to take hold of their circumstances and change them for the better. It proposes strategies for change that emphasize that process of self-generated change, and that promote capacity for 'participation' by ordinary people (or 'local people') in decision-making. It also emphasizes the priority that should be given to developmental change that 'empowers' people, and promotes sustainable development explicitly as an approach that achieves this.

Mainstream ideas about sustainable development were strongly influenced by 'neo-populist' ideas that came to prominence in development thinking in the 1970s (Byres 1979; Kitching 1982). Work on 'ecodevelopment' (I. Sachs 1979, 1980; Glaeser 1984) focused not only on *what* sustainable development should comprise, but also about *how* it should be undertaken. Eco-development involved decentralization of bureaucracy, disaggregation of development focus ('the achievement of sustainable development at local level'), self-reliance and self-sufficiency, the priority given to meeting basic human needs, public participation and equitable distribution (*Ambio* 1979). Singh (1980), writing from the perspective of a Malaysian non-governmental conservation organization, described eco-development as 'development of the people, for the people, by the people' (p. 1350).

Neo-populism emerged in Russia and Eastern Europe before the First World War. It embraced a range of ideas, which argue for 'a pattern of development based on small-scale individual enterprise both in industry and agriculture' (Kitching 1982, p. 19). Neo-populist thinking about rural society (classically by Kropotkin, Chayanov and Gandhi) has been widely applied in the developing world (Bideleux 1987). Kitching (1982) concluded that populism 'makes good social and moral criticism, and has often produced very effective political sloganeering, but on the whole makes rather flabby economic theory' (p. 140). Despite this flabbiness (or possibly because of it), these ideas have proved attractive to those approaching concepts of sustainable development and eco-development, and a number have been co-opted to strengthen the developmental credentials of thinking about sustainability.

First, there has been a focus on basic human needs (I. Sachs 1979; O'Riordan 1988). Debate about basic needs emerged in the Second Development Decade of the 1970s (e.g. Stewart 1985), and survives in the post-millennium concern with acute poverty as a global phenomenon (e.g. J. Sachs 2005). Critics pointed out the political weakness of the concept, resulting from the strength of vested interest and power that oppose wealth redistribution and decentralization (Lee 1981). Its adoption in sustainable development reflects a desire to address the problems of the poor without necessarily confronting the political economy of the development process.

Second, sustainable development picks up the ideas of appropriate and intermediate technology. This is hardly surprising, since *Small is Beautiful* (Schumacher 1973) was one of the definitive books of the 'environmental decade' of the 1970s as well as a successful critique of development. Schumacher wrote a first version of the key chapter of that book, 'Buddhist economics', in 1955, when he was economic adviser to Burma (McRobie 1981). In 1961 he spoke to an international seminar at Poona in India, and the following year wrote a report for the Indian Planning Commission, at the instigation of Nehru, developing the same line of argument. This was coolly received, but his ideas on intermediate technology received wider publicity at a seminar in Cambridge in 1964 (R. E. Robinson 1971). In May 1965 the Intermediate Technology Development Group (ITDG) was formed, its first major work being a guide to the availability of hand and small-scale tools useful in rural development. In *Small is Beautiful* the environmental movement found both icon and scripture for its crusade to forge a new approach to development.

Third, in mainstream thinking, sustainable development demands participation. I. Sachs (1979) called for 'participatory planning and grass-roots activation' (p. 113). Participatory approaches to development became a prominent element in development thinking in the 1980s, and have since become standard practice (Midgley *et al.* 1986; Cernea 1991; Wright and Nelson 1995). Indeed, Cooke and Kothari (2001) suggest that participation has become so dominant as to become 'the new tyranny' in development thinking. Participatory development is naive in implying the feasibility of non-hierarchical systems of organization and government. There is nothing magical in sprinkling the word 'participation' across the development process. White (1996) calls participation a 'hurrah' word, whose use 'brings a warm glow to its users and hearers', but she points out that this very quality prevents its detailed examination, and masks the fact that participation can take many forms and serve many different interests (p. 7). Participation is a highly political process, both within 'the community' and in the relations of the community to other agents (not least the eager development worker). However 'participatory' a development projects seeks to be, White argues that it cannot escape the grip of wider power relations within society. Development 'from below' provides no escape from hard distributional questions.

The idea in sustainable development that development should come 'from below', from the community and not the state, reflected a curious interlocking of opposing political ideologies in Europe and North America in the 1980s. On the one hand, political adoption of the conservative strictures of neo-classical economics led to a view of development as something driven by the market, not the state. This had a profound influence on ideas about national development strategies (Toye [1987] 1993). To achieve public policy goals (sustainable development, for example) the economic incentives for all the main actors must be set correctly, and that that was best done by the market mechanism. Right-wing political thinkers argued that the state had become bloated and inefficient, distorted markets and impeded efficient delivery of economic growth. State power had seemingly become too great and too centralized, generating a tidal wave of privatization of state-run services. Neo-classical economists in the World Bank and other international institutions, and industrial donor governments, therefore found the notion that 'the community' should take an active part in development reasonably palatable.

On the other hand, the notion of empowering people at the 'grass roots' was also attractive to political thinkers on the left and to communities themselves, threatened by the globalization of businesses and the social costs of rapid economic restructuring. By the 1980s new social movements of various kinds had become a political force, both in the former state socialist countries of Eastern Europe and in the West (notable among them, of course, the environmental movement), and governments faced challenges to their policies (including those of privatization and market dominance), and demands for new openness and improved local democracy.

Participatory development was therefore a hybrid of two contrasting sets of ideas. The first sought to expose more of public life to the discipline of the market. This meant a reduced role for the state and created spaces for 'communities' (villagers, private individuals, companies, groups of companies) to be more involved in development. These ideas about market, state and civil society formed the basis of a 'New Policy Agenda' for foreign assistance developed in the USA in the early 1990s (M. Moore 1993; M. Robinson 1993). The second sought to move power down from the state to more local levels, and emphasized the capacity of communities to organize themselves to manage development. This emphasized an enhanced role for civil society and democracy.

One reason for the centrality of populist ideas in sustainable development was their roots in critiques of the inhumane, monolithic and bureaucratic nature of the development process. The strongest insights of the *World Conservation Strategy (WCS)* are borrowed from critiques of development that emphasize basic human needs, and argue that those affected by development should participate in decisions that affect them. These 'populist' critiques stress the significance and importance of indigenous cultures (e.g. McNeely and Pitt 1987), indigenous knowledge (Brokensha *et al.* 1980; Chambers 1983; Richards 1985), the need for local participation in development, and 'development from below' (Stöhr 1981). The *WCS* suggested:

> Conservation is entirely compatible with the growing demand for 'people-centred' development, that achieves a wider distribution of benefits to whole populations (better nutrition, health, education, family welfare, fuller employment, greater income security, protection from environmental degradation);

that makes fuller use of people's labour, capabilities, motivations and creativity; and that is more sensitive to cultural heritage.

(IUCN 1980, para. 20.6)

These ideas were picked up and developed in the successors to the *WCS*, and became central to 'community conservation' strategies in the 1980s and 1990s (Adams and Hulme 2001a; see also Chapter 10). Ghai and Vivian (1992a) argued that 'people's legitimate interest in the conservation of their resource base must be recognised and supported – not only because this is their basic right, but also because it is a pragmatic course to take in the interests of achieving sustainable development' (p. 17). A participatory approach was incorporated into mainstream thinking in *Caring for the Earth* (IUCN 1991): one of the nine principles of sustainable development was to 'enable communities to care for their own environments' (Principle 7). As Holdgate (1996) commented, 'the fact is that sustainable development demands partnerships that involve all sectors of the community, from the individual and the small action group through industry to local and central government' (p. 117). In *Caring for the Earth* it was suggested that action in support of 'sustainable communities' should include giving individuals and communities greater control over their own lives (and resources), enabling them to meet their needs in sustainable ways and enabling them to conserve their environment. Achieving this would demand security of resource tenure, exchange of skills and technologies, enhanced participation in conservation and development processes, more effective local government and better financial and technical support (IUCN 1991).

This emphasis on participation and the empowerment of communities *vis-à-vis* the state was continued in *Agenda 21*. It was both advocated by and exemplified by Third World NGOs. Through the 1980s, locally focused grass-roots NGOs began to open up and organize and even federate internationally in search of influence over the wider institutional context for local change (Princen 1994a). It was these 'third-generation NGOs' that became engaged by the Rio process, and their agenda of participation and grass-roots empowerment became an important element within the wider mainstream (Chatterjee and Finger 1994).

The neo-populist project of 'participatory' development or 'development from below' is problematic (Kitching 1982; Harvey 1996b). Its neat rhetorical inversion awakes a promise that has often proved rather hard to realize in practice. It invites naive, simplistic and idealistic analyses of society, social engagement with nature, and the political economy of development. The idea of 'the community' as a source of legitimacy and a means of achieving effective and lasting developmental change touches on deeply wired Western romantic notions of communities as 'natural' organic social entities. These draw heavily on European works of fiction (for example, Tolkien's 'Middle Earth' and other fantastic versions of a rural European past), and were important elements in the opposition to modernity (industrialization, urbanization, pollution, specialization) in Northern environmentalism (Hays 1987; Veldman 1994). When reflected through lenses of paternalistic colonialism or idealistic postcolonial guilt, these ideas make it seem self-evident that rural people in the Third World live together in discrete 'villages', share common (and fixed) 'tribal' identities and are committed to each other by co-residence, kinship and shared poverty in a way that people in the urbanized, industrialized, 'developed' West have lost. This romanticism about 'community', and the associated vagueness about political conflict at local level (be it district, village or household), is characteristic of many of the documents of the sustainable development mainstream. This slightly odd romantic heritage by no means invalidates ideas about 'development from below', although it may explain why local people may have different ideas about the desirability of 'community action' from those held by development workers.

Environmental limits and the mainstream

It has been argued above that MSD is predicated on continuing economic growth. However, it differs in an important way from normal developmentalism, in that it does recognize the possibility of the biophysical limits within which the economy and global society function. There is an unresolved tension between continued growth and the existence of these limits. This issue is discussed further in Chapter 6.

Concern about the earth's supposed 'carrying capacity' has been a persistent element in environmentalist critiques of the state of the world (Cotgrove 1982; Pepper 1996; see also Chapter 2). Attacks on the possibility of sustainable economic growth have been a familiar element in environmentalist critiques of economics, and radical in the literal sense that they demand fundamental change in economic systems. The 1970s 'limits to growth' environmentalism (for example, the early computer models predicting resource exhaustion (Forrester 1971; Meadows *et al.* 1972)) occasioned considerable hostility from economists (e.g. Beckerman 1974, 1994; Rostow 1978; Kahn 1979; Simon 1981). Beckerman (1974), for example, stated the orthodox position: 'a failure to maintain economic growth means continued poverty, deprivation, disease, squalor, degradation and slavery to soul-destroying toil for countless millions of the world's population' (p. 9).

In some forms, 'zero-growth' ideas are naive, and are easily caricatured and dismissed. Other formulations have been better argued, notably perhaps Herman Daly's work (1977, 2007) on 'steady-state economics'. Daly's starting point in 1977 was his 'impossibility theorem': that a high mass-consumption economy on the US style was impossible (at least for anything other than a short period) in a world of four billion people. He pointed out that the rich countries of the world sought to place the burden of scarcity on the poor countries through population control, while the poor wanted it borne by the rich through limiting consumption. Both wanted to pass as much as possible of the burden onto the future. The simple (but, as he points out, hopelessly utopian) solution was for the rich to cut consumption, the poor to cut population and raise consumption to the same new reduced 'rich' level, and both to move towards a steady state at a common level of capital stock per person and stabilized or reduced population. This idea fails because of the lack of international goodwill, and internal class conflicts within Third World societies. Thus, in the case of Brazil, he said:

Now that the Brazilians have learned to beat us at our own game of growthmanship, it seems rather ungracious to declare that game obsolete. We can sympathise with Brazilian disbelief and suspicion regarding the motives of the neo-Malthusians. But the dialectic of change has no rule against irony.

(Daly 1977, p. 166)

Neo-Malthusian ideas about global population growth, in books such as *Population, Resources and Environment* (Ehrlich and Ehrlich 1970) and *The Population Bomb* (Ehrlich 1972), were fundamental elements of the 'futures' debate of environmentalism in the 1970s (Cole 1978), and were reflected in the *World Conservation Strategy*. Ehrlich and Ehrlich (1970) suggested that, because of population growth, many underdeveloped countries would 'never, under any conceivable circumstances, be "developed" in the sense in which the United States is today. They could quite accurately be described as the never-to-be-developed countries' (p. 2). In *The Third World Calamity*, May (1981) spoke of the 'dead-end societies' of the Third World: 'there is no prospect of change in the Third World that would substantially improve the lives of more than a few people' (p. 226). Neo-Malthusian ideas remain important features of MSD discourse (Kirchner *et al.* 1985). They continue to command attention in best-selling popular science books, such as Jared Diamond's *Collapse: how societies choose to fail or survive* (2006).

Of all the ideas current in the early 1970s, only that of population control caught on in terms of policy, largely because it proposed action where rates of growth were highest, in developing countries, and hence did not threaten the lives of people in the developed world, or fabric of advanced capitalist countries (Sandbach 1978). Support for neo-Malthusian ideas was particularly great in the USA. The 'positive programme' introduced slightly diffidently by Ehrlich and Ehrlich (1970) at the end of their book, was an American vision, based on a view of the world where US political and economic hegemony is assumed. It explored what the US government could and should do nationally and internationally to control global population. Other organizations – for example, the Environmental Fund (of which Paul Ehrlich and Garrett Hardin, among other things known for his paper on 'lifeboat ethics' in 1974, were founder members) – and various groups dedicated to population control such as ZPG and Planned Parenthood, focused on the population content of USAID giving, with significant impacts on USAID policy. These were only overturned by attacks in the 1980s by the anti-abortion lobby in the USA, ironically, of course, from a position to the far right of the political spectrum, which was precisely where liberal opponents had located those who had advocated population control.

Political agendas built on ideas of zero population growth tend towards authoritarianism (Enzensberger 1996). In the 1980s, some environmentalists seemed to despair of existing political structures for change, and called for a technocratic global government. Myers and Myers (1982) discussed the notion of technocratic political globalism, with supranational power exercised by some notional 'global community', and Polunin (1984) suggested the world needed 'saving from itself – from destruction perpetrated by Mankind, its uniquely intelligent component' (pp. 294–5). The superficially *non*-political call for 'impartial' or 'expert' government was, of course, highly political (Enzensberger 1996): scant room for effective democracy, for example, in a world run by an oligopoly of environmental experts and business leaders.

The politics of neo-Malthusian ideas were fiercely opposed by thinkers from the left in the 1970s and 1980s (e.g. Enzensberger 1974; Harvey 1974; Sandbach 1978). Criticism of arguments about resource exhaustion and disaster also came from conservative economists such as Beckerman (1974) and Simon (1981), and scientists with particular faith in the benign possibilities of technological change (e.g. Maddox 1972), but also from radical thinkers. Marxists made the obvious point that the relation between population and environment is not fixed: the density of people that is economically viable in Manhattan is different from that in the Sahel, and this difference is a function not of climatic conditions but of economic organization; the population size that can be sustained is therefore determined by social relations (Pepper 1993). Marxist analyses of the Sahel in the 1970s stressed the structural causes of famine (Copans 1983; Watts 1983a, b).

Radical critiques of neo-Malthusian analysis were accompanied by critiques of policy prescription. Harvey (1974) argued bluntly that

> the projection of a neo-Malthusian view into the politics of the time appears to invite repression at home and neo-colonialist policies abroad. The neo-Malthusian view often functions to legitimate such policies and, thereby, to preserve the position of a ruling elite.
>
> (p. 276)

Environmentalism offered an essentially determinist analysis based on the principle of unchanging limits on human action, and a pessimistic view of the potential impact of social reform (Pepper 1984, 1993). The 'new barbarism' (the phrase is from Commoner 1972) of neo-Malthusian environmentalism, probably most firmly identified with Hardin's 'lifeboat ethics' (Hardin 1974), therefore has the potential to generate conservative ideology and reactionary and repressive politics (i.e. 'eco-fascism', although Pepper (1996, p. 49) warns about applying this emotive label loosely).

In the early 1970s, Hans Magnus Enzensberger (1974) argued that the claimed social neutrality of neo-Malthusian ideas, derived from their ostensible roots in natural science, was a fiction (p. 9). Indeed, he saw ecology as taking shelter in global projection because it was overwhelmed by its inability to theorize sensibly about society, and surrendering 'in the face of the size and complexity of the problems which it has thrown up' (p. 17).

The neo-Malthusianism of the environmental debate of the early 1970s was taken up selectively in the sustainable development mainstream. Enthusiasm for the more pessimistic aspects of the neo-Malthusian 'limits' debate rapidly died away (Sandbach 1978). The Club of Rome picked up the calls for global stability from *The Limits to Growth* but not the notion of zero growth (Golub and Townsend 1977). The problem of population growth was a strong and central message of the *WCS*, but not the Brundtland Report. The concern was re-expressed, but in a much more muted form, in *Caring for the Earth*.

Over time, understanding about the links between population and development developed (Kiessling and Landberg 1994; S. Thomas 1995). Research has shown the inadequacy of earlier neo-Malthusian ideas. Thus it became appreciated that the benefits to a rural household of having more hands to work can outweigh the problem of having more mouths to feed – that there can be a strong economic logic in favour of large families; that fertility in most Third World countries is falling, and is related to wealth; and that people make strategic choices about numbers of children, especially where women have the power and education to control their own fertility (Bledsoe 1994; Tiffen *et al.* 1994). Analysts now have both a better understanding of the importance of political factors in food production and distribution, and a more sober appreciation of the prospects (Madely 1995).

The 'Programme of Action' agreed at the International Conference on Population and Development in Cairo in 1994 focused not on the problem of human numbers *per se*, but on reproductive health through better access to family planning and safe motherhood services, and investment in child survival, education and opportunities for women. The logic of this strategy was that social development and small healthy families would together allow a gradual decline in population growth, and this in turn would ease burdens of poverty and pressures on the environment. However, the annual cost of achieving this strategy through to the year 2000 was estimated to be $19 billion, and annual aid investment fell consistently short of this target (Rowley 1999).

There remain substantial barriers to the availability of even the simplest technologies for fertility regulation (Campbell *et al.* 2006). The contrast between the rapid global acceptance of the drug sildenafil (Viagra) and oral contraceptives speaks volumes for the unequal politics of reproductive agendas between men and women (Potts 2005).

In 2002 the United Nations projected a world population of 7.4 to 10.6 billion by 2050, with a middle estimate of 8.9 billion. These figures take account of fertility rates in industrialized countries falling below replacement levels, and of likely rises in mortality from HIV/AIDS. However, small variations in the level of fertility give dramatic differences in stable populations. By 2050 the global population will be much larger, but growing more slowly (and declining in industrialized countries). It will be more urban (especially in developing countries) and it will be older than it is now (Cohen 2007). Planners face high levels of uncertainty, and cannot be sanguine about either the best direction for policy or its effect on overall global population levels (Haub 1999). It is not neo-Malthusian to conclude that the prospects are daunting (Kates 2004).

In the successive reformulations of ideas about sustainable development, the more politically naive and unacceptable neo-Malthusian analyses of global population were progressively replaced with more sophisticated and more

Plate 5.2 Children in a village in eastern Bangladesh. Neo-Malthusian ideas about the dangers of population growth were an important feature of environmentalism and the 'futures' debate of the 1970s. Concern in the North for population growth in the Third World and the earth's supposed 'carrying capacity' has been a persistent element in environmentalist critiques of the state of the world, and of Northern government views of global environmental problems. Only in the 1990s did the dominant emphasis in Northern environmentalism shift to a concern about resource consumption, and the unequal demands on the biosphere made by wealthy countries. In the same decade, research such as that in rural Kenya by Mary Tiffen and Mike Mortimore built on the ideas of Esther Boserup to challenge the pessimism of neo-Malthusian analysis and reveal the possibility of beneficial interactions between population growth, environmental improvement and economic output (Tiffen *et al.* 1994; see Chapter 8). If they survived, these children will now be grown up, and will probably have families of their own.

empirically based analyses. In their place developed an analysis of the threats to global sustainability of the biological depletion and pollution of atmosphere and oceans by the world's gas-guzzling and hyper-consuming rich. Nonetheless, one does not have to dig far in environmentalist writing or popular concern to find neo-Malthusian thinking alive and well.

At Rio, the way the central message of MSD was cast in terms of the state of the global environment (biodiversity depletion and climate change) avoided explicit engagement with questions of limits (Brown *et al.* 1993). Such ideas were incompatible with business interests in an increasingly interconnected world economy, where market growth was the logic and global capital flows the lifeblood. They were also profoundly unattractive to politicians and policy-makers facing the challenge of providing for growing numbers of people in need of work, shelter and food. Whereas the notion of 'limits to growth' had been uncomfortable, challenging conventional capitalist growth, the problem of global environmental change was a problem 'amenable to mitigation' (Brown *et al.* 1993, p. 573). This allowed debate about sustainable development to be held within the bounds of conventional political and economic thinking. Furthermore, it could be tackled by the conventional weapons of technological innovation and the market.

Nonetheless, behind debates about market environmentalism and ecological modernization, there is still a tension about environmental limits to growth. In industrialized countries, MSD does not demand radical shifts in corporate or national wealth or power, or shifts in social or industrial organization. In fact, the path of sustainable development itself has been portrayed as offering dazzling market opportunities in clean technologies. The Organization for Economic Cooperation and Development (OECD) reported that the global market for environmental goods and services in the mid-1990s was $200–300 billion (OECD 1996).

MSD was environmentalism reinvented by the Reagan and Thatcher decades: free-market environmentalism (Low and Gleeson 1998). Hard questions, of poverty in developing countries and environmental degradation, remain. Ideas that address these problems are discussed in Chapter 7. First, Chapter 6 asks how it is envisaged that MSD should be delivered.

Summary

- There is a strong central stream to thought about sustainable development. This mainstream sustainable development (MSD) runs within dominant capitalist industrialism and developmentalism rather than challenging it. Two key elements within it are market environmentalism and ecological modernization.

- Market environmentalism suggests that problems of environmental management and degradation should be addressed by extending the institutions of the free market into further dimensions of the environment, setting prices for environmental 'goods' and 'services'. Market environmentalism assumes continued capitalist economic growth and rejects environmentalist ideas of limits to growth.

- The 'greening' of business corporations has become an important element in MSD since the Rio Conference.

- Ecological modernization proposes that capitalist modernization can be reformed through efficient regulation of markets, governance and technology. It is technocentrist, demanding improved planning, management, regulation and utilization of human use of nature. Ecological modernization assumes continued capitalist economic growth, within careful, in many cases technologically regulated, boundaries.

- MSD is populist, proposing to bring change about through the participation of citizens, use of appropriate technology and a focus on basic needs.

- Ideas about limits to growth do not form an explicit part of MSD, although they remain important in environmentalist critiques of development. Thinking about global population growth has become better informed, although neo-Malthusian ideas retain a persistent place in popular environmentalism.

Further reading

Atkinson, G., Dietz, S. and Neumayer, E. (2007) (eds) *Handbook of Sustainable Development*, Edward Elgar, Cheltenham.

Dresner, S. (2002) *The Principles of Sustainability*, Earthscan, London.

Elliott, J. A. (2005) *An Introduction to Sustainable Development*, Routledge, London (3rd edition).

Jacobs, M. (1991) *The Green Economy: environment, sustainable development and the politics of the future*, Pluto Press, London.

Jansson, A., Hammer, M., Folke, C. and Costanza, R. (1994) (eds) *Investing in Natural Capital: the ecological economics approach to sustainability*, Island Press, Washington.

Low, N. and Gleeson, B. (1998) *Justice, Nature and Society: an exploration of political ecology*, Routledge, London.

Mason, M. (2005) *The New Accountability: environmental responsibility across borders*, Earthscan, London.

Mol, A. P. J. (2001) *Globalization and Environmental Reform: the ecological modernization of the global economy*, MIT Press, Cambridge, MA.

Neumayer, E. (2003) *Weak versus Strong Sustainability: exploring the limits of two opposing paradigms*, Edward Elgar, Cheltenham (2nd edition).

Purvis, M. and Grainger, A. (2004) (eds) *Exploring Sustainable Development*, Earthscan, London.

Redclift, M. R. (2005) *Sustainability: critical concepts in the social sciences*, Routledge Major Works (four volumes), Taylor and Francis, London.

Stern, N. (2007) *The Economics of Climate Change: the Stern review*, Cambridge University Press, Cambridge.

Utting, P. (2002) (ed) *The Greening of Business in Developing Countries: rhetoric, reality and prospects*, Zed Books, London, for the United Nations Research Institute for Social Development.

Web sources

http://www.wbcsd.ch/ The World Business Council for Sustainable Development, a coalition of 140 international companies formed in January 1995, 'to provide business leadership as a catalyst for change toward sustainable development'.

http://www.bp.com BP

http://www.shell.com/ Shell International

http://www.exxon.com Exxon; for global oil corporations, see under 'environment and society'.

http://www.iucn.org/themes/business/index.htm The IUCN Biodiversity and Business Programme.

http://www.wwf.org.uk/business/ Worldwide Fund for Nature UK Business and Industry Unit.

http://www.cpi.cam.ac.uk/ Corporate Leaders Group on Climate Change: brings together business leaders from major UK and international companies who believe that there is an urgent need to develop new and longer-term policies for tackling climate change.

http://www.oecd.org Organization for Economic Cooperation and Development (OECD): sustainable development.

http://www.sd-commission.org.uk/ UK Sustainable Development Commission, 'the Government's independent advisory body on sustainable development'.

http://www.sustainable-development.gov.uk/ UK Government website on sustainable development.

http://www.forumforthefuture.org.uk/ Forum for the Future, UK environmental NGO: 'action for a sustainable world'.

http://www.iso.org/iso/home.htm The International Organization for Standardization: see ISO 14001, the environmental management system.

6 Delivering mainstream sustainable development

> The whole life of policy is a chaos of purposes and accidents.
>
> (E. J. Clay and B. B. Schaffer, *Room for Manoeuvre*, 1985)

Natural capital and sustainability

Economics is the most influential intellectual discipline of mainstream sustainable development (MSD). This chapter explores the way environmental and ecological economists think about sustainable development, and discusses some of the policy mechanisms that have been proposed to deliver MSD at a variety of scales.

Economic approaches to sustainability build on long-established ideas of maximizing flows of income while maintaining the stock of assets (or capital) from which they come (Munasinghe 1993a, b; Spash 1999). Ideas about natural resources have been established in neo-classical economics since the nineteenth century, and work in environmental economics that explored the interactions between economy and environment grew from the 1950s (Spash 1999). Environmental economics emerged as a distinct discipline in the USA in the 1960s, increasingly informed by work involving by the laws of thermodynamics and developments in mathematical modelling.

The critical development in terms of an economic engagement with the environmentalist critiques of development that gave rise to ideas about sustainable development (see Chapters 2 and 3) was the concept of natural capital. Åkerman (2003) traces this to two contrasting traditions. The first she ties to the work of the British economist David Pearce (e.g. Pearce *et al.* 1989), who drew on existing neo-classical economic theory to define sustainable development proposed in terms of a constant stock of natural assets. Pearce's aim was to stimulate more intelligent and sensitive approaches to ecology by economists, integrating ecosystem theory into mainstream economic theory. The second approach is that of ecological economics, which developed over the same period: the International Society for Ecological Economics was founded at a meeting in Barcelona (Spash 1999). Ecological economics sought to address the relationships between ecosystems and economic systems by combining both disciplines (Costanza 1989). Pearce's approach offered an accountant's view of nature: that of ecological economics sought to inject ecosystem thinking into economics through new and integrated work (Åkerman 2003; cf. Armsworth and Rough-garden 2001). In both, the distinction drawn between natural and human-made capital is critical (Common and Stagl 2005).

Economists have traditionally defined capital as things people have built that have value, such as roads or factories. Environmental economists define this as 'human-made capital'. 'Natural capital' is created by bio-geophysical processes rather than human action, and represents the environment's ability to meet human needs, whether through providing raw materials (fish or timber) or what in the functionalist term are called 'services'. Such ecosystem services include the role of global bio-geochemical cycles in maintaining ecological conditions suitable for human life, or the more mundane way in which wetlands moderate floods or absorb pollutants (Daily 1997; Millennium Ecosystem Assessment 2005).

Using these economic concepts, sustainability can be conceived of in various ways. The first, and most obvious, is to stipulate that constant stocks of both human-made and natural capital are maintained over time. Economic development has always implied increasing capital stocks, but the insights of environmental economics suggest the need to account for stocks of both human-made and natural capital separately. Thus development that increases stocks of human-made capital only by the depletion of natural capital of an equivalent value could be

Plate 6.1 A truck hauling fuel to the Democratic Republic of Congo. Road, truck and fuel are necessary building blocks of development, and conventional indicators of its achievement. Their creation involves exchange of natural for human-made capital in pursuit of development. Photo: W. M. Adams.

said to be not sustainable. This view of sustainability is commonly referred to as 'strong' sustainability (Beckerman 1994).

Techniques for the valuation of human-made capital are well established, although in practice even estimates of human-made capital can be problematic. However, one obvious problem with this whole economistic approach to the environment is that, while a value can be placed on most forms of human-made capital fairly readily (for example, the cost of constructing a road or factory, or the economic benefits that flow from it), it is much harder to place a value on natural capital (Holland and Roxbee Cox 1992). Where a market exists for privately owned facilities or products, valuation is fairly straightforward, and can build on the vast store of expertise in business practice and planning. Public benefits and costs are much harder to gauge; although techniques such as cost–benefit analysis (CBA) have been extensively developed (Sugden and Williams 1978; Brent 1990; Hanley and Spash 1994; see also later in this chapter), they are far from entirely satisfactory. Natural capital is often not privately owned; indeed, it is commonly shared by smaller or larger groups of people. Under the pressures of market forces and social change associated with development, locally recognized systems for allocating rights to benefits that flow from natural capital can become fiercely contested.

The development of techniques for the valuation of natural capital has proceeded apace (e.g. Turner *et al.* 1992; O'Connor and Spash 1999; Cleveland *et al.* 2001; OECD 2006), although it continues to present thorny conceptual and practical problems (J. Adams 1992). Most environmental goods are not subject to market relations, either because they are held in common (for example, clean air), because they have only recently become scarce (for example, clean groundwater, subject to slow and recent pollution penetration), because the structure of existing markets allows key actors to treat environmental costs as an externality, or because institutions for organizing a market do not exist.

Ecosystem services are not fully captured in commercial markets and are not adequately quantified (or in some instances adequately understood). However, attempts to place a money value on ecosystem services have become widely established (Daily 1997; Mooney and Ehrlich 1997; cf. Pimm 1997). Economists recognize direct and indirect use values and non-use values of various kinds – natural resources like fuelwood or fish are direct values, and ecosystem services like flood control are indirect values (Barbier 1998). They use a variety of techniques to capture these values (Gouldner and Kennedy 1997). Munasinghe (1993c), for example, used contingent-valuation, travel-cost and opportunity-cost approaches to measure the costs and benefits of a new national park in Madagascar. In a much discussed paper, Costanza *et al.* (1997b) attempted to estimate the current economic value of seventeen ecosystem services for sixteen biomes across the globe, using a variety of methods including contingent valuation and replacement cost. They calculated that the value of the entire biosphere was between \$16 and \$54 trillion (10^{12}), with an average of \$33 trillion per year. On a yearly basis, the total value from marine ecosystems was estimated to be \$577 per hectare per year, and from terrestrial ecosystems \$804 per hectare per year, the latter providing about a third of total global flow of value per year. For comparison, global GNP was about $18 trillion per year. Thus, in order for human-made capital to substitute for the natural capital of the biosphere, global GNP would have to grow by at least $33 trillion, but this would, of course, bring no increase in welfare.

This work stimulated extensive debate, not least in a special issue of *Ecological Economics* (Costanza 1998) about both whether it was broadly desirable that an attempt should be made to value environmental services in this way, and how it should be done (Costanza *et al.* 1998). Subsequent analyses have begun to explore consequent issues – for example, how the value of services at a local scale (for example, tourism based on scarce habitat) scale up globally, or the effect of growing scarcity of supply (as habitat is destroyed) on values, and the way human-transformed habitats can yield substantial benefits (Balmford *et al.* 2002).

Whatever the detail, calculations of this kind have demonstrated that nature has *some* value that can be measured in monetary terms, and that this value is extremely large (Balmford *et al.* 2002). As David Pearce hoped, the concept of 'natural capital' has changed the way people think about the biosphere, and the value of non-human nature. The conversion of the values of nature to a single enumerator of money worked, because this metaphor describes nature in the only language economists understand, and because economics has become such a dominant force in policy-making (Åkerman 2003). Natural capital has provided a simple and robust frame for debate within the sustainable development mainstream, and specifically the observation that, as natural capital becomes scarcer (as ecosystems are transformed and subject to greater demands), its value will rise. If human demands cause ecological thresholds to be passed, the value of ecosystem services may jump (conceivably to infinity, if life-support systems collapse). The work of the Millennium Ecosystem Assessment (2005) develops this pragmatic approach, translating complex issues about human demands on the biosphere into language that can be understood by policy decision-makers.

Weak and strong sustainability

The implications of the simple notion that sustainability demands maintenance of stocks of both human-made and natural capital over time has been fiercely debated. A distinction is now commonly drawn between weak and strong sustainability. 'Strong sustainability' places a special importance on natural capital and requires that the stocks of both natural and human-made capital be maintained. 'Weak sustainability' allows trade-offs between natural and human-made capital. Economists such as David Pearce have pointed out that such a requirement for zero or negative natural capital depreciation would place excessive constraints on economic growth (Pearce *et al.* 1989). If this requirement is imposed at the project level, it is likely to stultify development, since it effectively makes it impossible to do anything that damages the environment at all. This approach to development, 'strong sustainability', would suit environmentalists opposed to development and economic growth (and these tend to be those living comfortably enough in the

North), but is impossibly challenging for governments and business. It is unlikely to appeal to grass-roots environmentalists in the South facing the daily human tragedy of poverty.

This notion of 'weak sustainability' involves the principle of trade-offs between losses to natural capital in one project and gains elsewhere and the substitution of either human-made capital or human-induced 'natural capital' for lost natural capital (Barbier *et al.* 1990). If the requirement for zero natural capital depreciation is set at the *programme* level (that is, at the level of suites of projects of a particular region, agency or government), there is some flexibility to maximize economic returns from individual projects (Pearce *et al.* 1989). Lipton (1991) suggests that sustainability should be discussed at the level of the country as a whole, so that damaging and favourable effects of projects can be balanced out.

Economists differ as to how far trade-offs should be allowed to go. Beckerman (1994) argues that '"sustainable development" has been defined in such a way as to be either morally repugnant or logically redundant' (p. 192). He dismisses strong sustainability ('implying that all other components of welfare are to be sacrificed in the interests of preserving the environment in exactly the form it happens to be in today') as 'totally impractical' (p. 203). Jacobs (1995) argues that Beckerman's definition of strong sustainability is absurd, suggesting that 'sustainability is the injunction to maintain the *capacities* of the natural environment: its ability to provide humankind with the services of resource provision, waste assimilation, amenity and life support' (p. 62; emphasis in the original).

Beckerman (1994) is also scathing about 'weak' sustainability, and the welfare-based definition that allows substitution between natural and human-made capital. He argues that this adds nothing to 'the old-fashioned economist's concept of optimality' (p. 195). He suggests that the only efficient way to proceed is to disregard the distinction between human-made and natural capital altogether, and allow gains in the former to replace losses in the latter. Thus the replacement of an unproductive wetland with a productive irrigation scheme should be assessed quite simply in the different economic benefit streams they produce, and development should involve selecting the projects with the greatest economic return because it is this that will maximize human welfare.

Daly (1994) points out that natural and nature-produced services (for example, atmospheric regulation) and resources (for example, waste assimilation capacity) are non-replicable. Natural capital is, therefore, complementary to human-made capital and not a substitute for it. Barbier *et al.* (1990) recognize this problem of the non-substitutability of human-made for natural capital, and include it within their formulation of CBA as a constraint on the depletion or degradation of the stock of natural capital. El Serafy (1996) goes further in defending weak sustainability, pointing out that it is not a watering-down of strong sustainability but its precursor, as a concern for all capital that sustains future income. Strong sustainability represents a strengthening of this position to hold that *natural* capital specifically be kept intact. He sees the key issue as how much of the income derived from the exploitation of natural capital is *genuine* income – that is, available for consumption, and how much needs to be reinvested to sustain the same levels of consumption into the future. Weak sustainability, like strong sustainability, is arguably therefore a workable theoretical concept.

Among environmental economists who accept that sustainability demands maintenance of stocks of both human-made and natural capital, there is still debate about what kind of trade-offs should be made. One notion is that developments that reduce natural capital should be balanced by others that are specifically designed to compensate by creating new natural capital elsewhere. Thus a development programme should contain 'shadow projects' to compensate for degradation other projects cause (Pearce *et al.* 1989). This notion has proved attractive in industrialized economies such as the UK, where developers proposing a project that has adverse environmental impacts are now routinely expected to offer as part of the development 'ecosystem restoration' schemes such as new areas of woodland or wetlands to replace those lost (Eden *et al.* 1999).

This notion that trade-offs may be made between one piece of natural capital and another is opposed by conservationists who argue that natural capital is not like human-made capital. It is not made of bricks and mortar. Some forms of natural capital are more like precious pieces of architecture (such as a cathedral) than like a dispensable factory that can be taken down and replaced with another, better, one as circumstances change. Thus it might be argued that some species (perhaps the mountain gorilla or the blue whale) and some ecosystems (perhaps the highly biodiverse *fynbos* vegetation of South Africa or the Great Barrier Reef) are irreplaceable and should not be conceived of as 'capital' to be replaced or exchanged for some other form of natural capital.

There are actually two arguments here. The first is that some (or all) attributes of non-human nature have an 'intrinsic value' that cannot be captured by economic appraisal. This relates to a debate about ethics, and the intrinsic worth of nature, as opposed to the instrumental view of non-human nature as something to be valued only for its utility for humans (Low and Gleeson 1998). Economists try to capture such values using ideas such as 'existence value' – that is, the non-material value to people of the existence of species (Barbier 1998). However, the debate as to the validity of intrinsic worth is essentially philosophical and not methodological, and in the field of environmental ethics this approach is not regarded as a solution (Elliot 1997).

The other argument about the non-substitutability of natural capital holds that, regardless of the theoretical validity of replacing one piece of natural capital with another, this is in practice not possible. Thus if it were proposed to convert 1,000 ha of rainforest into ranchland, the complexity of the ecological interactions forged through co-evolution would make it impossible in practice to recreate it. Scientific knowledge is too slight for such exchanges to be feasible, and, even in the relatively intensively researched and simply structured temperate ecosystems, attempts to move or recreate ecosystems are still in their infancy (Buckley 1989; Cairns 1991).

The concept of 'critical natural capital' is used to refer to those parts of natural capital that cannot be replaced if lost (or at least, not within feasible time frames), and cannot therefore be substituted with human capital or compensated for by positive projects elsewhere (Buckley 1995).

An ecological economics?

The field of ecological economics developed through the 1990s (Costanza 1989, 1991; Costanza *et al.* 1997a; Daly 2007). It drew on earlier work by Kenneth Boulding (1966) and Herman Daly (1977), and recognition of the importance of material and energy flows (Åkerman 2003). It set out to be transdisciplinary, making explicit links with the natural science of ecology to explain the complex linkages between human and natural systems (Costanza 1991; Jansson *et al.* 1994; Berkes and Folke 1998; Åkerman 2003), and to apply those insights to policy that promoted sustainability, and the efficient allocation of scarce resources. It claimed to take 'a holistic systems approach that goes beyond the normal narrow boundaries of academic disciplines' (Folke *et al.* 1994, p. 3).

Plate 6.2 A settler family on the ill-fated Bura Irrigation Project in Kenya in the early 1980s. Settlers were brought to the Bura scheme from all over Kenya, but the remote location of the project and problems with water supply and distribution meant that the scheme failed (W. M. Adams 1992). The irrigation infrastructure (barrage, canals etc.) represents human-made physical capital, the transformed rural environment and water supply is natural capital, and the unfortunate settlers human, social and cultural capital (all, in this case, misused). The support and education of young people within and outside families is critical to human capacity to cope with development and manage the environment. Photo: W. M. Adams.

Ecological economics views socio-economic systems as inextricably linked to environmental systems, and hence seeks to escape the conventional notion that 'resources' can be understood (or exploited) in isolation. It focuses on the linkages between human and ecological systems, and the feedbacks between them. It is also explicitly policy oriented: its aim is the development of practical policies for sustainability (Folke *et al.* 1994; Berkes and Folke 1998).

In ecological economics, the zero-growth debate has grown up. Goodland *et al.* (1993) stress the continuing reality of environmental limits. The human economic system has to exist within the biosphere, on which it depends and of which it is part. The biosphere provides the source of all natural resources and is the ultimate sink for all wastes. The throughput of resources between biosphere and economic system is a function of population size and per capita resource consumption, and the capacity of the biosphere to provide the resources and sinks demanded of it is finite. Figure 6.1 contrasts the economic subsystem in some notional past 'empty-world' scenario with that of today. In the 'full-world' scenario, the economic system has already started to interfere with the biosphere through excessive demands on sinks – for example, through acid rain, green-house-gas accumulation and ozone depletion (Goodland *et al.* 1993).

Ecological economists add to the concepts of human-made and natural capital a third category, of 'cultural capital', referring to the factors that enable human societies to adapt to and modify the natural environment (Berkes and Folke 1994). The existence of natural capital is the basis and precondition for cultural capital, and is in turn regulated by cultural capital (through the institutions affecting the use of nature). Human-made capital is generated by both natural and cultural capital together. Human-made capital impacts on natural capital (for example, when a factory or a city discharges wastes into a river basin). However, it also affects cultural capital, as the creation of technologies (tools, skills and knowl-edges) affects human understanding of the dynamics of and human demands on nature, and hence the status of natural capital (Berkes and Folke 1994).

At its simplest, cultural capital can be conceived of as the interface between natural and human-made capital. It embraces the conditions and institutions under which collective action occurs, but also the philosophies, values and ethics that underlie them (Berkes and Folke 1994). The idea of cultural capital therefore relates very closely to two concepts, 'institutional capital' (Ostrom 1990) and 'social capital' (Coleman 1990).

The field of new institutional economics (NIE) developed in the 1980s and 1990s to address the economics of the institutions that govern human interac-tions in areas such as resource use (Bromley 1989, 1992; Ostrom 1990). Debates about property rights and the management of common property resources have been particularly influential in changing understanding of how people collaborate, or conflict, over the use of resources (Berkes 1989; Berkes and Folke 1994, 1998; Agrawal 2001). The concept of social capital was defined in its modern form by Pierre Bourdieu in his paper 'The forms of capital' (1986), and popularized by the American political scientist Robert Putnam in the 1990s, most notably in his book *Bowling Alone* (2000). The term refers to the norms and networks that

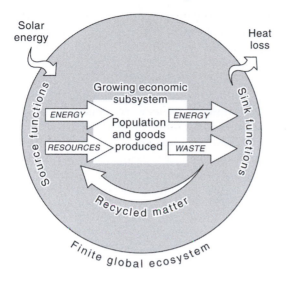

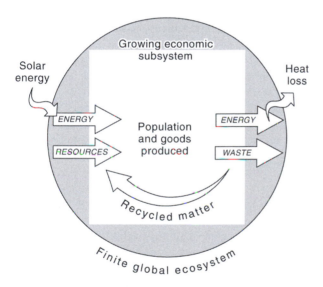

Figure 6.1 The finite biosphere relative to the growing economic subsystem in the 'empty world' (above) and 'full world' (below) scenarios (after Goodland *et al.* 1993).

enable collective action. It proved attractive to policy-makers who held that social cohesion was important for sustainable human and economic development and especially poverty alleviation (Bebbington 2004; Bebbington *et al.* 2004).

These areas of work address a serious problem of the reductionist way in which neo-classical economics treats natural capital. This downplays meanings of nature other than those that relate to production. The rich metaphor of natural capital has been narrowed to treat nature just as productive machinery (Åkerman 2003). Moreover, discussion of natural capital at the aggregate level pushes to the margin of debate the point that nature (and debate about what nature is) is contested. Environmental policy has to deal with conflicts over 'the distribution of environmental goods and bads' (Åkerman 2003, p. 442). Neo-classical economics smoothes these away behind a façade of optimality and the rhetoric of rational planning. As Spash (1999) observes, welfare economics has evolved as 'a method for removing any apparent need for moral discourse or politics from the agenda of the economic policy adviser' (p. 430). These and other shortcomings of neo-classical economics are increasingly widely recognized – for example, by the post-autistic economics network (www.paecon.net), and in wider academic critiques (e.g. Fullbrook 2004). Many are directly relevant to debates about mainstream sustainable development.

Trade-offs, equity and complexity

The question of sustainability can be built into economic appraisal of the costs and benefits of development in various ways (Barbier *et al.* 1990). In development economics, the relative importance of economic objectives of development are routinely weighed against the importance of social objectives such as poverty alleviation. Conventionally, development strategies involve a mix of economically efficient production (growth to make a larger cake) and targeted social programmes (aimed at more equitable division of the cake). The notion of sustainability adds a third dimension: maintaining the quality or capacity of environmental systems and the resources and benefit streams they generate (Munasinghe 1993b). Sustainable development can, therefore, be seen as involving trade-offs between environmental sustainability, economic sustainability and social sustainability. A variety of economic tools are conventionally used to aid decisions in development at different scales from the project to the national economy and internationally. These include CBA, the use of shadow prices and the manipulation of discount rates (Munasinghe 1993b). They demand sectoral or regional studies, national multisectoral economic analyses and analyses at the international scale of trade and financial flows (Barbier *et al.* 1990; Munasinghe 1993b).

Munasinghe argues that environmental economics is capable of providing a bridge between socio-economic and environmental systems, translating the results of environmental analysis into categories that fit the patterns of thinking and analysis used by economists. Unfortunately for economists, nature is no respecter of the neat categories of their models and analyses. Although ecological systems can be thought of as functioning at a hierarchy of scales, the fit with the units of markets, sectors and 'national economies' beloved of economists is poor.

The idea of trade-offs is fundamental to economic thinking about sustainability. A constant theme in accounts of sustainable development is equity: Jacobs (1995), for example, holds that it incorporates 'an inescapable commitment to equity' (p. 60). Thus it must involve not only the creation of wealth and the conservation of resources but also their fair distribution between rich and poor. Moreover, it demands not only this principle of *intra*generational equity but also a commitment to the future and *inter*generational equity. MSD has tried to take account of both these dimensions of equity, that between present and future generations and that between rich and poor in the present generation. As we have seen, in the thinking of market environmentalism, these concerns translate into the problems of making trade-offs.

The trade-offs that sustainable development demands are therefore complex, involving several different scales simultaneously. First, as has been seen, there can be difficult choices between natural and human-made capital or between social, environmental and economic dimensions of development. Second, trade-offs can take place either in space (between losses in natural capital or some other measure here and gains somewhere else) or in time (between losses or gains now, and those coming in the future).

Principles of inter- and intragenerational equity demand that a balance be struck between the needs of present and future generations. This is far from simple. One problem is that of uncertainty about the future (Dovers and Handmer 1992). The adoption of modest economic growth targets in the present generation may make it harder to tackle poverty now as well as having implications for the wealth or well-being of future generations (for example, by allowing existing inequalities to endure (Dovers and Handmer 1992; Munasinghe 1993a)). Uncertainty about future technologies and future resources makes it difficult to assess to what extent future generations will be able to provide different environmental goods. Uncertainty about the values future generations will hold makes it hard to know how they will value different environmental goods, and how they will view decisions made now on their behalf (Pasek 1992). As Beckerman (1994) points out, 'people at different points in time, or in different income levels, or with different cultural and national backgrounds, will differ with respect to what "needs" they regard as important' (p. 194).

There is also considerable uncertainty in predictions of the ways in which ecosystems will behave in the future (because of both scientific uncertainty and anthropogenic environmental change), and hence it is hard to predict the value of flows of benefits from existing and future natural capital. In semi-arid environments such as the Sahel, for example, rainfall variability and drought have severe implications for land use and yet are effectively impossible to forecast (see Chapter 7). Patterns of resource use that are analysed over a short time frame and judged sustainable may well not be sustainable over longer periods. Dixon and Fallon (1989) ask, 'how far into the future do we worry about? Is our concern next week, next year, next century?' (p. 81).

Because of this uncertainty about the future, a standard formula for sustainability is to demand that capital endowments are kept constant, such that each generation bequeaths a legacy of natural capital no smaller than the one it inherited. The snag with this is that the impacts of development can be delayed far into the future, either because the impact itself is delayed (for example, the impacts of poorly stored toxic waste that starts to leak after a few decades), or because the ecological repercussions of a development intervention take time to work their way through the ecosystem – for example, the problem of the responses of floodplain ecosystems to dam construction (Thomas and Adams 1997; see also Chapter 11).

Ecosystems function at a range of timescales, and ecological interactions and feedbacks can be complex and often delayed. Environmental response times may be so long as to conceal the link between past and future events, and the true severity of impacts may be revealed only when new economic opportunities come to be developed (for example, an attempt to irrigate using water from an aquifer contaminated with heavy metals). The importance of timescales in sustainability analysis can also be seen in reverse, in that degraded ecosystems can recover naturally over long time periods, or may be rehabilitated given appropriate management (Thomas and Adams 1999).

The socio-economic impacts of ecosystem change are likely to be even more complex, and may be further delayed. A further complication is that there are likely to be differences between the ways different impacts are viewed. People suffering serious short-term impacts (for example, floodplain farmers affected by an upstream dam) may have very short time horizons, while the governments that commissioned the dam may take a much longer view, arguing that eventual benefits to the national economy may outweigh short-term costs (Dixon and Fallon 1989; Thomas and Adams 1997). The definition of social time horizons is inherently political, because the ways people view their resources, and the relative merits of consumption in the present rather than the future, will inevitably vary (Dixon and Fallon 1989). Time horizons set according to political and economic expediency may be too short for the sustainable management of natural systems. As Dovers and Handmer (1992) note, 'a major implication of the moral principle of intergenerational equity is to force institutional systems to think over timescales that are somewhat closer to those of natural systems' (p. 219).

Spatial scale is also important in assessing sustainability (Fresco and Kroonenberg 1992), but again the judgement of the sustainability of a development decision is closely dependent on the scale chosen for analysis. Governments routinely trade off sustainability at one location to meet national goals. Thus, locally, the benefits of a development project (for example, a mine) might be at the cost of reduced sustainability elsewhere (for example, pollution downstream) (Low and Gleeson 1998). Internationally, industrialized countries seeking to make their policies sustainable may do so at the expense of other places by importing resources or exporting wastes. Both ecological and political boundaries (local, regional, national or international) are relevant to the assessment of sustainability. Debates about the sustainability of particular developments might very easily descend into arguments about boundaries, and different actors (for example, governments, non-government organizations (NGOs) and transnational corporations) may base conflicting assessments of the sustainability of controversial projects on different choices of boundaries for analysis. Sustainability can be determined at a range of scales from local to global (Thomas and Adams 1997).

Natural systems may provide appropriate boundaries for sustainability analysis, but these are not always easy to define. In particular, the ecosystem is effectively an arbitrary analytical category, not a natural entity whose characteristics are endogenously determined. Furthermore, in practice, development planning usually takes place within political jurisdictions, not within natural boundaries. Structures of governance are hierarchical, and human-made boundaries rarely fit the spatial patterns of natural systems (Conway 1985; Munasinghe 1993a). Ecosystems straddle political boundaries, and so do bio-geochemical processes such as trans-boundary acidification, international river flows and oceanic circulation.

Overlapping and conflicting political and environmental management institutions make the practical measurement of sustainability (let alone its promotion through environmental management) highly problematic. So too does the complexity of ecosystem behaviour.

Measuring sustainable economies

The logic of environmental economics in trying to express the value of nature in money terms is to incorporate environmental concerns into decisions about national development (Munasinghe 1993b). The definition of metrics for sustainable development has been a key element in ecological modernization (discussed in Chapter 4). The need for such metrics is widely recognized. Reliable measures of the supply of natural capital and human demand upon it are necessary to track progress and design policy to promote sustainability (Monfreda *et al.* 2004).

There are numerous candidate indicators of sustainability (Parris and Kates (2003) identify over 500). One such approach is the Ecological Footprint, incorporated into the annual *Living Planet Report* (Monfreda *et al.* 2004; Hails *et al.* 2006; for the Global Footprint Network, see www.footprintnetwork. org/). Another is the Environmental Sustainability Index (ESI), developed by the Yale Center for Environmental Law and Policy and the Center for International Earth Science Information Network of Columbia University, in collaboration with the World Economic Forum and the Joint Research Centre of the European Commission (see http://sedac.ciesin.columbia.edu). This ranks 146 countries on 21 aspects of environmental sustainability covering natural resource endowments, past and present levels of pollution, efforts towards environmental management, contributions to protection of the global commons and capacity to improve environmental performance over time. In 2005 Finland ranked first, followed by Norway, Uruguay, Sweden and Iceland. These are countries with substantial endowments of natural resources, low population density and good environment and development policy. The USA was in 45th place, just ahead of the UK (66th), brought low by problems such as waste generation and greenhouse-gas emissions. At the bottom of the table came a set of the world's poorest countries: North Korea, Iraq, Taiwan, Turkmenistan and Uzbekistan. These indicators are deliberately eye-catching, and of obvious value in focusing policy attention on the challenge of sustainability. However, none is universally accepted, or backed by rigorous theory and data and analysis (Parris and Kates 2003). More work is needed.

In recent decades there has also been considerable experimentation with new approaches to national accounting and the measurement of welfare (e.g. Daly and Cobb 1990), and integrated environmental and economic accounting (Bartelmus 1994). A key concern has been to replace established but deeply flawed measures such as gross national product (GNP). GNP is a measure of the way income flows in an economy, and does not take account of resource depletion or pollution or other environmental costs (Jacobs 1991). Any activity that involves the exchange of money contributes to GNP, so that the production of goods in a polluting factory contributes to GNP, but so do the costs of the resulting clean-up by government: pollution is, by this tunnel-visioned measure, deemed good for the environment. In this instance, the resulting GDP estimate is incorrect because harmful impacts such as pollution are ignored, and beneficial inputs related to environmental needs are undervalued (Munasinghe 1993b). Conventional presentations of national accounts also fail to take account of changes in stocks of natural capital, and the existence of hidden subsidies to certain activities and products, because their impacts on natural capital are not measured (ibid.). The existence of macroeconomic incentives for environmentally destructive behaviour (whether by corporations or small farmers) is of great importance in understanding unsustainable patterns of environmental management, and, of course, in adapting policy to change them (Barbier 1994).

Costanza and Daly (1992) emphasize the need to distinguish between growth in the size of an economy and growth in its capacity to deliver solutions to human needs. They write: 'Improvement in human welfare can come about by pushing more matter–energy through the economy or by squeezing more human want-satisfaction out of each unit of matter–energy that passes through' (p. 43). They suggest that increased throughput should be described as growth, and increases in the efficiency with which human needs and wants are met should be described as development. Beyond a certain point, as growth destroys natural capital, because it costs more than it creates, growth has become 'impoverishing not enriching' (p. 43).

GNP growth measures the size of the economy, but an assessment of the sustainability of that economy depends on what is growing and how. Atkinson *et al.* (1997) showed that growth in smaller resource-rich countries in the 1970s and 1980s was not environmentally sustainable because it was based on the depletion of natural capital. The scale of the economy is only one factor that determines environmental quality; others include structure (the mix of goods and services produced), the ability to substitute away from resources that are becoming scarce, the ability to use clean technologies and management practices to reduce damage per unit of input or output, and the efficiency of inputs used per unit of output (World Bank 1992).

One approach, therefore, is to address the economy's throughput of energy and materials, which on a finite planet are obviously limited (Ekins and Jacobs

1995). GDP measures value added in the economy. The relationship between this and the material throughput necessary to achieve it can be altered by structural economic change, substitutions between factor inputs, and more efficient use of the same input. Ekins and Jacobs (1995) suggest that it is perfectly possible for GDP growth and environmental sustainability to be compatible. If there were technological and structural changes in manufacturing and patterns of consumption such that GNP rose and yet the environmental impact coefficient (EIC) fell, economic growth could be accompanied by *reduced* rates of resource depletion (Jacobs 1991). What is needed is 'dematerialization', the reduction in the amount of natural resources and energy used to generate wealth. The concept lies behind the movement for radical improvements in resource productivity using technological advance, notably the 'Factor 10 Club' (www.factor10-institute. org/index.htm), founded in France in 1994 by Friedrich Schmidt-Bleek, whose goal is to dematerialize the economies of the industrialized countries tenfold on the average within 30–50 years (Hawken *et al.* 1999).

One possible measure of the environment impact of an economy is the material input per unit of service (MIPS) (Hinterberger *et al.* 1997). Another possibility is the calculation of an EIC of GNP (Jacobs 1991). This would involve measuring the amount of environmental consumption (raw materials and energy, the assimilation of wastes and the maintenance of life-support systems such as climatic regulation) created by each unit of national income.

Debates about environmental accounting have developed, taking into account conventional economic mechanisms such as taxation and monetary policy (Munasinghe 1993b), and issues such as depletion of natural resources, pollution and income distribution (Daly and Cobb 1990). It is clear that existing measures do not adequately measure environmental dimensions of human welfare. A graph of change in GNP and an Index of Sustainable Economic Welfare (ISEW) in the USA between 1950 and 1985 showed that, while GNP rose consistently, the ISEW did not (Daly and Cobb 1990; see also Figure 6.2). Similarly, calculations of net national income in the Netherlands show it to be 50 per cent higher than sustainable national income, a measure of the costs of bringing resource use back to a sustainable level (Gerlagh *et al.* 2002).

A range of practical attempts have been made to integrate the environment into national accounts, and therefore encourage transition towards the kind of 'green economy' that Jacobs (1991) outlines. One important initiative is the revision of the UN System of National Accounts (SNA). In the 1990s the United Nations Statistical Division (UNSTAT) developed a System of Integrated Environmental and Economic Accounting (SEEA) (Bartelmus 1994). This addressed the depletion of natural resources in production and final demand, and the changes in environmental quality resulting from production and consumption and natural events, and from environmental protection and enhancement. It did this by embracing the concept of natural capital, and seeking to measure its stocks and flows. These were measured in 'satellite accounts'. The SEEA therefore sought to account comprehensively for all impacts of development and link them to the activities and sectors causing them.

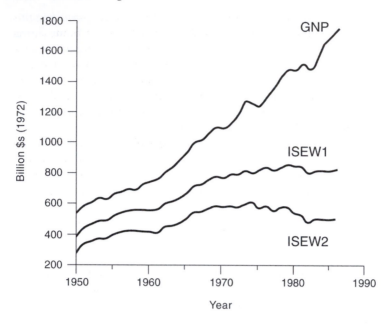

Figure 6.2 GNP and ISEW in the USA, 1950–85 (after Costanza and Daly 1992). Note: ISEW2 takes account of the depletion of non-renewable resources and long-term environmental damage, ISEW1 does not.

The SEEA approach was used in the 1990s by the World Bank in studies of Thailand, Papua New Guinea and Mexico. This revealed the potential of the approach, but also the problem of poor data (Bartelmus 1994). In the case of Mexico, two measures of environmentally adjusted net domestic product (EDP) were calculated. The first (EDP1) measured resource depletion (oil, forests, land and living water resources). This amounted to 94 per cent of net domestic product in 1985. The second (EDP2) attempted to measure externalities in the form of the costs of environmental quality degradation, and amounted to 87 per cent of NDP.

There was a surge of interest in environmental accounting at national level in the 1990s, but a variety of approaches were used. Countries with extensive natural resources (for example, Canada, Australia) focused on the development of resource stock accounts. Intensively developed countries like Germany emphasized land-use and land-cover accounting. The Netherlands developed relatively sophisticated measures combining economic statistics and pollution emissions data (R. Smith 2007). The SEEA 1993 provided some frame to such efforts, but national accountants were resistant to several core measures within it, particularly the need to adjust the central measure of GDP.

In 1992 the statistical offices of the Canadian and British governments and the European Union convened the 'London Group' of national accountants on environmental accounting, and in 1998 they undertook to revise the SEEA.

The United Nations Statistical Commission approved a new accounting hand-book in 2003. This is complex, incorporating four categories of accounts (flows of materials and energy, measures of good environmental management (for example, expenditures to maintain the environment), environmental assets and measures of resource depletion (R. Smith 2007). Some developed countries have implemented accounting systems that take a number of these into account (Australia, Canada, Denmark, Germany, Italy, New Zealand and Norway), but none has implemented the SEEA 2003 in its entirety.

Views of the SEEA vary. On the one hand, it is clear that the lack of national accounts that incorporate environmental information effectively is a major limitation of national and international capacity to achieve sustainability goals (R. Smith 2007). The SEEA 2003 does provide a consistent framework for data that can be used – for example, to measure weak and strong sustainability between countries (Dietz and Neumayer 2007). On the other hand, the complexity of the Handbook of National Accounting is likely to put off many government accountants, as its predecessor did, and they will turn to simple indicators instead (Bartelmus 2007). Critically, the SEEA does not adequately operationalize the concept of sustainability through the creation of environmentally adjusted accounts. It fails to grasp the nettle of adjusting central economic indicators by providing a full costing of environmental depletion and degradation. Without that, green accounting 'loses its capacity to compare the "goods" of production and consumption with the "bads" of waste, pollutants and natural resource depletion' (Bartelmus 2007, p. 614).

Sustainability at the project scale

The application of ideas about sustainability to policy also demands consideration at the project scale. At the scale of project or programme, sustainability is conventionally handled through CBA (Barbier *et al.* 1990). CBA is the established procedure by which economic decisions about project development can be made (Sugden and Williams 1978; Barrow 1997). Environmental sustainability was not recognized as an issue when the standard manuals of Third World project appraisal, using CBA, were written (Van Pelt *et al.* 1990). CBA is systematic, based on principles that are widely comprehended (people's choices as revealed by surveys and markets), and it produces a neat result that can feed directly into the planning process. Debate about the shortcomings of CBA is considerable, focusing particularly on the problem of determining social values, and the impact on public participation of the procedure's technical complexity (Barrow 1997). Nonetheless, CBA remains a central element in MSD.

CBA can be used to assess projects, programmes or policies. It is a decision tool that compares options in terms of present and future economic costs and benefits. There are many manuals of project assessment, setting out methods for CBA and warning against undue haste and narrow thinking (Bridger and Winpenny 1987; Brent 1990; OECD 1995). Key issues are the need to compare with- and without-project benefits and costs (doing nothing can sometimes prove an unexpectedly beneficial strategy), and the need to consider the opportunity costs of investments (that is, what the same investment could yield if spent on something different). A critical problem is time, since both benefits and costs tend to change as proposed projects develop and mature. Some projects can have high initial costs and slow payback (for example, pollution control technologies), or delayed payback (for example, afforestation). Time is conventionally dealt with by discounting future costs and benefits to calculate net present value. The 'discount rate' used is a critical figure, since it tends to cause projects to be favoured that yield benefits early and demand gradual investment. Environmental projects are often the reverse – expensive at the outset and yielding benefits slowly – and CBA tends to make them look unattractive. Furthermore, typical discount rates tend to give any cost or benefit that occurs more than thirty years in the future such a low value as to make it effectively irrelevant to the analysis. Thirty years may be a long time to a politician or a government planner, but it is little more than one human generation, less than a tenth of the lifespan of many trees, and an eye-blink in terms of evolution (Maser 1990). The discount rates conventionally used in CBA obviously raise significant issues as a basis for the discussion of sustainability.

There are serious questions about the extent to which it is possible to express social and environmental values in monetary terms in CBA calculations. For obvious reasons, many environmentalists and human-rights groups tend to distrust this aspect of CBA, for the way it reduces complex realities to a single metric, and transforms issues demanding ethical consideration to the single language of money. The question of intrinsic values of nature, and cultural and religious values, is highly problematic. Many techniques remain experimental and highly academic, and the whole field of valuation of ecosystem services is the subject of intense debate (Gouldner and Kennedy 1997).

However, if applied carefully, CBA has several useful things to offer. First, it provides a valuable discipline for gung-ho development planners. It can halt environmentally and socially destructive development that lacks even simple economic justification. Second, it can also provide a written record of the calculations that justified a project that is, at least in theory, available for subsequent scrutiny (without which any mistakes cannot be understood). Third, CBA should ensure that all externalities of a project are taken into account, including environmental and social impacts. Of course, the possibility of doing this is constrained by the problems of discount rates and the quantification of unquantifiable values discussed above; nonetheless, the structured approach of a CBA has some value. A clear distinction is drawn between economic and financial rates of return. Financial analysis measures the money costs and returns on a project (as a private corporation would do); economic analysis tries to measure the real or resource costs to the economy, which should include externalities (for example, health or environmental impacts), secondary and tertiary impacts, and questions of subsidy. Taking these into account, of course, makes project appraisal both harder and more expensive than it would otherwise be, but, in theory at least, a careful CBA will consider these things (or its failure to do so should be clear from the record).

Enthusiasts for CBA (mostly economists) argue that critics reject the technique too intemperately, and fail to see its potential. They argue that opponents of CBA criticize the technique of appraisal, when what is really at fault is the planning or political process that wields it. In this, CBA and environmental assessment (EA) techniques share the same weakness, as will be discussed in the next section.

The assessment of environmental and social impacts

A critical element in the formal assessment of the costs and benefits of projects is the ability to define, predict and manage environmental impacts. Environmental impact assessment (EIA) or environmental assessment (EA) techniques are used to insert environmental concerns into project and programme planning (Munn 1979; Wathern 1988; Barrow 1997). At their best, they could bring complex and unfamiliar environmental problems to the attention of planners as part of a holistic attempt to make development sustainable. At their worst, they can be little more than a short technical study carried out as an add-on to project planning simply to satisfy a bureaucratic requirement, and are completely ignored in project design. Which stereotype EA fulfils depends on the institutions and politics of the planning process and the individuals and organizations involved.

EA procedures are essentially qualitative, and therefore highly dependent on the skills, prejudices and perceptions of the analyst. There are a series of methodologies for determining weights or scores in a quasi-independent manner (Barrow 1997). Quantified methods may appear to confer enhanced legitimacy, particularly among a scientifically trained audience. However, as in the case of CBA, quantified evaluations in EA may be criticized because they make evaluation the preserve of the 'expert', remote from public comprehension and accountability. At the same time, they are attractive both to project developers wanting to present a clear case, and to politicians faced with difficult decisions. However, if EA procedures are used to palliate environmental concern without influencing project design, they are potentially deceitful and counterproductive: a waste of resources and a negation of their integrative and potentially transforming role.

Clearly, EAs are only as good as the policy frameworks within which they are carried out. The sequencing of tasks and the nature of the players among whom the EA will be created, assessed and acted on are particularly important (Munn 1979; Barrow 1997). Logically, national development goals define policies, programmes and projects in turn. Once an EA has been done, it needs to be reviewed by the competent body, the project implemented and a post-project audit carried out. C. J. Barrow (1997) identifies an ideal pattern of project planning as a 'helix', with environmental assessment taking place at both programme and project levels, its results being fed into ongoing cycles of planning (see Figure 6.3).

It need hardly be said that many of these steps are sensitive to critical pressures within government bureaucratic systems, and also that in many cases some or all of these procedures are skipped. It is, for example, one thing to have an EA commissioned and carried out, but quite another to integrate it into decision-making. The size of EA documents can be out of all proportion to the capacity

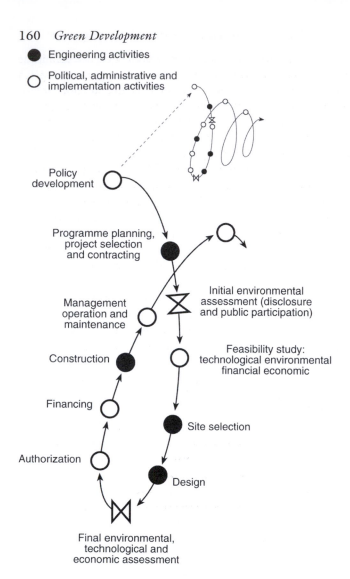

● Engineering activities

○ Political, administrative and
   implementation activities

Policy
development

Programme planning,
project selection
and contracting

Initial environmental
assessment (disclosure
and public participation)

Management
operation and
maintenance

Feasibility study:
technological environmental
financial economic

Construction

Financing

Site selection

Authorization

Design

Final environmental,
technological and
economic assessment

Figure 6.3 Environmental assessment and the project planning helix (after C. J. Barrow
         1997)

of agencies to digest data and reach informed conclusions. Indeed, the pursuit of
gigantism in an EA is one strategy open to the promoter of a major project who
is hopeful of drowning objectors in a flood of indigestible data.

The importance of carrying out assessments of both the long- and short-term
environmental impacts of projects in developing countries is clear (Ahmed and
Sammy 1985). However, a series of factors make EA potentially problematic in
these contexts.

One problem relates generally to the nature of environmental impacts themselves. In practice it is often far from easy to identify the nature of environmental

impacts at a particular point in space and time, let alone predict them. Significant impacts are often remote geographically from a project – for example, the impacts of logging or dam construction on sedimentation or river degradation down-stream. They may therefore be beyond the boundary of the specific development project, and therefore be unperceived by project developers. Such problems are compounded if, as in the case of downstream floodplain wetland environments, the place where the impacts occur is physically remote from centres of planning and decision-making, difficult to access and of marginal importance politically.

Environmental impacts can also be delayed, occurring some time after project development, and can involve complex sets of knock-on impacts – for example, ecological change in a floodplain following hydrological and geo-morphological change (see Chapter 11). Environmental impacts can also be increased by other developments or natural changes that have synergistic effects on ecosystems. A good example here is the impact of drought on the discharge of dammed rivers in the Sahel (Hollis *et al.* 1994).

There can also be problems knowing where to set the boundary for assessment of environmental impacts. The spatial and temporal scope chosen, and the range of development initiatives considered, can have significant effects on the outcome of an EA, as indeed on other aspects of project appraisal. For example, the contro-versial Sardar Sarovar Dam on the Narmada River in India made technical sense only if it were built to operate with the Narmada Sagar Dam upstream (to control inflows) and the smaller Omkareshwar and Maheshwar dams (Rich 1994). Without Narmada Sagar, the hydropower generation of Sardar Sarovar would be reduced by 25 per cent, and irrigation by 30 per cent. However, Narmada Sagar was expensive, would flood a large area of tropical forest and would force reset-tlement of an additional 100,000 people. The World Bank decided to appraise Sardar Sarovar as a discrete project and did not take account of its links with the other schemes (Rich 1994). The project's full costs were therefore not taken into consideration. Dam projects are particularly vulnerable to artificially narrow assessment in this way; for example, assessment of the viability of the Bakolori Dam and irrigation scheme in Nigeria specifically excluded any consideration of the considerable downstream impacts (W. M. Adams 1985a), and assessment of the Pangue Dam on the Bíobío River in Chile in the early 1990s ignored its dependence on other dam projects (Usher 1997b).

The technical challenge of assessing the nature and extent of environmental impacts is exacerbated by the enormous data hunger of EA procedures. Assess-ment of impacts demands large data-sets and long time-series, and such resources are unusual. The problem of 'planning without facts' (Stolper 1966) is familiar in developing countries. The problem is not simply the lack of data (although this is real enough), but a deeper ignorance of precisely what is not known. There are, of course, various technical tricks to extend short data-sets – for example, hydrologists correlate a short run of river discharge figures with a longer run of rainfall data and generate a synthesized data-set offering a long time-series. However, such methods have been devised for temperate environments, where the variability of the environmental conditions over time tends to be quite limited.

The enormous variability of rainfall in areas such as the Sahel makes it risky to extrapolate from short data-sets to derive longer-term averages, particularly if they include periods such as the 1960s (which were unusually wet compared to the long-term record (M. Hulme 1996)).

A lack of environmental data in developing countries is matched by a lack of expertise. Impact assessment demands knowledge of the dynamics of ecosystems, and an ability to make realistic predictions about their response to stresses of various kinds. Yet long-term ecological monitoring is unusual in developing countries, and experienced ecologists are few. There are few textbooks on ecology based on tropical examples (although see Deshmukh 1986). The usual response to the lack of indigenous expertise is to bring in outside 'experts' in the form of consultants or technical advisers. There may be many reasons for calling outside experts: the department in the host country may be empire-building; the civil servant may want a 'fall-guy', more time on his farm or access to foreign travel; or the experts may be the only way to obtain resources in the form of aid, coming as part of a tied aid package (Chambers 1983). Yet the insights to be derived from what Chambers (1983) calls 'rural development tourism' are limited. The employment of expatriate consultants and technical advisers can be effective, but it helps make planning remote, technocratic and untransparent.

Foreign experts may command a legitimacy denied prophets in their own country derived from their presentation of technology and experience, but this expertise may be less valuable than it at first appears. Thus Winid (1981) argued that the French consultants called in to plan development in the Awash Valley of Ethiopia lacked political, ecological, social, cultural, scientific and technical knowledge of the region, and in the field they had contact only with the local elite, lacked supervision and failed to transfer technology. There is a risk of foreign experts glibly repeating dominant preconceptions and myths about tropical ecosystems (for example, about overgrazing or desertification; see Chapter 8), which have misdirected development so often in the past (Leach and Mearns 1996).

The effectiveness of EA in project appraisal is affected by disciplinary bias inherent in the planning process (Chambers 1983). Project appraisal is dominated by the 'hard' technical disciplines such as engineering, hydrology and agronomy, and by the most technocratic of the social sciences, economics. Most project feasibility studies are undertaken by teams dominated by these disciplines, or by consortia typically dominated by engineering companies. The 'soft' disciplines such as ecology, geography, anthropology or sociology central to EA (and vital to social CBA) tend to be marginal to the planning process. Increasingly they have a place in project planning, but it is a small one. It is unusual for people from these disciplines to lead appraisal teams. Sociology, anthropology and ecology are typically slotted in with perhaps a one- or two-person-month input on a project where the total planning input is several person-years.

EA procedures can be costly, particularly if taken seriously. EA costs are sunk before it is known whether the project is technically, environmentally and economically viable, and will not be recouped if the project does not go ahead. The costs of EA need to be paid in foreign exchange, often financed by government loans that become part of national debt. EA therefore not only raises the cost of sunk investment in project appraisal, but also postpones the commencement of streams of possible benefits. There are obviously strong economic and financial incentives to truncate or speed up EA procedures, or even omit them altogether.

Towards sustainable projects

One way in which ideas from MSD have influenced development planning is through initiatives to broaden and strengthen techniques of project appraisal (Van Pelt *et al.* 1990). Alternatives proposed to CBA as a form of project appraisal (OECD 1995) include risk–benefit analysis, multicriteria analysis and decision analysis. Risk–benefit analysis effectively inverts CBA by looking at the risk of not taking action – for example, the risk of pollution if investment is not made in safety measures. Multicriteria analysis includes criteria in addition to quantifiable economic rate of return – for example, costs per beneficiary, or the number or characteristics of beneficiaries. Decision analysis drops the assumption in CBA that decision-makers are risk neutral, and seeks to assess their preferences, judgements and trade-offs. None of these will necessarily lead to development projects with smaller environmental and social impacts, but they do represent attempts to diversify and strengthen the range of tools available.

Three other approaches address sustainability more explicitly, strategic environmental assessment (SEA), integrated appraisal, and social impact assessment (SIA). SEA pushes the assessment of impacts upstream from the project, to the programme and policy arenas. Arguably, unless the environment is taken fully into account at the strategic level of planning, it will not be surprising if individual projects generate unforeseen and unwelcome impacts. This notion is not novel – it was, for example, a key recommendation of the *World Conservation Straegy* in 1980 (see Chapter 3) – but it is still not standard practice.

The idea of 'integrated appraisal' has several dimensions, including the need to bring different forms of appraisal together in project planning, to make sure they use consistent assumptions and to develop cross-disciplinary insights (Bond *et al.* 2001). In practice, 'weak' integrated appraisal involves the use of different forms of appraisal at different points in the project cycle, leaving the task of integrating them to the decision-making authority. In 'strong' integrated appraisal, the different forms of appraisal are integrated throughout. A key issue is the involvement of stakeholder groups in assessment and decision-making, something rarely achieved.

The most extensively developed attempt to broaden conventional EA processes is SIA (Vanclay and Bronstein 1995; Barrow 1997; Becker and Vanclay 2003). It is holistic, involving the analysis, monitoring and management of intended and unintended social consequences (both positive and negative), of planned interventions and any social change processes they bring about. Interventions can include policies, programmes, plans or projects. SIA seeks both to achieve better outcomes from specific developments in terms of ecological and social sustainability and to address the wider need for sustainability in development (Vanclay 1999).

This is likely to involve a participatory approach (working with communities and building their own capacities to plan), a recognition of the connections between social and biophysical impacts, and an explicit consideration of second- and later-order impacts (Vanclay 1999).

Unlike EIA, SIA seeks to address not only social impacts arising from biophysical impacts, but also those generated through social change that can both follow from and be created by planned development interventions. Thus people forced to resettle as a result of dam construction may be affected by a complex chain of social and environmental impacts associated with loss of resources, changed disease incidence, conditions in a resettlement location and social change resulting from translocation (loss of social cohesion, respect for elders, traditions or religious belief). SIA attempts to recognize and consider all these. It could be used as part of a state's regulatory process and address a single project, but it could also be used by communities or other actors as a means not only to appraise, but also to steer development. In that sense, SIA can be seen as a philosophy or paradigm rather than simply a methodology (Vanclay 1999).

The ability of improved project and programme appraisal methodologies to contribute to sustainability depends crucially on institutional capacity and governance. This capacity has been increasing, particularly since the Rio Conference in 1992, with the support of the GEF and the World Bank's insistence on National Environmental Action Plans (NEAPs). Provisions for assessing environmental impacts had become almost universal by the mid-1990s (Barrow 1997), although in many countries capacity to implement them still remained weak.

Aid agencies and environmental policy

In the absence of effective governance and environmental regulation, the effectiveness with which proper account is taken of sustainability depends to a considerable extent on the capacity of the aid donor organizations that fund so much development. The performance of both multilateral and bilateral aid agencies began to receive attention in the 1970s and 1980s (Stein and Johnson 1979; Johnson and Blake 1980; V. W. Kennedy 1988). Since that time there has been growing pressure by environmental groups for reform of aid agency environmental policy, particularly in the USA, aimed at both USAID itself and the World Bank, whose income is dominated by the size of the US contribution.

The key to the actual level of consideration of environmental aspects of project development in any aid bureaucracy is the attitude and perception of particular individuals within the normal planning structure. Bureaucracies such as aid agencies are extremely difficult to reform from within. There is often a wide gap between 'the increasingly alert concern of individuals and the official response of most institutions' (Stein and Johnson 1979, p. 133). In the 1970s and early 1980s this was a problem for the World Bank, despite its surprising record of employing staff popularly identified with the environmentalist cause such as Robert Goodland and the economist Herman Daly (Holden 1988). The World Bank (or more properly the International Bank for Reconstruction and Development and the

International Development Association, which are separate arms of the Bretton Woods family of organizations) is the largest multilateral aid donor. The sheer size of its lending makes its environmental policy of considerable importance, and of consuming interest to environmentalists (Fox and Brown 1998b).

In the 1970s, the centre of the World Bank's evaluation procedures was the Office of Environmental and Health Affairs (OEHA). In practice this had little influence on project design. It was represented on few World Bank project identification missions, and mission reports suffered from the constraint of 'paying almost unique attention to financial and economic considerations', rarely discussing environmental issues in enough detail (Stein and Johnson 1979, p. 17). While in theory project appraisal should have included environmental aspects of the project, in practice few projects were rejected on environmental grounds alone, although projects were seen by the OEHA before going to the Loan Committee of vice-presidents of the Bank and then the executive directors. By that time it was usually too late to redesign projects, and difficult to build in environmental safeguards. The pressure to keep projects on schedule was intense, and delays most unwelcome.

There is a danger that 'mavericks' within large bureaucratic organizations like the World Bank are 'good for the image but mouse-sized in impact' (Watson 1986, p. 275). As an ecologist working for the Bank, Watson found her ideas rejected as unrealistic and impractical every time they contained implications for practice. Similarly, environmental advisers with NORAD (the Norwegian aid agency) who were critical of dam projects on environmental grounds found their advice ignored. Usher (1997c) suggests that environmental and social issues are obstructions to the process within the aid bureaucracy to 'push development projects through the "pipeline"' (p. 69). Watson's experience (1986) was similar: she commented, 'by and large we were viewed by project and technical staff as an unessential office which at best put useless icing on the cake and at worst could halt or slow projects' (p. 269).

In 1984 the World Bank adopted its first official statement on environmental aspects of its work, an Operational Manual Statement (OMS). This drew attention to guidelines to be followed in Bank operations unless the borrowing country's standards were more strict. These principles include, for example, the provision that the Bank would not finance projects that 'cause severe or irreversible environmental degradation', that would affect the environment of neighbouring countries, or that would 'significantly modify' biosphere reserves, national parks or other protected areas (World Bank 1984c, p. 4). It was also stated that 'project designers, consulting firms and Bank project officers are expected to provide effective and thorough environmental input into project design construction and operation' (Goodland 1984, p. 13). With the World Environment Centre, the Bank organized a series of seminars for major consulting firms in the early 1980s to raise awareness of the need for proper environmental appraisal of projects (Goodland 1990).

Environmental groups maintained their critical pressure on the Bank, particularly US environmental NGOs, especially the National Wildlife Federation, the Sierra Club and the Environmental Defence Fund (e.g. Goldsmith 1987; Horowitz 1987; Fox and Brown 1998a). The US Congress held oversight hearings on the multilateral development banks in 1983, hearing in particular severe criticism of the World Bank's funding of Brazil's Polonoroeste Programme (Northwest Region Development programme), and the construction of Highway 364 into the heart of the remote rainforest region (Rich 1994). The barrage of environmental criticism had some effect. In 1987 the Bank created an Environment Department with forty new staff, and new scientific and technical staff in regional offices (Holden 1987). Behind these organizational changes lay some softening of the rigid doctrines of Bank economics, with the appointment to the Latin American office of the zero-growth economist Herman Daly. New systems of national accounting were developed to reflect the depletion of non-renewable resources and unsustainable exploitation of renewable resources, and to revise the discount rate, which biased appraisals in favour of projects with short-term pay-offs against longer-term cost/benefit considerations (Holden 1988).

The Bank also produced a series of policy papers in the 1980s addressing issues central to sustainability, including an overall environmental policy statement and papers on involuntary resettlement, wildlands conservation, pollution control and pesticides, and tribal people. These codified existing best practice, and had already existed in draft for several years. Such policy statements were valuable, but they did not change corporate culture. Goodland (1990) noted that 'their full implementation by the Bank's disparate staff of 6000, with many urgent priorities, cannot be achieved overnight, only through time' (p. 151).

By the early 1990s, environmental critique of the Bank had developed from specific issues (often concerning rainforests, roads or dams) into a broader attempt to make the Bank accountable to civil societies in donor and borrowing countries (Fox and Brown 1998b). The Bank's need for donor government contributions to the International Development Association every three years provided a cyclic window of opportunity for NGO pressure. The Bank instituted a range of reforms in the early 1990s, strengthening its policies on EA, involuntary resettlement and indigenous peoples. However, practice was publicly shown to be behind policy, particularly in the case of the Sardar Sarovar Dam on the Narmada River in India (discussed above). The Bank's board of directors set up a review of this project in 1991 under Bradford Morse, former director of UNDP (J. A. Fox 1998). The Morse Commission concluded in 1992 that the Bank had flouted its own environmental and resettlement policies, and recommended that it 'step back' from the project, which it eventually did (to no avail, since construction continued despite a storm of protest; see Chapter 13). The Bank also cancelled a proposed loan for Nepal's Arun II Dam (Usher 1997c). A review of resettlement in Bank projects followed directly from the Morse Report, recommending changes already proposed internally by Michael Cernea (J. A. Fox 1998). In 1993 a new water resources management policy was finally approved, including the stipulation that 'environmental protection and mitigation' would be integral parts of a comprehensive approach to water development (Moore and Sklar 1998).

The World Bank transformed its policies on the environment and sustainability in the 1990s, even creating a vice-presidency of environmentally sustainable development, and leading the development of economic ways to factor the environment into economic thinking (e.g. Munasinghe 1993b). Nonetheless, the Bank is introverted and narcissistic, learning primarily from itself and systematically ignoring outside knowledge, especially local knowledge on the ground (Goldman 2005). Its practice still falls short of its promises of reform (Fox and Brown 1998b).

Other multilateral and bilateral donors have not received the same level of environmental criticism as the World Bank, but most have felt the same pressures for reform. Most have adopted environmental appraisal procedures of some kind. The Asian Development Bank established an environmental unit in the early 1980s, and the UK's Department for International Development (DFID, formerly the Overseas Development Administration (ODA)) produced a manual of environmental appraisal in 1989 (Barrow 1997). The 1997 UK White Paper on international development explicitly addressed the challenge of sustainable development, committing the DFID to the promotion of 'sustainable livelihoods' and protection and improvement of the 'natural and physical environment' (Carney 1998).

Learning for change

The message of MSD, driven forward by ritualistic respect for the discipline of economics in a neo-liberal policy world, has penetrated governments, donors, business and the professional worlds of project planners. Like governments, the World Bank and other major donors changed their tune on sustainability in the 1980s and 1990s. To a variable (but mostly lesser) extent, they also began to change their practices. The degree of this change is critical, because of the enormous influence of donor organizations on planning in many Southern countries. However, government, donor and corporate business organizations are large and bureaucratic: hierarchical, autonomous and populated by people jealously aware of their professional skills and their established procedures. How can such organizations change?

Fox and Brown (1998b) distinguish between institutional adaptation and learning. If external political pressure is sufficiently strong, changes in organizational behaviour (adaptation) may take place without learning (changes in the way problems are perceived and explained). At the same time, learning can take place without adaptation, if staff lack the power to change organizational behaviour. Organizational learning that threatens dominant paradigms (for example, the hegemony of neo-classical economics) is likely to provoke resistance. However, over time, learning will take place, as a result of recruitment and staff training, the penetration of new ideas (for example, environmental and ecological economics) and changing internal institutional incentives. External pressure can also lead to the release into positions of influence of innovators who *have* learned. It is an interesting question, therefore, whether aid donor 'greening' is the fruit of internal learning or simply adaptation to external pressure from NGOs.

Fox and Brown (1998b) believed that World Bank staff did more adapting than learning.

Learning is not a challenge confined to the bureaucracies of aid organizations and governments. The evolution of thinking within the professional disciplines involved in development planning is also very important. The infiltration of environmental (and even ecological) economics into the economics faculties of less conservative universities represents a similar evolution (and in the case of ecological economics a potential revolution) within disciplinary boundaries.

Neither a general intention to improve environmental appraisal nor a superficial awareness that environmental impacts may occur is enough to guarantee that development will become sustainable, or that aid agencies will take the steps necessary to control the impacts of the developments they fund. Large bureaucracies are inherently conservative, and the 'greening' of development is bizarre theoretically (in the context of the established disciplines of development planning), troublesome in terms of policy and highly inconvenient administratively. Reforming the practice of development bureaucracies has to go a great deal deeper than the superficial transformation of rhetoric and terminology.

MSD commands an increasingly sophisticated range of methodologies that do much to make the environment a normal and integral issue in development planning. However, such managerialism does not by any means solve all the problems of sustainability. Behind the technical certainties of the mainstream lie other issues. These are addressed by a range of other ideas about sustainability – countercurrents to the mainstream of sustainable development. They are the subjects of the next chapter.

Summary

- A key element in mainstream sustainable development (MSD) is the development of environmental economics. Critical concepts include the distinction between natural and human-made capital. Natural capital includes stocks from which benefits flow, but which are not the product of technology or human action. 'Strong' sustainability demands that stocks of each are maintained; 'weak' sustainability allows some trade-off between natural and human-made capital.
- Ecological economics takes explicit account of relations between economic systems and ecosystems. An important concept here is 'cultural capital', relating to the institutions that regulate human use of the environment, a question also addressed by 'new institutional economics'.
- The delivery of MSD is being addressed through a number of policy initiatives, including the adjustment of national economic accounts to internalize the environment. Attempts to factor the environment into economic thinking faces significant problems of trade-offs between different people and interests in space and time, and in predicting future behaviour of environmental systems.
- The assessment of the social and environmental impacts of development

projects is an important element in promoting sustainability. Environmental assessment of projects faces numerous problems relating to the complexity of environmental responses to human action (for example, timescales, spatial boundaries), the difficulty of knowing about those impacts (lack of data or researchers), the nature of the organizations carrying out assessments (for example, their dependence on foreign expertise), haste and cost.

- Broader project appraisal methodologies have been developed, including social impact assessment (SIA). SIA addresses all social impacts, including direct impacts, those resulting from environmental impacts and the chain of impacts resulting from them.
- Aid donor organizations have a particular role in transforming the process of project appraisal. Most have 'greened' to some extent. The World Bank adopted a series of new policies – for example, on the environment, indigenous people and resettlement in the early 1990s. The process of institutional change, in both aid organizations and the academic disciplines from which their staff are recruited, is slow.
- Behind the technical virtuosity of MSD lies a fairly large degree of uncertainty. Alongside mainstream thinking flow other, more challenging and radical currents.

Further reading

Atkinson, G., Dietz, S. and Neumayer, E. (2007) (eds) *Handbook of Sustainable Development*, Edward Elgar, Cheltenham.

Barbier, E. B. (1998) *The Economics of Environment and Development: selected essays*, Edward Elgar, Cheltenham.

Barrow, C. J. (1997) *Environmental and Social Impact Assessment: an introduction*, Arnold, London.

Barrow, C. J. (2000) *Social Impact Assessment: an introduction*, Arnold, London.

Becker, H. A. and Vanclay, F. (2003) (eds) *The International Handbook of Social Impact Assessment; conceptual and methodological advances*, Edward Elgar, Cheltenham.

Berkes, F. and Folke, C. (1998) (eds) *Linking Social and Ecological Systems: management practices and social mechanisms for building resilience*, Cambridge University Press, Cambridge.

Brouwer, R. (2005) (ed.) *Cost–Benefit Analysis and Water Resources Management*, Edward Elgar, Cheltenham

Cernea, M. (1991) (ed.) *Putting People First: sociological variables in rural development*, Oxford University Press, Oxford, for the World Bank.

Costanza, R. (1991) (ed.) *Ecological Economics: the science and management of sustainability*, Columbia University Press, New York.

Daly, H. E. and Cobb, J. R. (1989) *For the Common Good: redirecting the economy towards community, the environment and a sustainable future*, Beacon Press, Boston, MA.

Dresner, S. (2002) *The Principles of Sustainability*, Earthscan, London.

Fox, J. A. and Brown, L. D. (1998) (eds) *The Struggle for Accountability: the World Bank, NGOs and grassroots movements*, MIT Press, Cambridge, MA.

Mol, A. P. J. (2001) *Globalization and Environmental Reform: the ecological modernization of the global economy*, MIT Press, Cambridge, Mass.

Neumayer, E. ([1999] 2003) *Weak versus Strong Sustainability: exploring the limits of two opposing paradigms*, Edward Elgar, Cheltenham (2nd edition).

Purvis, M. and Grainger, A. (2004) (eds) *Exploring Sustainable Development*, Earthscan, London.

Redclift, M. R. (2005) *Sustainability: critical concepts in the social sciences*, Routledge Major Works (four volumes), Taylor and Francis, London

Rich, B. (1994) *Mortgaging the Earth: the World Bank, environmental impoverishment and the crisis of development*, Earthscan, London.

Usher, A. D. (1997) (ed.) *Dams as Aid: a political anatomy of Nordic development thinking*, Routledge, London.

Vanclay, F. and Bronstein, D. A. (1995) (eds) *Environmental and Social Impact Assessment*, Wiley, Chichester.

World Commission on Dams (2000) *Dams and Development: a new framework for decision-making*, Earthscan, London.

Web sources

http://web.worldbank.org The World Bank: look under 'Topics' for environment, sustainable development, social development and Millennium Development Goals.

http://www.iaia.org/ The International Association for Impact Assessment: exists to advance innovation and development and communication of best practice in impact assessment.

http://www.dams.org Archive site of the World Commission on Dams: offers some carefully considered approaches to the planning, implementation and monitoring of controversial large projects such as dams.

http://www.unep.org/dams/ UNEP Dams and Development Project.

http://www.ecoeco.org/ The International Society for Ecological Economics (ISEE), an organization linking ecological economists and promoting research on and understanding of the relationships among ecological, social and economic systems.

http://www.paecon.net/ The Post-Autistic Economic network, the causes and implications of the academic monopoly of 'mainstream' neo-classical economics.

http://www.footprintnetwork.org/ The Global Footprint Network: data on global and national footprints.

http://www.factor10-institute.org/index.htm The Factor 10 Club, proposing tenfold improvements in resource and energy efficiency.

7 Countercurrents in sustainable development

Clearly eco-software will not save the planet if capitalist expansionism remains the name of the game.
(Martin Hajer, 'Ecological modernization as cultural politics', 1996)

Sustainable, ecologically sound capitalist development is a contradiction in terms.
(David Pepper, *Eco-Socialism*, 1993)

Beyond the mainstream

The philosophical basis of the environmentalism of the 1970s was complex, eclectic and confused. It combined modernism and anti-modernism, both a call for a better science and a critique of the rationality of science (Sachs 1992b). Sustainable development is the uncertain inheritor of this confusion. As we have seen in Chapters 5 and 6, mainstream sustainable development (MSD) has begun to acquire the intellectual scaffolding necessary to translate rhetoric into practical policy. The discipline of economics has furnished bridges between normal practice in development planning and concerns for environment. Environmentalism and human rights have been factored into the business spread-sheets of 'Earth plc', enabling trading and planning to continue very much as normal (Pearce 1992).

MSD has, therefore, transcended the uncomfortable claims of environmentalists and critics of development. The Rio Conference epitomized this mainstreaming of sustainable development. As was discussed in Chapter 4, some commentators saw Rio as a sell-out on critical environmental and development issues, arguing that the non-governmental movement was co-opted to a process that ultimately worked against its interests (Chatterjee and Finger 1994). Non-governmental organizations (NGOs) were invited in, indeed sucked in, by the lure of influence and by generous funding, to lend their support to the 'Rio process' – 'fed into the green machine', as Chatterjee and Finger (1994, p. 79) put it. While a small group of mostly US-based NGOs had some influence at Rio, most NGOs found the experience disorientating and disappointing.

The version of sustainable development expounded at Rio and Johannesburg demanded no radical changes in the relations between rich and poor countries, no systematic reorganization of the control of resources, no reining back of consumption of non-renewable or renewable resources that might harm the delicate constitution of the juggernaut of the world economy. As Katrina Brown (1997) suggests, 'Rio and the developments since have reaffirmed the South's suspicion that the North is simply not prepared to redefine the international division of labour or its economic, social and political relationship with the rest of the world' (p. 388). More generally, the United Nations Conference on Environment and Development (UNCED) reaffirmed the unwillingness of the international community to question the nature and direction of development or to consider an alternative to the dominant development paradigm. This is not surprising, for the inertia of the established models of development is very great. A transition from non-sustainable to sustainable development would inevitably create winners and losers (K. Brown 1997). Recasting of the global power game has not been part of MSD.

More radical voices have, however, been raised about environment and development. There are countercurrents within the mainstream of sustainable development that offer a significant critique of the dominant model. Thus, Martin Lewis (1992) outlines five schools of 'extremist' thought (anti-humanist anarchism, primitivism, humanist eco-anarchism, green Marxism and radical eco-feminism) in the radical environmental movement. He argues that they all reject representative democracy, and respond to US government institutions with contempt. Against these variously demonic 'extremisms' he makes a case for MSD, 'a new alliance of moderates from both left and right' (p. 250), building on the judicious use of market mechanisms.

Radical green ideas have not been completely swept away by the rising tide of MSD, and still deserve serious attention. Environmental organizations pushed radical approaches at the very meetings that codified the mainstream. Thus at Rio in 1992, a consortium of NGOs (including Greenpeace International, Friends of the Earth International and the Forum of Brazilian organizations), set out a '10-point plan to save the Summit' (Chatterjee and Finger 1994; see also Table 7.1). Few of these issues were addressed at Rio, and few subsequently found a place in the mainstream of sustainable development.

There are significant eddies within the mainstream ideas and differing and sometimes conflicting versions of ecological modernization. Thus Christoff (1996) distinguishes 'weak' ecological modernization, which is economistic, narrowly technical and focused within national boundaries, from 'strong' ecological modernization that is ecological, systemic and international (see Table 7.2; note the parallel with weak and strong sustainability; see Chapter 6). Spaargaren and Mol (1992) identify a more radical programme beyond the conventional conservative approach of compensation/impact minimization and that of classic ecological modernization (clean technologies, valuation of environmental resources and transformation of patterns of production and consumption). Their third programme involves 'a progressive dismantling or deindustrialization of the economy', and the transformation of industrial structure into small units that link production and consumption more closely (p. 339). This approach, so

Table 7.1 NGO ten-point plan to save the Summit

1	Set legally binding targets and timetables for reduction in greenhouse-gas emissions, with industrialized countries leading the way
2	Cut Northern resource consumption and transform technology to create ecological sustainability
3	Reform the global economy to reverse the South–North flow of resources, improve the South's terms of trade and reduce its debt
4	End the World Bank's control of the Global Environment Facility (GEF)
5	Regulate transnational corporations and restore the UN Centre on Transnational Corporations
6	Ban exports of hazardous wastes and dirty industries
7	Address the real causes of forest destruction, since planting trees as UNCED proposes cannot be a substitute for saving existing natural forests and the cultures that live in them
8	End nuclear weapons testing
9	Establish binding safety measures – including a code of conduct – for biotechnology
10	Reconcile trade with environmental protection, ensuring that free trade is not endorsed as the key to achieving sustainable development

Source: Chatterjee and Finger (1994, p. 40).

popular in the neo-Malthusian heyday of the 1970s, remains a potent element in more radical thinking about sustainable development. Thus, for example, ideas of this kind are integral to radical thinking about responses to the threat of rapid anthropogenic climate change – for example, in George Monbiot's *Heat: how to stop the planet burning* (2007). They are important elements, too, in calls for alternative patterns of production and consumption such as Lester Brown's *Plan B* (2006).

This chapter looks at some of the more radical ideas that run counter to the conformity of MSD. It considers green critiques of development, eco-socialism, eco-anarchism, deep ecology and eco-feminism. This is an admittedly eclectic list. It reflects the way that political thinkers of many persuasions have slowly begun to engage with problems of environment and development. It is also the result of the tendency for environmentalists, ill-informed about political ideas, to pick up fragments of ideologies that catch their eye. Environmentalists talking politics are

Table 7.2 Weak and strong ecological modernization

Weak ecological modernization	*Strong ecological modernization*
Economistic	Ecological
Technological (narrow)	Institutional/systemic (broad)
Instrumental	Communicative
Technocratic/neo-corporatist/closed	Deliberative democratic/open
National	International
Unitary (hegemonic)	Diversifying

Source: Christoff (1996, p. 490).

often magpies with a hoard of shiny ideas stolen out of context, and which they may not fully understand.

Green critiques of developmentalism

In his book *Green Political Thought*, Dobson ([1990] 2007) distinguishes between environmentalism and ecologism. The former, he argues, refers to a managerial approach to environmental problems: equivalent to the MSD discussed in Chapter 5 of this book. What he calls ecologism holds that sustainability demands radical changes to human relations to the non-human world and to social and political life. It is ecologism that is explored in this chapter, although the distinction between these concepts is not made by all authors, and both terms are used here. Dobson is at pains to emphasize that ecologism and environmentalism are quite different – and different in kind as well as degree. Environmentalism is not an ideology. It does not provide an analytical description of society, it does not prescribe a particular form of human society based on beliefs about the human condition and it does not prescribe a coherent programme of political change. Ecologism, or radical environmentalism, does these things.

Critiques of the standard Western model of industrial development have been important in Western environmentalist thought. Attitudes to industrialization began to shift in late Victorian England as the new industrial system began to look 'less and less morally and spiritually supportable' (Wiener 1981, p. 82). In Britain an eighteenth- and nineteenth-century tradition of romantic protest at industrialization and large-scale organization may be traced through the twentieth century to the Campaign for Nuclear Disarmament and the activism of the early Greens (Veldman 1994). At first changing ideas were linked to the rise of Victorian natural history focused on the loss of species and habitats, and to pastoral visions that contrasted a threatened countryside and dark satanic mills and expanding cities. However, the environmentalists of the 1960s and 1970s moved far beyond their predecessors' concerns for nature and countryside to argue for far-reaching social, economic and political change: 'they condemned not only the environmental degradation but also the society that did the degrading' (Veldman 1994, p. 210). Opposition to industrialism, or at least to its particular local manifestations in polluted or destroyed ecosystems, was a major element in the new environmentalism in other industrialized countries, notably the USA (Hays 1987; Guha 2000), as it became in later industrializers such as India (Gadgil and Guha 1995).

David Pepper (1996) argues that environmentalism is more than a localized romantic opposition to industrialism, but a rejection of modernism itself – that is, a rejection of the whole project of science, technology and organization that was ushered in by the Enlightenment of the eighteenth century. He points out that Greens not only offer a critique of industrialization's claim to control nature (stressing instead the prevalence of high-technology risks), but also express mistrust of the grand political theories of the 'modern' period, liberalism and socialism. Such critiques are broad. They underlie the ideas of the

German sociologist Ulrich Beck (1992) about risk society (see Chapter 12), while from a different (postcolonial) perspective Escobar (2004) argues that modernity is compromised by its increasingly obvious failure to provide answers to modern problems. In understanding how to move beyond ideas of development, or the Third World, it is therefore necessary to look beyond interpretations of modernity as globalization of European thought, to marginalized (subaltern) knowledges and practices, new forms of social action and networked social movements.

Some analysts have tried to set out a green alternative to both capitalism and Marxism. Against the 'Blue' (market, liberal, capitalist) and 'Red' (state, socialist) strategies a 'Green counterpoint' could be identified that opposes the institutionalization of the 'modern complex' (that is, bureaucracy and the industrial, urban, market and techno-scientific systems, and the military-industrial complex (Friberg and Hettne 1985, p. 207)). This green position is obviously a hybrid, incorporating elements of romanticism, anarchism and utopian socialism, but arguably its commitment to a just world order means that it cannot be interpreted simply as nostalgic conservatism.

Green political ideas are therefore marked out by this opposition to the conventional political strategies of both left and right, and the industrialism and consumerism that support them. Thus in *A New World Order*, Paul Ekins (1992) identified a single systemic 'global *problématique*' of great complexity, comprising four interlinked global crises: militarization, poverty, environmental destruction and human repression. This *problématique* was maintained by modern technology, world capitalism and state power, all the fruits of the modern project. They were in turn sustained by three forces: 'scientism' (an exclusive trust in the scientific worldview), 'developmentalism' (a belief that economic and human progress depends on an expanding consumer society) and 'statism' (a belief that the nation state is the ultimate form of political authority) (Ekins 1992, p. 207). Such an analysis suggests very different strategies from the incremental technocentric ecological rationalization that epitomizes MSD.

Radical green critics of development reject the possibility that capitalism can deliver just, equitable, humane and sustainable conditions of human life. Addo *et al.* (1985) argue that

> the bankruptcy of dominant development models, the deterioration of living conditions virtually everywhere, the sharpening of conflicts within and between nations, and the destruction of the foundations of existence should overwhelm the illusion held for so long of the possibilities of developmental transformation within the capitalist world-system.
>
> (p. 2)

Friberg and Hettne made a classic statement of a green critique of the dominant development paradigm in 1985. The processes that contribute to the developmentalist world system are modernization, economic growth and nation-state building. Its core metaphors are progress, growth and development (Aseniero 1985, p. 51). Now that the 'market' triumphalism of the end of the cold war has turned sour, it is perhaps more obvious that capitalism and state socialism were two varieties of a common developmentalist corporate industrial culture based on the values of 'competitive individualism, rationality, growth, efficiency, specialisation, centralisation and big scale' (Friberg and Hettne 1985, p. 231).

The evolutionary assumptions of 'developmentalism' imply that development is directional and cumulative, that its path is predetermined and that it is by its nature progressive (Sachs 1992a; Cowen and Shenton 1995; Watts 1995; see also Chapter 1). Peet and Watts (1996) describe development as 'modernity on a planetary scale' (p. 19). By the end of the nineteenth century the concept of development had brought about a pervasive ordering of ideas, drawing on 'universal' concepts of science, linearity, modernization and progress, which 'carried the appeal of secular utopias constructed with rationality and enlightenment' (Cowen and Shenton 1995, p. 19). In the twentieth century, the ideology of development reflected the desire of colonial and postcolonial states to control territory, ecology and subjugated people (Scott 1998; Drayton 2000; Demeritt 2001). After 1945 development also provided the discursive and practical strategies necessary to negotiate the end of European colonialism, and the rise of US political and economic power and of neo-liberalism.

Critique of these ideologies and their policy outcomes has been a fundamental element in postcolonial analysis of development (e.g. Radcliffe 2005a). It has also been a feature of green critiques of development and developmentalism. Thus Friberg and Hettne (1985) reject the evolutionary notion of progress. People make development, and human agency is therefore a decisive element within the process. 'So-called developed societies' are therefore neither model nor forerunner for those yet to industrialize (p. 219).

In place of evolutionary developmentalism, Friberg and Hettne (1985) propose a green 'endogenous development' based on communitarianism (with development rooted in the values and institutions of a culturally defined community); self-reliance (at different scales within society, not autarky or national self-reliance); social justice; and 'ecological balance' (implying an awareness of local ecosystem potential and local and global limits). Development is therefore to be sought in each country's own ecology and culture, not in the supposed 'model' of a developed country. Furthermore, as development is to be through 'voluntary cooperation and autonomous choices by ordinary men and women', the unit of development is not the state, but people and groups of people defined by culture (p. 221). Their notion of social justice goes beyond the established idea of redistribution with growth, and monetary compensation for marginalization and alienation, to embrace access to wealth, knowledge, decision-making and meaningful work.

From this green perspective, the modern world system, dominated by capitalism and embarked upon first in Western Europe, has been imposed upon the periphery through geographical expansion and socio-economic penetration in association with colonialism. It aimed at control, expansion, growth and efficiency, and was legitimized by evolutionist thinking. The logic of the modern project is eventually to eradicate all pre-capitalist social formations through continued modernization in an expanding world economy, whether through the 'Blue road' of capitalism or the 'Red road' of socialist world government. In contrast, Friberg and Hettne (1985) propose a green strategy of 'demodernization' (see Figure 7.1). This would involve gradual withdrawal from the modern capitalist world economy and the launch of a 'new, non-modern, non-capitalist development project' based on the 'progressive' elements of pre-capitalist social orders, plus their successors, avoiding the exploitative and dehumanizing aspects of some small-scale pre-capitalist societies (p. 235).

The green project demands circumvention (or in some cases subversion) of the nation state (which they see as 'one of the greatest obstacles to "the Greening of the world"' (Friberg and Hettne 1985, p. 237)) and the elites that sustain the modern project. They drive increasing involvement in the world economy to increase their access to wealth, and at the same time foster the political, military and bureaucratic development of the nation state as a power base. Only counter-mobilization by 'counterpoint movements', non-party politics, spontaneous networks and voluntary organizations can stand against them. Friberg and Hettne (1985)

The 'Red' or 'Blue' road of continued modernization

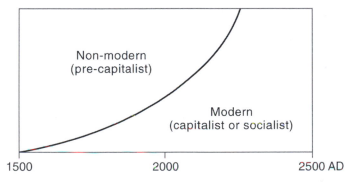

The 'Green' road to demodernization

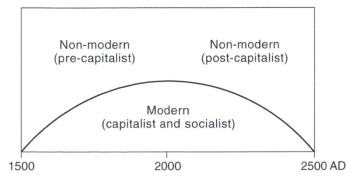

Figure 7.1 Friberg and Hettne's 'Green' project (after Friberg and Hettne 1985)

identify three sources of such a counterpoint. First, there are 'traditionalists' who wish to resist capitalist penetration in the form of state-building, commercialization and industrialization (including non-Western civilizations and religions, old nations and tribes, local communities, kinship groups). Second are 'marginalized people', including unemployed people, those who are mentally ill, handicapped people, and people in dehumanizing jobs who have lost 'a meaningful function in the mega-machine' (p. 264) through pressures for increased productivity, rationalization and automation. The third group consists of the 'post-materialists' who dominate Western environmentalism (Cotgrove 1982), the mostly young and well-educated members of 'the ecological, solidarity, peace, feminist, communal, regional, youth, personal growth and new age movements' (Friberg and Hettne 1985, p. 248).

Reactions to capitalist penetration can be violent and revolutionary, but most developing world counterpoint movements have been small scale and concerned with specific issues of protest. Friberg and Hettne (1985) argue that alternative movements, from the 1950s anti-bomb protests onwards, have tended to converge towards a small-scale logic of 'functionally integrated communal societies based on direct participation and self-management' (p. 258), in which capitalist growth, central bureaucracy and 'techno-science' have a small place. Arguably these localized movements can coalesce and challenge the power of the state. There has been increasing discussion of the political significance in development of new social movements, and the effectiveness of the 'weapons of the weak' (Scott 1985; Ekins 1992; Ghai and Vivian 1992a). Thus Ekins (1992), for example, believes that 'another development' (a phrase coined by the Dag Hammarskjöld foundation in 1975) can arise out of networks of resistance against exploitation and struggles for justice represented by grass-roots activism. Thus the waning effectiveness of bureaucratic authoritarianism in Latin America has made space for new forms of social action in pursuit of equity and empowerment. New popular movements, operating within civil society rather than through armed struggle, can be widely identified – for example, in Mexico, Ecuador, Bolivia and Peru and Brazil (Vanden 2003). Manuel Castells (2000) urged the potential for social movements to provide the engine for social change through new forms of political engagement in the contemporary Information Age.

Issues of democratic decentralization (Ribot and Larson 2005) and environmental democracy (Mason 1999) are central to the questions of the politics of environment and development. Environmental movements in the developing world have often stemmed from actions in defence of specific resources such as land, water or forests against the claims of wealthy elites and corporations (Dwivedi 2005). However, the power of local grass-roots organizations has increasingly been exercised through their capacity to network internationally and represent their concerns within global arenas: the Internet, and the possibility of accessing international summit meetings, are critical to this connectivity. The importance of social action to pursue sustainability will be discussed further in Chapter 13.

There is no single grand theory of green development to compare with Marxism or capitalism. However, there is a persistent core of ideas that comprise a critique of conventional modernism and developmentalism. Some of these ideas are echoed within MSD (for example, in their engagement with neo-populism, or their concern for equity and justice); others are beyond the pale. Many of these more radical green ideas are, however, strongly reflected within socialist thought, in a set of ideas referred to as 'eco-socialism'. These are discussed in the next section.

Eco-socialism and sustainability

The attempt by Friberg and Hettne (1985) to demonstrate the distinctiveness of green and red development strategies (see Table 7.3) is not entirely successful. It depends on a simplistic and one-sided caricature of 'Red', socialist or Marxist thinking. Having pointed out the existence of green strands within radical thought (for example, utopian socialism), they proceed to ignore these in claiming the distinctiveness of green ideas. They also more or less dismiss the labour movement as an element in their 'counter-structure', arguing that it has become 'wedded to the large-scale industrial system' and incorporated into the formal-rational complex (pp. 253, 259). They argue that the labour movement has lost its momentum as a creative force and itself reproduces the features of the modern system. This dismissal of socialism is unhelpful, for eco-socialism is, in fact, both diverse and more interesting in its engagement with green critiques of developmentalism.

There are strong traditions of political thought poised between green and red. Thus, in Germany, Rudolf Bahro occupied a distinctive position on this boundary,

Table 7.3 Comparison of Red and Green strategies of development

	Red strategy	*Green strategy*
Oppressive system	The capitalist economy	The technological culture
Enemy	Capitalists	Technocracy
Social vision	Socialist society	Communal society
Method	Macro-revolution, socialism from above by working-class revolution	Micro-revolution, small groups withdraw from system to defend autonomous ways of life
Spatial frame	No socialist islands in a capitalist sea	Local experiments, liberated zones
At stake	Material interests, collective identity, ownership	Existential needs, personal identity, autonomy
The new person	Social transformation before personal	Simultaneous personal and social transformation
Leadership	Intelligentsia	Post-materialistic elite
Social base	Working class	Marginalized people
Institutional base	Big industrial sites	Small neighbourhood communities
Organization	Centralized, formal	Decentralized, informal
Ideology	Abstract, rational	Concrete, intuitive, open

Source: Friberg and Hettne (1985).

both personally and intellectually. He left East Germany after the publication of *The Alternative in Eastern Europe* (Bahro 1978). In West Germany he became involved in the Green Party, because he believed that the ecological crisis would bring about the end of capitalism (Bahro 1984). In *Socialism and Survival* (Bahro 1982) he offered a double critique, of capitalism and 'actually existing socialism'. To Bahro (1984), socialism had failed to address the ecological crisis, which brought questions onto the agenda that were 'already there before the first class society took shape' (p. 148). He attacked industrialization in both capitalist and socialist systems, arguing that 'the increase in material consumption and production, with the inbuilt waste, pollution and depletion of resources … is enough to destroy us in a few generations' (p. 179). Bahro criticized the 'huge structures' of industrialization and bureaucracy, and questioned not only the kind of products made (that is, the organization of industrial output), but also the 'reproductive process itself' (that is, the nature of industrial society) (p. 147). Bahro concluded that 'the time has come when the utopian communist and socialist visions are no longer utopian. We have reached the limit. Nature will not accept any more, and it's striking back' (p. 184).

Eco-socialists believe that socialism can explain the cause of development and environmental crises, and that it alone can deliver solutions. Historically, it can be argued that in many ways the labour movement in Europe *was* an environmental movement, concerned with the living and working conditions, health and life chances of the poor. David Pepper (1993) comments: 'Marxism reminds us that for most people, nineteenth century environmental problems were *clearly* socially inflicted' (p. 63; emphasis in the original). Later-twentieth-century environmentalism in industrialized countries such as the UK emphasized more bourgeois concerns about species extinction and loss of landscape heritage, and exported these concerns to the Third World through international ideologies of conservation. However, the core concern with the quality of the environment in which people live was common both to the environmental struggles of the developing world (Gadgil and Guha 1995; Guha and Martinez-Alier 1997; Ho 2001; Dwivedi 2005) and to the arguments of analysts of industrialization concerned about the 'Risk Society' (Beck 1992, 1995; see further discussion in Chapter 12).

The links between environmentalism and socialism are complex. Pepper (1993) points out the rather confused blend of Marxist and anarchist ideas in green thought. Some ideas are compatible with capitalism, while others follow the logic of socialist analysis. Pepper is unimpressed with eco-socialists' enthusiasm for 'new social movements' rather than labour as the basis for a revolution in both consciousness and material social relations. He argues that these movements (the green movement prominent among them) are 'idealistic and superstructural' (p. 136), meaning that they do not address the material basis for exploitation of people or nature; he concludes that 'they have more to do with ahistorical postmodernism than with Marxism's historical materialism' (p. 136). In particular, he argues that, in the developing world, environmental struggles are still about the basic requirements for an environmentally secure life. This is a distant world from the obsession with aesthetic aspects of the environment, the politics of 'NIMBYism' and the rhetoric of 'sustainable development planning' in industrialized countries.

Socialist thought offers a significant radical critique of development, moreover one that is in many ways at least as green as it is conventionally red. Indeed, Pepper (1993) argues that eco-centric thought is inherently socialist in that it is anti-capitalist (p. 70). There is certainly common ground between green visions of the future and ideas of decentralism, communalism and utopian socialism that are part of the socialist tradition (for example, the ideas of William Morris, with his vision of production for use and not exchange value, and production to meet human need). These grade into (although they are distinct from) social anarchism or anarcho-communism (Pepper 1984, 1993). Environmentalism ('ecologism', as Pepper (1993), like Dobson [1990] 2007, prefers) is idealist rather than materialist in its approach, viewing humankind as part of a global ecosystem and subject to 'natural' laws. Eco-centrics start with a view of nature, and attempt to develop a human response to it. Socialists start with social concerns, particularly wealth distribution, social justice and quality of life, and see the environment as an issue that vitally affects those concerns, and is in turn affected by social action.

Green critiques of Marxism tend, as we have seen, to argue that Marx, and early Marxists, assumed that resources were inexhaustible; they have taken the state capitalism of the Soviet Union in particular as proof that Marxism has been woefully blind to the environment (Pepper 1993). However, as Michael Redclift (1987) points out, some Marxist writers have addressed environmental and resource depletion issues. Thus Hans Magnus Enzensberger (1974) argued that there were real scientific problems lying behind the bourgeois packaging of the environmental movement, and reiterated the 'commonplace of Marxism' that environmentalists had highlighted, the 'catastrophic consequences' of the capitalist mode of production (p. 10). The editorial in that 1974 volume of *New Left Review* asserts that 'to identify and combat these has become a central scientific and political task of the socialist movement everywhere'. In France, the work of eco-socialists André Gorz and René Dumont in the 1970s linked ecological destruction and its social and political causes, and subsequent writers such as Félix Guattari and (in the 1990s) Alain Lipiez have continued to developed a diverse tradition of eco-socialist thought (Whiteside 2002).

It is probably true that Marxist thinkers historically underplayed the importance of the environment and any 'environmental crisis'. Enzensberger (1974) argued that the left in Europe remained sceptical and aloof from environmentalist groups, simply incorporating selected elements of the environmental debate in their repertoire of anti-capitalist agitation (p. 9). Arguably the rise of green thinking itself reflected the failure of the strategies of the European left over several decades (Amin 1985). However, there was awareness among Marxists of environmental dimensions of the impact of capitalism, particularly in the work of Friedrich Engels (Parsons 1979). As the environmental revolution flowered, and ideas of MSD began to be laid out, the theoretical framework for a Marxist theory of nature was duly discussed – for example, by Burgess (1978), N. Smith (1984) and Benton (1996). The question of the relevance of Marxism to environmentalism has been extensively explored – for example, by Pepper (1984, 1993), Redclift (1984, 1987) and Gimenez (2000). Michael Redclift (1984) called for 'a fundamental revision of Marxist political economy, to reflect the urgency of the South's environmental crisis' (p. 18).

The environmentalist consciousness offers a potentially radical and enduring critique of the social effects of developmentalism. Arguably, in fact, Marx and Engels were forerunners of human political and social ecology, favouring 'active and planned intervention in nature but not its triumphant and ultimately irrational destruction' (Pepper 1993, p. 62). Marx was no eco-centric, but his view of the instrumental values of nature embraced aesthetic, scientific and moral values as well as straightforwardly economic or material values (Pepper 1993, p. 64). Marxists take explicit account of the historical conditions that shape human lives and relations with nature (Gimenez 2000), and many have now pointed out that capitalism is inherently incapable of dealing fully with ecological problems, or even that capitalist management of the earth will necessarily bring about ecological catastrophe (Vlachou 2004).

Marxist analysis of capitalism underlies most green critiques of economic development. Capitalism emerged in Europe out of feudalism as a means to allow new wealth (from slavery, agricultural production, mining and simple manufacture) to be invested:

> Capitalism was made possible by the raiding of stored wealth, the reorientation of trade routes, the imposition of unequal exchange, the forceful movement of millions of people in world space, and the conversion of the people and territories of whole continents into colonies where all aspects of existence were subject to the purposes of the Europeans.
>
> (Peet 1991, p. 145)

Merchant capitalism gave way to industrial capitalism, and increasingly this has evolved through Fordism (the division of labour into specialized tasks and their integration and routinization on a production line), and into various more flexible forms of capitalism (Harvey 1990; Peet 1991; Pepper 1993).

Capital is the result of the surplus derived from the employment of labour (that is, the difference between the value of what labour produces and the price that has to be paid to workers to persuade them to work). This 'surplus value' (profit) accrues to whatever individual or group owns the production process and the 'means of production'. The motive force for capitalism is therefore the accumulation of wealth derived from profits (Johnston 1989). For capitalism to work, the desire to accumulate profits has to be made to seem 'natural'. Ideas about how society should be organized (forms of 'social consciousness') are therefore closely related to (and influenced by) the way production is organized. Capitalist relations of production therefore correspond to a particular way of understanding how the world works (a particular 'form of consciousness' (Pepper 1996, p. 68)).

Institutions (both formal institutions such as laws and informal institutions such as ideas and values) emerge that support the capitalist system, and also support the class interests that chiefly benefit from that system – that is, the owners (and increasingly the elite managers) of capital and the means of production (Pepper 1996, p. 69).

An important feature of Marx's account of the transition to capitalism is that it links the removal of people from the land (an economic alienation) with their separation from nature and loss of awareness of human dependence on the environment or impacts upon it (Pepper 1993, p. 72). The romanticism about nature and 'the countryside', which in due course provided a powerful root of environmentalist thinking in industrialized countries (e.g. Bunce 1994; Veldman 1994), has to be understood as itself the fruit of capitalism. Under capitalism, people not only sell their labour power, but relate to nature as an object, to be bought and sold. Increasingly nature is also a product, physically refashioned by state or business corporation and paid for at point of consumption, or is packaged as an image or a product in cyberspace (Wilson 1992). Capitalism therefore commodifies both labour (and hence relations between people) and nature (and hence relations between people and non-human nature). There is more Marx in environmentalist thought than might at first appear.

Marxism suggests that, in various ways, the capitalist system is unsustainable: it 'contains within itself the seeds of its own destruction' (Johnston 1989, p. 58). One problem is the power of large corporations to create monopolies (which negate the claimed 'efficiency' of the market). The growing internationalization of capital and the vast size of the largest global corporations have severely restricted the capacity of national governments to restrain the profit-seeking behaviour of capital. A second problem is that increased productivity can lead to production outstripping the capacity to consume, resulting in overproduction and reduced profitability. Creative destruction is embedded within the circulation of capital itself: 'innovation exacerbates instability and insecurity, and in the end, becomes the primary force pushing capitalism into periodic paroxysms of crisis' (Harvey 1990, p. 106). This means that, as modern industry goes through periods of boom and stagnation, labour is subjected to instability and uncertainty as a 'normal' part of life – the inevitable result of what Margaret Thatcher described as 'the laws of economic gravity' (Pepper 1996, p. 89). In response, enterprise managers seek to reduce costs (through cheaper raw materials, cheaper labour and more efficient machines), stimulate demand by advertising and find new markets and products.

The search for cheaper materials, cheaper labour and new markets is, of course, the engine that drives the development/modernization process in the developing world. The post-Second World War 'crisis' of Fordism in Europe and North America led to both the transformation of industrial production to more flexible systems of labour organization in the developing world (deskilling, extension of automation, longer and flexible working hours, de-unionization, reduction of job security) and the extension of Fordist production systems to the developing world, creating 'peripheral Fordism' (Harvey 1990, p. 186). Relocation of production to the Third World was a strategy aimed at reducing production costs and maximizing profit. Important elements in these costs were the costs of employment (wages and related living costs, and measures to protect employees from sickness and to provide job security) and the costs of taking account of 'externalities' of production such as pollution. Relocation to countries with low wage rates, weak labour and environmental laws, and weak enforcement of those laws made perfect sense in terms of global business strategies of maximizing returns on investment (see Chapter 12). David Harvey (1990) argues that, while this capitalist penetration of the periphery promised emancipation from want and full integration into Fordism, it delivered instead destruction of local cultures, oppression and various forms of capitalist domination in return for rather meagre gains in mass living standards and services (except, of course, for the small and soon super-affluent indigenous elites, who collaborated with and profited from the penetration of capital (p. 139)).

Many analysts have lost confidence in the capacity of capitalism to 'develop' the developing world to the level of the industrialized West. In his book in 1972, Sutcliffe argued for industrialization in large-scale units under autarchic conditions. By 1984 he had changed his view. He now suggested that capitalism produced inappropriate products (cars, weapons, obsolescent goods) and it used the wrong methods (centralized, deskilled, totalitarian and alienating) (Sutcliffe 1984). Capitalist industrialization could not, therefore, be the material basis for what we might now call a sustainable world. Sutcliffe added that it had also created its own mirror image in the 'centralised, statist and bureaucratic' view of state socialism. In his 1984 paper Sutcliffe urged a recapture of utopian traditions by socialist thinking on industrialization and development. He admitted certain strengths in the critiques of industrialism by populists and intermediate technologists, and called for a search 'for a more humane alternative to economic development than the rocky path represented by actually existing industrialisation' (p. 133). Rudolf Bahro (1984) also regarded established development strategies for the developing world (through increasing trade and industrialization) as 'a tunnel without exit' (p. 211), arguing that industrialism in the developing world would mean 'poverty for whole generations and hunger for millions' (p. 184). The poverty created by capitalist industrialization in eighteenth- and nineteenth-century Europe was 'made bearable' by the prospect of escaping it through exploitation of the periphery. However, 'for the present periphery there is no further periphery to be exploited, no way of attaining the good life of London, Paris or Washington' (p. 211), making the prospect of proletarianization in the contemporary Third World 'a horrific vision' (p. 184).

Pepper (1993) argues that capitalism is '*inherently* "environmentally unfriendly"' (p. 91; emphasis in original): it 'continuously gnaws away at the resource base that sustains it' (p. 92). It externalizes its costs, leaving them to be met by the state or in the bodies of the poor in terms of sickness or reduced life expectancy, or by future generations. Thus Löwy (2005) argues that 'the insatiable quest for profits, the productivist and mercantile logic of capitalist/industrial civilization is leading into an ecological disaster of incalculable proportions' (p. 15). Capitalism endangers or destroys the conditions of its own systems of production, among them the natural environment.

Capitalism as a system (like individual enterprises) can reward those in a structural position to profit from it, but only at the expense of others elsewhere. Low and Gleeson (1998) discuss the problem of 'environmental racism', and the way the distribution of polluting industries and the dumping of hazardous waste within the USA reflects the structural powerlessness of class and race. The same kinds of inequalities in the distribution of environmental risk and other externalities of production exist at a global scale. The capitalist system, involving the circulation of money, products and risks, allows developed countries to externalize risk by shifting hazardous forms of production beyond their borders, to countries with lower environmental or employment standards.

The 'environmental crisis' is, therefore, far from uniform, for it affects some people much more than others, at all scales from local to global. The impacts of environmental degradation are socially and spatially differentiated: 'they may end up affecting the global environment, but first they damage small parts of it' (Low and Gleeson 1998, p. 19). As capital is increasingly managed globally, with investment and disinvestment to maximize profit, its beneficiaries are, first, the super-rich elite of the international banking and finance system and, second, the employed of the industrialized world, particularly the self-regenerating class of managers. The losers are the unemployed workers of 'rustbelt' regions in the North and the vast numbers of the Third World poor, without a chance of access to a share of global wealth. As work on transnational commodity chains shows, 'sustainable development' for the First World (and for wealthy areas within those countries) is all too easily built on selective unsustainable extraction of resources, unequal trade of commodities and inhumane and polluting manufacture of products in the South (A. Hughes 2001; Fold 2002).

Socialism clearly provides a powerful critique of the environmental and developmental impacts of capitalism. Developing without strong institutions of civil society, the profoundly corrupt version of the 'free market' that developed in these 'countries in transition' gives much food for thought. Eco-socialism remains an important influence on debates about sustainable development.

Eco-anarchism

Environmentalist critiques of development also draw extensively if not always explicitly on anarchist thinking – for example, in the work of people such as Peter Kropotkin and his ideas of social anarchism or anarcho-communism (Kropotkin [1906] 1972, [1899] 1974; Galois 1976; Breitbart 1981), or in the work of writers such as Murray Bookchin (1979) and Theodore Roszak (1979). In particular, Murray Bookchin's 'social ecology' was influential in the development of green thought – for example, in *Post-Scarcity Anarchism* (1971) and *The Ecology of Freedom* (1982). Bookchin argues that the domination of non-human nature by humans arises directly from the existence of hierarchy in society, and the domination of humans by humans (Eckersley 1992). There is a particular focus on the coercive power of the state: without the state and other structures of exploitation associated with hierarchy among people and structure of domination, environmental problems would not arise (Pepper 1993). Carter's attempt to devise a 'green political theory' drawing on anarchist (and other) thought maps tight links between centralized state, inegalitarian economic relations, 'hard' technologies and militarism (see Figure 7.2). As he comments, this dynamic is 'environmentally hazardous in the extreme' (Carter 1993, p. 45). An alternative dynamic can be imagined, drawing on ideas of decentralization, participatory democracy, self-sufficiency, egalitarianism, alternative technology, pacifism and internationalism (see Figure 7.3). Such a shift would be resisted, the suggested route towards it being through non-compliance with the state.

A central feature of anarchism is the importance of viewing development from the perspective of the individual. Indeed, Pepper (1993) suggests that anarchism can be viewed as a form of extreme liberalism. Kropotkin believed that 'true individualism can only be cultivated by the conscious and reflective interaction of people with a social environment which supports their personal freedom and growth' (Breitbart 1981, p. 136). Pepper (1984) points out the difference between this starting point in the individual and that of environmentalist thinking such as *Blueprint for Survival* (Goldsmith *et al.* 1972), which suggests an ecological imperative for action to achieve utopia, the human dimensions of which are secondary. Bookchin notes the environmentalist urge to protect nature from destructive societies, but argues that the social root of the destruction of nature is

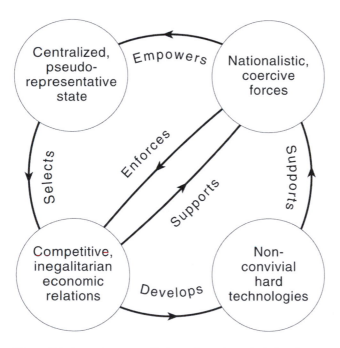

Figure 7.2 An environmentally hazardous dynamic (after Carter 1993)

our particular civilization, with its hierarchical social relations, which pit men against women, privileged whites against people of colour, elites against masses, employers against workers, the First World against the Third World, and, ultimately, a cancer-like "grow or die" industrial capitalist economic system against the natural world and other life forms.

(Bookchin in Bookchin and Foreman 1991, p. 31)

In *Person/Planet*, Roszak (1979) argued that, as long as Western society remains locked into 'the orthodox urban-industrial vision of human purpose', there is no hope that poverty, and the injustice it brings with it, can be 'more than temporarily and partially mitigated for a fortunate nation here, a privileged class there' (p. 317). Bookchin (1979) argues that 'in the final analysis, it is impossible to achieve a harmonisation of people and nature without creating a human community that lives in a lasting balance with its natural environment' (p. 23).

Galtung (1984) differentiated between 'alpha' social systems, which are hierarchical, unlimited in size and tending towards uniformity, and beta systems, which are horizontal, limited in scale and inclined towards diversity. Eco-development requires beta structures, but set within a matrix of a benign, flexible, communicating and restraining alpha system. Galtung's ideas owe little to social theory and a great deal to the diversity–stability debate in ecology in the 1960s (Margalef 1968). Nonetheless, they reflect the element of anarchism in environmentalism that provides one of the central themes of 'green' development

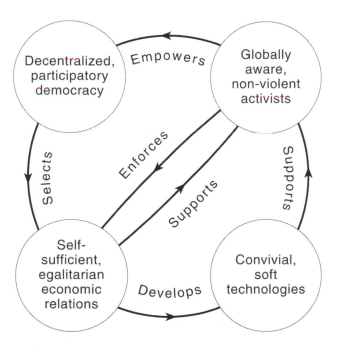

Figure 7.3 An environmentally benign dynamic (after Carter 1993)

thinking (e.g. Bookchin 1979; Roszak 1979; Pepper 1993), as well as notions of participation in planning and 'development from below' (Stöhr 1981).

Bookchin (1979) based his arguments quite extensively on ecological ideas, seeing ecology as one science that might avoid assimilation by the established social order, and he mourned the absorption of environmentalists into governmental institutions, warning that 'ecological dislocation proceeds unabated, and cannot be resolved within the existing social framework' (p. 22). Ecology is integrative and reconstructive, and this 'leads directly into anarchist areas of social thought' (p. 23).

Clearly, 'nature' is being treated here as a source of values, and political ideas are being drawn from particular interpretations of the organization of nature (Dobson [1990] 2007; see also Table 7.4). To eco-anarchists, collaboration (Kropotkin's 'mutual aid') is 'natural' (Pepper 1993). In this, eco-anarchism shares with other forms of ecologistic thinking the fallacy of treating human understandings of non-human nature at a particular point in time (including the notion of separation of 'human' self and the non-human 'other') as true, and using that understanding as the basis for political argument. If understanding of nature (whether through science or through other means) is seen to be contingent on particular historical moments and social processes, the 'naturalness' of that arrangement is at once seen to be also a social construct.

Pepper (1993) argues that eco-anarchists tend to be vague about why the problems they identify in society (and its treatment of nature) emerged. For example, if states are unnatural, what forces brought them into being; if people 'naturally' cooperate, why is altruism so rare? He believes that green political writers often fudge the boundary between anarchism and socialism. Eckersley (1992) distinguishes between two strands in eco-anarchism, social ecology and 'ecocommunalism'. Pepper (1993) suggests that fuzzy green thinkers fasten loosely on to the latter, proposing a form of anarcho-communalism that claims to attempt to bypass or subvert the state through communal living and lifestyle change. Such ideas have played a major role in the mainstream Western environmental movement (captured perhaps in the now tired slogan 'think globally, act locally' and in the dubious logics of 'green consumerism'), although, as environmental organizations in the North have adopted increasingly complex corporate structures in the 1990s, it is more than ever clear that anarchism is uncommon and perhaps unsustainable in the structure of environmental organizations, whether environmental pressure groups or Green political parties. Dave Foreman comments: 'I guess if you organize yourself

Table 7.4 Political ideas drawn from nature

Perceived attribute of nature	Political idea or principle
Diversity	Toleration, stability and democracy
Interdependence	Equality
Longevity	Tradition
Nature as 'female'	A particular conception of feminism

Source: Dobson (1990, p. 24).

like a corporation, you start to think like a corporation' (in Bookchin and Foreman 1991, p. 38). Of course, most Northern countries can show peaceful (and sometimes long-lived) 'alternative' communities, but Pepper (1993) reminds us to be suspicious of utopias, anarchist or otherwise, and the belief that changing minds and vocabularies can change material relations (noting that 'the world cannot be restructured by moral example' (p. 150). He also notes the danger of anti-urban (and anti-working-class) romanticism.

Anarchists also commonly advocate political tactics that include strikes, boycotts and demonstrations (predominantly non-violent, although commitment to pacifism and non-violence varies, and is hotly debated). It is easy to see the influence of anarchist ideas in the work of developed-world environmental organizations such as Greenpeace and Earth First!, and the self-avowedly 'eco-anarchist' opposition to roads and other projects in countries such as the UK (Rawcliffe 1998). It is also possible to trace related traditions elsewhere – for example, in the techniques of peaceful direct action developed by Mahatma Gandhi in India, and in more recent struggles such as the Chipko movement and Narmada Bachao Andolan in India (Guha and Martinez-Alier 1997). These will be discussed further in Chapter 13.

However, Pepper admits that anarchism, anarcho-communism and decentralist socialism lie close alongside each other, and they have been drawn on extensively by those developing radical green critiques of society. All demand drastic changes to 'business as usual', and all provide a very different picture of sustainability from MSD:

> The anarchist concepts of a balanced community, a face-to-face democracy, a humanistic technology and a decentralised society – these rich libertarian concepts – are not only desirable, they are also necessary. They belong not only to the great visions of a human future, they now constitute the preconditions for human survival.
>
> (Bookchin 1979, p. 27)

To Bookchin (1979), the 'ecological crisis' was just part of a larger malaise: 'Humanity has produced imbalances not only in nature but, more fundamentally, in relations amongst people and in the very structure of society.' Therefore, 'what we are seeing today is a crisis in social ecology', in which Western society is 'being organised round immense urban belts, a highly industrialized agriculture and, capping both, a swollen bureaucratised, anonymous state apparatus' (p. 25). Bookchin's 'ecological anarchism' is therefore essentially anti-industrial, anti-bureaucratic and anti-statist. In some ways it shares ground with Schumacher's *Small is Beautiful* (1973), but for Bookchin that approach demands adaptation to the norms of society, not the 'revolutionary opposition' he believes is required.

Deep ecology

Much of the environmentalist critique of development has drawn on 'eco-centric' or 'bio-centric' ideas, which contrast with the techno-centrism of MSD (O'Riordan

[1976] 1981). The core of bio-centrism is the ascribing of intrinsic values or moral status to non-human nature. Dobson (1990) describes the political ideology of 'ecologism' as seeking 'to persuade us that the "natural" world has intrinsic value: that we should care for it not simply because this may be of benefit to us' (p. 49). A conventional distinction is drawn between 'deep ecology', which is bio-centric in this way, and 'shallow ecology', which is anthropo-centric, and concerned about the values of nature for the human species. A parallel distinction has commonly been drawn between 'dark' or 'light' green thinking (e.g. Wissenburg 1993). It should be noted that the word 'ecology' here does not represent in any direct way the practice of the experimental natural science of ecology (e.g. McIntosh 1985), but rather the looser meaning of an 'ecology movement' (cf. Dobson [1990] 2007). The MSD described in Chapter 5 draws chiefly on the 'shallow' end of the ecology movement. The 'deep long-range ecology movement' (Devall 2001) offers a radical alternative.

This distinction, and the phrase 'deep ecology', come from the writings of the Norwegian philosopher Arne Naess. From the first, deep ecology was conceived of as both a call to activism and a normative worldview about the place of humans in the world (Katz *et al.* 2000). Arne Naess began to consider ecological philosophy in the late 1960s, drawing on philosophy of Spinoza and following his own previous work on Gandhian thought. He first wrote about deep ecology in 1973 in a paper explicitly contrasting 'shallow ecology' and 'deep long-range ecology' (Naess 1973). He suggested that the shallow ecology movement had as its central objective the health and affluence of people in developed countries, and fought against pollution and resource depletion. Deep ecology differed in two ways. First, it rejected this anthropo-centrism and the separation of 'human' and 'environment' (and indeed the separation of 'thing' and 'its milieu') in favour of a 'total field model', of organisms as knots in a field of intrinsic relations. Second, deep ecology was based on the principle of 'biospherical egalitarianism', and recognition of the equal rights of organisms to live and blossom. Naess argued that the anthropo-centric restriction of this right to humans was detrimental to life quality for humans themselves, and that the attempt to ignore the interdependence between humans and other organisms and to establish a master–slave role had contributed to the alienation of humans from themselves (Reed and Rothenberg 1993).

Naess has sought to set out a philosophical system that relates self to nature, which he calls an 'ecosophy', a personal philosophy or a code of values and a view of the world that guides personal decisions about relations with the natural world (Reed and Rothenberg 1993). He called his own version 'Ecosophy T', and it was offered not as a finished system of thought, but as a means for other people to develop their own personal ecosophies. His ideas continued to evolve, and in 1984 he produced the 'deep ecology platform' (see Table 7.5), to establish a common eco-philosophical ground for deep ecology (Sessions 1995).

This platform was not an attempt to define a 'deep-ecology' dogma. George Sessions, and other commentators, have pointed out their various proposed modifications or elaborations. This is important, for the very openness of Naess's

Table 7.5 The deep ecology platform of Arne Naess

- The well-being and flourishing of human and non-human life on earth have value in themselves (synonyms: inherent worth; intrinsic value; inherent value). These values are independent of the usefulness of the non-human world for human purposes.
- Richness and diversity of life forms contribute to the realization of these values and are also values in themselves.
- Humans have no right to reduce this richness and diversity except to satisfy vital needs.
- The flourishing of human life and cultures is compatible with a substantial decrease of the human population. The flourishing of non-human life demands such a decrease.
- Present human interference with the non-human world is excessive, and the situation is rapidly worsening.
- Policies must therefore be changed. The changes in policies affect basic economic, technical and ideological structures. The resulting state of affairs will be deeply different from the present.
- The ideological change is mainly that of appreciating life quality (dwelling in situations of inherent worth) rather than adhering to an increasingly higher standard of living. There will be a profound awareness of the difference between big and great.
- Those who subscribe to the foregoing points have an obligation directly or indirectly to participate in the attempt to implement the necessary changes.

Source: Reed and Rothenberg (1993).

account of deep ecology has allowed an enormous diversity of thinking to develop and claim this label. By the mid-1980s, discussion of 'deep ecology' had begun to develop through a varied mix of writings by people such as Warwick Fox (1984, 1990) and Bill Devall and George Sessions (1985). This writing tended to divide into a development of ecosophical thinking, notably in the 'transpersonal ecology' of Warwick Fox (1990), and 'a range of normative and sometimes radical visions of the human relationship with nature' (Reed and Rothenberg 1993, p. 2) – for example, in the work of Devall and Sessions (1985).

Deep ecology emphasizes the transcendental attributes of nature. Graber (1976) suggested that the 'wilderness ethic' is strongly religious in character (p. 111). 'Wilderness purists' draw on the works of Thoreau, Muir and Aldo Leopold for inspiration and group definition. Deep ecologists too reference themselves by the writings of such people and their sense of moral order in nature, and of the continuity between humans and other organisms (and indeed inanimate nature). Deep ecology calls for a new relation with nature that challenges both established anthropo-centric utilitarian ideas (that is, conventional 'development') and the managerialist reformism of MSD.

In his book *Simple in Means, Rich in Ends*, Bill Devall (1988) attempted to outline the basis of a *practice* of deep ecology, taking the phrase to refer to 'finding our bearings, to the process of grounding ourselves through fuller experience of our connection to earth' (p. 11). One element in such thinking is the notion of bio-regionalism, a deliberate focusing on the 'homeland of ecological self' (p. 58). Devall and Sessions (1985) argued that 'many individuals and societies throughout history have developed an intuitive mystical sense of interpenetration with the landscape and an abiding and all-pervading "sense of place"'

(p. 241). Katz and Kirby (1991) speak of 'constructs of the Native American lifeworld' – a system in which 'there exist no dualities between humans and nature, or necessarily between animate and inanimate' (p. 262). This shift in consciousness reflects the wider angst at modernity and globalization that has fed Northern environmentalism, and reflects the relatively narrow social and political base of that movement in terms of education, wealth and employment (see Cotgrove 1982). It has significance for debates about sustainable development because of the continuing influence of Northern environmentalism on ideas about nature (and its 'development') in the Third World, and because of the increasing global exchange of such ideas (Guha and Martinez-Alier 1997).

Deep ecology is an important element in a wider stream of radical environmentalist thought that can be described as fundamentalist ecology (J. Shantz 2003). Neo-Malthusianism is important in framing radical calls for change in relations between people and environment. The radical 'eco-warriors' of the USA in particular have alarmed many observers because of the anti-social and anti-human elements in their ideas (Bookchin and Foreman 1991; J. Shantz 2003). The practice of 'eco-defence' –also referred to as 'monkey-wrenching' or 'ecotage' (Devall 1988) – is named from Edward Abbey's classic novel *The Monkey Wrench Gang* (1975). This is epitomized in the industrialized world by the eco-radicalism of Earth First!, and the development of direct-action protests against environmentally destructive development – for example, clear-felling of old-growth forests in western Canada and the north-west USA, or against roads and other developments in the UK (Devall 1988; Bookchin and Foreman 1991; Rawcliffe 1998). However, strategies such as 'tree-spiking' (driving nails or metal rods into timber trees as a protest against clear-fell logging) threatened the lives and safety of timber workers, and took such actions well beyond conventional 'monkey-wrenching' (cf. Abbey 1975).

Non-violence has been an important element in deep-ecology-inspired protest against ecosystem transformation: Arne Naess himself took part in peaceful non-violent protests against the construction of a dam at the Mardøla Falls in Norway in the 1970s (Reed and Rothenberg 1993). The resonance between such protests and those of developing world environmental organizations such as Narmada Bachao Andolan in India (Guha 1989; Guha and Martinez-Alier 1997) is obvious, although the depth of the similarity between environmentalist direct action in the developed and developing world is less clear. Developing world environmental movements are often rooted in human needs and represent, at least in the first instance, localized responses to threats to human welfare (Ho 2001; Dwivedi 2005). On the other hand, fundamental ecology movements in developed countries typically lack both a consistent and clear social analysis of ecological crisis (that, for example, an eco-socialist analysis would give them), and a consistent commitment to humane social ethics. The eventual renouncement by Earth First! of tree-spiking in 1990 and the search for collaboration with timber workers against the unsustainable pursuit of profit by logging companies took them back towards pacifism and non-violence, although the move was itself met with violence by the US state (Rowell 1996; J. Shantz 2002).

Pepper (1993) notes the dangers of reactionary political ideas as part of bio-regionalism. The extreme right-wing politics associated with 'survivalist' groups in the American West, and the aggresion of the US 'Wise Use' movement towards radical environmentalists, mark a significant departure from the more demure symbolic protests that have characterized environmental movement in the past (Rowell 1996). Critics of deep ecology have been quick to criticize the misanthropy and glorification of violence that have been part of the history of (and remain an element within) this strand of radical green thought (Lewis 1992).

The bio-centrism of deep ecology potentially undermines the moral basis of most development action. It is easy to identify a possible alliance between the neo-Malthusian science-based critique of population growth (with its apparently sound concepts such as 'carrying capacity') that sees famines as somehow 'natural', and bio-centric ideologies that identify people as organisms with no special rights and that see intervention to sustain human lives at the expense of other organisms and inanimate objects as unacceptable. This kind of green thinking is perhaps what Amin (1985) has in mind when he speaks of green ideas as 'a form of religious fundamentalism' (p. 281). Such ideas can generate the deeply conservative

Plate 7.1 800-year-old Douglas fir trees at Cathedral Grove, British Columbia, Canada. MacMillan Provincial Park on Vancouver Island comprises a small area of huge Douglas fir trees. The park covers 136 hectares. Outside it all the surrounding forest has been clear-felled. Such contrasts inspire deep ecologists in North America; see, for example, Ecotrust (www.ecotrust.org). Deep ecology draws on the work of the Norwegian philosopher Arne Naess, and rejects anthropocentrism and the separation of 'human' and 'nature'. Photo: W. M. Adams.

ideology and reactionary and repressive politics of 'ecofascism' (Pepper 1984). Lewis (1992) excoriates 'harsh deep ecology' as 'primitivism', accusing it of advocating 'the active destruction of civilisation' (p. 28).

Eco-feminism

The bio-centrism of deep ecology is both echoed and challenged by eco-feminism (Plumwood 2000). Awareness of the importance of gender in relations between people and non-human nature grew apace through the 1980s and 1990s (Nesmith and Radcliffe 1993; Jackson 1994). In particular, 'eco-feminism', or 'ecological feminism' (Warren 1994), became an important challenge both to MSD thinking, and at the same time to other radical streams of thought such as deep ecology. However, the world of eco-feminism is (once again) rather complex. Rocheleau *et al.* (1996a) identify a series of schools of thought within which environment and gender engage, including feminist environmentalism, feminist poststructuralism and socialist feminism. Environmentalists have begun to engage with liberal feminist agendas, to consider women as actors in environmental management and (in terms of policy) as active agents in environmental projects. This approach mirrors wider changes in development thinking and recognition of the significance of gender as a factor in the management of the environment (argued, for example, by Townsend 1995). However, in practice a simplistic and functionalist focus on gender can lead to a policy straitjacket that simply sees women as cost-effective 'target groups' in development (Elmhirst 1998), and the erroneous notion that policy can be devised that is synergistic in addressing problems of population, environment and development (C. Jackson 1993).

Feminist poststructuralism (Rocheleau *et al.* 1996b) builds on poststructuralist ideas about development (for example, the ideas of Arturo Escobar (1995)) and on feminist critiques of science (for example, by Donna Haraway (e.g. 1991)). Haraway writes about research on primate behaviour, and shows that, over time, competing scientific 'stories' in the academic literature (based on 'scientific' data) mirror the social and political world in which the scientists moved: ideas about what women were like, and how they ought to behave, were inextricably linked to the conclusions scientists drew about non-human apes. Haraway's analysis suggests that science cannot produce meanings that are free of their context. Nature cannot exist without social meaning (Fitsimmons 1989).

Socialist feminists have addressed the importance of gender in an understanding of political economy, and hence have focused on gender divisions of production and reproduction (Rocheleau *et al.* 1996a), and issues of access to and rights over land and other environmental resources (e.g. Carney 1993). Thus Bandarage (1984) fiercely criticized liberal feminism and the Women in Development (WID) school, which, while it correctly identifies the systematic impoverishment and disempowerment of women, fails to escape the Western modernization model of development: WID is a movement *for* women, not *of* women. By contrast, she argued that a 'Marxist–feminist synthesis' could 'situate sexual oppression historically as it interacts with class oppression and imperialism' (pp. 504–5).

Eco-feminists argue that the coercive relations between humans and non-human nature are the result of an essentially gendered process of exploitation. Eco-feminist analysis argues that the same structured oppression of women by men (patriarchy) is reflected in patterns of imperialism and capitalist accumulation; there is a dual subjugation of both women and nature (Mies 1986; Shiva 1988). Furthermore, this oppression extends to science and other forms of hegemonic 'Western' knowledge. In her influential book *Staying Alive*, Vandana Shiva (1988) portrayed science as a 'masculine and patriarchal project' that 'necessarily entailed the subjugation of both nature and women' (p. 15). She argues that a gender-based ideology of patriarchy underlies ecological destruction, and the definition of nature as passive and requiring taming and 'development':

> From the perspective of Third World women, productivity is a measure of producing life and sustenance; that this kind of productivity has been rendered invisible does not reduce its centrality to survival – it merely reflects the domination of modern patriarchal economic categories which see only profits, not life.
>
> (p. 5)

Thus Shiva (1988) argued that, in the forests of India, reductionist Western science, driven by capitalist profit maximization, had marginalized women and degraded their environment. The destruction of the forest and the displacement of women were both structurally linked to the reductionist (and capitalist) paradigm of science. Attempts to deal with deforestation that stem from the same patriarchal science and capitalism exacerbate both the crisis of human survival and that of environmental degradation. Salvation, the recovery of 'the feminist principle in nature', and of the view of the earth as sustainer and provider, is to be found in the capacity of women, specifically Third World women, to challenge established male-dominated modes of thinking about knowledge, wealth and value. Shiva argues that 'the intellectual heritage for survival lies with those who are experts in survival' (p. 224). Only they have the knowledge and experience 'to extricate us from the ecological cul-de-sac that the western masculine mind has manoeuvred us into'. She called for a 'non gender-based ideology of liberation' (p. xvii).

However, at the core of Shiva's analysis is an essentialist proposition: that men and women are essentially different, and that women are closer to nature than men. Thus 'the reductionist mind superimposes the roles and forms of power of Western male-oriented concepts on women, all non-Western peoples and even on nature, rendering all three "deficient", and in need of "development"' (p. 5). What Nesmith and Radcliffe (1993) call 'spiritual environmental feminists' have affirmed the female–nature connection. However, many other feminists have challenged that essentialism, while agreeing with the history of patriarchy and the twin domination of nature and women (see also Sargisson 2001). Plumwood (2000), for example, criticizes the failure of deep ecology to deal adequately with difference, among other problems.

Cecilia Jackson (1994) argues that eco-feminism, and ecocentric environmentalism as a whole, is essentialistic in its understanding of both women and environments. The problem is that women are conceived of as a category with universal characteristics that transcend place, time and material circumstances. Thus many eco-feminists attempt to revalue the feminine – for example, emphasizing 'feminine' values such as caring and nurturing, and identify them as universal attributes of women, and the possible basis for a strategy for reworking human relations with non-human nature. However, if 'feminine' has been defined in distinction from 'masculine', can a feminine essence be defined? The acceptance of a fundamental similarity between women and nature invites the continued oppression of both.

Val Plumwood (1993) critiques the notion of 'special' connection between women and non-human nature, and argues for an anti-dualistic 'critical ecological feminism' that challenges, among others, the dual categories of human/nature and man/woman. She suggests that the fierce debates within 'green theory', particularly the two-sided argument (largely between male protagonists) between social ecology and deep ecology, have obscured the potential for explanations of human domination of nature that are compatible with older critiques of hierarchy and human social domination. Plumwood describes a 'web of oppression' whose source is the ideology of the control of reason over nature: oppressed groups have been counted as part of the 'chaotic and deficient' realm of nature by the mastering and ordering 'reason' of Western culture (p. 74). Eco-feminism needs to understand masculinity and femininity as 'relational, socially constructed, culturally specific negotiated categories' (Jackson 1994, p. 125). Plumwood (2003) identifies a series of rationalist dualisms in dominant modes of thought that oppose, for example, reason to nature, mind to body, emotional female to rational male and human to animal. 'Progress' is then understood as 'the progressive overcoming, or control of, this "barbarian" non-human or semi-human sphere by the rational sphere of European culture and "modernity"' (p. 52).

Feminist views on the environment and development are broad and continuing to evolve. Almost all, however, share one characteristic, which is the fierce challenge they offer to developmentalism and the status quo of the world system, and hence to the conformism of MSD.

Towards a political ecology

Radical streams in writing on the environment are rich and strong. They are of great importance to understanding sustainability. A coherent understanding of how society and nature relate must go beyond the simple oppositionism of conventional Western environmentalism, and the limited reformism of MSD thinking. That thinking provides no answers to the proposed 'environmental crisis' beyond reform of the procedures and organization of development planning. Its view of society and environment is restricted, untheorized and naive.

In order to move beyond this, it is necessary to take account of political economy, and move outside environmental disciplines, and outside environmentalism, to approach the problem from political economy and not environmental science. There is great power in an approach to the understanding of environmental aspects of development that uses the insights of both natural and social science. In the 1980s, work by Piers Blaikie and Harold Brookfield (Blaikie 1985; Blaikie and Brookfield 1987) made this innovation, in the process building bridges between environmental issues and radical social studies. The specific focus of their work was the study of soil erosion. It led to the establishment of the burgeoning field of political ecology (Haila and Levins 1992; Rocheleau *et al.* 1996b; Bryant and Bailey 1997; Bryant 1998; Stott and Sullivan 2000; Zimmerer and Bassett 2003; Peet and Watts 2004; Robbins 2004).

Political ecology is diverse and trans-disciplinary. Political ecologists understand environmental or ecological conditions as the product of political and social processes, linked at a number of nested scales from the local to the global (Bryant and Bailey 1997). Their work seeks to tie the logics, dynamics and patterns of economic change to the politics of environmental action and to actual ecological outcomes (Peet and Watts 2004). They typically engage in field-based empirical research (often case-study research), a localized or regional approach with roots in geography, anthropology, sociology and environmental history (Zimmerer and Young 1998).

The field of political ecology explicitly addresses the relations between the social and the natural, arguing that social and environmental conditions are deeply and inextricably linked. Moreover, it emphasizes not only that the actual state of nature needs to be understood materially as the outcome of political processes, but also that the way nature itself is understood is also political. Ideas about nature, even those that result from formal scientific experimentation, are formed, shared and applied in ways that are inherently political (Escobar 1999). There is particular interest in the place of the apparatus of the state in directing, legitimizing and exercising power and control (Forsyth 2003; Peet and Watts 2004; Robbins 2004).

At the heart of political ecology is the observation of the centrality of politics in attempts to explain the interactions between people and the environment, or the 'dynamics and properties of a "politicised environment"' (Bryant 1998, p. 82). Specifically, political ecology emphasizes the importance of asymmetries of power, the unequal relations between different actors, in explaining the interaction of society and environment (Bryant and Bailey 1997).

Blaikie (1995) suggests that, since the 1970s, political ecology, like the wheel, has been repeatedly reinvented. Bryant and Bailey (1997) identify two phases in its development. The first, from the mid-1970s to the mid-1980s, was built on neo-Marxism and emphasized structural explanations of human–environment relations. Development in this period was slow and piecemeal, partly because of the lack of interest by Marxist scholars in the environment, and partly because of radical aversion to the neo-Malthusian explanations of environmental 'problems' associated with ecological approaches to understanding environmental change (see Chapter 6). Nevertheless, in the 1980s, critiques of neo-Malthusianism emphasized the political economy of the environment – for example, work on famine in the Sahel (e.g. Copans 1983; Watts 1983a, b) and made a major contribution to the development of political ecology (Robbins 2004).

This work is discussed further in Chapter 8. The second period in the development of political ecology, from the later 1980s, has been more complex, with a
greater focus on the role of grass-roots actors and social movements (Bryant and
Bailey 1997); it has seen in particular a greater awareness of discursive dimensions of environment–society interactions (Peet and Watts 2004).

The critical innovation of political ecology has been the search for explanations that take account of both the natural and the social sciences (Blaikie and
Brookfield 1987). The appeal of such a synthesis is obvious, although the intellectual
coherence is more questionable (Watts and Peet 2004). Bryant (1998) suggests
two key themes running through the political ecology literature. The first is
the way in which unequal power relations relate to conflicts over access to and
use of resources; the second concerns the ways in which power relations are
reflected in conflicting discourses and knowledge claims about the environment
and development. Peet and Watts (1996) pointed to new work in four areas.
The first sought to make explicit the causal connections between the logics
and dynamics of capitalist growth and specific environmental outcomes. The
second specifically addressed the politics of social action about the environment.
The third explored social movements and civil society. The fourth addressed
discursive dimensions of the way the environment is defined and its dynamics
described (Peet and Watts 1996).

All relations between environment and people are political, just as all development is ideological. In a preface to the 1979 reprint of his paper 'Ecology and
revolutionary thought' (originally written in 1965), Murray Bookchin argued
that 'the domination of human by human lies at the root of our contemporary
Promethean notion of the domination of nature by man' (p. 21). The way people
relate to non-human nature around them (their environment) – as well as the
way they understand it – is created by culture and bounded by social relations,
by structures of power and domination. Development itself is a product of power
relations, of the power of states, using capital, technology and knowledge, and
the market to alter the culture and society of particular groups of people. States
co-opt cultures while the world system engages indigenous economies. We call
the result of this process development.

Development is about control, both of nature and of people. It is at this level
that MSD thinking is seen to be so profoundly limited. For, although sustainability seems to offer a moral critique of the development process, in the end it is
simply co-opted into an adapted version of the same paradigm. Indeed, Hartwick
and Peet (2003) suggest that sustainable development serves as a conceptual
device that promises, but fails, to bridge the impassable divide between growth
and environmental destruction. The word 'sustainable' needs to be understood
ideologically as 'the effects most people can be persuaded to find tolerable, as
the necessary environmental consequences of an even more necessary growth
process' (p. 209).

This view of sustainable development as essentially a managerial process, where
reform of procedures will ensure some 'optimal' outcome, fails to address the
central issue of the ideology of the 'modern project' (to use Friberg and Hettne's

1985 phrase). Michael Jacobs (1995) reminds us not to forget that sustainability is necessarily about ideology:

> It needs to be remembered after all that sustainable development and sustainability were not originally intended as 'economic' terms. They were, and remain, essentially political objectives, more like 'social justice' and 'democracy' than 'economic growth'. And as such their purpose, or 'use', is mainly to express key ideas about how society – including the economy – should be governed.
>
> (p. 65)

Development creates losers as well as winners. It is something that is done to people, by governments, corporations, civil society and other groups of people and individual actors. Those who drive change co-opt or reflect dominant ideologies and often draw on financial capital from outside interests. Behind all change is the blind neo-liberal engine of the market. To the target of change, 'development' holds many terrors, significant costs and risks as well as possible benefits. Governments make decisions on behalf of citizens regarding proper balances of costs and benefits, and about where those costs should be borne and those benefits enjoyed. Very often (as environmentalists devastatingly have been pointing out for almost half a century now) the two do not balance. Even more often, the costs and benefits are unequally shared.

Development is a double-edged phenomenon, a tiger that governments ride and to which they attach – sometimes by force (Crummey 1986; see also Chapter 12) – the lives of individuals and groups of people. In development the domination of nature is part of wider political and economic processes. It is impossible to understand the relation between development and nature without considering and comprehending political economy. However, once we escape from the straitjacket of evolutionary thinking about 'development', which argues that it necessarily involves a progression towards 'better' conditions, it is possible to start to consider anew the persistent questions raised by environmentalist critiques of development practice.

These are to be considered in the remaining chapters of this book, looking at issues of sustainability and political ecology across a range of specific contexts. The next chapter looks at the political ecology of debates about environmental degradation, particularly dryland desertification. What do we know about threats to drylands? How are ideas about degradation affected by ideology? Is it possible for farmers or pastoralists to manage their environments sustainably, and, if so, what impacts do policies to 'combat desertification' have on them?

Summary

- Radical green critics of developmentalism reject the possibility that capitalism and industrialism can deliver justice, equity and humane conditions of human life. They suggest that the modern world system is flawed, generating interlinked crises (militarization, poverty, human repression and environmental destruction); mainstream sustainable development (MSD) cannot bring about the changes required.

- There is an important stream of radical thought in socialism, although historically much socialist writing (particularly in Marxism) underplayed the importance of the environment. Eco-socialism includes a range of political ideas, including decentralization, communalism and utopian socialism, and the contribution of socialist ideas to green thinking is greater than is commonly recognized. Socialism provides a powerful critique of the environmental and developmental impacts of capitalism.

- Anarchist thinking also provides a basis for radical green ideas about development, particularly in Murray Bookchin's 'social ecology'. Ideas with anarchist roots such as decentralization, participatory democracy, self-sufficiency, egalitarianism and alternative technologies are important elements in environmentalism, and of direct relevance to debates about sustainable development.

- Biocentrism (the recognition of intrinsic values or moral status in non-human nature) is an important theme in environmentalism. Deep ecologists reject anthropocentrism, and their ideas offer a profound challenge to many of the assumptions on which MSD is based, although they also raise critical political questions.

- Eco-feminism offers a diverse range of ideas that challenge MSD – for example, in emphasizing the patriarchal roots of developmentalism and capitalism. There are significant differences between separate strands of feminist thought – for example, the nature of any special connection between femininity and nature.

- Radical ideas about environment and development have by no means been suppressed by the routinization of environmental concern in MSD. They emphasize in particular the political dimensions to decisions about society and nature, and challenge the notion that sustainable development can be achieved through technical and managerial responses alone.

Further reading

Bookchin, M. (1982) *The Ecology of Freedom: the emergence and dissolution of hierarchy*, Cheshire, Palo Alto, CA.

Brown, L. R. (2006) *Plan B. 2.0: rescuing a planet under stress and a civilization in trouble*, W. W. Norton, New York, for the Earth Policy Institute.

Dobson, A. ([1990] 2007) *Green Political Thought*, HarperCollins, London (4th edition).

Dresner, S. (2002) *The Principles of Sustainability*, Earthscan, London.

Eckersley, R. (1992) *Environmentalism and Political Theory: toward an ecocentric approach*, UCL Books, London.

Ekins, P. (1992) *A New World Order: grassroots movements and global change*, Routledge, London.

Guha, R. and Martinez-Alier, J. (1997) *Varieties of Environmentalism: essays North and South*, Earthscan, London.

Katz, E., Light, A. and Rothenberg, D. (2000) (eds) *Beneath the Surface: critical essays in the philosophy of deep ecology*, MIT Press, Cambridge, MA.

Lewis, M. W. (1992) *Green Delusions: an environmentalist critique of radical environmentalism*, Duke University Press, Durham, NC.

Mason, M. (1999) *Environmental Democracy*, Earthscan, London.

Monbiot, G. (2007) *Heat: how we can stop the planet burning*, Penguin Books, Harmondsworth.

Pepper, D. (1993) *Eco-Socialism: from deep ecology to social justice*, Routledge, London.

Pepper, D. (1996) *Modern Environmentalism: an introduction*, Routledge, London.

Redclift, M. R. (2005) *Sustainability: critical concepts in the social sciences*, Routledge Major Works (four volumes), Taylor and Francis, London.

Redclift, M. R. and Benton, T. (1994) (eds) *Social Theory and the Global Environment*, Routledge, London.

Redclift, M. R. and Woodgate, G. (1997) (eds.) *Developments in Environmental Sociology*, Edward Elgar, Cheltenham.

Reed, P. and Rothenberg, D. (1993) (eds) *Wisdom in the Open Air: the Norwegian roots of deep ecology*, University of Minnesota Press, Minneapolis.

Sessions, G. (1995) (ed.) *Deep Ecology for the 21st Century: readings on the philosophy and practice of the new environmentalism*, Shambhala, Boston.

Whiteside, K. (2002) *Divided Natures: French contributions to political ecology*, MIT Press, Cambridge, MA.

Web sources

http://www.deepecology.org/index.htm The Foundation for Deep Ecology, Sausalito, California, established 'to support education and advocacy on behalf of wild Nature'.

http://www.greenspirit.org.uk/ GreenSpirit, a 'movement which celebrates all life as deeply connected and sacred': follow links for ecofeminism.

http://www.ecotrust.org/ Bioregionalism in action: Ecotrust is a non-governmental organization based in Portland, Oregon, USA. It was created in 1991 'to bring some of the good ideas emerging around sustainability back to the rain forests of home': to build 'salmon nation' along the western North American coast, from San Francisco to Anchorage.

http://www.conservationeconomy.net/ The 'bioregional pattern language' of the coastal region.

http://www.social-ecology.org/ The Institute for Social Ecology, founded by Murray Bookchin.

http://www.worldincommon.org/ The World In Common group, founded in 2002 to help inspire a 'vision of an alternative way of living where all the world's resources are owned in common and democratically controlled by communities on an ecologically sustainable and socially harmonious basis'.

http://www.tlio.org.uk/ The Land is Ours: a UK-based group that 'campaigns peacefully for access to the land, its resources, and the decision-making processes affecting them, for everyone, irrespective of race, gender or age'.

8　Dryland political ecology

Desertification is caused by the excessive pressures of overuse on productive ecosystems that are inherently fragile.

(Mustafa Tolba, 'Desertification in Africa', 1986)

We can no longer separate the natural phenomenon and the necessary political translation of its effects.

(Jean Copans, 'The Sahelian drought', 1983)

The political ecology of degradation

When the forester Edward Stebbing toured West Africa in 1934, he visited the Emir of Katsina in the arid far north of Nigeria. Here he encountered conditions that seemed familiar from his work in the Indian drylands. The Emir's representative showed him evidence of serious and recent environmental deterioration conditions. To Stebbing the problem was obvious. The dry lands of West Africa were, he said, undergoing progressive desiccation and the Sahara was moving southwards, a 'silent invasion of the great desert' (Stebbing 1935, p. 518). Stebbing saw the savannah as a form of open deciduous forest, progressively degraded by burning and shifting cultivation, grazing, browsing and pollarding by pastoralists. The result was inevitable. Under such management 'the final extinction of the savanna forest takes place, when the weakened roots and vanishing rainfall result in the death of the trees' (p. 513).

What Stebbing described in that now famous (or infamous) passage is the phenomenon that has become known as desertification, the degradation of productive land in dry regions. Concern about desertification has evolved a great deal since Stebbing's work in the 1930s, but it remains vigorous. In 1980 the United Nations Environment Programme (UNEP) estimated that about 35 per cent of the terrestrial globe was vulnerable to erosion (about 4.5 billion hectares); this land supported about a fifth of the world's population, some 850 million people. Of this area, 30 per cent was severely or very severely desertified; about 80 per cent of rangelands were affected by 'overuse' (Tolba 1986). The extent and severity of desertification were seen to be increasing in every arid region in the developing world (Mabbutt 1984). Analysts have suggested close links between drought, environmental conditions and methods of land management (for example, in the context of rain-fed farming in the Gambia (Baker 1995; see also Figure 8.1)).

For three decades the existence and significance of desertification as a global process have been taken for granted by policy-makers and scientists in environmental organizations, bilateral and multilateral donor organizations and national governments. Desertification, and soil erosion more generally, were key themes of environmentalist writing on the developing world (e.g. Brown and Wolf 1984; Timberlake 1985). The *World Conservation Strategy* (*WCS*) (IUCN 1980) saw desertification as 'a response to the inherent vulnerability of the land and the pressure of human activities' (para. 16.9). It described the way in which farmers had moved onto land marginal for agriculture and displaced pastoralists onto land marginal for livestock rearing in Africa's Sahelian and Sudanian zones, while in the Himalayas and Andes 'too many improperly tended animals remove both trees and grass cover ... and erosion accelerates' (para. 4.13).

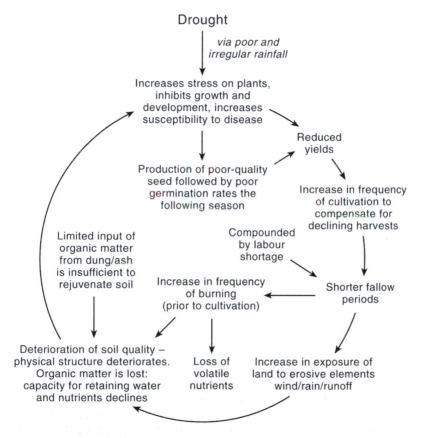

Figure 8.1 Drought and dryland agriculture and environmental degradation in the Gambia (after Baker 1995)

The Brundtland Report (1987) was more alarmist, stating that 'each year another 6 million hectares of productive dryland turns into worthless desert' (p. 2). Desertification was included among the environmental trends that threaten 'to radically alter the planet, that threaten the lives of many species upon it, including the human species' (p. 2). *Agenda 21* devoted a chapter to the problem of 'combating desertification and drought', as one of two under the rather emotive title 'Managing fragile ecosystems' (the other was on 'sustainable mountain development').

In the literature of mainstream sustainable development (MSD) the natural environment is repeatedly portrayed as suffering degradation at the hand of the teeming multitudes of the poor. The spectre of desertification has swept many observers and commentators along. The desert seemingly stalks green fields, livestock nibble inexorably away at the basis of pastoral subsistence, and burgeoning human and animal numbers threaten the long-term sustainability of soil productivity and rural economies. Such has been desertification's prominence that Thomas and Middleton (1994) described it as 'a concept out of hand' and 'an institutional myth' (pp. 63, 50). Yet similar stories about human overuse are told of other environments – for example, where small farmers cling to steep hillsides in Nicaragua (Ravnborg 2003), or where poor farmers and herders in the Altiplano of Bolivia and Peru face declining soil fertility (Swinton and Quiroz 2003; Zimmerer 2004). Repeatedly, the poor are portrayed teetering on a slender cusp between survival and both economic and ecological disaster (Swinton *et al.* 2003; Gray and Mosely 2005).

However, the links between poverty and environmental degradation are far from simple. It is often not the poorest land-users who cause the most degradation, but their richer neighbours, as Ravnborg (2003) showed in Nicaragua. Social, political and economic issues are central to questions of environment and development, a set of relations central to the field of political ecology, introduced in the previous chapter (Dietz 1996; Rocheleau *et al.* 1996b; Bryant and Bailey 1997; Stott and Sullivan 2000; Peet and Watts 2004). Political ecology has diverse roots (Bryant 1998), but came to prominence in Piers Blaikie's classic book *The Political Ecology of Soil Erosion in Developing Countries* (1985). He commented, 'small producers cause soil erosion because they are poor, and in turn soil erosion exacerbates that condition. A set of socio-economic relations called underdevelopment is at the centre of this poverty' (p. 138). Thus farmers or pastoralists enduring drought and poor soil fertility in the Sahel, or women forced to travel miles to collect fuelwood, are all experiencing environmental degradation that relates directly to their ability to achieve sustainable livelihoods and an acceptable quality of life.

Environmental degradation is integral to the hazards of life experienced by the poor. Very often those affected have neither the freedom to stop causing degradation, nor the opportunity to move elsewhere. Poverty and environmental degradation can form a trap from which there is little chance of escape. Farmers in the highlands of Ethiopia or Nepal farm steep and eroded hillsides, and agropastoralists in the Sahel who live and work degraded landscapes have nowhere else to go (Blaikie 1988; Ives and Messerli 1989; Mortimore 1998).

The environment is not neutral in its effects on the poor: environmental quality is mediated by society, and society is not undifferentiated. Access to and the distribution of environmental 'goods' (be they land, water, grass or fuelwood) are uneven. Piers Blaikie (1985) showed the need for a political–economic understanding of the phenomena of environmental degradation:

> It is when the physical phenomenon of soil erosion affects people so that they have to respond and adapt their mode of life that it becomes a social phenomenon. When this response affects others and brings about a clash of interests ... it becomes a political problem as well.
>
> (p. 89)

Political ecology also seeks to integrate explanations across spatial (and temporal) scales (Blaikie and Brookfield 1987). Thus local 'environmental problems' (soil erosion is now the classic example) need to be seen as the product not only of local processes (farming practices) but also of political economy at local, national and international scales. Thus soil erosion might be said to happen because it rains hard just after fields have been dug, but is also influenced by a series of other factors – among others, perhaps, because labour shortages prevent timely land preparation, because men have migrated out in search of wage employment owing to a depressed agricultural economy, because national fiscal policy subsidizes imported agricultural products, or because national debt and 'structural adjustment' have cut funding for agricultural extension or health clinics. An explanation that leaves out the wider political economy leaves out vital parts of the story. In 1981 Blaikie argued the importance of linking the circumstances of decision-makers to micro-level, national and worldwide political economy. In 1995 he developed this into the concept of a 'chain of explanation' (see Figure 8.2). Thus, in considering soil erosion, Piers Blaikie suggested that physical changes in soils and vegetation were linked to economic symptoms at particular places at particular times, and in turn to land-use practices in that place, to the resources, skills, assets, time horizons and technologies of land-users, to the nature of agrarian society and finally to the international political economy (see Figure 8.3).

The links between environment, economy and society are complex, and the process of development is both a response to and a cause of environmental change: 'land degradation can undermine and frustrate economic development, while low levels of economic development can in turn have a strong causal impact on the incidence of land degradation' (Blaikie and Brookfield 1987, p. 13). More importantly, because the development process involves the transformation of social and economic relations, it relates to the ways in which individuals and groups within a society experience their environment, and the ways in which they use it.

Conventional analyses of environmental 'problems' tend to draw heavily on scientific explanations, with social and political dimensions ignored or downplayed. And yet, questions of epistemology (of theories about knowledge) make the assumption of the possibility of impartial or neutral definitions of environmental problems or analyses of environmental change problematic. The definition of

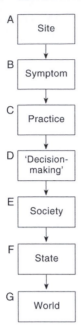

Figure 8.2 Political ecology's chain of explanation (after Blaikie 1995)

environmental problems is also socially mediated. As Jarosz (1996) points out, environmental 'problems' tend to be defined in ways that make them amenable to technical solutions; and technical interventions do not appear from nowhere, but are themselves highly political (Blaikie 1985, 1995; Forsyth 2003).

Before 'environmental issues' can be articulated, the environment is experienced by people and made the subject of conscious thought (whether by scientists or by pastoralists). Moreover, the ways in which different people experience the environment, derive their understandings and develop discourses about it vary, and differences of view (whether between people with different bundles of rights, or between those with different claims to environmental knowledge such as scientists and lay people) interact in a political process. The status of actors will be reflected in the power of their arguments: 'hard' natural scientists may disparage the insights of 'soft' anthropologists; bureaucrats or urban businesspeople may seek to ignore the views of Indian *adivasis* or African pastoralists; men may downplay the environmental knowledges and understandings of women. In political ecology the exercise of power (over nature, over other people) therefore has to be understood at the discursive as well as the material level (Bryant and Bailey 1997).

The social construction of knowledges about the environment can itself be the means by which power is exercised over both nature and society. Recognition of the depth and complexity of debates about the social construction of nature – and about the material reconstitution and 'invention' of life-forms possible through genetic engineering – suggests that political ecology should be

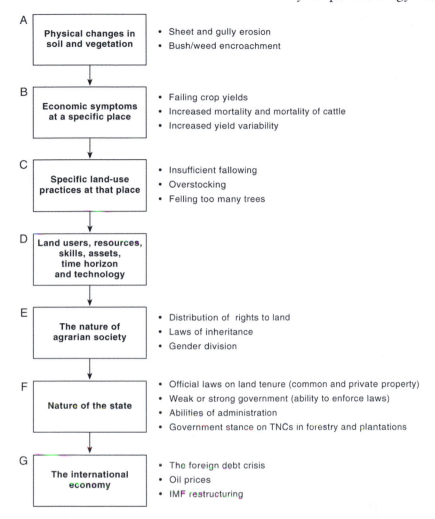

A **Physical changes in soil and vegetation**
- Sheet and gully erosion
- Bush/weed encroachment

B **Economic symptoms at a specific place**
- Failing crop yields
- Increased mortality and mortality of cattle
- Increased yield variability

C **Specific land-use practices at that place**
- Insufficient fallowing
- Overstocking
- Felling too many trees

D **Land users, resources, skills, assets, time horizon and technology**

E **The nature of agrarian society**
- Distribution of rights to land
- Laws of inheritance
- Gender division

F **Nature of the state**
- Official laws on land tenure (common and private property)
- Weak or strong government (ability to enforce laws)
- Abilities of administration
- Government stance on TNCs in forestry and plantations

G **The international economy**
- The foreign debt crisis
- Oil prices
- IMF restructuring

Figure 8.3 The chain of explanation of land degradation (after Blaikie 1995)

open to new understandings of culture and nature, and embrace 'poststructural Political Ecology' (Escobar 1996). The importance of discursive dimensions of political ecology has been widely noted (Peet and Watts 1996; Stott and Sullivan 2000; Watts and Peet 2004; cf. Forsyth 2003; Robbins 2004).

In their book *Liberation Ecologies*, Peet and Watts (2004) emphasize the importance of the politics of meaning and the construction of knowledge. They argue that political ecology has moved forwards along two axes. The first relates to work on the politics of knowledge, the second to the politics of social action, or 'ecological democracy'. Rather than 'environmental problems' generating single understandings to which people respond, there are in fact 'multiple realizations about all levels of environmental problems' (Peet and Watts 1996, p. 37).

What is known and understood about the environment and people's relation to it is highly political. The ideas of different local people and groups, and the ideas of scientists and other experts, are often different, and, when they clash (for example, in debates about whether particular development projects should go ahead), the outcome can have profound implications for both nature and people. Different understandings stimulate creative reactions that may (or may not) emerge as social movements. While the struggles that emerge may be material struggles about survival or livelihoods, they are also struggles about the ways in which people speak about and think about and organize understandings of human and non-human nature.

In the politics of ideas about nature and society, some sets of ideas are more powerful than others, particularly the ideas of development planners. Emory Roe (1991) drew attention to the significance of the way environmental ideas become standardized in development, in what he calls policy narratives. Development planners create self-referencing stories, convincing each other that their understanding of problems is correct and their choice of solutions appropriate. Policy narratives are remarkably persistent, and cannot be overturned by simply showing that they are untrue in a particular instance, but only by providing a better and more convincing story (E. Roe 1991, 1994; Leach and Mearns 1996). Such narratives structure thinking about sustainable development in many contexts, as far apart geographically and substantively as debates about soil erosion in Bolivia (Zimmerer 2004) or flooding in Bangladesh (Bradnock and Saunders 2000). The issue of desertification in the Sahel is a casebook example of the power of environmental narratives (Swift 1996; S. Sullivan 2000). In the Sahel it is possible to trace a continuous thread of environmental concern from early in the twentieth century to the present day. The ideas of Edward Stebbing about northern Nigeria have persisted to a remarkable degree. Their many contemporary equivalents are just a couple of mouse-clicks away (see, e.g., http://en.wikipedia.org/wiki/Desertification).

Desertification and the scientific imagination

If it is accepted that environmental change is understood in different ways by different actors, it is possible to look anew at concern about environmental degradation. By the time Stebbing visited West Africa, the fact that deserts came and went over geological time was well established (e.g. Hobley 1914). However, the issue that he picked up was rather different, a fear of a rapid increase in aridity and the encroachment southwards of the Sahara. Others had already expressed this concern (e.g. Bovill 1921). Droughts in West Africa in the early decades of colonial rule (for example, in Hausaland in 1913 (Grove 1973)) had significant effects on the thinking of colonial administrators. For example, low rainfall and low river floods in Sokoto Province in northern Nigeria in 1917 and 1918 led to a remarkable level of colonial government concern at 'desiccation' and consequent interest in small-scale flood irrigation in the valley of the River Sokoto between 1919 and 1921, although these experiments ultimately failed (Adams 1987).

Stebbing's hypothesis about the spread of the Sahara desert southwards was rapidly refuted by Brynmor Jones (1938) on the strength of the fieldwork of the Anglo-French Forestry Commission to northern Nigeria and Niger in 1936 and 1937. Jones argued that there was no evidence of southward encroachment of sand, retrogression of vegetation, permanent reduction of rainfall, recent shrinkage of streams or lakes, or lowering of the water table. He concluded, 'there is no reason to fear that the desiccation through climatic causes will impair the habitability of the West African colonies for many generations to come' (p. 421). Subsequently, Stebbing himself (1938) went some way towards this position, stressing instead the spoliation and erosion of land owing to population increase and to agricultural and pastoral practices. The idea of the spreading desert did not, however, go away.

While the geographer Dudley Stamp concluded that there was 'no need to fear desiccation through climatic causes', he argued (with Stebbing) that 'the spread of man-made desert from within is quite another matter' (Stamp 1940, p. 300). This concern about the links between the management of drylands and *in situ* land degradation and soil erosion was linked to global alarm at the American Dust Bowl (Beinart and Coates 1995). Soil erosion and dust storms in the American Midwest, and particularly the experience of dust over the eastern seaboard,

Plate 8.1 Erosion gully, northern Nigeria. It may seem obvious that a gully of this kind indicates overgrazing or some other inappropriate use of land, yet without a clear understanding of soil conditions and vegetation and rainfall change over time any such assumption would be rash. The narrative of dryland degradation encourages observers to jump to conclusions. Knowing whether they are right demands a little more time and care. Photo: W. M. Adams.

generated huge public concern in the USA. The Soil Erosion Act was passed in 1932, and the Soil Conservation Service in 1935. The report of the Great Plains Committee in 1936 ushered in 'a radically new environmental outlook' (Worster 1985, p. 232) based on the ecological thinking of scientists such as Frederick Clements, whose ideas of vegetation succession and climatic climax (see Chapter 2), came to dominate ecological ideas about land use and indeed the US national imagination.

News of this environmental crisis in the USA was rapidly spread internationally through newspapers and other routes (including the Imperial School of Tropical Agriculture in Trinidad (D. M. Anderson 1984)) to feed concern about soil erosion in arid and semi-arid environments across the world (Jacks and Whyte 1938; Furon 1947). These concerns took firm root in both British and French territories in Africa (Aubréville 1949; Harroy 1949; Anderson 1984; Beinart 1984), although the scope of concern was essentially worldwide. Stamp (1940) suggested, 'there now seems little doubt that the problem before West Africa is not the special one of Saharan encroachment but the universal one of man-induced soil erosion' (p. 300). Similarly, Vogt (1949) concluded: 'Whether or not Africa is actually suffering a climatic change, man is most effectively helping to desiccate the continent' (p. 248).

To some extent at least, desertification is in the eye of the beholder. Historically, people have seen what they wanted to see. Preconceptions about ecological processes, an exaggerated ability to infer process from form, and therefore make assumptions about what is happening in a landscape on the basis of what can be seen at any given moment, and above all the ideological baggage of the observer all contribute to the perception of desertification. David Anderson (1984) describes four factors that encouraged a move by the state in colonial Kenya to undertake direct intervention in African agriculture over the 1930s: the Depression, the American Dust Bowl, population growth and drought.

The Depression threatened both African and settler farmers in Kenya. African farmers responded to the slump in agricultural prices by expanding the area under production of maize. White farmers facing bankruptcy responded defensively, seeking to entrench their position, in the process claiming (notably to the Kenya Land Commission 1932–4) that African farming practices were damaging to the environment and unproductive. The impacts of the Depression and the political response to it were compounded by reports of the American Dust Bowl, which 'caused Agricultural Officers all over British Africa to examine their own localities for signs of this menace' (Anderson 1984, p. 327). The prevention of erosion, particularly as this was perceived to occur primarily on African land, was a major stimulus to intervention by the colonial state in African agriculture (which had previously been ignored). Meanwhile, by the 1930s the growth of the African population had become a matter of official concern, because of problems of overcrowding and landlessness in the Kikuyu 'reserves' (Throup 1987). European settlers farming in the White Highlands perceived soil erosion both on the reserves and on their own farms, where 'squatters' exchanged plots of land for their labour. Finally, the years from the mid-1920s to the mid-1930s were dry in

East Africa, and drought created periodic local food shortages and, sometimes widespread, mortality in cattle herds. Famine relief by the state was expensive, and drought heightened the perception of environmental degradation caused by African husbandry.

By 1938 soil conservation had become a major concern of government throughout the East African colonies, the cutting edge of a policy of state intervention in African agriculture. In Kenya, there were progressive attempts to persuade, and to compel, Kikuyu farmers to adopt soil conservation measures such as crop rotation, compositing and grass leys. From 1938 the Soil Conservation Service under Colin Maher promoted the construction of terraces by hand in African reserves. The policies were deeply unpopular, particularly among women, who bore the brunt of the forced labour, and generated significant opposition (MacKenzie 1998, 2000). Discourses of conservation, environmentalism and betterment served as what Foucault (1975) describes as 'political technologies', allowing political issues such as the control and use of land to be translated into technical issues, where formal science could be drawn upon to diagnose land health and prescribe treatment (Mackenzie 2000). Interventions and reactions were similar elsewhere (for example, in southern Africa (Beinart 1984)). Elsewhere, similar soil conservation policies fell on more favourable ground – for example, in Kigezi, Uganda, where proposals for strip cropping and terraces built on indigenous techniques (Carswell 2003, 2007).

The powerful and emotive image of environmental decline in semi-arid regions of Africa under the pressure of agricultural misuse led to a persuasive and self-confirming perception of environmental degradation, or what would now be called desertification. There are few data with which to assess scientifically the severity of the problems to which colonial administrators reacted. Such data were not collected. They were not deemed necessary; the ideology of degradation was sufficient to generate policy. Where such data do exist, they can tell a very different story. Areas such as Machakos in Kenya were portrayed in the 1930s as degraded wastelands, where human survival was at risk from soil erosion. When historical data from this period were used to examine changes in land use over time, they revealed a remarkable phenomenon, one of progressive improvements in soil conservation (Tiffen *et al.* 1994). This work is discussed further below.

It was this same fear of degradation that resurfaced in the 1970s under the new label of desertification (Ribot 1999). The severity of the 1972–4 drought led to a debate on the floor of the United Nations. As a result of this, the United Nations Conference on Desertification (UNCOD) was held in Nairobi in 1977, organized by UNEP. In that year, UNEP was made responsible by the UN General Assembly for coordinating implementation of a Plan of Action to Combat Desertification (PACD). UNEP established a desertification branch, an Interagency Working Group and a Consultative Group on Desertification Control, and began publication of the *Desertification Control Bulletin* and various editions of the *World Atlas of Desertification* (Middleton and Thomas [1992] 1997). This institutional interest, and the persistence of Sahelian and African drought through

Plate 8.2 Agricultural terracing in Kigezi, south-west Uganda. Land in Kabale District is in short supply and rarely flat. Indigenous soil conservation practices created unusually receptive conditions for a colonial terracing programme (Carswell 2007), and land in the district is now everywhere carefully and thoroughly terraced. Photo: W. M. Adams.

the 1970s and 1980s, kept desertification on the international agenda (Mabbutt 1984; Tolba 1986; Warren 1993, 1996).

Swift (1996) argued that the desertification narrative became so widely accepted because it served the interests of specific groups of powerful policy actors: national governments in Africa, international aid bureaucracies (especially UN agencies) and scientists. In the 1970s, recently independent African governments were restructuring their bureaucracies and strengthening central control over natural resources. Drought, and the assumptions about human-induced environmental degradation linked to them, legitimated such claims and made centralized top-down environmental planning seem a logical strategy. Pastoralism, in particular, could be portrayed as doomed and self-destructive, and its replacement by sedentary agriculture made to seem necessary and beneficial. Aid donors, meanwhile, found in desertification a problem that seemed to transcend politics and legitimated 'large, technology-driven international programmes' (Swift 1996, p. 88). To scientists developing new fields such as remote sensing, desertification offered fertile terrain for expansion: satellite imagery seemed to offer answers without the need for lengthy and tedious fieldwork, and desertification became a source of funding and legitimacy for new cadres of technicians and researchers. The role of science in the analysis and boosting of desertification are discussed in the next section.

At the Rio Conference in 1992, a resolution was passed to draw up a convention to combat desertification. This was adopted in 1994 and entered into force in December 1996, the laboriously entitled 'UN Convention to Combat Desertification in Those Countries Experiencing Severe Drought and/or Desertification, especially in Africa' (International Institute for Sustainable Development 1997).

Given this international attention, it is interesting that the exact definition of desertification has been the subject of extended debate among researchers and policy-makers (Verstraete 1986; Thomas and Middleton 1994). The term was coined by the French ecologist Aubréville in 1949, who used it to refer to an extreme form of the process of 'savannization', the conversion of tropical and subtropical forests into savannahs as a result of severe soil erosion, changes in soil properties and the invasion of dryland plant species. The geographer A. T. Grove (1977), conscious of the fact that over geological time deserts come and go without human influence, took a broad view, defining it as 'the spread of desert conditions for whatever reasons, desert being land with sparse vegetation and very low productivity associated with aridity, the degradation being persistent or in extreme cases irreversible' (p. 299). The UN Convention to Combat Desertification (in 1994) pragmatically defined desertification as 'land degradation in arid, semi-arid and dry subhumid areas resulting from various factors including climatic variations and human activities' (N. Robinson 1993).

The science of desertification

Climate variability is central to an understanding of desertification. It is now clear that drought is a persistent feature of tropical Africa. Its occurrence is complex in both space and time. Yet, at the time of the Sahel drought of 1972–4, relatively little was known about climate change in Africa. There had been sequences of dry years in the Sahel and Sudan zones at the start of the twentieth century (Grove 1973), but few remembered them, and their causes were little understood (Lamb 1979). Years of lower than average rainfall actually began in 1968, before the particularly poor years between 1972 and 1974 (Kowal and Adeoye 1973), and it is now clear that the 1950s and 1960s were rather wet (Hulme 1996). Moreover, drought did not end in 1974. The recurrence of drought and hunger in the Sahel and the Horn of Africa in 1984 reawakened international concern, as did the drought of 1992 in southern Africa (Hulme 1995). There have been numerous subsequent drought (for example, in the Sahel in 2005 and in the Horn of Africa and northern Kenya in 2006).

In the 1970s climatologists hypothesized that there might be links between human action on the land surface and climate through the impacts of albedo

Plate 8.3 An active sand dune, Kaska Village, northern Nigeria. The commonest popular description of desertification is 'the advance of the desert'. Sand dunes do move, but this is a poor basis for understanding arid ecosystem degradation. This dune, slowly overwhelming a small village near the northern border of Nigeria with Niger in West Africa, has become something of a PR celebrity, much frequented by people with visitors whom they wish to convince about the perils of desertification. In fact, the dune forms part of the Manga Grasslands, a relict of the former sub-Saharan dunefields active at the height of the last glaciation over 12,000 years ago. Mike Mortimore has studied them intensively, and points out that the grassland landscape today is much as described by the Anglo-French Boundary Commission in 1937, when it had much the same borders, and was also treeless and with active dunes (Mortimore 1989). Photo: W. M. Adams.

(reflectance of the land surface), vegetation cover and atmospheric dust on earth surface temperature and cloud formation. Otterman (1974) reported links between surface albedo and climate, describing a temperature difference of 5°C between the heavily grazed Sinai desert and the less grazed Negev. He suggested that bare soils caused by overgrazing might be a possible mechanism of deser-tification, with higher albedo causing lower surface temperatures and reduced atmospheric convection. These observations were subsequently backed up by modelling experiments (e.g. Charney 1975; Charney *et al.* 1975) that produced interesting results, and some controversy (Idso 1977).

Researchers mapped changes in vegetation in the Sahel using satellite imagery (Henderson-Sellers and Gornitz 1984; Tucker *et al.* 1984, 1985), and discussed links between albedo and drought (Courel et al. 1984). Others measured atmospheric dust and considered its possible effects in creating warm inversion layers at altitude that suppressed convective rising of humid surface air (Prospero and Nees 1977; N. J. Middleton 1985).

The hypothesis that, in Charney's memorable phrase (1975), 'the desert feeds on itself' was widely taken as fact. However, the evidence for it remained incon-clusive. Subsequent work showed that links between human agency and climatic variation are complex, and it is now clear that they must be understood to embrace the global atmospheric system rather than any small local impacts. The impacts of CO_2 and other greenhouse gases, as well as emissions of soot, carbon nitrogen and ozone (from tropical burning as well as from urban industrial sources) are impor-tant (Williams and Balling 1995). There are, for example, links between Sahel rainfall and anomalies in global sea temperatures (Folland *et al.* 1986), implying possible impacts of atmospheric carbon dioxide and global warming. Droughts in the Sahel are associated with warm sea surface temperatures in the southern oceans and Indian Ocean, and cool temperatures in northern oceans (Hulme 1995). Rainfall anomalies in Southern Africa (and, for example, maize yields in Zimbabwe) are linked to the El Niño/Southern Oscillation (Hulme 1996).

On timescales of centuries, it has become increasingly clear that prolonged droughts are an established feature of African drylands (Grove 1981; Nicholson 1996). Historical records showed that the margins of the Sahara were wetter throughout the sixteenth to eighteenth centuries, but experienced major droughts in the 1680s and between 1738 and 1756 (Nicholson 1978). From the end of the eighteenth century to the third quarter of the nineteenth, conditions were generally drier, with another drought between 1828 and 1839. Then, between 1875 and 1895, it was wetter than the present on both edges of the Sahara, in the central Sahara and in much of East and North Africa. This, of course, was the period of colonial expansion: the Europeans arrived at the end of a period excep-tionally blessed with rainfall. Similar periods of wetter and drier conditions than the present have been demonstrated by historical research in southern Africa, although the timing of these periods is regionally specific. Maps of rainfall anoma-lies in twentieth-century Africa reflect those of the nineteenth century and earlier; regional patterns are complex, and changes between wet and dry conditions can occur rapidly (Nicholson 1996).

The longer-term palaeoclimatic record shows much larger climatic variations over timescales of thousands of years (Hulme 1996; Nicholson 1996; Roberts [1989] 1998). The tropics were arid during the last glacial maximum (Williams 1985), with dunefields far into the Sahel and other now wetter regions in Africa (Grove 1958; Grove and Warren 1968; Thomas 1984). Prior to that, fossil lake shorelines show previously much wetter periods (Washbourn 1967; Street and Grove 1976). Palaeoenvironmental studies in South America and Asia support the view of variability in past climates in the tropics, and widespread (but not universal) aridity at the peak of the last glaciation (Roberts [1989] 1998).

Perceptions of the severity of recent desertification are to a large extent self-confirming, but have a scanty scientific base, despite the use of satellite remote sensing over several decades (e.g. Justice *et al.* 1985; Tucker *et al.* 1985, 1986). Agencies such as UNEP lack the money or the remit to undertake scientific research on an international scale, and the urgency of policy concern about 'desertification' is based on loosely conceived concepts that provide no clear and consistent theoretical basis on which scientists might be attracted to build their research programmes and careers. Estimates of the global extent of desertification have continued to be problematic. In 1992 UNEP estimated that between 1,016 and 1,036 million hectares of land were experiencing soil degradation, a total of perhaps 2,556 million hectares if vegetation change is included (see Figure 8.4). This was less than one-third of the area estimated at UNCOD in 1977, or in the previous 1984 survey (Thomas and Middleton 1994; Middleton and Thomas 1997).

Attempts to measure current 'desertification' based on actual observations of real places over time are few. Measures used have included the expansion of areas of moving sand, the deterioration of rangelands, the degradation of rainfed croplands, waterlogging and salinization of irrigated areas, deforestation and declining ground or surface water supplies (Mabbutt 1984). The classic account of the rate at which deserts 'spread' was a study by Hugh Lamprey in 1975 on vegetation change in the southern Sudan (Lamprey 1988). A comparison of the boundary between desert and sub-desert grassland and scrub between a 1958 vegetation map and a 1975 satellite image suggested that the boundary had shifted 90–100 kilometres in the intervening seventeen years (5.5 kilometres a year). This statistic was taken up, repeated and reinvented by politicians and many others in the ensuing decades, and the process that it implies (of desert expansion) still dominates debate about desertification, despite a careful follow-up study by Hellden (1988) that questioned Lamprey's findings. The 1958 vegetation boundary was not surveyed but interpolated from a rainfall isohyet (itself likely to be notional in such a remote area, and of course subject to considerable inter-annual variation); comparison of 1975 and 1979 satellite imagery showed no change in vegetation boundaries. Other Swedish studies, including field surveys, showed short-term impacts of drought between 1965 and 1974, but no systematic decline in crop production, no shift in dunefield positions and no major changes in vegetation cover (Thomas and Middleton 1994).

However, it is all too easy to over-react and misinterpret the lack of clear evidence of desertification as evidence that environmental degradation does not

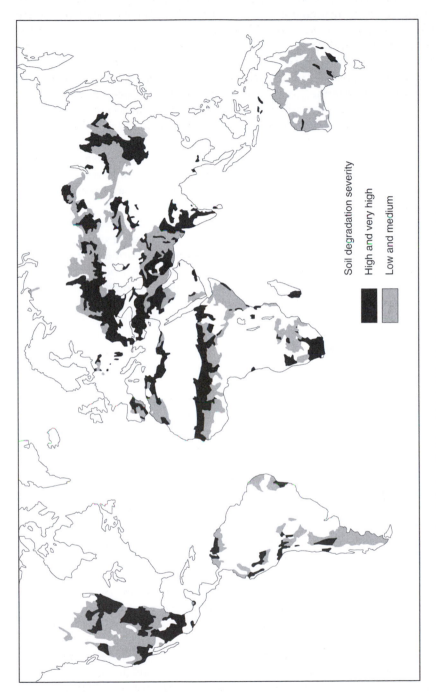

Figure 8.4 Desertification in the world's susceptible drylands after Middleton and Thomas (1994)

Soil degradation severity

High and very high

Low and medium

happen. A better conclusion would be that stereotyped narratives of degradation are dangerous, and that careful empirical research in specific locations is needed before pronouncements are made on environmental trends. There is, for example, a significant problem of soil erosion in Ethiopia, and projects to restrict access to hillsides to allow recovery work (Nyssen *et al.* 2004). In northern China, researchers are confident (admittedly from a comparison of 1980s printed maps and 1990s satellite imagery, the same methods Lamprey used) that desertification is developing rapidly owing to 'intensified and irrational human activities', meaning reclamation, overgrazing and over-cutting (Wu and Ci 2002). On the other hand, in north-west Namibia, Sian Sullivan (1999, 2000) demonstrates that (despite entrenched official fears of resource overuse in the savannah woodland) degradation was limited to small areas within settlements themselves. In southern Morocco, long-term surveys show little evidence of permanent overgrazing, or significant decline in vegetation over time (Davis 2004).

The truth seems to be that desertification is not a distinctive problem in its own right (Warren 1996). It is the result of the interlocking of a complex set of phenomena in semi-arid landscapes whose ecosystems are highly variable in space and time. There is a valuable distinction between drought, desiccation and dryland degradation (Warren and Khogali 1992). 'Drought' here means a dry period that lies within the range of ecosystem response: ecosystems (and the economic systems that depend on them) are affected, but survive, and can return to their former state when the dry spell is over. 'Desiccation', on the other hand, is a longer dry period, long and intense enough to bring about permanent (destructive) change in ecological and/or human communities. While one or two years of drought will be accompanied by loss of annual vegetation cover, this will return quickly from buried seed and deep-rooted trees, and with it will return the economic productivity of the rangeland (or fields). More prolonged drought will kill seed-bearing plants, and have a deeper destructive signature on the landscape. The third element of the triptych of dryland troubles is 'dryland degradation', defined as 'a process that reduces land productivity to the extent that natural recovery can only happen over many decades or where artificially accelerated recovery is beyond the capital and technical resources of existing communities' (Warren 1996, p. 346).

Of course, this neat separation is something of an intellectual sleight of hand, for, while to a climate change analyst the difference between drought and desiccation may be clear, to a pastoralist or Sahelian farmer looking at ruined crops, hungry cattle and intruding neighbours, problems are indivisible, and have a holistic power perhaps lost on an outsider (Mortimore 1989, 1998); as Andrew Warren (1996) points out, these three problems never come singly.

Neo-Malthusian narratives

Underlying the desertification narrative, with all its weight of scientific and pseudo-scientific argument, lies the familiar logic of neo-Malthusianism (see Chapter 5). A large fraction of the power of the narrative derives from the wider support for neo-Malthusian thinking within environmentalism, and within debates about people and environment in certain parts of the globe, particularly Africa. As long ago as 1949 Vogt (1949) wrote:

> The European in Africa has temporarily removed the Malthusian checks. He has put down tribal wars, destroyed predators, moved enough food about the continent to check famine – but he has not substituted constructive measures to balance his destruction of the old order.
>
> (p. 260)

Vogt's central concern is carrying capacity, and he argues that 'man has moved into an untenable position by protracted and wholesale violation of certain natural laws' (p. 264). This idea has persisted in thinking about desertification. For example, forty years later, Curry-Lindahl (1986) argued that it was an 'ecological rule' that land had a carrying capacity 'beyond which it cannot be utilised without causing damage, deterioration, and decreased productivity' (p. 125).

Neo-Malthusian thinking about overpopulation, overgrazing and land degradation was present in muted form in the *World Conservation Strategy*. Here desertification was seen to be caused primarily by human transgression of ecologically defined limits. Although there are problems of 'unwise development projects' and the 'inadequate management' of irrigated areas, the key problem is seen to be 'pressure of human numbers and numbers of livestock' (IUCN 1980, para. 16.9).

The neo-Malthusian view is readily found in starker form in writings on desertification, particularly in studies of Africa. Thus governments and aid agencies drew the wrong conclusions about the Sahelian drought of the 1970s: 'they called the effects of the drought a "natural disaster" and blamed the climate! The real root of the problem, that Man himself was responsible, was not recognised or admitted: it was too unpopular a message' (Curry-Lindahl 1986, p.107).

Neo-Malthusian blinkers do not help the environmental analyst make sense of the Sahel. Sinclair and Fryxell (1985) explained the 1984 crisis in the Sahel and Ethiopia, not in terms of drought (a failure of rain exacerbated by warfare), but in terms of a 'settlement-overgrazing hypothesis'. They argued that until about the middle of the twentieth century the 'normal' land-use pattern in the Sahel had been based on migratory grazing using seasonally available resources. They suggested that this system had been operating in a 'balanced and reasonably stable' way for many centuries, possibly since domestic cattle first appeared in the Sahel 5,000 years ago. It broke down 'through well-intended but short-sighted and misinformed intervention through aid projects'. Problems began after the Second World War exacerbated by population growth, overgrazing and agricultural practices aimed at short-term profit not sustained yield (Sinclair and Fryxell 1985, p. 992): overgrazing brought about the 'regression of plants' around boreholes and wells, and, as these 'piospheres' (Warren and Maizels 1977) expanded and joined up, extensive areas of the Sahel became desertified. Arguments about the possible feedback effects of bare desertified soil on climate then suggest that the Sahelian ecosystem 'is being pushed into a new stable state of self-perpetuating drought' (Sinclair and Fryxell 1985, p. 992). It is was stark story of humans degrading the land, although in their analysis the real blame is laid on the aid agencies, which fund projects that break down the older and sustainable migratory pattern. Their conclusion is that short-term food aid by itself will 'only make the situation worse', since 'simply feeding the people and leaving them on the degraded land will maintain and exacerbate the imbalance and not allow the land to recover' (Sinclair and Fryxell 1985, p. 992).

Similar neo-Malthusian narratives exist in other environments. In their book *The Himalayan Dilemma*, Ives and Messerli (1989) question alarmist reports of massive environmental degradation in the Himalayan region. The degradation narrative suggests that population growth (2–3 per cent per year in Nepal), plus immigration from India into the Terai lowlands, which have increased demand for fuelwood, timber, fodder and land, have brought about a reduction in forest cover on the hills (half of Nepal's forest reserves being lost between 1950 and 1980), and the extension of agricultural terraces onto steeper and more marginal slopes, unprotected by trees. This has caused more soil erosion, loss of land to landslides and disruption of hillslope hydrology. In turn, this has increased run-off in the monsoon and caused flooding and sedimentation downstream in the Ganges valley (burying fertile land, filling reservoirs and reducing dry-season flows because less water is retained higher up river basins). As land is eroded, agriculture reaches yet further onto more marginal land, and, as trees are eliminated, people turn to animal dung for fuel, and soil fertility declines. This tightens a spiral of environmental overuse and degradation.

There is little doubt that many, perhaps most, of the problems comprising the 'Himalayan environmental degradation theory' are real. However, in *The Himalayan Dilemma*, Ives and Messerli (1989) argue that the way in which they have been wrapped up together into a 'supercrisis' is unhelpful: it homogenizes spatially specific issues, and it paralyses policy. Floods in Bangladesh did not increase in frequency through the twentieth century, but there were more large floods. This, however, was not due to land-use change in the Himalaya, but to variations in rainfall, high groundwater levels and spring tides and lateral embankments along river channels in Bangladesh itself (Hofer and Messerli 2006).

An inability to understand the processes going on means that any attempt to develop policy is liable to have unforeseen and perhaps deleterious effects on the poor. Thompson and Warburton (1988) suggest that the real problem is that there is not simply *one* problem, but 'a plurality of contradictory and contending problems' (p. 46). Even the best-planned projects face a sea of uncertainty, endless problems of implementation and many surprises (Ives and Messerli 1989, p. 255). Degradation narratives give policy-makers a wholly false sense of security.

The political ecology of famine

In the neo-Malthusian frame, a direct link is made between environmental degradation that is due to unwise overuse of land, and long-term food shortage and famine. In the case of African drylands (especially the Sahel, the Horn of Africa and Darfur) commentators are quick to deploy a determinist language of crisis (Watts 1989). Mortimore (1989) criticizes the 'doomsday' scenario so often painted for African drylands, blamed on drought and poor soils, pointing out that exogenous political–economic challenges to the livelihood security of the rural poor are just as important as drought or climate change. He asks: 'what are the effects on environmental management of world markets, global economic recession and the impoverishment of African governments through debt repayments, diminishing revenues, inefficiency, corruption and war?' (p. 4).

Individual households facing a shortfall in rain in a drought year are linked through the market and the state to the international economy. They are also part of a global information economy. The two Sahel famines of the 1970s and 1980s occurred in the television age, with the suffering of refugees relayed nightly to living rooms in the North by television, stimulating a surge of humanitarian concern (most notably, perhaps, in the Live Aid Concert, organized by Bob Geldof in response to an item on the UK evening news (Geldof 1986)). The outrage of televised famine demanded a response; the prevention of famine obviously demanded an explanation of what caused it.

The simplest, and perhaps most obvious, explanation of the famine was that it was caused by a decline in food availability; and in semi-arid regions the most obvious causes of food availability decline (FAD in the literature) were drought and environmental degradation. In a dry area, such as the Sahel, it was easy to picture farmers at risk from fickle rains in every year, and in years when the rains were delayed or too small, being tipped over into hunger; failure of the rains on a large scale would lead to famine. A neo-Malthusian argument would suggest that excessive numbers of people caused land to be overused, reducing soil cover and soil fertility, vegetation cover and productivity, and setting the preconditions for catastrophic failure of production when the next drought arrived. This explanation of the 'drought–degradation nexus' was widely accepted in the governments of Sahelian countries, in the international agencies that provided capital for relief and rehabilitation and in the universities that claimed to understand how the Sahel worked (Mortimore 1989, 1998). These ideas proved convenient to many parties. In Ethiopia, for example, the government and Western aid donors found that a neo-Malthusian explanation of the causes of famine (too many people causing a degraded environment) could be expressed in technical terms (free of political ideology and the awkwardness of dealing with an avowedly Marxist regime) and provided a rationale for a large food-aid programme (Hoben 1995).

However, several bodies of work suggested that the explanation of famine, and its putative link with environmental degradation, was somewhat more complicated. Field research on how farming families coped with drought and hunger gave a very different picture from that so readily adopted by well-meaning outsiders. In Nigeria, for example, work by Mortimore (1989) on villages in the Sahelian north of the country demonstrated the enormous depth of indigenous capacity to plan and cope with variable rainfall. The variability of rainfall both within and between years was considerable, and had great significance for production, but people had developed strategies to cope with it, spreading risk, diversifying (within farming into livestock and other economic activities) and maintaining mobility (for example, seasonal labour migration). Furthermore, farmers were well aware of the importance of soil fertility, and had (in areas such as the Kano close-settled zone) long since established permanent and sustainable systems of agriculture. Sahelian farmers were far from the passive victims of fate, driven to a marginal and famine-stalked exposure to famine by their own fecundity (Mortimore 1989, 1998; Mortimore and Adams 1999).

A second, and highly influential, approach to famine was that taken by Amartya Sen, in his book *Poverty and Famines* (1981). Here, and in the subsequent *Hunger and Public Action* (Drèze and Sen 1989), Sen argued that famine and hunger were caused by a collapse in entitlements to food, and not in food availability. Entitlements derive from trade, production, labour and inheritance or transfer. In a series of case studies, including accounts of the Great Bengal Famine of 1943 (when people starved outside the doors of grain warehouses), and of famine in the Sahel and in Ethiopia in the 1970s, Sen argued that it was a breakdown in the ability of people to obtain food that led to widespread starvation, not a physical shortage of food as such. Sen's work on famine has been the subject of intensive academic debate (Watts and Bohle 1993). It has provided an important framework for many studies – for example, de Waal's account (1989) of famine in Darfur in the Sudan.

The entitlements approach views famine within an economic lens. Diverse actors contribute to acute food shortage. While many starve in a famine, not all will do so: 'different groups typically do have very different commanding powers over food, and an over-all shortage brings out the contrasting powers in stark clarity' (Sen 1981, p. 43). However, the entitlements approach is less adequate in explaining the politics and political economy of famine – the 'long-term structural and historical processes by which specific patterns of entitlements and property rights come to be distributed' (Watts and Bohle 1993, p. 48). Political ecology provides a better understanding of famine, whether in the massive mortalities of the Victorian colonial world in China, India and Brazil or late-twentieth-century Africa (M. Davis 2004).

Copans (1983), writing about the social and political context of the 1970s drought in the Sahel, pointed out that the phenomenon of famine must be understood in the context of a series of scales: first, the conditions of production (for example, rainfall); second, the social organization of production (for example, arrangements for agriculture); third, the national political economy (for example, issues of inequality of both class and region); and, fourth, the international political economy (dependency and neocolonialism). Copans highlighted the lack of work during the research 'boom' that followed the 1972–4 Sahel drought on the links between social, economic and political history and related changes in the natural environment that these produced.

This challenge has been addressed by several authors, notably in the Sahel by Franke and Chasin, in their book *Seeds of Famine* (1980), and Michael Watts (1983a, b, 1984, 1987) in his studies of the village of Kaita in Katsina Emirate in northern Nigeria (especially the book *Silent Violence* (1983a)). Franke and Chasin describe the impact on the Sahel of the expansion of European economic power and trading influence in the seventeenth century, and latterly, at the end of the nineteenth century, by colonial expansion. The French colonial governments in French West Africa brought about the expansion of groundnut (peanut) cultivation through a mixture of incentives and coercion (particularly taxation and forced labour), and the commoditization and monetization of the economy. Regular shipments of groundnuts to Marseilles began in 1884, and they were exempted from duty in 1892. Railways and roads, and after 1913 agricultural research, credit and extension, were all marshalled to promote groundnut production. Production rose through the colonial period and after independence. Production in Senegal was 45,000 tons in 1884–5 and over 1 million tons in 1965–6. By 1964 groundnuts represented 58 per cent of exports by value in Mauritania, 59 per cent in Mali, 63 per cent in Niger and a staggering 79 per cent in Senegal.

Franke and Chasin argued that individual farmers were caught in a production trap as groundnut yields declined, demanding the expansion of production into fallow areas and into areas used for growing staple food crops of millet. As a result, cash was required to purchase foodgrains, demanding more groundnut production. Furthermore, intensification of groundnut cultivation led to the abandonment of former practices: rotating groundnuts and millet, and a fallow period of six years to allow nutrients to recover in the soil (Franke and Chasin 1979, 1980). Continuous sole cropping with groundnuts, and the violation of fallow, led to soil deterioration and declining yields. The expansion of groundnut cultivation northwards, and the cultivation of fallow lands further south, in turn disrupted the pastoral economy, creating conflicts over dry-season pastures and additional pressure on seasonal northern grasslands.

The one-crop economies of West Africa suffered competition in the post-war period in the European oilseed market, particularly from American soya oil. After independence, expansion of groundnut production became a national goal, with production doubling in Senegal, for example, between 1954 and 1957, but it brought soil exhaustion and eroded food self-sufficiency. It was this that created conditions in which Sahelian producers were exposed to famine during the drought of the early 1970s. Thus cash cropping, driven initially by colonial policy and latterly by the demands of the national economy and by international political economy, created both degraded croplands and desertified rangelands in West Africa.

The sweep of this argument is compelling, although Franke and Chasin do not analyse the politics of the colonial state, or the possibility of winners and losers among African producers. Nonetheless, the links between political economy and environment are clear. The impacts of the actions of colonial and postcolonial states on the actions of farmers and pastoralists, and the effects of their actions in turn on the ground, must be the centrepiece of accounts of the political ecology of environmental degradation.

In his work on farming and famine in northern Nigeria, Michael Watts (1987) offered a detailed account of the relations between producers and non-producers, and the social mechanisms for surplus extraction, arguing that they are the means of understanding the connections between material circumstances and ecological conditions. His analysis started, therefore, from the peasant household, locked into the local economy and ecology. He discussed the impact of capitalism on production systems in northern Nigeria, and the articulation of the pre-capitalist mode of production with the global capitalist system through the agency of the colonial state. With colonial rule came taxation, cash cropping (in the north of Nigeria chiefly cotton and groundnuts) and railways. The railway reached Kano in 1913 and generated a rapid rise in groundnut exports. While some may have profited from this trade, it is argued that these changes reduced the 'margin of security' that poorer Hausa farmers had previously enjoyed (Watts 1983b, p. 251), both from the highly adaptive ecology of their production system (based on intercropping and the skilled exploitation of the diversity of upland and wetland environments open to them) and from the range of sources of livelihood outside farming (crafts, farm labour, livestock, seasonal migration and sale of land).

Hausa farmers became progressively more involved in cash cropping. Falls in groundnut prices caused a 'reproduction squeeze' where farmers either increased production or reduced consumption. This squeeze was felt unequally by the poor in the differentiated Hausa society of northern Nigeria, and it promoted (among other things) the decline of the moral economy – the patterns of reciprocity between households, particularly between rich and poor, that had provided a safety net in times of drought or disease. Poor farmers, 'shackled by their own poverty, are largely powerless to effect the sorts of changes which might mitigate the debilitating consequences of environmental hazards' (Watts 1983b, p. 256). With drought, in the twentieth century as before, came famine (van Apeldoorn 1980). Michael Watts (1983b) saw drought 'refracted through the prism of community inequality' (p. 256), and the environmental crisis of drought as something that revealed the structure of the social system reworked by colonialism and maintained by international political economy.

It is now accepted that famine is a phenomenon with complex causes, and that many of these are political. Thus Olsson (1993) concludes that the idea that environmental degradation was a significant cause of famine in the Sudan in 1984–5 should be abandoned. Famine was the result of drought and market failure. Severe reductions in rainfall in 1984 triggered widespread speculation in food, which pushed prices beyond the reach of ordinary rural people. At a national scale there was sufficient food, but poor distribution allowed famine to develop. Colonialism reduced the autonomy of West African rural communities (Mortimore 1989); decades of development have done little to increase their capacity to withstand drought and have in many instances exposed them to new risks that have eroded their flexibility and adaptability (Mortimore and Adams 1999).

Risk and vulnerability are critical concepts in understanding who is threatened by environmental change (Blaikie *et al.* 1994). The prevention of famine, hunger and environmental degradation must start from an understanding of both environmental variability and political economy – that is, from a political ecology (Davis 2004).

The politics of desertification policy

The influence of the dominant narratives of desertification and overgrazing, and the neo-Malthusian threat of famine, are not simply of academic interest. They have provided the foundation stone of policy in areas such as the Sahel. Many analyses of ecological change were (and still are) based on problematic assumptions about livestock management or indigenous agriculture. The conventional view of rangeland management and mismanagement has been built around ideas of range conditions, class and carrying capacity. The logic was that the environment is capable of supporting a certain fixed number (or biomass) of livestock, and that for any given ecosystem this could be calculated primarily as a function of rainfall. There is a general relationship between rainfall and the productivity of herbivores (Coe *et al.* 1976), and similar plots of livestock biomass against annual rainfall can be drawn (e.g. Jewell 1980). If these regressions are taken to represent 'carrying capacity', it can be argued that, at stocking levels lower than the solid curve, pasture resources are being underused, and that, at stocking levels above the solid curve, resources are being overused, such that there is likely to be adverse ecological change (for example, extinction of palatable species and eventually loss of vegetation cover), and eventually the death of excess stock.

The concepts of overgrazing and carrying capacity have underpinned conventional pastoral policy in many parts of the Third World, and particularly in Africa. Policies aimed at confining, controlling and often settling nomadic pastoralists in sub-Saharan Africa have reflected what Horowitz and Little (1987) call an 'intellectual tradition of anti-nomadism'. Governments have tended to distrust people who are mobile and difficult to locate, tax, educate and provide with services. Conventional rangeland science has added to this a particular distrust of their apparently feckless management of seemingly fragile rangelands. Recommended management strategies have been stereotyped (Swift 1982; Lane 1998). They have involved, first, the control of livestock numbers to match range conditions, through destocking and the promotion of commercial offtake, and to improve stock health and weight; second, fencing and paddocking to allow grazing pressure on particular pieces of land to be closely controlled, and provision of watering points to allow optimal livestock dispersal; third, manipulation of range ecology through controlled burning, bush clearance and pasture reseeding; and, fourth, disease control and stock breeding. None of these strategies fits with nomadic or semi-nomadic subsistence livestock production, so government pastoral policy has tended to emphasize sedentarization, formal (that is, freehold or leasehold) land tenure and capitalist production. Alternative ideas about pastoral ecology, and development strategy, are discussed in the next section.

In political terms, dryland policy that tries to impose solutions to presumed degradation can be draconian. Hare (1984) concluded that desertification 'harshens the microclimate', and proposed land-use control (meaning fencing and the control of the movement of animals) to 'restore plant cover and soil conditions' and 'repair the microclimate' (p. 19). Sinclair and Fryxell (1985) also called for social control (see Table 8.1). They suggested that the regeneration of degraded environments could take place only if the land were 'rested' until vegetation could recover, and this required population resettlement and control, destocking and the establishment of a new migration or rotational grazing system (see Table 8.1). The political, economic and human costs of extensive involuntary government resettlement in the Sahel would be vast. Sinclair and Fryxell suffer the blindness common to many commentators to the ecological and economic logics of African grazing systems, and assume that destocking is ecologically necessary, and economically and practically feasible. They propose turning the Sahel into a massive controlled grazing scheme, a breathtakingly misconceived proposal.

Repeatedly, state intervention in rangelands, often triggered or legitimized by concerns about desertification, has involved attempts to make nomads settle, often as part of a wider attempt to transform the economy of semi-arid areas and the livelihoods of their people. Aid donors responded to the African droughts of the 1970s with commitments of funds to support and transform the region. In 1973 seven Sahelian nations established the Comité Inter-États de Lutte contre la Sécheresse Sahélienne (CILSS). The US Congress supported the notion of a long-term development programme in the Sahel in December 1973 under the Foreign Assistance Act (Derman 1984). In 1976 CILSS and certain aid donors established the Club du Sahel, to coordinate aid giving; Henry Kissinger made a public commitment of $7.5 billion to 'roll back the desert' (Franke and Chasin 1980, p. 137).

The thrust of the new aid was to solve the problems of the Sahel using development investment and direct environmental management, not by addressing underlying political economic structures. A strategy for drought control and development in the Sahel produced by the Organization for Economic Cooperation and Development (OECD) was adopted by CILSS and the Club du Sahel in 1977 (Derman 1984). Its aims were to alleviate drought and re-establish food production and 'ecological recovery' and to enable the Sahelian countries to achieve food self-sufficiency by raising agricultural production (Franke and Chasin 1980, p. 148). The programme would require massive increases in

Table 8.1 Policy recommendations for the Sahel

To move people from the degraded land to new areas
To educate and help people in those new areas to establish a new and suitable rural economy
To institute education and family planning to control population growth
Severely to restrict and cull cattle herds
To monitor vegetation succession on degraded land
When recovered, to establish a modified migration or rotational grazing system
To construct wells only if they do not harm the migration system
To encourage African governments to institute these measures

Source: Sinclair and Fryxell (1985).

the production of wheat (seventyfold), rice (fivefold), cattle (twofold) and fish (twofold) on 1970 figures. In large part this was to be achieved by improved rain-fed farming (doubling production by the end of the century, working on drought-tolerant crops and off-farm employment), but a massive fivefold increase in the area irrigated (to 1.2 million hectares) was also planned. Other initiatives were to include transport infrastructure, well construction, animal health, work on crop storage and reforestation.

Franke and Chasin (1980) criticized this plan on a number of grounds, including its excessive focus on 'cash' rather than 'staple' crops (and a failure to appreciate the fluidity of such categories), and the excessive speed with which development was proposed, particularly in the field of irrigation. The plan accepted too readily simplistic assumptions about the nature and causes of overgrazing ('traditional' livestock-raising practices) and failed to recognize the potential impacts of initiatives such as well-digging. Derman (1984) concluded that, despite the New Directions mandate, new lending by the United States Agency for International Development (USAID) in the Sahel simply turned the 'familiar arsenal of modernization' on the problem of small farmers. Policy sought simply to get the commercial structure right, with no effective appreciation of the links between ecology and the way production is organized. Franke and Chasin (1980) concluded that the vaunted 'new' approach to development in the Sahel would simply reproduce existing vulnerability to drought. Development, in this instance at least, would bring no solution to environmental degradation.

The PACD, agreed at UNCOD in Nairobi in 1977, was designed to 'prevent and to arrest the advance of desertification, and, where possible, to reclaim desertified land for productive use' (Walls 1984). The PACD envisaged both translational projects (for example, a 'translational green belt in North Africa') and action by national governments (see Table 8.2). It was not a success, largely because the aid needed from donor nations did not materialize. UNEP estimated in 1980 that about $4.5 billion would be needed every year for twenty years to finance the core of the PACD. Yet in its first six years the UN General Assembly's 'special account' attracted only $48,500. Overall only $7 billion was spent from all sources between 1978 and 1983 ($1.17 million per year). Furthermore, of that $7 million, only about $400 million was actually spent directly on projects aimed at 'desertification control', the rest going on infrastructural projects such as roads, water supplies, buildings, research and training (Dregne

Table 8.2 National government actions under the 1977 Plan of Action to Combat Desertification

---

To establish or designate a government authority to combat desertification
To assess desertification problems at national, provincial or sub-provincial levels
To establish national priorities for actions against desertification
To prepare a national plan of action against desertification
To select priorities for national action
To prepare and submit requests for international support for specific activities
To implement the national plans of action

1984; Walls 1984). While such infrastructure might be useful, it had little to do with desertification control as such, and appeared to result in existing projects in semi-arid areas being relabelled as 'desertification' projects, either by aid agencies eager to show they were doing something or by governments hoping to make projects more attractive by jumping on the latest bandwagon. National planning to combat desertification also faltered. By 1984 only two countries (Sudan and Afghanistan) had drawn up anti-desertification plans, although there were nine others in draft (Walls 1984). Few countries had designated agencies for desertification control.

Projects to achieve 'desertification control' have been both ambitious and expensive. For example, in Mauritania the UN Sudan–Sahelian Office has explored physical and biological methods of controlling sand movement and encroachment on fields and facilities (Grojean 1991). Work began in 1983, identifying biological protection techniques (live fences of introduced species such as *Prosopis juliflora*, or local species such as *Leptadenis pyrotechnica*), and 'raising awareness among the insufficiently motivated, newly sedentary population' (Grojean 1991, p. 6). On average, mechanical stabilization cost $1,000 per hectare and biological stabilization $400 per hectare. The aim of this 'curative sand encroachment control project' was technical, not developmental; the technology could be made to work, but not at a feasible economic cost.

There is a great deal of evidence of the failure of external plans to transform dryland production systems, and the lives of farmers and pastoralists. One example is the history of attempts to settle pastoralists in northern Kenya destituted by drought in the 1970s and 1980s, often on irrigation schemes. These were expensive, unpopular and unsuccessful (Hogg 1983, 1987a, b). In the 1980s substantial numbers of Turkana pastoralists were settled on irrigation schemes along the seasonal Kerio and Turkwel rivers, and on the shores of Lake Turkana as fishermen (Hogg 1987b). A Ministry of Agriculture appraisal in 1984 calculated that the total development cost of the Turkana cluster of small-scale irrigation schemes was fifteen times the cost of setting up each family with a herd of replacement livestock, and the equivalent to famine relief for 200 years (Adams 1992). Hogg argues that irrigation contributed to the further marginalization of already poor pastoralists and significantly increased pressure on key grazing lands through clearance of riverine forest vegetation near settlements (a process Hogg (1987a) sees as 'policy-induced desertification'). The root causes of desertification in Turkana are, then, first poverty, second induced settlement and third modernization. Development 'solutions' made all of this worse.

Overgrazing and new range ecology

There are wide gaps between pastoral policy prescriptions and the ways pastoralists actually manage their herds and rangelands. By implication, conventional pastoral policy has tended to emphasize the production of animal products from the slaughter of stock (that is, meat and hides) rather than the products from live animals (milk or blood) that typify many indigenous pastoral systems. Similarly, pastoral development planning tends to focus on cattle, whereas indigenous production systems typically involve a mix of species. Indigenous pastoral ecosystems seem well adapted to exploit the spatial and temporal variability in production, adapting herd composition and using movement to maximize survival chances. The Turkana in Kenya, for example, have mixed herds, with camels, which use browse resources that are available even in the dry season, and cattle, which are more productive in the wet season but have to move out of the plains into the hills in the dry season (Coughenour *et al.* 1985; Coppock *et al.* 1986; McCabe 2004). Such systems offer a relatively low output compared to modern capitalist systems such as ranching. However, they are remarkably robust in terms of providing a predictable, if limited, livelihood.

Researchers have increasingly expressed reservations about the universal applicability of the concept of overgrazing, and the unreflective links drawn between it and desertification (Sandford 1983; Horowitz and Little 1987; Mace 1991). Judgements about carrying capacity are subjective, although that subjectivity is rarely admitted (Hogg 1987a). Estimates of 'carrying capacity' take no account of seasonal variations in fodder availability or annual variations (Homewood and Rodgers 1984, 1987). They concentrate on absolute numbers of livestock and not densities, and rarely consider spatial mobility. They are, therefore, of little value in understanding either rangeland ecology or pastoral practice. Although they have become both entrenched and self-reinforcing, they are an unsatisfactory and sometimes dangerous basis for management.

The productivity of semi-arid rangelands varies both seasonally and between years. The primary cause of this variation is rainfall. The high spatial and temporal variability of precipitation, particularly in sub-Saharan Africa, has increasingly been recognized. Other factors affecting productivity also vary over time and space, notably soil nitrogen and phosphate (Breman and de Wit 1983) and the impact of fire (Norton-Griffiths 1979). Ecological studies tend to be of short duration (often confined to the fieldwork period associated with a Ph.D. thesis) and localized. Yet semi-arid grazing ecosystems undergo considerable and important fluctuations from year to year and place to place. Analyses of real or predicted degradation tend to be built on estimates of regional stocking rates. Not only are such estimates notoriously unreliable (because livestock are hard and expensive to count, particularly if their owners do not want you to); they also fail to take account of spatial and temporal variations. More seriously still, the assumption of mechanistic relations between stock numbers and biological productivity completely fails to take account of the social processes of herd management, such as, for example, the boom in cattle numbers that took place in the 1960s in the north-west part of the Niger Inland Delta in Mali (M. Turner 1993).

Biologically based estimates of carrying capacity tend to be arbitrary, based on limited data and deriving from 'rule of thumb' and the experience of range ecologists rather than empirical scientific fieldwork (Mace 1991). Arguably, the concept of carrying capacity is not appropriate in areas with great annual variation in primary productivity, and most estimates fail to take account of the variability and resilience of savannah ecosystems (Homewood and Rodgers 1987).

Plate 8.4 Groundwater-fed acacia trees along a stream, Turkana, northern Kenya. Rangeland resources are not uniformly distributed in either space or time. This photograph shows mature *Acacia tortilis* trees along seasonal drainage channels. These trees can obtain water from the shallow aquifer in the sands below the streambed long after the rains have ended. These trees produce seed pods that are an important and nutritious food resource for Turkana livestock. Riparian acacia woodland is an important rangeland resource, and access to it is carefully controlled. The survival of the woodlands of the main river of north-west Kenya, the Turkwel, is threatened by construction of the Turkwel Gorge Dam (W. M. Adams 1992). Photo: W. M. Adams.

Furthermore, it is fairly meaningless to seek to identify a single 'carrying capacity' for an ecosystem. Carrying capacity should be defined in terms of the density of animals and plants that allows the managers to get what they want out of the system; in other words, it is possible to speak of a carrying capacity only in the context of a particular management goal. What suits a nomadic pastoralist may not suit a rancher; many African systems have a subsistence stocking rate higher than commercial ranchers would adopt, giving low rates of production per animal but high output per unit area (Homewood and Rodgers 1987). Actual stocking levels can and do exceed 'carrying capacity' for decades at a time (Behnke and Scoones 1991).

Overgrazing is a classic 'institutional fact'. As Ruth Mace (1991) comments, 'sometimes we are so sure of something we don't need to see the evidence' (p. 280). Despite the volume of literature on overgrazing and carrying capacity, researchers now conclude that there is no one simple ecological succession towards an overgrazed state, but complex patterns of ecological change in response to exogenous conditions (especially rainfall) and stock numbers and management.

Such ecological changes can take many forms, not all of them serious, and they can proceed by diverse routes, some of which can be reversed more easily than others, and some of which are more sensitive to particular management than others. There are no 'naturally' stable points in semi-arid ecosystems that can usefully be taken to define an 'equilibrial' state.

Through the 1980s and 1990s, conventional thinking about carrying capacity and overgrazing began to be challenged by so-called new range ecology (Behnke *et al.* 1993). In drier rangelands, with greater rainfall variability, ecosystems exhibit non-equilibrial behaviour. Ecosystem state and productivity are largely driven by rainfall, and pastoral strategies are designed to track environmental variation (taking advantage of wet years and coping with dry ones), rather than being conservative (seeking a steady-state equilibrial output). This awareness of the non-equilibrial nature of savannah ecosystem dynamics reflects a wider understanding of the importance of non-linear processes in ecology as a whole (e.g. Botkin 1990; Pahl-Wostl 1995).

Whatever the implications for ecological thought, it is certainly the case that non-equilibrial behaviour is characteristic in the dry tropics. Once this is appreciated, much of what appeared to be perversity or conservatism on the part of pastoralists is revealed to be highly adaptive (Behnke *et al.* 1993; Prior 1993; Scoones 1994).

The argument of new range ecology is that in dry rangelands with a strong inter-annual variability in rainfall there is no single 'carrying capacity' of the ecosystem, represented by an equilibrium number of livestock (Scoones 1991). Instead, the balance of livestock and range resources changes over time, with drought years first reducing the condition of stock and then (through disease, death and destitution-forced sales) reducing stock numbers. Good rain years then allow pastures to recover, allowing a lagged recovery of herd numbers as pastoralists track environmental conditions. Not only do herd managers need extensive knowledge of environmental conditions and opportunities in different areas open to them, and resilient multi-species herds, to survive under such conditions, but they also need institutions for the exchange and recovery of stock through kinship networks. Development strategies must support indigenous capacity to track rainfall and maintain social and economic networks, rather than demand a shift to a static, equilibrial capitalist form of production.

Recognition of the non-equilibrial nature of savannah ecology presents a considerable challenge to policy-makers (Scoones 1999). It has cast increasing doubt on the validity of traditional management of rangelands aimed at maintaining a range in a specified condition. Alternative strategies have been proposed that emphasize the opportunism of pastoral management (Behnke *et al.* 1993; Sandford 1983; Scoones 1994). New ideas recognize that opportunistic strategies are long established, that husbandry systems may well not need drastic reform (let alone abandonment) and that development strategies can be gradual and fully participatory, leading to piecemeal and not wholesale change (Scoones 1994). The insight that ecologists can offer is confirmation of the herders' need to balance fodder supply and stock numbers. Strategies to help herders track environmental change include a focus on enhancing

Plate 8.5 Fulani cattle, northern Nigeria. New understandings of range ecology in
savannah areas of low rainfall emphasize the important of mobility and
flexibility in livestock management. Mixed herds of cattle, sheep, goats and
camels use available graze and browse resources in different ways. Social
interactions are critical to the sustainability of livestock husbandry.
Photo: W. M. Adams

feed supply (maintaining exchanges with farming communities, supplying feed),
supporting mobility (supporting tenure of key dry-season grazing sites and access to
trekking routes) and promoting human rights. Animal health is important to stock
survival in drought, and mobile vaccination facilities can be important, while there
is still a role for the stock-breeding beloved of government livestock researchers,
but the focus needs to be on the capacity of animals to survive disease, drought
and poor dry-season grazing, in preference to milk or meat yield under favourable
conditions.

It is also now widely recognized that pastoralists need help to endure crises such
as drought. Innovative policies include provision for purchasing stock at reasonable
prices in droughts (when the supply of poor animals rises and prices crash) and
for helping pastoral families restock, and communal grain banks for pastoralists
(thus enabling them to weather spiralling grain prices during droughts). Most
important of all is the provision of security to rights in key areas of rangeland,
particularly wetlands patches that support communities in surrounding drylands,
and particularly in drought years (Scoones 1991).

Finally, there is a need for more support for herders to move into and out of
stock-keeping; not through mass resettlement and retraining campaigns (of the
kind that have, for example, sought to turn herders into coastal fishermen), but
by supporting a diversity of livelihood options between which people can choose.

Diversity and flexibility are cornerstones of survival in both pastoral and agricultural production in drylands, and policy-makers must recognize and foster these, rather than seeking to sweep them away in the pursuit of higher productivity and a cash income (Mortimore and Adams 1999).

Overpopulation or intensification?

Empirical research in Africa in the 1990s has called into question neo-Malthusian assumptions about the inevitability of environmental degradation as population density rises, and neo-Malthusian policy narratives are increasingly under fire (Roe 1991, 1995; Leach and Mearns 1996; Forsyth 2003; Robbins 2004). Rural population densities in Africa are low compared to those in equivalent drylands in Asia, and historically the lack of labour for agriculture has been a critical factor in the evolution of farming systems and environmental management (Iliffe 1995). Comparative study of agricultural farming systems in a range of countries shows increases in agricultural output per head, quite contrary to the customary wisdom of agrarian crisis and falling food production per capita (Wiggins 1995). As more careful studies have been undertaken, it has become clear that, under some circumstances, population growth in sub-Saharan Africa is leading to sustainable intensification of agriculture, not degradation. In northern Nigeria high population densities have been maintained for centuries in the close-settled zone around Kano City. This agricultural landscape is referred to in the literature as 'farmed parkland', with closely packed fields set with economic trees. By 1913 no more than one-third of the land was fallow, and by 1991 87 per cent of it was cultivated, and rural population densities were 348 people per square kilometre (Mortimore 1993). The farming system is complex, with several crops (particularly millet, sorghum, cowpeas and groundnuts) of a wide range of local varieties grown together in different intercropping and relay cropping mixtures (Adams and Mortimore 1997; Mortimore and Adams 1999; see also Plate 8.6). The key to the sustainability of cultivation without prolonged fallow periods, however, lies in the maintenance of soil fertility through the close management of nutrient cycles, use of legume crops and the integration of agriculture and livestock-keeping, particularly in the use of crop residues as fodder for small stock, such as sheep and goats (Harris 1998). Some soil nutrients also arrive in the form of dust deposits.

It could be argued that the Kano close-settled zone is a remarkable and untypical place, and that circumstances there are unlikely to be repeated elsewhere. However, studies in drier Sahelian farming systems further north-east in Nigeria suggest that similar patterns of intensification may be developing as population densities rise (Harris 1999; Mortimore and Adams 1999). For rural households the allocation of household labour to different tasks in cultivation, livestock-keeping, off-farm activity and household work is a critical factor in their ability to achieve sustainable livelihoods (Mortimore and Adams 1999).

The possibility of a positive relationship between rural population growth and environmental sustainability has started to become conventional wisdom, as the implications of a major historical study of Machakos District in Kenya have been

Plate 8.6 Sustained intensive agriculture in the Kano close-settled zone, Nigeria. A field
    intercropped with late millet, cowpeas and sorghum. Land around Kano City
    has been farmed continuously for several centuries. Cropping systems are
    diverse, and population densities are high. Soil fertility is maintained through
    careful nutrient management (particularly the integration of livestock keeping
    and crop production, and use of compound sweepings as manure). Photo:
    W. M. Adams.

disseminated within the policy community. This book, provocatively titled *More People, Less Erosion* (Tiffen *et al.* 1994), presents a detailed and empirically backed argument that there has been a beneficial interaction between population growth, environmental improvement and economic output per capita: they reiterate the optimism of Esther Boserup's arguments (1965) and challenge the pessimism of neo-Malthusians.

What gives this story its significance is the way in which Machakos was held up in the 1930s, both within colonial Kenya and more widely, as epitomizing the threat of soil erosion under relentless pressure of population. It was the very public target of late-colonial projects to tackle the menace of land degradation, an 'experiment in colonialism' (E. Huxley 1960). Not only was disaster averted, but Tiffen and Mortimore argue that population growth allowed an astonishing level of investment in land (particularly terracing) and the wholesale transformation of agriculture to highly intensive production systems (Tiffen and Mortimore 1994; Mortimore and Tiffen 1995; Tiffen *et al.* 1994).

Machakos lies south-east of Nairobi, and stretches from relatively high and well-watered land (2,000 metres above sea level, 1,200 mm rainfall) to lower dry range-lands (600 metres above sea level, 700 mm rainfall). Like the rest of Kenya, it has been subject to a high rate of population growth, ranging up to 3.7 per cent per year in the 1970s. The population of the district was 240,000 in 1930 and 1.4 million in 1990. The value of agricultural output rose three times per capita and eleven times per unit of area between 1930 and 1990 as farmers invested off-farm incomes in land, intensified production (see Plate 8.7), turned to cash crops such as coffee, harnessed labour to terrace hillsides and made use of the denser networks of contacts to learn new ideas and sell their produce. Tiffen *et al.* (1994) conclude that

> The Machakos experience between 1930 and 1980 lends no support to the view that rapid population growth leads inexorably to environmental degra-dation. It is impossible to show that a reduced rate of population growth might have had a more beneficial effect on the environment. On the contrary, it might have made less labour available for conservation technologies, resulted in less market demand and incentives for development, and reduced the speed at which new land was demarcated, cleared and conserved.
>
> (p. 284)

This is a remarkable and sustained argument, and it has had a significant impact on the way development policy-makers think about African agriculture. There are, inevitably, other arguments that need to be heard. For example, Murton (1999) demonstrates that not all Machakos households can cope with and effec-tively harness extra labour: those with buoyant off-farm income (particularly in nearby Nairobi) can accumulate land and innovate as farmers; those dependent on agricultural labour opportunities struggle. Terraces on the hillside may repre-sent a control of erosion, and perhaps environmental sustainability, but they do not necessarily translate into sustainable livelihoods for all, especially in the longer run. Samantha J. Jones (1996) draws similar conclusions about the ability of

Plate 8.7 Smallholder coffee, Machakos District, Kenya. Machakos, south-east of the city of Nairobi in Kenya, was known in the 1930s as the classic threatened environment, where overpopulation was causing drastic soil erosion. Sixty years later, slopes were terraced and cloaked with trees and cash crops such as coffee. Stall-fed cattle provide manure for intensively managed fields. Photo: W. M. Adams.

richer and poorer households to sustain soil fertility in the Uluguru Mountains in Tanzania.

There is a need, also, to be careful about running away with policy conclusions (Siedenburg 2006). Rocheleau *et al.* (1995) look at Machakos in a longer time frame, and quite rightly draw attention to the way in which for over a century it has been the subject of repeated study and intervention by outsider 'experts' (and clearly even this most recent enthusiastic endorsement of indigenous skill and enterprise falls into that category). They emphasize the dangers of the policy search for rapid diagnosis and the 'quick fix', pointing out how this tends to produce problems further down the line. They emphasize the importance of taking account of the multiple stories that local people have about places and people. Murton (1999) showed how the changes in Machakos have been accompanied by increasing inequality and a reduction in food self-sufficiency.

Despite these caveats, there is no doubt that the pessimistic neo-Malthusian narrative about dryland farming has been convincingly and widely challenged. Comparable findings to those in Nigeria and Kenya have been reported from other areas. Lindblade *et al.* (1998), for example, examined land-use change in Kabale District in Uganda between 1945 and 1996, finding that, despite population growth (and population densities of 265 per square kilometre), and a history of colonial concern about overpopulation, a higher proportion of land was now being fallowed, and evidence of land degradation was limited. Farmland was carefully terraced (see Plate 8.2). Steep slopes had been turned over to woodlots, and valley-bottom wetlands drained for grazing, while soil fertility was being maintained by using animal manure, household compost and mulching.

Desertification as policy fact

Desertification underwent an institutional renaissance in the run-up to the UN Conference on Environment and Development in Rio de Janeiro in 1992. In a manner reminiscent of the dissatisfaction of non-industrialized countries at Stockholm twenty years before, Southern countries resented the sidelining of the environmental problems relevant to them, and desertification came to embody their dissatisfaction. Southern Africa was also in the grip of severe drought. As a result, the issue was discussed at length in the final PrepCom meeting before the Rio Conference, and a chapter on desertification was included in *Agenda 21*. A formal commitment was made at Rio to negotiate and agree a Convention on Desertification by 1994, although this did not go through without opposition (Carr and Mpande 1996). Arguments included the question of whether desertification was actually a global issue, and the question of whether Southern demands for a desertification convention would be traded off against the desire from the USA and the EU for a forest convention.

Following Rio, an Intergovernmental Negotiating Committee was established rapidly, meeting in Geneva in 1993. It worked through a series of issues, including scientific uncertainty about the definition of desertification, and the extent to which it was a global problem (Carr and Mpande 1996). After five sessions, a text of the Convention with four regional annexes (on Africa, Asia, Latin America and the northern Mediterranean respectively) was complete for signature by the deadline in June 1994 (although the Intergovernmental Negotiating Committee continued to meet to clarify the meaning and implication of certain articles). The final convention is an interesting reflection of both the politics of the Rio process (discussed in Chapter 4) and several decades of confused thinking about environmental degradation and development.

The convention came into force in December 1996, the first Conference of the Parties taking place in Rome in 1997. A permanent secretariat was established in Bonn, Germany, and (unusually) the conference included a plenary meeting for dialogue with non-governmental organizations (NGOs) (Toulmin 1993; European Commission 1997; International Institute for Sustainable Development 1997). By 1997 the convention had been ratified by 113 countries (twice as many as the other two Rio conventions, on Biological Diversity and Climate Change), although several key countries had yet to ratify it, notably Japan, Russia and the USA.

The Convention to Combat Desertification was proposed by Southern countries primarily as a way to focus financial resources on real problems of some of the world's poorest people, but in the end it contained only the weakest of commitments by donor countries to provide extra funds (Carr and Mpande 1996). As with the PACD, lack of funds is a key constraint on implementation. A 'global mechanism' (administered by the International Fund for Agricultural Development (IFAD)) was agreed to mobilize and channel funds (a mechanism similar to that of the Framework Convention on Climate Change), but the flow has been slight. It is not clear whether, in the long run, Africa will be able to persuade other regions to let it be a 'special case' deserving privileged attention, and whether the broad focus of the convention (embracing environmental management, poverty, democratization and governance) will prove workable, or will actually have any impact on the lives of the poor in arid areas (International Institute for Sustainable Development 1997).

Some aid donors geared up to support the convention, notably perhaps the European Union, which is recognized as having desertification within its own region, in the Mediterranean. Between 1990 and 1995–6 some 524 million ECU were dedicated to 237 desertification projects in developing countries through the European Development Fund, cooperation agreements with Asian and Latin American countries and thematic budget lines (European Commission 1997). The European Community (EC) spent 280 million ECU between 1990 and 1996 in twenty-six countries of sub-Saharan Africa, 51 per cent of which went to West Africa.

Since the 1970s, anti-desertification projects have become increasingly multidisciplinary and diverse, reflecting the growing perception that the problem of 'desertification' is not simple, and certainly not conducive to narrow technical solutions. In the 1990s the EC spent 23 million ECU on a project to rehabilitate common lands in the Aravali Hills, Haryana, India. The work of the project ranged from tree- and grass-planting, contour-trenching and wall-building through to work on land-management institutions (to encourage effective community control of these lands and the involvement of women in land-management decisions) and the introduction of new technologies such as fuel-efficient stoves, grass-harvesting and silage-making (European Commission 1997).

This chapter has argued that experts and planners have a very mixed track record in their attempts to define and identify environmental degradation, and often a frankly poor record in trying to overcome it. The poor experience degradation not as an aggregate phenomenon of ecological change, but directly, in the form of challenges to welfare and livelihood sustainability. Poverty, economy and social organization are an integral part of the challenge facing development planners, and those they seek to help. To Bo Kjellén (2003), a Swedish diplomat involved in the Rio Conference, the United Nations Framework Convention on Climate Change (UNFCCC) and the United Nations Convention to Combat Desertification, the fundamental point is the responsibility of the international community to 'combat unacceptable conditions for the more than one billion people who live in the vast drylands of the planet' (p. 132). That requires that 'the whole context of sustainable development debate must be brought to bear on the problems of the drylands'. To that end, liaison between the Conventions on Desertification and the Conservation of Biological Diversity began in 1998, to avoid duplication of efforts and to promote complementarity and synergy. A joint work programme was agreed in 2003 (Zeidler and Mulongoy 2003). The UN General Assembly declared 2006 the 'International Year of Deserts and Desertification' (Fisher 2006).

Concern about desertification, which started by addressing 'the inexorable advance of the desert', has ended up addressing questions of poverty and sustainable livelihoods within the broader frame of sustainable development. It shares with that wider field its combination of science and social concern, in its weakness for vague rhetoric in place of material analysis, and in its reluctance to engage with the politics of the relations between people and environments.

Summary

- Environmental degradation is a central issue in sustainable development. While references to degradation (and particularly 'desertification') in dryland areas abound in the literature and in policy, considerable caution is needed in thinking about them. Definitions are confused, and strong scientific evidence on long-term environmental change is often lacking. Discussions of desertification in particular are often more dependent on commonly accepted wisdom among so-called experts than on hard field evidence.
- Poverty and environment change are linked in a close and complex way. Political ecology offers a challenge to established approaches to understanding social action and environmental change. Writing in political ecology is diverse, embracing the links between the logics of capitalist growth and environmental change, the politics of social action for the environment, and

the discursive power of social constructions of nature (including scientific explanations of environmental change).

- Climate change is a long-established feature of regions such as dryland Africa, and rainfall varies in space and time in complex ways. Although scientific understanding is growing, evidence does not encourage simplistic conclusions about causes of drought or the links between climate and land-use change.
- Fear of desertification and soil erosion has been a repeated theme in African development thinking, but neo-Malthusian analyses of population growth and environmental degradation have been challenged by studies (for example, in Machakos, Kenya) of agricultural intensification and sustainability.
- Studies of dryland pastoralists also now challenge conventional wisdom about the inevitability of environmental degradation. New range ecology involves a recognition of the way indigenous pastoral systems are adapted to variations in rainfall and grazing productivity in space and time.
- The causes of famine are complex and deeply political, demanding an explanation that addresses both the environmental and the political–economic context of rural production and consumption.
- Policies devised to 'combat desertification' are unlikely to be successful if they are based on an erroneous understanding of the relations between people and environment (which they often have been). The Convention to Combat Desertification was negotiated following the Rio Conference, and came into force in 1996. It embodies much of the confused thinking about dryland degradation that has dominated international debate since the 1970s.

Further reading

Behnke, R. H., Jr, Scoones, I. and Kerven, C. (1993) *Range Ecology at Disequilibrium: new models of natural variability and pastoral adaptation in African savannas*, Overseas Development Institute, London.

Blaikie, P. (1985) *The Political Economy of Soil Erosion in Developing Countries*, Longman, London.

Blaikie, P. and Brookfield, H. (1987) *Land Degradation and Society*, Methuen, London.

Blaikie, P., Cannon, T., Davis, I. and Wisner, B. (1994) *At Risk: natural hazards, people's vulnerability and disasters*, Routledge, London.

Botkin, D. B. (1990) *Discordant Harmonies: a new ecology for the twenty-first century*, Oxford University Press, New York.

Cutter, S. L. (2006) *Hazards, Vulnerability and Environmental Justice*, Earthscan, London.

Devereux, S. (2006) *The New Famines: why famines persist in an era of globalization*, Routledge, London.

Ives, J. and Messerli, B. (1989) *The Himalayan Dilemma: reconciling development and conservation*, Routledge, London.

Leach, M. and Mearns, R. (1996) (eds) *The Lie of the Land: challenging received wisdom on the African environment*, James Currey/International African Institute, London.

Mortimore, M. (1998) *Roots in the African Dust: sustaining the sub-Saharan drylands*, Cambridge University Press, Cambridge.

Mortimore, M. and Adams, W. M. (1999) *Working the Sahel: environment and society in northern Nigeria*, Routledge, London.

Pahl-Wostl, C. (1995) *The Dynamic Nature of Ecosystems: chaos and order intertwined*, Wiley, Chichester.

Roberts, N. ([1989] 1998) *The Holocene: an environmental history*, Blackwell, Oxford (2nd edition).

Thomas, D. S. G. and Middleton, T. (1994) *Desertification: exploding the myth*, Wiley, Chichester.

Tiffen, M., Mortimore, M. J. and Gichugi, F. (1994) *More People, Less Erosion: environmental recovery in Kenya*, Wiley, Chichester.

Scoones, I. (1994) *Living with Uncertainty: new directions in pastoral development in Africa*, IT Publications, London.

Stott, P. and Sullivan, S. (2000) (eds) *Political Ecology: science, myth and power*, Arnold, London.

Rocheleau, D., Thomas-Slayter, B. and Wangari, E. (1996) (eds) *Feminist Political Ecology: global issues and local experiences*, Routledge, London.

Web sources

http://www.fao.org/desertification/ The FAO desertification website: desertification data.

http://www.undp.org/drylands/ The UNDP Drylands Development Centre.

http://www.environment.gov.au/events/iydd/index.html International Year of Deserts and Drylands – 2006.

http://www.unccd.int/main.php The United Nations Secretariat of the Convention to Combat Desertification, including information on the meetings of the Conference of the Parties.

http://www.iied.org/NR/drylands/index.htmll IIED Drylands Programme home page: sensible research and networking on drought and land degradation.

http://www.ilri.org/ The International Livestock Research Institute (ILRI) in Nairobi, CGIAR (Consultative Group on International Agricultural Research) centre for livestock research: information on its ideas about issues such as livestock nutrition, health and livestock policy.

http://www.fews.net/ USAID Famine Early Warning System network: reports, monthly updates, information on vulnerability and hazard monitoring (climate, prices, vegetation).

9   Sustainable forests?

All critical examinations of the relation to nature are simultaneously critical examinations of society.

(David Harvey, *Justice, Nature and the Geography of Difference*, 1996)

Tropical deforestation

It has been taken for granted since the 1980s that tropical forests are uniquely precious and threatened. Environmentalists and many scientists have painted a stark and insistent picture of rainforest destruction (e.g. N. Myers 1984; Caufield 1985; Groombridge 1992; E. O. Wilson 1992; Terborgh 1999), and a 'doomed forest' narrative has been widely accepted (Hecht 2004). Thus, in the 1970s, Denevan (1973) suggested that development was bringing about 'the imminent demise of the Amazonian rainforest' as a result of a 'pell-mell destructive rush to the heart of Amazonia' (p. 137). In the 1980s, the loss of rainforest became a significant theme within environmental concern in the developed world. Statistics on the rate of loss of forest land were stated and extrapolated to emphasize the scale and speed of crisis. An advertisement for the World Wide Fund for Nature (formerly the World Wildlife Fund (WWF)) in *The Times* in 1980 showing cleared forest in Sumatra accompanied a six-page special report on the *World Conservation Strategy* (*WCS*) (IUCN 1980) that stated (rather wildly) that 'the earth's lungs are being destroyed at the rate of fifty acres [20 hectares] a minute'. The Friends of the Earth Tropical Rainforest Campaign, which began in 1985, claimed that 7.5 million hectares of undisturbed tropical moist forest were being destroyed or degraded annually, with 14 hectares being cleared every minute, so that by 1990 the rate of extinction of species would have risen globally to one per hour (Secrett 1985).

Although there can be no doubt about the rapidity of forest-cover change in many humid and sub-humid tropical areas, particularly from the 1980s onwards, there remains much debate about its rate and extent. Consistent and reliable data on rates of forest loss are surprisingly difficult to find. This is partly because of difficulties of definition, both of forest types (dry forests like the Atlantic forests in Brazil share many of the same threats as 'rainforest'), and of what is meant by 'deforestation'. Conservation scientists are chiefly interested in tropical forests for their diversity of species, and define them floristically. Thus Olson and Dinnestein (1998) define sixty-five tropical forest eco-regions in five bio-geographic realms, as a basis for conservation action. Foresters, who see forests as a resource, define them by the amount of timber they comprise.

In the 1980s the UN Food and Agriculture Organization (FAO) (Lanly 1982) distinguished between closed forest (of which there were some 1.20 billion hectares globally) and open forest (0.73 billion hectares). However, the FAO also recognized fallows in both closed and open forest (0.24 billion hectares and 0.17 billion hectares respectively), and scrubland (0.62 billion hectares). The FAO's *Forest Resources Assessment 1990* suggested a total of 3.4 billion hectares of forest, but categories were not fully compatible with previous surveys (Grainger 1996). The *Forest Resources Assessment 2000* estimated a larger total of 3.9 billion hectares under forest (almost 30 per cent of the earth's land area), but this reflected new data and changed categories rather than physically more trees. Of this total, a mere 5 per cent was plantation forest (Siry *et al.* 2005).

If there is confusion over the extent of different kinds of forests, it is perhaps unsurprising that estimates of the rate and extent of deforestation differ quite widely (Allen and Barnes 1985). There is no clear and universal definition of deforestation (Melillo *et al.* 1985), and the data are confusing. Ickowitz (2006) points out that forest cover in Cameroon (a forest loss hotspot (Mertens and Lambin 2000)) can be shown to have fallen sharply or to have increased over the last two decades of the twentieth century, depending on which set of figures is used. Where large tracts of land are cleared permanently of forest and turned over to a wholly different form of land use, definition is easy. However, it is more difficult to classify land-cover change in areas with a complex mix of old growth and successional forest – for example, associated with shifting cultivation or forest farmland abandonment (Armitage 2002; Hecht *et al.* 2006; Ickowitz 2006).

Data on deforestation have improved, but have also been made more complex since the 1980s by the use of satellite imagery (Myers 1980; Turner and Meyer 1994; Smith 2004b). Thus Tucker *et al.* (1984) used data from the Advanced Very High Resolution Radiometer (with a coarse spatial resolution but available on a daily basis) with some success to study strip clearance of forest in Rondônia in Brazil for roadside settlement, and Fearnside (1986) analysed Landsat data to derive estimates of the extent of forest clearance in the Brazilian Amazon, showing rapid acceleration in the rate of clearance between 1975 and 1980 in certain key areas, notably strategic highways – for example, those through Mato Grosso and Rondônia. However, technical problems such as cloud cover and the problem of detecting vegetation change from forest to other land covers (involving subtle changes in spectral signature) in small areas of irregular shape (typically the result of forest clearance by small farmers) are significant (Singh 1986), as are practical problems of cost and data availability (Smith 2004b).

Despite these problems of definition and measurement, the general pattern of global forest-cover change is now broadly accepted (Groombridge 1992; Williams 1994, 2003; Rudel and Roper 1997). The Millennium Ecosystem Assessment

(MEA) reported that global forest cover had declined by 50 per cent in the last three centuries, and twenty-nine countries had lost 90 per cent of their forest cover (Millennium Ecosystem Assessment 2005). Broad regional variations in patterns of forest use are clear (N. Myers 1980). Logging for timber production (largely for European and South-East Asian markets) is important in South-East Asia, particularly Malaysia, Indonesia, Papua New Guinea and the Philippines. The international timber trade is much less important in Latin America, although there is a considerable internal trade in timber: in the 1980s logging took over ranching as the leading cause of deforestation in the Brazilian Amazon (Parayil and Tong 1998). In Africa, internal markets for timber are also large, and rates of logging (usually clear-felling) are locally high. Elsewhere, settlement schemes are important in Amazonia, Indonesia and Malaysia; ranching and pasture development in Central and Latin America; fuelwood and charcoal collection in Africa and India (Williams 1989, 1994, 2003).

However, behind these general trends, considerable uncertainty remains about the true extent of remaining moist forest cover, and rates of loss. National patterns of forest-cover reduction are increasingly well documented. Many countries experienced substantial deforestation during the twentieth century – for example Thailand (Delang 2005), Mexico (Bray and Klepeis 2005) or El Salvador (Hecht *et al.* 2006). Thus 67 per cent of Costa Rica was under primary forest in 1940, and only 17 per cent by 1983 (Groombridge 1992). In Sumatra the area of unlogged forest declined dramatically in the last two-thirds of the twentieth century, as large areas were cleared for industrial plantations and subsistence agriculture, for settling transmigrants from elsewhere in Indonesia and for logging (Groombridge 1992; see also Figure 9.1).

Debate about tropical forest loss therefore remains high on international environmental and conservation agendas, and a key issue in sustainable development. This concern has three main dimensions. The first is the loss of species (particularly in recognized centres of high species diversity (Groombridge 1992)), which has been particularly significant since the signing of the Convention on Biological Diversity (Prance 1991; Smith *et al.* 1993; Hails *et al.* 2006; Secretary of the Convention on Biological Diversity 2006). It was with a compelling description of the vibrant living diversity of the rainforest that Edward Wilson began his book *The Diversity of Life* (1992), with its classic celebration of biological diversity and warning about the implications of the looming anthropogenic extinction spasm. Rainforest maintains a central place in the iconography of global environmentalism. The second dimension of concern about forest loss is in many ways the most significant, and it relates to the impact of deforestation on the rights and needs of forest people. Organizations like the World Rainforest Movement and Survival International have led a debate about forests that centres not on questions of environmental sustainability, but on human rights (Byron and Arnold 1999).

The third concern about deforestation is of growing importance in debates about sustainable development. This is the role of forest loss in the context of the global carbon balance, and hence anthropogenic climate change (Gash *et al.* 1996). This issue was central to debate over the successor to the post-Kyoto

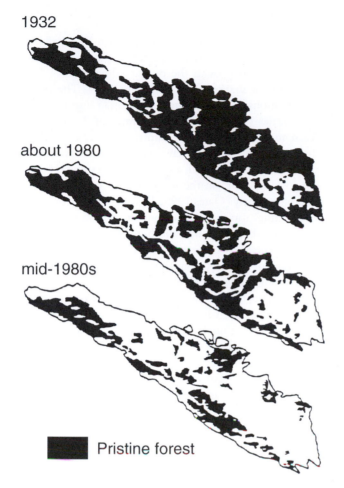

Figure 9.1 Deforestation in Sumatra (after Groombridge 1992)

climate regime post-2012, particularly at the UN Climate Change Conference in Bali in December 2007. Interest focuses on attempts both to re-establish cleared forest, funded by developing markets in carbon, and to preserve standing forest for the carbon locked in its ecosystem (Stier and Siebert 2002; Bonnie and Schwartzman 2003). The argument for strategies based on reduced emission from deforestation and degradation (REDD) is that forest loss (to all causes) releases more carbon to the atmosphere than the fossil-fuel-intensive global transport industry. Keeping trees standing locks up carbon, maintains biodiversity and provides the energy-guzzling industrialized countries with a cost-effective way to offset their carbon without changing their lifestyles. In order to work, of course, REDD strategies need to be based on a clear understanding of the causes of forest loss (Kanninen *et al.* 2007).

Explaining forest loss

Conventional explanations of deforestation (or forest conversion to other land uses) have tended to present it in simplistic terms, either as a direct result of pressure on resources, with population growth the primary driving force (e.g. N. Myers 1980), or as the result of market forces (Hecht 2004). The population pressure explanation summons well-worn neo-Malthusian arguments in its support. Thus Allen and Barnes (1985) showed that rates of deforestation in twenty-eight African, Asian and Latin American countries between 1968 and 1978 were correlated with population growth, fuelwood production and wood export, and agricultural expansion. However, such analyses tend to suggest purely technical responses to forest loss, focusing on the need for preservation and the exclusion of people to protect biodiversity (Oates 1999; Terborgh 1999), or trying to establish 'rational' and 'scientific' forest management to achieve sustainable harvests, while tackling human population growth and human needs separately.

Economic analysis of forest conversion, on the other hand, suggests that deforestation happens because the flow of benefits from the forest is not matched by benefits from other uses (Pearce and Brown 1994; Barbier *et al.* 1995). However, a simple market analysis is incomplete, for strong political economic processes are at work. Those who gain the benefits are often different (and more powerful) people than those who lose benefits from uncleared forest (Hecht 2004). This is clearly the case where logging companies obtain felling licences and clear forest land occupied by shifting cultivators or hunters, or ranchers obtain rights to clear forests for large-scale stock raising. It is also true where small farmers wish to move to the forest frontier as settlers, and are able to do so and clear the forest without the consent of existing forest people.

Most forest is owned by the state (87 per cent globally according to Siry *et al.* (2005)). The legal basis of such claims by the state often flies in the face of the rights of indigenous people and ignores their systems of tenure and rules of resource management. Moreover, forest land is often treated as a de facto open-access resource – that is, either one to which access is wholly unregulated, or one where regulations exist but are not implemented (for example, where forest or land concessions are given corruptly). In many developing countries, the state has taken authority to regulate the use of forest land, timber and other forest products, but lacks the authority and power to implement them. Forests are thus wide open to any entrepreneur able to negotiate unofficial access to forest land independently of an official legal regime (for example, by bribery), and convert a slow, sustainable trickle of economic benefits into a one-off windfall of timber or cleared land. Much commercial logging in rainforests takes place in contravention of formal plans and regulations, again often because logging contractors are able to arrange de facto access even to reserved forests.

The fact that forests are treated as open-access resources can be explained in terms of market failure (Pearce and Brown 1994). First, local market failure means that those who convert forest land take the benefits from doing so (for example, timber revenues) but do not have to pay the costs (loss of subsistence livelihoods, soil erosion, loss of biodiversity, loss of future revenues from timber if forests do not regenerate). Second, governments either fail to address this market failure (and make those who gain from forest conversion pay the costs), or offer perverse incentives (for example, failing to tax or regulate logging companies, or allowing inefficient logging practices that encourage wastage), so that the economic balance is pushed towards forest conversion. These problems are exacerbated by poor governance, with inefficiency and corruption tending to favour the short-term profits of large corporations at the expense of forests (and very often the people who live in them). Third, the global values of forests are poorly represented by the market. These include carbon storage, water supply and biodiversity (whether this is seen to have intrinsic value or use value as a reservoir of pharmaceutical products or a destination for future ecotourists). There is no direct market for many of these values (although a growing global system of trade in carbon – for example, at the Chicago Climate Exchange – is starting to produce one), and as a result the values of standing forests are under-represented when choices are made about clearance.

The policy framework has huge significance for deforestation. Thus, in West Africa, forest loss is related to the agricultural policy environment. Overvalued exchange rates, and pricing that effectively taxes agricultural exports, subsidizes agricultural imports and discourages market production, have prevented farmers from investing in their land, and particularly from adopting new technologies and inputs (Cleaver 1994). Poor rural services and infrastructure (lack of roads to get to market, of clinics to tackle sickness, of access to inputs and new agricultural knowledge) have further hampered innovation. As a result, farmers have not intensified, but have extended production, expanding onto new forest land, and clearing trees to plant crops.

Socio-economic models of the causes of deforestation have grown in sophistication. However, it is still not easy to explain forest loss satisfactorily in a statistical sense in response to factors such as population growth or debt (Gullison and Losos 1993; Pearce and Brown 1994). Cross-national data on forest loss for the period 1975–90 suggest that rural population growth and poverty (lack of other economic options) were key factors where forest extent was already limited, while, where forests were extensive, loss was due to the actions of entrepreneurs, small farmers and companies working in concert (Rudel and Roper 1997). However, it is increasingly clear that single-factor explanations of tropical deforestation are inadequate: multiple causative factors operate. A detailed analysis of 152 case studies showed that different combinations of proximate factors and underlying drivers are important in different geographical and historical contexts (Geist and Lambin 2002). To understand forest-cover change, it is necessary to understand the specific political, economic, social and environmental processes at work there.

Deforestation narratives

The explanation of forest conversion may be complex and as yet inadequate, yet powerful narratives are widely accepted that explain deforestation in simple neo-Malthusian terms (Fairhead and Leach 1995b). As discussed in Chapter 8, policy-makers faced by uncertainty and an urgent need to make decisions fall back on conventional understandings of the nature and causes of problems and the action needed to solve them. Once established, such policy narratives are persistent in the face of rebuttal and counter-evidence (Roe 1991; Leach and Mearns 1996). Established narratives of deforestation are now being widely challenged (Fairhead and Leach 1996, 1998; Kull 2004).

Forests are understood in different ways by different actors. Ecologists and conservation scientists tend to explain forest dynamics in terms of species and ecological dynamics (Struhsaker 1999), foresters in terms of their capacity to supply a certain number of board feet of timber over a specified period (Demeritt 2001; Barton 2002) and local people in terms of soils, plants, animals, place and their cultural associations. Forests are complex and contested spaces, not fixed entities whose nature can be stated in an absolute way. The way forests are understood, and the way they are valued, is inextricably linked to the ideas of the diverse actors who view or lay claim to them.

The best-known study of an established, yet flawed, deforestation narrative is by James Fairhead and Melissa Leach, who worked in Kissidougou Prefecture in Guinea, West Africa, in their study *Misreading the African Landscape* (1996; see also 1995a, b). Kissidougou lies on the ecological transition between what ecologists classify as moist forest and savannah. Rainfall is variable, but about 1,900–2,000 mm per year. The current landscape is one of dense forest patches and corridors, around villages and along streams, set in a matrix of grassland. Throughout the twentieth century a succession of outside experts and administrators have understood the Kissidougou landscape as the product of human clearance of forest, and thought that the mosaic landscape was subject to rapid degradation, particularly from fire. The French colonial botanist Auguste Chevalier reached this conclusion in 1909, as did Kissidougou's senior administrator and the staff of a European Commission watershed rehabilitation project in the 1990s. The forest patches around villages and along streams were presumed to be the fragments of a once-continuous forest cover, requiring urgent and draconian measures for their conservation (for example, prohibiting tree-cutting and fires, and attempting to persuade local people to plant trees in open patches).

Fairhead and Leach demonstrate that exactly the reverse is the case. They show from historical written accounts, maps and sequential air photographs that, if anything, forest cover has increased in the past century; that the landscape that greeted the first French colonial officers was substantially the same as it is today; that agriculture improves and does not degrade land; that high population densities allow fire management and control rather than increasing wildfires. The savannah was not 'derived', as so much research in Guinea and elsewhere in Africa had assumed for so long, and the forest islands were not 'relics' of former forest cover. Kissi informants made clear that they did not see the area as a forest landscape that was progressively losing its trees, but as a savannah landscape filling with forest: policy-makers were 'reading forest history backwards'. This reading of landscape led to and legitimized the intrusion of the state into the lives of Kissi villagers, taking away control of resources, imposing taxes and fines, and diverting funds better spent in other ways. Inspired by this discourse of degradation, outsiders to Kissidougou have

> accused people of wanton destruction, criminalised many of their everyday activities, denied the technical validity of their ecological knowledge and research into developing it, denied value and credibility to their cultural forms, expressions and basis of morality, and at times denied even people's consciousness and intelligence.
>
> (Fairhead and Leach 1996, p. 295)

French colonial foresters and their successors in independent Guinea have systematically misunderstood the role that people play in relation to the creation and destruction of forest (Fairhead and Leach 1996). Subsequent work has extended this analysis to other parts of Africa (Fairhead and Leach 1998; Ribot 1999; Cline-Cole and Madge 2000; Ickowitz 2006). Indeed, as research in environmental history and political ecology develops, the phenomenon of 'false forest histories' (Fairhead and Leach 1995b) seems to have a global scope. In Madagascar, for example, Kull (2000, 2004) traces a narrative of catastrophic deforestation as a result of slash-and-burn agriculture from the first days of French colonial annexation in the 1890s to the present day. The deeply held belief that Madagascar was relatively recently covered in forest was refuted only in the 1990s with the advent of palaeoclimatic research on Holocene vegetation, but even now it is widely parroted in popular and scientific accounts of conservation in Madagascar. In fact, the highlands and the West of Madagascar were never continuous closed forest like the moist forests of the east, but a mosaic of riparian forest, heath and grassland (Kull 2000): indeed eighteenth-century travellers commented on the open grasslands of the centre of the island.

The enthusiasm of the scholars analysing the flaws in classic uniform stories about deforestation is such that a counter-narrative that automatically questions their empirical basis is now firmly established. Certainly, accounts of forest loss are often historically weak and unsupported by data. Moreover, generalized accounts of the inevitability of forest loss also fail to take account of the phenomenon of forest recovery. Mather (1992) identified a 'forest transition', linking deforestation and industrialization and the growth of a national economy. He pointed out that forest area fell in industrialized countries in the past (as it is doing in non-industrialized countries now), but that it subsequently grew again (for example, in North-West Europe and North America). This pattern is common in developed countries: no country with a per capita GNP greater than $5,600 in 2005 was experiencing falling forest area. The phenomenon is also now widely reported from the developing world – for example, in China (where it dates to the late 1970s), India and Vietnam (from about 1990 (Kauppi *et al.* 2006)). Rudel *et al.* (2000) describe the recovery of forest in Puerto Rico in the second half of the twentieth century, and Bray and Klepeis (2005) note the same of south-east Mexico in the two decades from 1985. There are many reasons for such reversion

Plate 9.1 Clear felling in the Chocó rainforest, Ecuador. Little biodiversity and few other economic values survive clear felling of tropical forest. In some countries forest recovery is taking place. The biodiversity of regenerated forest is the subject of considerable research. Since 1999 Fauna & Flora International and its partners have been developing the Awacachi Corridor to save an area of Chocó rainforest from oil palm and logging interests in north-west Ecuador (www. fauna-flora.org). Photo: Juan Pablo Moreiras/FFI.

to forest – in Puerto Rico, it was the result of out-migration to urban employment to escape rural deprivation; in southern Mexico a much wider range of social, economic and political processes were at work. It is important to note, too, that some increases in forest cover involve plantations on former grassland, as in the case of the formerly biodiverse *páramo* grasslands of Ecuador: however attractive economically, some forest transitions may have significant negative environmental impacts (Farley 2007).

The important message here is that generalizations about forest loss based on loose but compelling narratives are highly misleading: what is needed is specific knowledge of forest-cover change in particular places, and the social and economic processes that provide its context. Hecht *et al.* (2006) explain how conservationists, obsessed with their 'Malthusian nightmare' of deforestation in El Salvador, fail to see or understand the extent of forest cover. Forests in El Salvador are not the pristine landscape of the American conservation imagination, but a mosaic of patches of successional, anthropogenic and old growth forests. Many are the result of the recovery of abandoned pastureland, others comprise complexes of forest and areas of coffee or fruit production. They argue that such fragments are dismissed by most analysts because they are small, heavily influenced by human action and often inhabited. Yet Hecht *et al.* argue that many of these 'secret forests' contain significant biodiversity and are worthy of conservation attention, as well as being of enormous local economic value. The way forests are imagined and defined is central to the way they are used, and the kinds of rules that get made about who can enter and draw sustenance from them.

The political ecology of deforestation

Political ecology is central to the issue of deforestation. Forest conversion cannot be explained simply in neo-Malthusian terms (as a product of rural population growth), or in terms of economic demand for export crops, or the inappropriateness of agricultural technology (Hecht 1984, 1985). Explanations of deforestation are 'socially and politically constructed to the advantage of powerful people' (Jarosz 1996, p. 148). In eastern Madagascar, colonial discourse linked forest loss to 'irrational' peasant farmers, whose shifting cultivation (*tavy*) was duly banned under a 'rational' approach to forest management (Jarosz 1996). Forest reserves were created and commercial timber exploitation began in 1921. However, suppression of *tavy* removed indigenous institutions that regulated how and where the forest could be cleared. Forest cover fell dramatically in response to the spread of cash cropping (for coffee), demands for timber (for example, by the railways) and an effectively uncontrolled mixture of cultivation, grazing, burning and extraction forest products. Population, initially, remained static. Forest loss was driven, not by population growth and irrational peasants, but by an entirely rational commitment by Malagasy farmers to new market opportunities. Government forestry failed to understand that, and their policies made things worse (in terms of forest cover) rather than better.

Forest management (including forest clearance) reflects the material interests of powerful actors, working either directly or through the apparatus of the state, and must be analysed in this political frame. In Thailand, for example, Delang (2005) describes the political decisions in the first decades of the twentieth century that led to the construction of railways into remote forested regions, and the links between agricultural colonization and the demand of the industrial manufacturing sector for cheap rural labour. After many decades of deforestation, a logging ban was eventually imposed in the 1980s. However, by then powerful allies of the military had established de facto cross-border access to logging operations in Burma, Cambodia and Laos, where illegal logging continued unabated.

Structures and decisions by actors at a range of levels are important in forest-cover change. Wood (1990) suggests that the politics of deforestation can be imagined as an upside-down pyramid of increasing scale (local, national, multilateral and global) and increasing ecological and political complexity (see Figure 9.2). At the local scale, from the neighbourhood to first-order administrative units (for example, provinces), drivers of deforestation are source specific and directly concern local economies – for example, the links between deforestation and fuelwood shortage, or the downstream effects of soil erosion. However, while local, these impacts relate to complex and disintegrated government bureaucracies, and are

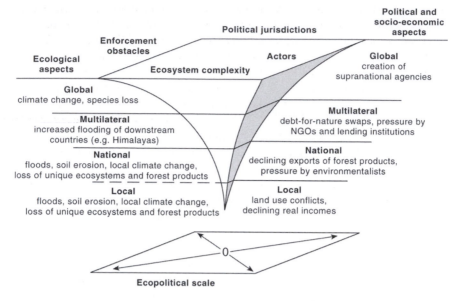

Figure 9.2 Ecopolitical hierarchy of tropical deforestation (after W. B. Wood 1990)

often beyond the influence of local people themselves. Any benefits from unsustainable resource use are lost to remote cities, or internationally, or the pockets of business or an often corrupt bureaucratic elite.

At the national level, deforestation reflects different factors – for example, loss of ecosystems and desire for cash revenue from standing forests. National politics will reflect strategic differences between government ministries, and between national and regional or local government. Even where laws are passed to promote sustainable forest management, capacity for implementation may be very limited. Internationally, agreements on forest management can be controversial, and the strategic interests of countries may differ (for example, between forest countries such as Pacific states and countries hosting major logging interests such as Japan). Globally, issues such as climate change and species loss become dominant, and, as the Rio Conference demonstrated, political debate can focus on differences between poor rainforest countries and the environmental sensibilities of delegations from rich countries.

The interplay between these levels can be complex. Internationally, the governments of industrialized economies can seek to apply pressure on national policies through 'environmental conditionality' and targeting in aid giving. National governments can also seek to change policy regimes, as Brazil, for example, did in 1989, with a new programme to remove tax incentives for ranching, to ban timber exports and to establish new reserves (W. B. Wood 1990). These interactions will in due course have some kind of impact on the ground, moderated by complex political and economic feedbacks at intervening levels.

In the Brazilian Amazon, the social, economic and environmental dimensions of settlement in the Amazon are diverse and complex (Furley 1994). The penetration of settlers and ranchers into the forest must be seen in the context of the political economy of Brazil (Hecht and Cockburn 1989; Parayil and Tong 1998). Deforestation has been driven primarily by subsidy provided by the Brazilian state (e.g. Goodland 1980; Moran 1983; Caufield 1985; Hecht and Cockburn 1989).

Brazil's region of Amazonas covers 5 million square kilometres (58 per cent of the country), but only 3.5 million square kilometres are (or recently have been) actually forested, and only 70 per cent of that is tropical moist forest (Barraclough and Ghimire 1995). Extensive forest clearance began to occur in the early 1970s, and has occurred at about 20,000–30,000 square kilometres per year. Following the military coup in 1964, the state sought legitimacy through economic growth. However, the astonishing growth in the Brazilian economy in the 1960s and 1970s (9 per cent annually from 1965 to 1980) was accompanied by increasingly industrialized methods in the rural sector, and a crisis of access to land by the poor. The opening-up of Amazonia (through the construction of roads to central Brazil and the coastal cities and other infrastructure) obviated the need for land reform by providing land for unemployed agricultural workers. Operation Amazonia was launched in 1965. The population of Amazonia was about 1.5 million in 1940, but this had grown to 8.6 million by 1989, mostly through immigration from the south and north-east, the settlement of small farmers being actively promoted by the National Institute for Colonization and Agrarian Reform (INCRA) (Ghimire and Pimbert 1997).

Laws passed in 1966 provided tax reductions for companies (including foreign corporations) investing in Amazonia, and additional credit was available both from within Brazil and internationally. These tax incentives, and cheap credit, encouraged the annexation of land by ranchers and speculators (Hecht and Cockburn 1989; Ghimire and Pimbert 1997). Land allocation was chaotic and disputes became common, their outcomes determined by 'political connections, economic power, bribes and sheer physical force' (Ghimire and Pimbert 1997, p. 56). New technologies of seeded pastures opened up profitable possibilities for industrial agribusiness to move into beef ranching, thus providing an important (and largely new) export product, new forms of production and an outlet for investment. The incentives for the development of Amazonia were supported by international concerns and interests – for example, the views of the World Bank and the FAO.

In 1970 the New Integration Programme brought a new focus on small-scale settlement, but the policy was again reversed in 1975, and ranching predominated once more. Cattle ranching achieved notoriety through critical accounts of the 'hamburger connection' (Myers 1981), and the linkage between Amazonian forest clearance and American fast-food outlets. American beef prices rose in the 1970s, and by 1978 Central American beef was less than half the wholesale price of beef raised in the USA. American beef imports rose rapidly, the lean grass-fed beef of rainforest pastures being suited to the fast-food industry, which comprises 25 per cent of US beef consumption. Between 1961 and 1978 beef exports rose by five times in Costa Rica and by fifteen times in Guatemala (Myers 1981). In Brazil the dry *terra firme* became the subject of intensive development both by transnational companies such as Volkswagen, King Ranch and Armour-Swift, and by large investors from Brazil (Fearnside 1980). Deforestation was driven by both domestic structural factors ('possibly the most extensive, destructive and chaotic private land enclosure in history' (Dove and Noguiera 1994, p. 492) and macroeconomic disequilibrium (inflation and foreign debt).

Environmental myths and narratives also legitimate forest clearance and settlement. In Brazil there was a pervasive belief among both scientists and policy-makers through the 1980s that cattle ranching was a valid and effective long-term land use, and indeed that the creation of pasture improves soils (Fearnside 1980). This was based on the argument that clearance of forest releases nutrients held in rainforest vegetation, and there is a rapid increase in nutrient availability. In fact, critics of cattle ranching in rainforest-derived pastures argue that this view ignores serious longer-term ecological impacts that make cattle ranching far from sustainable (Fearnside 1980; Hecht 1980, 1984, 1985). Forest clearance does generally bring about short-term falls in soil acidity, and a rise in the amount of calcium, magnesium, potassium and phosphorus, although there is a great deal of variability, and larger samples tend to show that increases are less dramatic than is often claimed. However, the pulse of nutrients is short lived. In particular, phosphorus levels decline after about five years to something very close to the levels in soils under undisturbed forest, and over longer-term five- to fifteen-year periods nutrient status and pasture productivity decline, sometimes drastically. Soil erosion can also become a serious problem: cattle ranching is not usually a sustainable land use in rainforest areas (Fearnside 1980). Ranch economics dictate the sale of pasture after about five years; indeed, S. B. Hecht (1985) reported that 85 per cent of ranches in Paragominas had failed by 1978. In ecological terms this seems to make little sense, but, of course, such short-term exploitation of land can be good economics. Entrepreneurs are attracted to rainforest ranching not by the long-term productivity of the land but by short-term returns on investment. As Hecht (1984) comments, 'if the productivity of the land itself is of low importance, cautious land management becomes irrelevant and environmental degradation is the inevitable result' (p. 393). Lack of secure land tenure meant that landowners, tenants and squatters had no incentive to do anything but mine soil fertility, forest resources and minerals as rapidly as possible, a process exacerbated by the way in which land clearance was taken as evidence of land improvement and effective occupancy (Ghimire and Pimbert 1997).

Forest clearance and forest people

Although frequently referred to by conservationists (especially from the USA) as 'wilderness' (see Chapter 10), almost all tropical forests are or have been inhabited. Many forests that appear to be primary have a deep history of occupation (Ickowitz 2006). In the mid-1970s it was estimated that forest farmers numbered perhaps 140 million people (N. Myers 1980).

Small forest farmers have long been regarded as villains in accounts of forest clearance, and 'slash-and-burn agriculture' has been held up for ritual condemnation,

Plate 9.2 Forest clearance for agriculture on the edge of the Bwindi Impenetrable
    National Park, south-west Uganda. Forest clearance for small-scale agriculture
    can be rapid and permanent, although without appropriate farming systems
    cultivation may not prove sustainable. Bwindi Impenetrable National Park
    was declared in 1991 to protect a range of species, including most notably
    the mountain gorilla. The agricultural frontier advanced in Kigezi throughout
    the twentieth century; rural population densities in the region are still high
    and poverty a major problem. The interests of conservation and farmers are in
    many ways diametrically opposed, although the generation of foreign-exchange
    income from gorilla-watching tourism provides some prospect of common
    ground for those who see a sustainable future for 'community conservation'
    strategies (see Chapter 10). Photo: W. M. Adams.

first by colonial agricultural officers and later by the international consultants who
followed them (Jarosz 1996; Ickowitz 2006). The social processes of resource
definition, extraction, control and distribution are central to the understanding
of deforestation (Jarosz 1996). The attitudes to indigenous resource users on the
part of colonial foresters in Madagascar that Jarosz describes, and their resort to
a policy of rational forest exploitation and reservation, were repeated throughout
the colonial world – for example, in India (Gadgil and Guha 1995; Jewitt 1995;
Barton 2002). The carrot of the market and the stick of colonial taxation stimulated forest clearance for cash crops in many places (for example, South-East Asia
in the nineteenth century (Parnwell and Bryant 1996)). In Burma a discourse of
'forestry as progress' supported a view of the forest that focused on teak and the
revenue that could flow from it, effectively making alternative uses invisible (and
illegal). In India the Forest Department was established in 1864, and the Forest
Act of 1878 allowed for the closing or 'reservation' of forests to allow 'scientific
forestry' to concentrate on efficient timber production. Existing use rights for

timber and non-timber forest products were extinguished. In the Uttarakhand (the Himalayan hills of Kumaon and Garhwal), demand for railway sleepers led to the closure of vast tracts of forests to subsistence use, and the banning of practices such as burning. The resulting hardship began a long history of protest against state forest policy (Guha 1989). The idea that forests should be places for timber production alone, and dismissive or hostile attitudes to the wider range of uses typical of forest-dwellers, persist today in many parts of the Third World.

Some colonial forest-management models were built around forest farmers (for example, the Malayan Taungya system, introduced by the government forest department in southern Nigeria (Hellerman 2007)), and others (notably the Forest Department in India) gazetted vast areas of inhabited land as 'forest' regardless of its land use – or, in some instances, tree cover (Vira 1999; Barton 2002). Whether conscious of forest people or not, most forestry departments have tended to regard rural people as the self-deluded destroyers of a valuable resource of timber, cutting minor forest products (canes, rattans, and so on), hunting and gathering, and above all clearing forest for farmland, only to abandon it and seek to move on. In Thailand, indigenous forest-dwellers have been particularly unfairly targeted, the natural scapegoat for forest loss in Thailand (Delang 2005), and in Madagascar forest farmers have been blamed for poor agricultural practices while, paradoxically, also being celebrated as the source of indigenous community management institutions (Kull 2000).

The ecology of forest clearance for agriculture is complex. As mentioned above, forest clearance leads to a flush of nutrients that supports one or more growing seasons, followed by a fairly rapid decline in soil nutrient status and yields (Nye and Greenland 1960). Soil fertility can be replaced by a period of fallow, or by cropping systems that mimic the physical structure of rainforest. Shortening fallow cycles are conventionally blamed for fertility decline. However, evidence that there was ever a 'golden age' during which ideal practices were followed, or that fallow intervals have widely declined in recent decades, are both scant for regions such as Africa (Ickowitz 2006).

The image of the forest farmer is either of the rapacious chancer, knowingly clearing land and managing it unsustainably, or of the ignorant victim, condemned by lack of education to gnaw away at the edge of the forest and turn it to worthless fallow behind him (*sic*: such myths often carry a curious gender bias). Thus, with characteristic tunnel vision, forestry staff in Indonesia lament illegal cutting by people 'who do not understand the ecological consequences of forest destruction' (Soerianegara 1982, p. 82). However, between apparent irrationality and greed lies a whole world of decision-making by forest farming households (Townsend *et al.* 1995). Institutional and economic factors affect both the productivity and the sustainability of farming of frontier settlers in Amazonia. Even where forest farming is a significant cause of forest loss, as in south-east Nigeria, the decisions of farmers are entirely rational at the household scale (Ite 1996, 2001).

Forest farmers themselves generally have a very clear understanding of the ecology of fallow plots in the forest. They leave trees in cleared plots (they have many recognized uses, but are also hard to remove), and are aware of the economic and ecological options in cropping systems (cassava in intensive cleared land near the house, and banana and plantain in regenerating forest further away), and of the implications of the total loss of forest cover. Forest farms tend to be small, diverse, and integrated in space and time with the forest around them (see Plate 9.3). Patterns of forest clearance, and post-clearance field management and forest regeneration, are therefore the result of both necessity and choice at household scale. Forest farmers are intelligent and active agents of forest management, far from the unwitting agents of degradation so often portrayed in accounts of deforestation. Tropical forests are often profoundly managed ecosystems, even when they are rich in biodiversity and to the untutored eye look 'natural'.

Much forest loss to 'small farmers' in fact involves schemes to settle farmers into the forest zone, usually from densely populated regions elsewhere. Sometimes this is done to attract and organize such settlers (as in Indonesia), sometimes to manage spontaneous settlement processes (as in Amazonia). Settlement schemes in rainforest areas tend to have limited success even in narrowly economic terms (Schmick and Wood 1984). Hiraoka and Yamamoto (1980), for example, showed that new settlements in north-east Ecuador were ill suited to the environment. Many such projects, such as the enormous Transmigrasi Project of Indonesia to resettle people from Java (Hardjono 1977; Otten 1986), appear to derive their justification from powerful political support (Budiardjo 1986) rather than from any substantive success shown by any actual projects themselves. Tony Allan (1983)

Plate 9.3 Land cleared for a farm in the dry forests of the Madhupur Tract, Bangladesh. Fertility is high on newly cleared land, and a mixture of crops is grown without extensive cultivation, achieving rapid coverage of bare soil. Photo: W. M. Adams.

argued that the idea of the Amazon as a 'virgin land' was sustained by fantasies woven and sustained by government rhetoric. Skillings (1984) suggested that the psychological effect of the sheer extent of Amazonia, which formed 57 per cent of Brazil, held only 4 per cent of the population and generated 2 per cent of GDP, acted as a challenge to political leaders, one supported by a geopolitical imperative to secure remote forested regions by showing an effective state presence.

Policy for forested land is often developed as if forests were empty and unclaimed, an untapped resource for the state and its contractors to exploit. As a result, the most destructive impacts of forest policy are often on forest people themselves. There is increasing awareness of the impacts of deforestation on indigenous people and other forest-dwellers – for example, in Amazonia (Arvelo-Jiménez 1984; Vickers 1984; Survival International 1985). The World Bank's policy paper *Tribal Peoples and Economic Development* (1984a) stated the Bank's policy not to develop areas presently occupied by tribal people, with the proviso that, if this was unavoidable, the Bank would ensure that 'best efforts have been made to obtain the voluntary, full, and conscionable agreement (i.e. under prevailing circumstances and customary laws)' of the tribal people or that of their advocates, and that project design and implementation 'are appropriate to meet the special needs and wishes of such peoples' (p. 1). Many observers remain unconvinced by policy statements of this kind. Maybury-Lewis (1984) argued that 'we have to come up with strategies of development that offer reasonable and equitable opportunities to participate to all members of society, both Indians and non-Indians' (p. 134); it should not be assumed that Indian societies are doomed just because of the strength of the ideologies behind the second conquest.

On the ground, forest 'development' has destroyed the environment and economy of forest-dwelling people in many parts of the world, their small but sustained flow of revenue from the forest swept away in the pursuit of larger short-term benefits for the state or (more usually) private business interests. Thus, in Médio Amazona, large-scale ranching on *terra firme* had a serious impact on small peasant producers (Bunker 1980). Peasant producers often lacked title to land, and were unable to sustain production in the face of competing demands from ranchers for accessible land as land prices rose. Ranching, therefore, was socially regressive as well as environmentally damaging. Indigenous people in Rondônia (Brazil) lost land to gold prospectors (whose mercury polluted rivers far from their scene of operation), ranchers, loggers and land speculators (Barraclough and Ghimire 1995). Riverine cultivators, established Portuguese-speaking farmers of the seasonally flooded *varzeas*, suffered from ecological changes that were due to pollution as well as to dam construction (see Chapter 11).

The close link between forest people and their land, and their interest in defending their way of life, have led to various kinds of resistance (Colchester 1994). Conflict and violence between those who hold land and power and those without either have long been a feature of the south of Pará in the Brazilian Amazon, partly the result of ambiguous land laws (Simmons *et al.* 2007). Such ambiguity favours the powerful, and allows the institutions of the state to be captured to foster private gain, and public resources to be annexed: a process of 'privatization of the public purse'. The Chipko movement in the Garhwal Himalaya in India provides one example of such protest (Guha 1989), although there is much debate about the politics of the Chipko movement (Rangan 1996; see also Chapter 13). The work of the National Council of Rubber Tappers in Brazil in the 1980s is also much cited as an example of grass-roots attempts to retain rights to forest resources. When the Amazonian rubber boom collapsed, production from wild trees persisted (Coomes and Barham 1994). However, as land colonization, deforestation and mining gathered pace, those living in the forest collecting rubber and Brazil nuts steadily lost land to speculators and ranchers. The rubber-tappers began to organize and protest, and met with violent oppression from powerful economic interests. The plight of the rubber-tappers became part of international protest at Amazonian 'development' following the murder of the head of the recently formed National Council of Rubber Tappers, Chico Mendez, in 1988 (Barraclough and Ghimire 1995). Extractive reserves were eventually established in 1990 (C. Campbell *et al.* 1996). By that time, indigenous reserves covering 270,000 square kilometres had been set aside in Amazonia, and the rights of indigenous people had been recognized in international law, although in practice and on the ground these rights were still widely abused (Barraclough and Ghimire 1995).

Conservationists worry about the loss of biodiversity associated with tropical forest clearance, yet the impacts of that clearance on the livelihoods of the poor are often more alarming. Clearly, there is potential for common ground to be established between those advocating forest conservation because of its diversity, and forest people and their supporters (Redford and Stearman 1993; Sanderson and Redford 2003a). Both deplore unsustainable logging and privatization of public assets. However, conservation can itself be a direct challenge to the immediate livelihood needs of the rural poor in tropical forests, and more generally. In its desire to 'protect' forests from all forms of human depredation, conservation can become, to forest people, just another form of state-sponsored development. There is a political ecology of conservation, as will be discussed in Chapter 10.

Sustainable forestry?

Consumer campaigns by environmental pressure groups and environmentalist–business partnerships both seek to bring about sustainability, albeit in different ways (see Chapters 5 and 6). Such efforts are wholly dependent on the success with which sustainable resource management strategies can be defined and made technically feasible. The notion of 'sustainable management' of tropical forests has, therefore, become an important element in debates between environmental pressure groups and the global timber industry. Unsurprisingly, there has been a great deal of research to establish what 'sustainable tropical forest management' would look like, and there have been a succession of guidelines for development in rainforest regions (e.g. Poore 1976; Poore and Sayer 1987; Gómez-Pompa *et al.* 1991; Tumaneng-Diete *et al.* 2005). The *FAO Model Code of Forest Harvesting Practice* (Dykstra and Heinrich 1996) attempted to contribute to the expectations of the Rio Conference about sustainable forestry, and particularly the Forest Principles. It provided a model for national forest harvesting codes that sets out best practice in the areas of harvest planning, forest road engineering, extraction, landing operations, transport operations, harvesting assessment and the forest harvesting workforce.

Logging of natural forest can broadly be classified as selective or clear-felling, but within these categories there are distinctions. Selective logging intensities can vary from two to three stems per hectare taken out in some African forests to twenty or more stems per hectare in parts of South-East Asia. This difference is caused less by the attitude of foresters to conservation than by the density of timber of suitable size and quality. Both selective felling and clear-felling have extensive impacts on forest ecology, although, of the two, clear-felling is obviously the more destructive. The two may be operated together in different zones of the same forest, and worked forests may exhibit a complex and fragmented structure, with some areas nearly 100 per cent logged and others with a canopy reduced by 30–50 per cent. Selective logging can take place on a monocyclic basis (a single operation to remove all saleable trees) or in a polycyclic system, where trees are removed in a series of felling cycles as they reach suitable sizes (A. D. Johns 1985).

The broad technical requirements for sustainable forest management have been identified and are now well established (e.g. Gómez-Pompa *et al.* 1991). The environmental impacts of logging vary considerably with the logging regime adopted. Typical problems of most large-scale commercial forms of timber extraction include soil compaction and loss of internal structure, effects of leaching and run-off on soil nutrients and micro-organisms, the loss of nutrients in timber and changes to the internal microclimate of the forest (Shelton 1985).

The most critical factor in determining the ecological impacts of logging is subsequent land use. This can range from abandonment following timber clear-felling through replacement with a pasture for cattle-raising, or establishment of a plantation of tree crops, to an agroforestry system. In many cases forest can regenerate, and, with it, eventually, timber value, non-timber forest products and biodiversity. For example, a local operating company of the Japanese company Honshu Paper, part of the Mitsui group, began felling for woodchips and saw logs in the Madang and Naru Valleys in Papua New Guinea in 1973 on a logging concession of 68,000 hectares (Lamb 1980; Seddon 1984). By 1983 more than half had been logged. Logging had caused the removal of humus and topsoil, leading to high rates of soil erosion and leaching, deposition of sediments and waterlogging on valley floors and flats, with associated soil acidity and the loss of phosphorus. However, by 1984 regeneration of secondary forest was taking place. There were significant effects on the diversity of the forest, which regenerated, partly because of the loss of soil seed banks. However, where logged land was cultivated or disturbed (for example, in log loading areas), tree regeneration was prevented, and grassland took over (Saulei 1984).

This is not an untypical story. It can be read from the standard idealist conservation perspective (of biodiversity lost as natural forest is logged). However, the 'naturalness' of the forest is not in fact so straightforward. Despite appearances, the forest cleared was itself a secondary growth resulting from drought and fires in the 1930s and during 1944–5 (Saulei 1984). Many areas of tropical forests have a long history of human use. In Marovo Lagoon in the Solomon Islands, the apparently 'natural' forest is in fact rich in cultural sites of fields, settlements and fruiting trees (Hviding and Bayliss-Smith 2000). The diversity of such forests may make them a legitimate target of concern for conservationists, but their apparently pristine nature is a product of the romanticism and ignorance of outside observers. Arguably, a sustainable logging regime might be perfectly possible, and perfectly acceptable, in such a location.

The most acute impacts on biodiversity are associated with clear-felling, followed by permanent conversion (see Plate 9.4). Even in this situation the ecological impacts are complex. Studies of the ecology of forest fragments in Amazonia are increasing knowledge of exactly what may survive in refugia left after widespread deforestation (Lovejoy *et al.* 1983). Shankar Raman *et al.* (1998) demonstrate that shifting cultivation with a typical five- to ten-year cycle in north-east India has significant implications for birds and plant species richness: a fallow interval of twenty-five years for birds and fifty to seventy-five years for plants would be

Plate 9.4 Bwindi Impenetrable National Park, Uganda. The mountain gorilla survives in the wild only in four remaining areas of forest in central Africa, Bwindi Impenetrable National Park in Uganda, with Mgahinga Gorilla National Park (Plate 9.2), the Parc National des Volcans in Rwanda and the Parc National des Virungas. These parks are islands set in a sea of recently cleared and now intensively managed farmland. Gorillas and farms do not mix, presenting a sharply focused choice between retaining the forests and perhaps keeping the gorillas, and clearing the remaining forest for farms. Photo: W. M. Adams.

necessary to maintain community composition. Inappropriate management of clear-felled land can lead to widespread environmental impacts such as the large-scale fires that afflicted large areas of Indonesia in the late 1990s during dry spells associated with El Niño events (Tacconi and Ruchiat 2006). On clear-felled land, ground vegetation is lost through desiccation, and animal and bird species dependent on shaded forest conditions are also lost. Wildlife species such as primates are unable to survive in heavily logged forests, although a limited number of species can survive selective logging (Johns 1985). Territorial species such as gibbons are tenacious in surviving forest fragments (Shelton 1985). Long-term responses of primate populations (and by implication other species) depend on the availability of refugia, the nature of hunting pressure and the willingness of foresters to adhere to silvicultural rules designed to achieve sustainability (Johns and Johns 1995).

Tropical forest loss presents huge challenges in terms of biodiversity conservation in many countries. In West Africa, Oates (1995, 1999) suggests that anything other than strict protection will fail to preserve biodiversity (certainly for easily hunted primate species). This 'empty-forest' problem (Redford 1992; Struhsaker 1999) has been widely recognized as demanding strong protection of forest protected areas (Redford *et al.* 1998). However, others hold that the biodiversity

Plate 9.5 Logging road, Ecuador. Logging roads provide access for bushmeat hunters, illegal loggers and informal settlers in previously remote regions, to the detriment of local forest people and biodiversity. Since 1999 Fauna & Flora International and its partners have been working in the Awacachi Corridor in north-west Ecuador to protect rainforest from oil palm and logging interests (www.fauna-flora.org). Photo: Juan Pablo Moreiras/FFI.

value of human-transformed forests has been underestimated (e.g. Hecht *et al.* 2006). Data on birds and plants from Puerto Rico suggest that models based on island biogeography theory tend to overestimate the rate of species extinction associated with deforestation. There may be potential to maintain and enhance biodiversity through appropriate management (Lugo *et al.* 1993), and there is a need to understand the relationships between diversity, ecological dynamics, social ecology and political economy (Hecht *et al.* 2006).

There is active discussion of the extent to which hunting of forest animals ('bushmeat') can be sustainable, and if so how (Milner-Gulland *et al.* 2003). In Africa, the market for bushmeat is considerable and growing, as prosperous city-dwellers turn away from mass-produced farmed meat (Fa *et al.* 2006). There is a considerable international trade, especially to Europe, all of it illegal. At the same time, studies suggest that, where the main target species are common and reproduce relatively rapidly (such as porcupine or cane rat), sustainable harvests might be possible. This may be the situation in parts of West Africa, where larger and rarer forest species have already been hunted out (Cowlishaw *et al.* 2005). If so, asking hunters and traders to regulate a legal harvest of common species might be easier than trying to stop a rampant illegal harvest of all species. Sustainable hunting of primates is effectively impossible, as well as unwise (because of cross-species disease transmission risks) and morally fraught.

Most proposals for sustainable forestry are built on the idea of selective logging. This does less environmental damage than clear-felling, although impacts are not negligible. Many non-target trees can be damaged by falling timber trees and drag paths or other works. In East Kalimantan, intensive selective felling of fourteen stems per hectare by mechanized methods caused damage to 41 per cent of residual trees; skid tracks, haul roads and log yards occupied 30 per cent of the logged area (Kartawinata *et al.* 1981). The recovery of these compacted and scraped areas is slow: water infiltration rates are low and erosion is high. Selective logging also involves the loss of the best specimens of commercial species, and potentially the extinction of species, forms of 'genetic erosion' with potential impacts on the ecology of the forest.

Although claims are now widely made that logging practices in rainforest countries have made substantial shifts towards sustainability, much doubt remains about whether that is the case. In Sarawak, for example, selective logging and an annual cut below estimated maximum sustainable yield are used to justify a claim of sustainable forestry. However, the state forest output (15.8 million cubic metres in the mid-1990s) was 70 per cent over the estimated sustainable cut, and there are serious problems of collateral damage to non-target trees, soil erosion and slow revegetation of skid trails; the planned cutting cycle (twenty-five years) is too short for dipterocarp trees. Furthermore, logging takes place without regard to the needs of forest-dwellers, for example, the Penan people. In Sarawak, as in many other places, sustainability is but a 'thin veneer' (Pearce 1994, p. 30).

The social dimensions of forest policy are also an essential component of sustainability. Conventional forestry usually involves a company (often a transnational) paying a government for a licence to extract timber, in the form of chips or sawn logs. It may employ a second company as felling contractor. The benefit of logging to the local economy is usually limited (some low-paid work, but loss of non-timber forest products). Benefit to the national economy is restricted, because, while value is added to the timber when it is sawn and made into products, this typically takes place elsewhere. Moves to promote sustainable forest operations therefore not only address the ecological impacts of timber extraction but also seek to capture more of the value of wood extracted to supply the local economy. This might involve increases in local involvement in (or even ownership of) commercial forestry operations, and increases in the extent to which timber is processed locally.

One early example of an alternative approach to logging was the Gogol Valley development in Madang Province, Papua New Guinea. From 1975 to 1978 the Gogol Project was the subject of Man and the Biosphere (MAB)-funded research, and was seen in some quarters as a showcase of forestry management (Lamb 1980). Initially the project was intended to integrate sawn timber, veneers and woodchips for pulpwood. The plan was for 48 per cent of the land to be clear-felled, 22 per cent selectively felled and 30 per cent unlogged, and the project was to include an element of reforestation. In practice, reforestation fell far short of the 800 hectares per year target, only 1,000 hectares being planted in the first five years, mostly with species of *Eucalyptus, Terminalia* and *Acacia*. The rate of planting subsequently increased, but the extent of plantation remained too little to sustain the pulp mill after the fifteen- to twenty-five-year life of the project (Seddon 1984). The potential sustainability of the Gogol Valley model depended not only on the profitability of the timber operation, but also on the economic benefits and costs to residents of the valley. The Gogol Valley is home to 2,000–4,000 indigenous people who lived by shifting agriculture, and for whom forms of permanent agriculture were sought. Royalties paid to the indigenous people injected cash into their economy, but experience of other promised benefits (access, education, jobs) was mixed. For example, logging roads have not survived the extraction operation. Against this must be balanced the trauma of the loss of traditional forests (although areas around villages and river corridors were spared), and the economic fact that the holding company JANT failed to show a profit, and thus escaped taxation. Profits were written off against capital borrowing costs, and (it is alleged) transfer pricing allowed profits to be repatriated (Seddon 1984).

There is increasing awareness of the multiple functions of forests, and the need to look beyond the timber resource. Thus, for example, the FAO's *Mangrove Forest Management Guidelines* (1994b) argued that 'the traditional "*management paradigm*" implying that if forests are well managed then, *ipso facto*, the non-wood ecosystem components will remain stable, is notionally flawed' (p. xxiii). The products of mangroves are legion, including capture and culture of fish, salt production, honey, agriculture, charcoal, firewood, poles, tannin, palm and wildlife (Barbier 1998). Mangroves are also vital in many instances to coast protection. Social and economic benefits need to be maximized through multiple-use of resources, involving both integration with management of the wider coastal zone, and an approach to planning that allows for the participation of the rural poor (FAO 1994b). In Indonesia, as in many other countries, dominant policy narratives support intensive aquaculture production at the expense of local common pool resources and local people (Armitage 2002).

Multiple-use forest management that takes account of the needs of existing local populations approaches is commonly labelled 'social forestry', although that label has been applied loosely to a great many different kinds of projects and initiatives. It has been incorporated into the rubric of most international agencies involved in forestry, particularly the FAO and International Tropical Timber Organization (ITTO) (Barraclough and Ghimire 1995). In a number of countries – for example, in India – conventional forestry strategies based on narrowly defined commercial timber production have been broadened to embrace a wider range of non-timber forest products. Moves to relate the needs of tribal people to forestry management – for example, in Gujarat (Murdia 1982) – have seen institutional innovations such as Joint Forest Management in India, where to some extent at least technical forestry bureaucracies begin to share power with local communities.

However, it should not be expected that communities will necessarily share the opposition to commercial clear-cutting expressed by Northern environmentalists. In northern New Georgia (Solomon Islands), local people (backed by Northern environmentalists) originally opposed logging by Lever Pacific Timbers (part of Lever Brothers) in the 1980s, causing the company to withdraw. However, despite this apparent victory for conservation, and explicit local support for the idea of sustainable forestry, the community soon awarded a new contract to another company, which promptly moved back into the old Lever camp and carried on much as before. The reasons for this are complex (Hviding and Bayliss-Smith 2000), but it serves as a reminder that sweeping assumptions should not be made that 'the community' will choose the same interpretation of sustainability as international environmental groups.

Forest products are of enormous importance to both forest-dwellers (hunters and farmers) and those living on the forest margin. There are important distinctions to be drawn between those who are dependent on the forest and those whose use is the subject of choice (Byron and Arnold 1999). Both can be adversely affected by the arrogation of rights in timber by the state, and the closure of the forest to alternative uses (grazing, fuelwood and poles, charcoal-making, collection of medicinal plants and hunting). The removal of forest cover can create new frontiers for settlement, but at great cost to existing communities at the forest edge, and most particularly to groups economically dependent on the forest itself. Patterns of forest-product demand, use and supply are changing almost everywhere, particularly as forests are cleared and product supply is limited to bush fallow and farm trees. The balance of individual and communal tenure of forest resources tends to change, and the social arrangements surrounding forest-product extraction are complex and fluid (Coomes and Barham 1997).

The state has a vital role in supporting institutional arrangements that maximize sustainable livelihoods, and the matching of those institutions to social and economic change is a key challenge for the future. In Costa Rica, rates of deforestation have been very rapid – over 7 per cent of forest area per year was lost in the 1980s (Gottfried *et al.* 1994). The Golfo Dulce Forest Reserve (613 square kilometres) lies on the Osa peninsula, which contains the largest area of rainforest on the Pacific coast of Central America. Some 8,000 people live within the reserve, where forest loss has run at 5 per cent per year. The Costa Rican government has sought to maintain forest cover and enhance sustainable livelihoods through natural forest management and small-scale timber enterprises. However, this strategy faces the major problem of insecure land tenure (forcing farmers to clear land to prove effective possession) and lack of credit (forcing farmers to seek a quick return through cattle-raising, even though this is ecologically unsustainable). There are also tax incentives and government loans for land clearance (Gottfried *et al.* 1994). The pursuit of sustainability in forest environments therefore depends crucially on the institutions governing social, economic and environmental change.

Regulating tropical forests

If it is possible to identify new ways to manage forests that avoid the worst environmental impacts and maximize social benefit, the next question is whether they can be made to happen. New methods, new forms of social ownership, new approaches to people and nature, all require new institutions. How do you get established ways of thinking and acting to change? Critically, how do you persuade those with vested interests in current ways of working to accept change?

There is now a growing literature on attempts to transform forest management policies in Third World countries to include the designation of forest reserves, the creation of sustainable agriculture in rainforest regions, the managed exploitation of natural forests and the restoration of logged and degraded forest lands (see, e.g., Gradwohl and Greenberg 1988), and/or the creation of systems of incentives for sustainable production systems (Repetto 1987). The development of policy regimes in practice can be problematic, because of the number of possibly conflicting policy goals, including stimulating the forestry industries, generating revenue for government services, spreading economic benefits across society and achieving environmental aims. Thus in the Philippines, for example, a partial log export ban, a commodity tax on logs and taxes on the exportation of logs and lumber and the importation of wood pulp had different aims and effects (Tumaneng-Diete *et al.* 2005).

One approach, in line with the idea of ecological rationalization and MSD (see Chapter 5) is regulation. This is extremely challenging because of the complexity and international connectivity of the tropical timber trade. One approach was the Tropical Forestry Action Plan, released at the time of the World Forestry Congress in Mexico (Poore and Sayer 1987). It was the work of the World Resources Institute (a powerful and wealthy environmental think tank based in Washington), the IUCN, the FAO, the United Nations Development

Programme (UNDP) and the World Bank. It argued for increased investment in the forestry sector, proposing to make sustainable forestry in the humid tropics a reality by doubling spending on timber and fuelwood plantations. Its target, however, was unreservedly industrial forestry, and environmentalists argued that it would neither promote the protection of remaining rainforest areas nor protect the interest of rainforest people. It was, they said, a 'top-down' approach to preserving forests (Caufield 1987).

Another approach to the regulation of the international timber trade was more deeply embedded within the private timber industry. The International Tropical Timber Organization (ITTO) was formed in 1983 following the UN Conference on Tropical Timber held under the auspices of the UN Conference on Trade and Development (Johnson 1985). The ITTO is a joint organization between producing and consuming countries, with votes on the Tropical Timber Council divided equally. It has been dogged by financial shortages because member countries have been slow to pay their dues. There were complaints of under-funding at its first meeting, in Yokahama in March 1987. The ITTO is different from other commodity agreements, because it specifically promotes reforestation and national polices of sustainable utilization and conservation. Its aims are broad, embracing the improvement of tropical forest management, the improvement of marketing and distribution of tropical timber and the promotion of wood processing in the producing countries. In 1991 the ITTO eventually set the goal making the global timber trade sustainable by the year 2000, through the establishment of demonstration plots, the development and dissemination of guidelines and the promotion of the case for sustainable management to member governments. Some observers suggested that the ITTO offered an opportunity for conservationists to promote conservation 'in the maximum degree of harmony with the tropical timber trade that is consonant with sound conservation objectives' (Johnson 1985, p. 44). However, its progress has been slow, and its achievements very modest (WWF 1991). While the word 'sustainable' is attached to many tropical hardwood products, and claimed for many production systems, the majority of tropical forestry fails to sustain the full range of values of forests, and much fails even to sustain the timber resource itself. The institutional challenge of creating a forestry industry capable of something other than timber mining is considerable.

Attempts to regulate international trade in timber and forest products have taken place through decades dominated by neo-liberalism, and against the background of attempts to promote 'free trade'. The World Trade Organization has not been friendly to attempts to regulate trade on environmental grounds (Hartwick and Peet 2003). Support for neo-liberal strategies (deregulation, privatization and structural adjustment) has limited the World Bank's capacity to promote sustainable forestry. Organizations like the United Nations Forum on Forests are talking shops without 'teeth' (Humphreys 2006).

A critical challenge for those who seek to establish sustainable forestry through regulation is the problem of illegal logging. This is a major problem in most forest-rich timber-producing countries (W. Smith 2004a, b). In some countries more than half the timber in trade is estimated to have been harvested illegally, with estimates up to 70 per cent in Indonesia, 80 per cent for the Brazilian Amazon and 90 per cent in Cambodia (W. Smith 2004a). Illegally logged timber is used to meet developed world demand for such 'vital 'products as hardwood garden furniture (Alley 1999). In so doing, the economic values of the standing forest to local people are liquidated and privatized. Because of its illegality, the benefits of such trade are lost to the public purse and recycled abroad or through elite consumption and the black economy.

Timber certification

A major factor in the attempt to make forestry sustainable has been the drive towards the certification of the origin of tropical timber. As described in Chapter 5, environmental NGOs in Europe led vigorous campaigns against unsustainable logging in the 1980s. Campaigning strategies evolved from defining a global 'environmental problem' (such as 'rainforest destruction') and linking it to specific events around which public protest could be focused, into wider campaigns involving consumer boycotts of specific products (drawing explicit links between environmental issues and specific products on sale in the industrialized world, such as hardwood garden furniture). In the early 1990s various environmental groups (Friends of the Earth, rainforest action groups and eventually Earth First!) began direct-action protests against the six largest DIY superstore chains in the UK.

One problem with this campaigning strategy was that there was no immediately available alternative to the unacceptable and unsustainably produced tropical timber products – no carrot to combine with the stick of consumer protest. An attempt was therefore made in the 1990s to provide this through certification of the origins of products and the conditions under which they were produced. The idea was that, if Northern consumers could distinguish products produced sustainably from those that were not, their purchasing power could drive production towards more sustainable methods.

NGOs campaigned for the adoption of a code of conduct by UK and EC timber traders, under which they would stock only timber from concessions with a government-approved management plan that stipulates post-logging management, where annual timber extraction does not exceed the concession's sustainable yield and where logging impacts are minimized by sympathetic extraction methods. Under the code of conduct, traders should not stock wood from plantations established in virgin forest areas, and they must label country and concession of origin of wood products. Furthermore, traders would be asked to devote 1 per cent of profits towards a fund to promote the sustainable use of rainforest.

In 1991 the WWF for Nature UK challenged the timber industry to make the world's forest production sustainable by 1995. Ten companies committed themselves to meeting this target, and formed the '1995 Group' (Murphy and Bendell 1997). The 1995 Group was effectively an informal environmental management system accreditation process with set targets. WWF-UK did not verify company

Plate 9.6 Timber depot in the Chocó rainforest, Ecuador. The value in the tropical
timber commodity chain is added as logs are processed into sawn wood and
end products. Many developing countries export sawn logs, and derive minimal
benefit from the exploitation of their forests. Since 1999 Fauna & Flora
International and its partners have been working in the Awacachi Corridor in
north-west Ecuador to protect rainforest from oil palm and logging interests
(www.fauna-flora.org). Photo: Juan Pablo Moreiras/FFI.

policy, and companies did not gain use of the WWF panda logo. Members of
the group committed themselves to the Forest Stewardship Council (FSC) as a
source of certification and labelling of 'sustainable production', and to moving
towards phasing out wood not accredited in this way. They committed themselves to identifying the source of wood and wood products, and they agreed to
appoint a named senior manager to implement this commitment, and to monitor
progress every six months. In return they could use the FSC logo on appropriate
products (WWF 1996). The FSC, which came into existence formally in 1993,
is a membership organization, with both industry and environmental interests
represented in two chambers. It adopted ten principles of forest management
(see Table 9.1).

These developments took place against the background of an industry backlash against the partnership and certification process, driven by organizations
such as the Canadian Pulp and Paper Association. In 1996 the timber industry
organization the Timber Trade Federation opposed this arrangement as contrary
to the principles of free competition and free movement of goods under UK
and EU law, and WWF-UK was forced to water down the group's membership
requirements (Murphy and Bendell 1997). This was one dimension of a wider
and more systematic assault on environmentalism in the 1990s, as industries and

Table 9.1 Forest Stewardship Council (FSC) principles of forest management

---

1   *Compliance with laws and FSC Principles.*   Forest management shall respect all available laws of the country in which they occur and international treaties and agreements to which the country is a signatory, and comply with all FSC Principles and Criteria.

2   *Tenure and use rights and responsibilities.*   Long-term tenure and use rights to the land and forest resources shall be clearly defined, documented and legally established.

3   *Indigenous peoples' rights.*   The legal and customary rights of indigenous peoples to own, use and manage their lands, territories and resources shall be recognized and respected.

4   *Community relations and workers' rights.*   Forest management operations shall maintain or enhance the long-term social and economic well-being of forest workers and local communities.

5   *Benefits from the forest.*   Forest management operations shall encourage the efficient use of the forest's multiple products and services to ensure economic viability and a wide range of environmental and social benefits.

6   *Environmental impacts.*   Forest management shall conserve biological diversity and its associated values, water resources, soils and unique and fragile ecosystems and landscapes, and, by so doing, maintain the ecological functions and integrity of the forest.

7   *Management plan.*   A management plan – appropriate to the scale and intensity of the operations – shall be written, implemented and kept up to date. The long-term objectives of management, and the means of achieving them, shall be clearly stated.

8   *Monitoring and assessment.*   Monitoring shall be conducted – appropriate to the scale and intensity of forest management – to assess the condition of the forest, yields of forest products, chain of custody, management activities and their social and environmental impacts.

9   *Maintenance of natural forests.*   Primary forests, well-developed secondary forests and sites of major environmental, social or cultural significance shall be conserved. Such areas shall not be replaced by tree plantations or other land uses.

10   *Plantations.*   Plantations shall be planned and managed in accordance with Principles and Criteria 1–9, and Principle 10 and its Criteria. While plantations can provide an array of social and economic benefits, and can contribute to satisfying the world's needs for forest products, they should complement the management of, reduce pressure on, and promote the restoration and conservation of natural forests.

Source: WWF (1996).

right-wing political activists in a number of countries (but particularly in the USA) attempted to reverse the influence of several decades of populist environmentalism, and challenge the growing hegemony of MSD (Rowell 1996). Despite this setback, and the broader anti-environment backlash that it reflected, a number of environmental organizations have continued to try to extend partnerships with business to promote sustainability. The drive for free trade and the increasing awareness of the global ramifications of business organization have made it clear that national government regulation is of limited power, and that businesses must be central to any significant move towards sustainability (Murphy and Bendell 1997).

The 1995 Group had some success in increasing penetration into the UK market of sustainable timber, although only 4 per cent of the timber sold by members of the group was certified by 1995, and only twenty-three of the forty-seven companies in the group had purchased some certified wood or wood product (although fourteen of the remaining twenty-four companies used paper, for which there was no certified source at this time (WWF 1996)). However, the group had no rules to prohibit members from marketing illegally harvested timber, and in 2003 several prominent members were shown to be doing so (Gulbrandsen 2006). WWF restructured the network, forming a 'UK Forest and Trade Network' in 2004, members of which would commit themselves to use only certified timber, and to reducing proportions of timber of unknown origin. In 2005 WWF announced that 56 per cent of members' wood supplies were certified, representing 16 per cent of UK wood production (Gulbrandsen 2006).

This is obviously a substantial achievement, but the numbers hide as much as they reveal (Gullison 2003). First, the move to certified timber has been largely confined to industrialized countries. Globally, there are 121 million hectares of certified forests. However, 93 per cent of this area is in the northern hemisphere (Siry *et al.* 2005). By 2002 the FSC had accredited 500 forestry operations in 56 countries, but the only countries with a significant proportion of their forest cover certified were Estonia (52 per cent), Poland (38 per cent) and Sweden (37 per cent). The next largest proportions were in South Africa (11 per cent) and New Zealand (8 per cent). In Indonesia, 0.1 per cent were certified, in Malaysia 0.4 per cent (Gullison 2003). Second, there is a confusing range of certification schemes, some far less strict than others. The FSC is the only global scheme, but large areas are outside this system – for example, under the Pan European Forest certification and the American Forest and Paper Association's Sustainable Forestry Initiative. Third, there are questions about the extent to which certification actually corresponds to improved environmental management on the ground. Gullison (2003) concludes that FSC certification does improve the value of certified forests in terms of biodiversity, although in many instances the forests that are certified are those that are managed in a way sensitive to biodiversity anyway. In Poland, for example, 6 million hectares of forest became FSC certified, but management did not change (Siry *et al.* 2005).

A final issue with timber and forest certification is whether it will continue to expand and transform the whole industry. Many observers conclude that the economic incentives are currently too small to attract the majority of forest managers to shift to sustainable forestry: costs are considerable and the benefits limited. Developed country consumer demand for certified products is too small significantly to reduce logging pressure on standing tropical forests (Gullison 2003).

The other important aspect of timber certification is its potential to deliver local development benefits by maximizing local capture of timber values (Upton and Bass 1995).

The price premium for certified timber can be important, if it can reach its market. The price of certified timber in the Solomon Islands in 1991 was 3.7 times that for unprocessed round logs (Hviding and Bayliss-Smith 2000). Relatively simple and affordable technology (chainsaws and milling frames) can go some way to making micro-scale logging feasible. A project called SWIFT (Solomon Western Islands Fair Trade) was established in 1994 by a Dutch group associated with the Solomon Islands United Church's Integrated Human Development Programme. Its aims were to produce FSC-certified 'eco-timber', and make the vital connection to European consumers.

However, the structure of the global timber market, and the diversity of products from unprocessed raw logs to highly value-added consumer products such as furniture, do not make it easy to extend 'Fair Trade' strategies to tropical timber production. Unlike coffee, for example, the timber commodity chain currently lacks an equivalent to the small coffee roaster who might have strong economic incentive to promote market access by small community producers (Taylor 2005). It is notable that the very success of timber certification has arisen through bringing large high-street DIY corporations on board. The scale of their operations makes accessing small-scale or community-based timber products problematic. The success of FSC was precisely that it did not challenge the structure of the established trade (as Fair Trade coffee did), but worked within it.

Hviding and Bayliss-Smith (2000) are dubious about the financial feasibility of the small-scale timber model being developed by SWIFT in the Solomon Islands, and suggest that it is too dependent on top-down management and Dutch expertise and money. Nonetheless, it is clear that there are opportunities to use certification to deliver direct community benefits as part of an alternative strategy for forest management to the 'strip and run' of so much conventional tropical forestry (Taylor 2005). It has been demonstrated – for example, in Mexico or Ecuador (Becker 2003; Bray *et al.* 2003) – that community ownership and management of forests is capable of delivering significant gains in terms of social and economic justice, good forest management and the conservation of biodiversity. These are essential elements in any future for forest management that can be described as sustainable.

Summary

- Tropical deforestation is a classic issue that demands a rich explanation drawing on political ecology. Data on tropical forest loss are poor, although the fact of rapid reduction in area is undisputed. Conventional explanations include population growth, economics (chiefly market failure) and poor governance.
- Forest clearance – for example, in Amazonia – needs to be understood in the light of political economic structures and decisions by actors at a range of levels.
- Environmental myths or narratives can be enormously important in justifying policies towards forests and forest clearance, and can be profoundly erroneous.
- Forest policy has serious implications for forest-dwellers. Much policy is based on a limited understanding of the way people use forests and their reasons for managing forest resources as they do

- There has been much interest in strategies for timber exploitation that are sustainable, and there is growing understanding of the environmental impacts of clear-felling and selective logging.
- Attempts to regulate forestry have met with mixed success. The control of illegal felling and international trade in timber is an essential but problematic component of sustainable forestry. The certification of forests is promising, but its impact in the tropics is still limited. Certification needs to take account of social as well as environmental conditions in forests.

Further reading

Barraclough, S. L. and Ghimire, K. B. (1995) *Forests and Livelihoods: the social dynamics of deforestation in developing countries,* Macmillan, London.

Brown, K. and Pearce, D. (1994) (eds) *The Causes of Tropical Deforestation: the economic and statistical analysis of factors giving rise to the loss of the tropical forests,* UCL Press, London.

Fairhead, J. and Leach, M. (1996) *Misreading the African Landscape: society and ecology in a forest savanna land,* Cambridge University Press, Cambridge.

Hecht, S. and Cockburn, A. (1989) *The Fate of the Forest: developers and defenders of the Amazon,* Verso, London (repr. Penguin Books, Harmondsworth, 1990).

Humphreys, D. (2006) *Logjam: deforestation and the crisis of global governance,* Earthscan, London.

Hviding, E. and Bayliss-Smith, T. (2000) *Islands of Rainforest: agroforestry, logging and ecotourism in Solomon Islands,* Ashgate, Aldershot.

Ite, U. E. (2001) *Global Thinking and Local Action: agriculture, tropical forest loss and conservation in South-East Nigeria,* Ashgate, Guildford.

Kull, C. A. (2004) *Isle of Fire: the political ecology of landscape burning in Madagascar,* University of Chicago Press, Chicago.

Oates, J. F. (1999) *Myth and Reality in the Rain Forest: how conservation strategies are failing in West Africa,* University of California Press, Berkeley and Los Angeles.

Parnwell, M. J. G. and Bryant, R. L. (1996) (eds) *Environmental Change in South-East Asia: people, politics and sustainable development,* Routledge, London.

Peet, R. and Watts, M. (2004) (eds) *Liberation Ecologies: environment, development, social movements,* Routledge, London (2nd edn).

Tacconi, L. and Seymour, F. (2007) *Illegal Logging: law enforcement, livelihoods and the timber trade,* Earthscan, London.

Williams, M. (2003) *Deforesting the Earth,* University of Chicago Press, Chicago.

Web sources

http://www.fsc-uk.demon.co.uk/ *The Forest Stewardship Council (FSC) is an international* non-governmental organization founded in 1993 to promote good forest management worldwide through certification of forest products.

http://www.survival-international.org/ Survival International is a major international campaigner against the threats to indigenous people.

http://www.wrm.org.uy/ The World Rainforest Movement: a global network of non-governmental organizations, founded in 1986, which campaigns on threats to tropical rainforests; the site has information on deforestation, mining, colonization and forest people.

http://www.unep-wcmc.org/forest/homepage.htm UNEP-World Conservation Monitoring Centre forests page for data on forests and deforestation.

http://www.fao.org/forestry/en/ Food and Agriculture forestry programme, including material on the state of the world's forests.

http://www.un.org/esa/forests/ The United Nations Forum on Forests

http://www.afandpa.org/ American Forest and Paper Association; see links to Sustainable Forestry Initiative (SFI).

http://www.weyerhaeuser.com/ Weyerhaeuser, 'an international forest products company with annual sales of $21.9 billion'. See environmental policy and sustainability report.

http://www.illegal-logging.info/ Illegal Logging: a site run by Chatham House in London, providing 'background information on the key issues in the illegal logging debate'.

http://www.cifor.cgiar.org/ Centre for International Forestry Research (CIFOR): 'science for forests and people'.

http://www.bushmeat.org/portal/server.pt The Bushmeat Crisis taskforce, founded in 1999 by leading international biodiversity conservation organizations.

10 The politics of preservation

Poor people should not pay the price for biodiversity protection
(Dilys Roe and Joanna Elliott, 'Poverty reduction and
biodiversity conservation', 2004)

Conservation and sustainability

Concern for the conservation of species and ecosystems played an important role in forging mainstream sustainable development (MSD) (see Chapters 2 and 3). While the first document of the mainstream, the *World Conservation Strategy* (*WCS*) (IUCN 1980), addressed questions of development, poverty alleviation and wider environmental management, it grew from an initiative by wildlife conservationists in the International Union for the Conservation of Nature and Natural Resources (IUCN) and the World Wide Fund for Nature (formerly the World Wildlife Fund (WWF)) to redirect development in ways that were more benign for nature conservation (Dasmann *et al.* 1973; Farvar and Milton 1973). Two arguments were made in the *WCS*. First, that all truly sustainable development depended on environmental conservation, specifically on the sustainable use of living organisms and ecosystems. Second, that development could be reconfigured to promote conservation. With the concept of sustainable development, conservationists began to claim that conservation and development objectives could be achieved together at global, national and local scales. In particular, the argument gained ground that conservation could help meet the true interests of poor people, and particularly the rural poor, who were themselves often the victims of inappropriate development.

The idea that environmental conservation underpinned development was the basis for a substantially increased flow of funds into conservation work in the 1990s – for example, through the Global Environment Facility and the work of bilateral donors such as the United States Agency for International Development (USAID). Such funding was the primary fuel for extensive experiments with 'community' approaches to conservation (Western and Wright 1994; Hulme and Murphree 2001), such as integrated conservation–development projects or programmes (ICDPs) and community-based natural resource management

(CBNRM), described later in this chapter. Advocates of 'sustainable use', or 'incentive-based conservation', propose that conservation can best be achieved by giving rural people a direct economic interest in the survival of species, thus literally harnessing conservation success to the issue of secure livelihoods (Hutton and Leader-Williams 2003).

This can be achieved through consumptive use (where wild organisms are hunted or harvested) or non-consumptive use (for example, wildlife-viewing tourism). Outside protected areas, such conservation strategies based on the consumptive use of natural resources by local people are often cautiously supported by conservationists. Inside protected areas such resource use is regarded as highly problematic, because of fears of over-hunting, over-harvesting or overgrazing. Strategies based on income from sport hunting are also problematic, even if such hunters can be made to obey rules, because of sympathy for ideas about animal welfare and rights among supporters of Northern conservation non-governmental organizations (NGOs). However, non-consumptive use of wildlife fits with the ethical and ecological predispositions of conservationists rather better. The idea of parks as a foundation for economic development is long established. In the early decades of the twentieth century, the Society for the Preservation of the Fauna of the Empire extolled the virtues of hunting safaris as a source of government revenue in colonial Africa (Adams 2004). Tourism, by train and latterly by motor car, was central to arguments for the development of national parks in the USA and Canada (Runte 1987; A. Wilson 1992; McNamee 1993), and a little later in South Africa (J. Carruthers 1995; S. Brooks 2005).

By the late twentieth century, the potential of wildlife or landscape-based tourism to contribute to national economies (for example, in Costa Rica, South Africa or Kenya) was not in doubt, although the capacity of protected areas, even with such activities, to contribute to the livelihoods of poor people was less clear. However, in the 1970s, the notion that parks should be socially and economically inclusive became part of mainstream conservation thinking (e.g. Western and Wright 1994; Ghimire and Pimbert 1997; Hulme and Murphree 2001; W. M. Adams 2004). In 1975 the United Nations Educational, Scientific and Cultural Organization (UNESCO) World Heritage Convention made provision for protection of areas of historical and cultural significance, including rural landscapes, with the UN protected area system.

By the 1980s the dominant conservation paradigm had changed to feature social inclusion rather than exclusion (Adams and Hulme 2001a). On paper at least, the needs of local people were established on the conservation planning agenda and community-based approaches had become important in debates about conservation strategies in the developing world (Wells and Brandon 1992; Western and Wright 1994; Ghimire and Pimbert 1997; Brosius *et al.* 1998, 2005; Hulme and Murphree 1999, 2001; Adams and Hulme 2001a, b; Wilshusen *et al.* 2002).

However, despite the continuity of efforts by conservationists to put flesh on the bones of the argument that wildlife (or in the new parlance 'biodiversity') conservation could contribute to development, conservation was no longer at the core of MSD debate. As described in Chapter 4, its place had already begun to slip

by the time the Brundtland Report was published in 1987, and this was accelerated by debates at Rio and Johannesburg, and the new emphasis on an international effort to eliminate poverty, reflected in the Millennium Development Goals.
Previously obvious links between conservation and poverty alleviation began to be
questioned. Some conservationists expressed concern that they were losing their
grip on the development agenda, that 'poverty alleviation has largely subsumed or
supplanted biodiversity conservation' (Sanderson and Redford 2003a, p. 390) or
that conservation had 'fallen off the bandwagon' in the new emphasis on development to alleviate poverty (Sanderson 2005, p. 326).

Debate about poverty and conservation has become more sophisticated
and complex (Adams *et al.* 2004). It is recognized that the linkages between
conservation and poverty are dynamic and context-specific, reflecting social
and political factors and issues of geography and scale (Kepe *et al.* 2004). There
are calls for new approaches to protected areas, and alternatives to protected
areas (Roe and Elliott 2004). Broad arguments have continued to be made that
the conservation of biodiversity can and should contribute to poverty alleviation (e.g. Brockington and Schmidt-Soltau 2004; Roe and Elliott 2004). The
argument that ecosystem services underpin the welfare and livelihoods, particularly (although not exclusively) of the poor, was central to the Millennium
Ecosystem Assessment (2005). Major programmes such as the United Nations
Development Programme's Equator Initiative aim precisely to reduce poverty
through the conservation and sustainable use of biodiversity (Timmer and Juma
2005). In September 2005 a statement from the Secretariats of the five biodiversity conventions (the Convention on Biological Diversity, the Convention
on International Trade in Endangered Species of Wild Fauna and Flora, the
Bonn Convention on the Conservation of Migratory Species of Wild Animals,
the Ramsar Convention on Wetlands and the World Heritage Convention)
argued that biodiversity underpinned all the Millennium Development Goals.
Biodiversity could, they suggested, help alleviate hunger and poverty, promote
good human health and 'be the basis for ensuring freedom and equity for all'
(http://www.biodiv.org/programmes/outreach/press/default.aspx).

This chapter explores the political ecology of conservation. It analyses the costs
and benefits of conservation and the way they are experienced by different people.
It reviews the three main strategies used to make conservation deliver benefits to
local people: people and parks programmes, projects to integrate conservation
and development, and projects to promote CBNRM. First, however, it considers
the politics of ideas about nature and the conservation of the 'wild'.

Protecting the wild

The argument that conservation and human needs can be combined as part of
sustainable development has been important, but it has always been an aspiration
rather than a description of conservation in practice. Much of the history of conservation in the Third World is not one of happily shared interests between rural people
and state conservation bodies, but one of exclusion and latent or actual conflict. The

historical evolution of conservation across the developing world has been described briefly in Chapter 2. The conventional approach followed the experience of indus- trialized countries by establishing protected areas (PAs), land set aside for 'nature' or 'wildlife', where human use could be either prevented or severely constrained. This approach, often called 'fortress conservation' (Brockington 2002), or a 'fences and fines' strategy (because it keeps people out with fences and fines them if they cross (Wells and Brandon 1992)), has tended to place conservation in direct conflict with those people with rights to, or need for, resources in protected areas.

PAs were the mainstay of conservation strategies globally through the nine- teenth and twentieth centuries (Adams 2004). The PA network expanded rapidly following the Second World War in an unprecedented 'conservation boom' (Neumann 2002). IUCN established a Provisional Committee on National Parks in 1958 (now the World Commission on Protected Areas (Holdgate 1999)). A 'World List of National Parks and Equivalent Reserves' was adopted by the

Plate 10.1 The gate of the Mgahinga Gorilla National Park in south-west Uganda symbolizes the separation of the 'wild' and the 'tame', and separates intensively managed farmland from both residual forest and land used for agriculture until cleared of farmers to create the national park, within the past forty years. The park is small, consisting of the northern slopes of three volcanoes forming the international border with Rwanda, and contains giant heather forests and bamboo thickets visited by groups of gorillas moving between Rwanda, Uganda and the Democratic Republic of Congo. Treks to see habituated gorilla groups attract international tourists and generate significant revenue, much of which is used for conservation in the park and through 'revenue-sharing' in infrastructure development for local communities in the form of school buildings. Photo: W. M. Adams.

United Nations General Assembly in 1962. This list was standardized by IUCN, whose categories of PA have been repeatedly refined over the years (Ravenel and Redford 2005). The list now includes both highly exclusionary Category I and II PAs (including classic National Parks), and a variety of other categories that are more inclusive of human activities, such as protected landscapes and reserves intended to maintain flows of products and services for human society (see Table 10.1). By 2005 there were over 100,000 PAs, covering over 2 million square kilometres, more than 12 per cent of the Earth's land surface (Chape *et al.* 2005). By the end of the twentieth century, systems of PAs had become universal in every country, rich and poor.

The approach taken to conservation following the Second World War in developing countries blended experience in North America, Europe and European colonial territories (W. M. Adams 2003; Neumann 2004a). From Europe came the idea of exclusive royal or aristocratic hunting grounds, where the unlicensed

Table 10.1 IUCN Protected Area categories

Category	Description
Category Ia: Strict Nature Reserves	Managed mainly for science
Category Ib: Wilderness Area	Land retaining its natural character and influence, without permanent or significant habitation, which is protected and managed so as to preserve its natural condition
Category II: National Parks	Areas of land or sea, designated to protect the ecological integrity of one or more ecosystems for present and future generations, to exclude exploitation or occupation inimical to the purposes of designation of the area and to provide a foundation for 'spiritual, scientific, educational, recreational and visitor opportunities' (all of which must be environmentally and culturally compatible)
Category III: Natural Monuments or Landmarks	Specific natural or natural/cultural feature of outstanding or unique value for its inherent rarity, its representative or aesthetic qualities or its cultural significance
Category IV: Habitat and Species Management Areas	Areas subject to active management intervention to maintain habitats or to meet the requirements of specific species
Category V: Protected Landscapes or Seascapes	Areas where the interaction of people and nature over time has produced distinct character with significant aesthetic, ecological or cultural value, and often with high biological diversity (safeguarding of the integrity of this traditional interaction is vital to the protection, maintenance and evolution)
Category VI: Managed Resources Protected Areas	Areas that contain predominantly unmodified natural systems that are managed to maintain a sustainable flow of natural products and services to meet community needs while ensuring long-term protection and maintenance of biological diversity

Source: www.unep-wcmc.org/protected_areas/categories/

killing of game (by rural people marked down as 'poachers') was closely policed. For the British Victorian elite, the preservation of wild 'game' for hunting was an obsession, both at home and in the Empire (Mackenzie 1988; Adams 2004). The British tradition of privately owned nature reserves, where non-proprietors lacked rights of access and use, was transferred to colonies, where the colonial state designated game reserves for the use of sporting gentlemen in the colonial service or on safari. This became the mainstay of British colonial conservation, a resort for gentleman hunters, whether traveller or colonial servant (MacKenzie 1988; Neumann 1996; Prendergast and Adams 2003; Adams 2004). Colonial conservationists maintained and policed game reserves in British colonies as a version of the Victorian sportsman's country estate long after that world had disappeared at home (Neumann 1996).

In Europe itself, state PAs took different forms. In the UK, for example, small national nature reserves were established by the government conservation agency from 1949, while the conservation of fragments of biodiverse habitat on private farmland and other economic landscapes was promoted by a mix of planning restrictions and financial incentives and grants (Adams 1996). It was understood that nature was not pristine, but deeply affected by human management (Sheail 1987). British national parks, finally designated after the Second World War, were selected to protect beautiful lived-in landscapes, and to provide citizens with access to them. Their scenic beauty was seen to depend on continued human use of land. They were not tracts of empty state-owned 'wilderness', but made up of private landholdings, mostly active farms (Evans 1992; Bunce 1994; Adams 1996). The British model of national parks, and indeed the wider experience in Europe of integrating people and wildlife in a landscape context, had limited influence on conservation in the developing world, although the IUCN Commission on National Parks and Protected Areas now recognizes the category of 'protected landscape', and there is interest in the applicability of the UK model in the non-industrialized world.

The dominant model in terms of its influence on the global ideas about conservation was that developed in the USA, where national parks were created in remote and sparsely populated areas, where the human costs of eviction were lost in wider distinction and disruption of indigenous people, and the voice of those excluded faint. Behind the US national park model, epitomized by Yellowstone and Yosemite (Runte 1987, 1990), was a conception of nature as something pristine and separate from lands transformed by people: nature as wilderness (Cronon 1995; Schama 1995). Wilderness was an important element in emergent national identity in the USA and in other colonial settler countries, notably Australia (Griffiths and Robin 1997; Dunlap 1999). Previous human occupation of parks, and indeed the extent to which the pre-Columban American West was far from 'pristine wilderness', were not widely recognized until the late twentieth century (Denevan 1992). Indian people were excluded from US parks (often forcibly evicted by the US army) and their presence forgotten or erased: places in the new parks were rechristened to describe 'natural' wonders (Runte 1990; Jacoby 2001). The burgeoning park-based tourist industry in the USA, ranging from the automobile-based mountain viewing from newly made roads to the specialized lightweight 'back-country' camper, was built on the appeal of accessible 'wilderness'. These people, and the staff who serve them and clean up after them, are permitted presences in the wilderness (Runte 1987; Wilson 1992). Neither indigenous people nor settlers could be tolerated (Jacoby 2001).

The political ecology of parks

In the developing world, the exclusionary approach to protected areas drawn from US national parks has dominated. In colonial Africa, government development plans allotted nature its fixed place. Protected areas were deemed to be 'natural', and people were excluded (Adams 2004). At least, as in North America, certain categories of people were excluded, for, while farmers, hunters and other uneducated resource users were unacceptable, tourists with cameras, hotel proprietors and tour operators, scientists and sometimes big game hunters were allowed. Unlike North America, most areas of tropical forest and savannah were not emptied of people upon colonial annexation and settlement, yet for the purposes of conservation large tracts of land have routinely been adjudged to be empty, or empty enough to be treated conceptually as 'wilderness' (Neumann 1998).

The ideas about previous human occupation that were part of the Yellowstone model went with it, and gave rise to the removal of people from 'natural' parks in many countries, as discussed below (Colchester 1997, 2002; Neumann 1998; Poirier and Ostergen 2002; Langton 2003). Thus conservationists saw Africa as an 'unspoiled Eden' (Anderson and Grove 1987, p. 4), or 'a lost Eden in need of protection and preservation' (Neumann 1998, p. 80) and planned parks accordingly. The idea is still in use (e.g. Quammen 2003). Ironically, as parks spread, the eviction of people to create them have indeed created wilderness from previously inhabited lands (Neumann 1996, 2001; Brockington 2002).

The exclusion of people from protected areas needs to be understood in the context of the way the modern state operates (Neumann 2004a). In both colonial and postcolonial periods, the priorities of national development have justified population displacement for projects such as dams (Howarth 1961; Scudder 2005) or agricultural development schemes (Lane 1992), in very much the same way as they have for parks. Colonial maps expressed the Enlightenment conceptual divide between natural and human on the land, between empty and settled lands, between space for wild nature and for civilization. This conceptual distinction led to physical separation on the ground. Thus the Tanganyikan colonial government separated spaces for wildlife and for people in Liwale District in Tanzania, creating the 'wilderness' of the Selous Game Reserve by displacing some 40,000 people towards the coast, away from crop-raiding elephants and sleeping sickness, and from their homes (Neumann 1998). Sleeping sickness provided the reason for other draconian clearances, which created empty lands for conservation – for example, in the Belgian Congo, where the Parc National Albert expanded onto land cleared in 1933 by the colonial state as part of its drastic sleeping-sickness campaign (Lyons 1985; Fairhead and Leach 2000).

It is a feature of attempts by the state to displace people in the name of conservation that many have involved coercion (Peluso 1993). Neumann (2004b) analyses the use of extreme force in conservation, including the bizarre character of 'shoot-to-kill' policies against poachers in countries where poaching is not a capital offence. The military style of national park management that arose in the USA (where the large western parks were managed by the US Army until 1918) has been important in many countries – for example, in Kenya with its aggressive armed pursuit of poachers (Leakey and Morell 2001). In their self-image, conservationists are heroic soldiers, literally fighting to protect nature (Peluso 1993). Such ideologies arise from the idea that 'nature' has to be protected against humans (Neumann 2004b). They can lead in strange directions. Terborgh (1999) calls for stronger defence of protected areas, and he proposes an internationally financed elite force legally authorized to carry arms and make arrests to achieve this aim (p. 199). In the rainforests of the Central African Republic, private paramilitary security forces already exist, undertaking counter-poaching activities against Sudanese gangs (Clynes 2002).

Most enforcement is less draconian. It more often consists of taking action against unarmed women collecting firewood or herders straying over unmarked park boundaries. Here the heroic game guard has a less glamorous role. The people most effectively excluded from the fortress are the rural poor. This is where the critique of the negative impact of conservation on poor rural people has its purchase. As Neumann (1998) argues, 'parks and protected areas are historically implicated in the conditions of poverty and underdevelopment that surround them' (p. 9).

The costs and benefits of conservation

Protected areas have a range of social and economic impacts on local people (Emerton 2001). Direct costs to neighbours include the depredations of crop-raiding wild animals (elephants, pigs, primates or a host of smaller species (Naughton-Treves 1997; Sekhar 1998; Woodroffe *et al.* 2005)). Problems include direct crop damage, the opportunity costs of crop defence (for example, effects on human capital when children cannot go to school) and risk of injury or death. Park neighbours can also suffer corrupt behaviour by conservation staff, particularly linked to minor infringements of regulations (for example, bribes to avoid arrest or fines for cutting fuelwood) or park boundaries (for example, impoundment of stock alleged to be grazing illegally). These things do not happen everywhere, but the problem of corruption in conservation is well recognized (Smith *et al.* 2003).

The greatest social impacts of PAs relate to population displacement, the loss of rights to residence and to use of land and resources now and in the future (Emerton 2001). The value of lost agricultural production from land set aside for conservation can also be important to local and even national economies (for example, in Kenya (Norton-Griffiths and Southey 1995)). The problem of loss of access to land of religious or cultural value is also significant. In 2004 the World

Bank extended the definition of 'involuntary displacement' in its guidelines on resettlement to include 'the involuntary restriction of access' to legally designated parks and protected areas, which has resulted in 'adverse impacts on the livelihoods of the displaced persons', even where no physical removal occurs (Cernea 2006). The term therefore covers restrictions on the use of resources imposed on people living outside a PA as well as those living inside it.

Displacement from parks has direct effects on livelihoods (West and Brechin 1991; Colchester 2002; Brechin *et al.* 2003). Social and economic impacts are felt both by those displaced and by those in communities that receive them (Cernea and McDowell 2000). Impacts include landlessness, joblessness, homelessness, marginalization, food insecurity, increased morbidity and mortality, loss of access to common property and services and social disarticulation (Cernea 1997).

There are numerous studies of particular cases of evictions from protected areas (e.g. Neumann 1998; Ranger 1999; Brockington 2002; Poirier and Ostergen 2002), although some widely quoted cases are inaccurate, such as Turnbull's account of the Ik people following removal from Kidepo National Park in Uganda (Turnbull 1974; Heine 1985). In Tanzania and Kenya, early colonial administrators developed the view that the Maasai were 'predators terrorising neighbouring groups', accumulating and refusing to sell stock and overgrazing pastures (Collett 1987, p. 144). It seemed axiomatic that they should be excluded from conservation areas (Homewood and Rodgers 1991). These ideas legitimated the eviction of pastoralists from the Mkomazi Game Reserve in northern Tanzania (Brockington and Homewood 1996; Brockington 2002). Conservation planners feared the present and future impacts of their livestock on biodiversity (Brockington and Homewood 1996). When Parakuyo and Maasai were eventually evicted in 1988, a full four decades after the game reserve was first designated, the area became 'wilderness' for the first time (Brockington 2002).

In 1979 a Biosphere Reserve was established in the Lufira Valley in south-eastern Shaba, in what is now the Democratic Republic of Congo. In this area of savannah woodland, Lemba people used to integrate shifting cultivation with more intensive practices, including mounding and ridging using mulch, and hand irrigation of streamside gardens. Biosphere reserves are zoned, with a core where all productive activities are prohibited, surrounded by a buffer zone where existing activities may continue, but innovations causing environmental change are banned. The Lufira Valley reserve became caught up in efforts by the state to increase agricultural production through obligatory cultivation of cassava and maize, which was to be enforced in the experimental zone. Moreover, while the reserve boundary was adjusted to exclude the charcoal-cutting areas of parastatal and private firms, and certain large farms, the central zone was said to be uninhabited. The chiefs of the Upper Lufira Valley complained in 1980 about the establishment of the reserve, and listed 2,000 people who lived or cultivated there. The chiefs were dismissed as uncooperative, obstructionist and anti-government.

In Ethiopia, Turton (1987) discusses the impact of state conservation on the Mursi of the Omo Valley. The Omo National Park was declared in 1966.

It was perceived by the Wildlife Conservation Department, quite incorrectly, as 'wilderness'. In fact it is an anthropogenic ecosystem created by the Mursi. Their economy is based on cattle herding, dry-season cultivation and flood-retreat farming along the Omo River. All three have to be combined to achieve subsistence. The Mursi also hunt in the hungry season before harvest, and trade ivory, leopard skins and other products. A second park east of the Omo River, the Mago National Park, was proposed in the 1970s. Over this period, the Mursi had been driven south into the Mago area by drought, and some land uses had intensified. The 1978 report saw the Mursi as a threat to conservation, although without defining in what way, and proposed resettlement. Turton (1987) ridiculed the policy of exclusive conservation, and drew a sharp contrast between attitudes to the conservation of wildlife and those towards the survival of the Mursi themselves. He comments that, if the integrity of the Omo and Mago Park boundaries had been successfully defended against human use, there would simply have been no Mursi economy left. In 2006 the Mursi were again threatened with displacement before the Omo National Park was taken over to be run by the Dutch NGO the African Parks Network.

Protected areas also generate economic benefits. Like costs, these too are rarely equitably distributed. The benefit of ecosystem services may be enjoyed locally, but economically larger values are often generated at regional or national levels. The existence value of species and habitats in protected areas are mostly enjoyed by tourists and conservation supporters from developed countries (Balmford and Whitten 2003). Local economic benefits include the opportunity to gain from tourism activities through employment, or related economic activities (for example, selling curios, food or cultural performances), and (more rarely) equity share or even ownership of lodges. Many local people are ill-equipped to take advantage of these activities, and benefits often end up concentrated in a few hands. The significant environmental costs associated with global tourism (not least the production of excessive CO_2 from air flights) are often conveniently ignored by conservation planners proposing tourism enterprises in remote parks.

There are also questions about the sustainability of tourist income. Tourism is a fickle industry, typically developing and wearing out destinations on a short cycle, and vulnerable to issues of security. Many parks lack the pulling power of, for example, the charismatic megafauna of the African savannah, or the mountain gorillas of the central African mountains. The sustainability of global wildlife tourism remains highly debatable. Funds from tourism or other enterprises need to be shared between many claimants, and the portion for local communities may be quite small (Adams and Infield 2003). Moreover, access to these benefits tends to be controlled by employees of the PA authority, and can be somewhat arbitrary, the product of administrative convenience rather than considerations of equity or justice.

There are also illegal benefits to be obtained from PAs – for example, from hunting or charcoal production. Like legal benefits, these tend to be unequally available with local communities (if only because not all households have the ability to undertake illegal activities and willingness to take risks). Benefits from conservation tend to reproduce existing economic inequalities within local communities (Paudel 2006).

Parks for people?

Growing awareness that there could be negative impacts of population displacement from protected areas was important in the shift that took place in the 1980s and 1990s in ideas about people in PAs (Western and Wright 1994; Hulme and Murphree 2001). Could the benefits of parks be better used to offset costs, and thereby build support for conservation? The resulting 'community conservation' strategies fell along a continuum from PA outreach to programmes building on local capacity for sustainable resource management (Barrow and Murphree 2001).

The shift towards people-friendly parks came about gradually. In the 1970s and 1980s, experience in Canada and Australia stimulated changes in the way indigenous land title and resource rights were understood globally (Colchester 1997; Morrison 1997; Langton 2003). In 1975 the IUCN General Assembly passed the Kinshasa Resolution on the Protection of Traditional Ways of Life, calling on governments not to displace people from PAs, and to take specific account of the needs of indigenous populations (Colchester 2004). Biosphere reserves introduced under the Man and the Biosphere (MAB) programme in the 1970s tried to deal with potential conflict by being large enough to include both a strictly protected 'core zone' and a 'buffer zone' where compatible resource use was allowed (UNESCO 1973; Batisse 1982). The relations between PAs and local people were debated at the Third and Fourth World Congresses on National Parks and Protected Areas in Bali in 1982 and Caracas in 1992 (McNeely and Miller 1984; McNeely 1993).

A new language of partnership was forged (McNeely 1996; see also Table 10.2). Park managers needed to start to address the needs of local communities by providing services such as education and healthcare, and by allowing local people to participate in park management and allowing consumptive and non-consumptive resource use (for example, hunting and gathering, agriculture, religious practices and pastoralism (Brandon and Wells 1992; Western and Wright 1994)). PAs should provide direct benefits to local people, should have a positive benefit–cost ratio to local people and should be positive if they are to prosper, and should allow those people to be involved in their planning. PAs should be planned and managed so that they meet local needs as well as biodiversity conservation goals and in a way that is integrated with surrounding human uses (McNeely 1996; see also Table 10.2).

The World Parks Congress in Durban in 2003 took as its theme 'benefits beyond boundaries' and one theme was 'communities, equity and protected areas'. Over 100 indigenous people attended. Events included an open meeting between leaders of some of the major international conservation NGOs and representatives of indigenous peoples (Brosius 2004). Issues of social exclusion from lands declared as protected for biodiversity, and marginalization from policy decisions

Plate 10.2 Community Game Guard, Uganda. The work of speaking to community groups and schools is combined with anti-poaching patrols in Mgahinga Gorilla National Park Uganda. Such mixed roles can be challenging for protected-area staff and local communities. Photo: W. M. Adams.

Table 10.2 Ten principles for successful partnerships between protected-area managers and local people

1	Provide benefits to local people
2	Meet local needs
3	Plan holistically
4	Plan protected areas as a system
5	Plan site management individually, with linkages to the system
6	Define objectives for management
7	Manage adaptively
8	Foster scientific research
9	Form networks of supporting institutions
10	Build public support

Source: McNeely (1996).

about conservation, were widely aired and fiercely debated. The 'Durban Accord', agreed at the conference, spoke of a new paradigm for PAs, 'equitably integrating them with the interests of all affected people', such that they provide benefits 'beyond their boundaries on a map, beyond the boundaries of nation states, across societies, genders and generations' (World Conservation Union 2005, p. 220).

The Durban Action Plan, agreed at the end of the conference, included numerous provisions long demanded by NGOs representing indigenous people, including Key Target 10, which called for participatory mechanisms for the restitution of lands incorporated into PAs without 'free and informed prior consent' (World Conservation Union 2005; see also Table 10.3). It also recognized a diversity of forms of PA governance, including co-managed and community-managed PAs ('community conserved areas').

It is commonly argued that there is substantial common interest between indigenous people, who wish to retain their rights to land (particularly forest land) in the face of competing demands, and conservationists, who wish to maintain habitat for its biodiversity. Others point out that such arguments tend to trade on essential and romanticized images of the non-Western primitive 'other', the 'ecologically noble savage', living in harmony with nature (Redford 1990; Conklin and Graham 1995). The interests of indigenous people (or other forest dwellers) in development even within the broad frame of a forested landscape can be different from those of biodiversity conservationists concerned to promote the survival of all species regardless of indigenous valuation (Redford and Sanderson 1992; Redford and Stearman 1993). Thus in the Brazilian Amazon, the way native peoples relate to their environment changes over time as populations grow and inevitable cultural changes take place (Seeger 1982). The romantic belief that such peoples are a 'natural' part of the ecosystem has led to proposals for the establishment of multiple-use reserves, creating national parks for native peoples and allowing native peoples in national parks. Seeger (1982) argues that this is unworkable and that, 'where resources are limited, the conflict between Indians and forest management has no real solution' (p. 188), if only because of the proven capacity of others to use indigenous people to exploit park resources.

Table 10.3 Outcomes of the Durban Action Plan

Outcomes		Key targets	
1	Protected areas' critical role in global biodiversity conservation fulfilled	1	Specific action by the Convention on Biological Diversity to improve the role of protected areas in biodiversity conservation
		2	Specific action by all signatories to the World Heritage Convention to improve the role of World Heritage sites in biodiversity conservation
2	Protected areas' fundamental role in sustainable development implemented	3	Action taken to ensure that protected areas strive to alleviate poverty and in no case to exacerbate poverty
3	A global system of protected areas linked to the surrounding landscapes and seascapes achieved	4	System of protected areas representing all of the world's ecosystems to be completed by 2010
		5	All protected areas to be linked into wider ecological/environmental systems on land and at sea by 2015
4	Improved quality, effectiveness and reporting of protected area management in place	6	All protected areas to have effective management in existence by 2015
5	The rights of indigenous peoples, mobile peoples and local communities recognized and guaranteed in relation to natural resources and biodiversity conservation	7	All protected areas to have effective capacity to manage
		8	All existing and future protected areas to be managed and established in full compliance with the rights of indigenous peoples, mobile peoples and local communities
		9	Protected areas to have representatives chosen by indigenous peoples and local communities in their management proportionate to their rights and interests
		10	Participatory mechanisms for the restitution of indigenous peoples' traditional lands and territories that were incorporated in protected areas without their free and informed consent to be established and implemented by 2010
6	Empowerment of younger generations achieved	11	Greater participation of younger generations in the governance and management of protected areas to be ensured and action taken to strengthen their capacity to contribute to and expand the conservation community as a whole
7	Significantly greater support for protected areas from other constituencies achieved	12	Support to be achieved from all major stakeholder constituencies
8	Improved forms of governance, recognizing both traditional forms and innovative approaches of great potential value for conservation, implemented	13	Effective systems of governance to be implemented by all countries
9	Greatly increased resources for protected areas, commensurate with their values and needs, secured	14	Sufficient resources secured to identify, establish and meet the recurrent operating costs of a globally representative system of protected areas by 2010
10	Improved communication and education on the role and benefits of protected areas	15	All national systems of protected areas to be supported by communication and education strategies by the time of the next World Parks Congress

Source: www.iucn.org, 28 November 2005.

Strategic alliances, based on conservation support for securing indigenous land rights, are therefore possible, but not automatic and not necessarily easy (Redford and Stearman 1993; McSweeny 2004). Attempts to broker partnerships need to start from the recognition of indigenous people as 'equals at the discussion table', not (as so often in the past) as subaltern groups to whom rights might be conditionally ceded by pragmatic conservation proprietors (Alcorn 1993). Moreover, such partnerships must address the widely embedded intolerant and coercive approaches of park planners and managers to indigenous residents in parks (Colchester 1997, 2002).

Community conservation demands new thinking by conservation managers, reform of the organizations in which they work and the legislative and policy framework. It is asking a lot of game guards and wardens trained in a military or policing role, armed and uniformed, to begin to see people they have been harrying (however unsuccessfully) for their law-breaking poaching or other activities as 'partners'. The reassignment of rangers from anti-poaching patrols to 'community rangers' can be highly confusing for both staff and local people (particularly if they still sometimes bear weapons and join anti-poaching patrols). Conservation staff are often under-trained, poorly equipped and badly paid. Government conservation organizations are frequently immobilized for lack of operating expenses or transport. Conservation laws and policies are frequently divorced from other policy sectors (for example, in agriculture, forestry, river-basin planning or tourism).

New partnerships with local people also have implications for conservation planning. Very often, regulations (and boundaries) are set from outside the community, written in plans that local people never see by faceless government officials, advised by scientists whose expertise is untestable, remote and not always sound. Research and technical assistance are often funded from overseas conservation NGOs. Conservation 'experts' bring assumptions from previous work that may not be appropriate, stay for short periods and suffer from the same seasonal, urban and tarmac biases as their economic and engineering counterparts (Chambers 1983). Repeatedly, such expert missions identify the current actions of local people as a threat to the survival of some feature of conservation interest. The policy proposals of these conservationists are likely to be as alien to local people as any proposed by conventional development planners (dams or irrigation schemes, for example; see Chapter 11), and potentially as adverse to their interests. Invoking 'the community' as a solution to the problem of the social impacts of PAs ignores the huge complexity of the concept. The terms of local people's involvement in conservation are still set by outsiders. As Murphree (1994) comments, 'imposed community-based conservation is a contradiction in terms, and implies an exercise in futility' (p. 404).

Conservation with development

Application of the concept of sustainable development in conservation outside PAs has given rise to widespread experimentation with programmes of two kinds.

The first consisted of projects that attempted to combine both conservation and development under a single project umbrella; these are often labelled 'integrated conservation–development projects [or programmes]' (ICDPs) (Wells and Brandon 1992; Barrett and Arcese 1995), or 'conservation-with-development projects' (Stocking and Perkin 1992). The second took the form of community-based natural resource management (CBNRM). Both are attempts to make wildlife 'pay its way' (Eltringham 1994).

The first generation of ICDPs had mixed success. As Wells and Brandon (1992) note, 'linking conservation and development objectives is in fact extremely difficult, even at a conceptual level' (p. 567). Conservationists may have been naive in assuming that a commitment on paper to sustainability and participation or 'bottom-up' planning would yield successful projects where more conventional development projects have a poor record. Stocking and Perkin (1992) provide a case study of ICDPs in action in the East Usambaras Agricultural Development and Environmental Conservation Project in Tanzania. The East Usambaras reach an altitude of 1,500 metres and support submontane forests with a very high level of endemic species. The IUCN project began in 1987 with three aims: to improve the living standards of the people; to protect the functions of the forest (particularly its role as a catchment for downstream water supply); and to preserve biological diversity. Traditional conservation objectives were deliberately de-emphasized to stress revenue generation and development. After four years, achievements were modest. A vast range of project activities had been begun, from agricultural extension to attempts to control illegal pit sawing, most with limited success.

The problems of the East Usambaras Project included lack of funds, leading in turn to a lack of breadth in technical expertise, and the way in which capital and energy were dissipated in too wide a range of activities. Behind many of these problems lay the lack of a proper feasibility study, a common failing in conservation projects (Caldecott 1996). Conservation organizations have discovered (like developers before them) that development plans are hard to transfer from paper to reality (Stocking and Perkin 1992). As projects, ICDPs are inherently highly complex and demand high levels of skill on the part of project staff. They also demand substantial funds and a realistic (that is, slow) timescale. Their chances of success depend on local perceptions of the project, and these are vulnerable to the public failure of particular components. Clear and precise objectives, careful evaluation of the costs and benefits of project components at the level of the individual household, long-term commitment to funding and strong local participatory linkages are essential. Projects of this sort will not be cheap to implement, and will not yield results quickly. Furthermore, there is a real risk that any positive impacts of the project on the local economy will be transient and dependent on the maintenance of flows of project revenues (Adams and Infield 2003). Barrett and Arcese (1995) conclude that ICDPs are 'no more than short-term palliatives' (p. 1081).

Most CBNRM involves 'consumptive use', killing or harvesting wild species (Campbell 2002b). This approach to conservation is built on the idea of wildlife simply as an economic resource that should be exploited in an effective and sustainable way. It may take the form of hunting by local people (for example, for bushmeat), killing in return for a licence fee by big game hunters, or collecting marketable or consumable natural products (for example, rainforest rattans or turtle eggs). In 1990 IUCN established a Specialist Group on 'sustainable use', and began to develop guidelines for utilization of wild species (Adams 2004).

A key issue with sustainable use is the feasibility of identifying a 'sustainable' level of harvesting. This is a complex scientific task conventionally requiring good data over long periods and regular monitoring, things often not available in practice in many developing countries. Sustainable-use projects also require effective institutions to enforce that harvest (involving rules, agreement by potential hunters that these are fair and reasonable rules, and measures to deal with those who break them). There are both monetary and non-monetary reasons why people harvest illegally, whether they defy national laws or local conventions. Hunting is not always done by 'local' people, and, even if it is, they often do so to supply an organized national trading network and an urban market in bushmeat (Milner-Gulland *et al.* 2003) It may, therefore, be hard for CBNRM projects to provide sufficient incentives to decouple livelihoods (for example, of hunters and local or national traders) from unsustainable patterns of wildlife harvest. It can be equally hard to regulate commercial hunting and the associated flow of trophies and products such as ivory. Legalization of ivory trading has been pursued strongly by the countries of Southern Africa such as Zimbabwe (Hill 1995) and Botswana, although opponents of elephant hunting (chiefly motivated by ethical considerations, or fears that institutional failure in measures to control hunting would allow indiscriminate and illegal killing to continue, particularly in East and West Africa) won their case for a total ban on international trade in ivory from African elephants at the meeting of contracting parties to the Convention on International Trade in Endangered Species (CITES) in 1989 (Princen 1994b). The ban was subsequently partially lifted for certain elephant-range states (Namibia, Botswana, Zimbabwe) at an emotional CITES meeting in Harare in 1998.

CBNRM programmes co-evolved in several different southern African countries in response to a range of historical, political, social and economic experiences, conditions and challenges (Fabricius *et al.* 2004). In Zimbabwe, under the Communal Areas Management Programme for Indigenous Resources (CAMPFIRE) programme, the same benefits from wildlife use that were enjoyed by landowners on leasehold and title-hold land were extended to residents of communal lands. In Zambia, CBNRM was a response to the challenges of engaging traditional authorities in the management of the benefits of hunting in state 'game management areas'.

CAMPFIRE granted *de facto* authority over wildlife resources of power to district authorities, such that they could profit from hunting revenues (Metcalfe 1994). The CAMPFIRE model has been seen internationally by conservation policy-makers to offer a form of conservation that is both popular and affordable (Olthof 1994). However, problems are emerging from the fact that authority (and hence revenues) is devolved only to district level, not to communities themselves (Murombedzi 1999). While CAMPFIRE has worked quite well in some areas

(Murphree 2001), in others, particularly those less rich in high-value trophy species such as elephant, and with rapid rates of immigration, it has not (Murombedzi 1999, 2001). Even where it was initially successful, there have been problems of lack of local leadership and sustainability of outside institutional support under difficult political and economic circumstances (Balint and Mashinya 2005).

In Zambia, the Lwangwa Integrated Resource Development Project (LIRDP) (adjacent to South Lwangwa National Park) and the wider ADMADE programme (begun in 1987) are often reported as a success, and their economic (consumptive use) and benefits-sharing approach is held to be a valuable model for conservation elsewhere (Swanson and Barbier 1992). However, research suggests that neither the LIRDP nor ADMADE has been effective in changing the level of hunting by local people (Wainwright and Wehrmeyer 1998). Gibson and Marks (1995) argue that the community benefits generated by ADMADE fail to compensate for the economic, social and political benefits of hunting: hunters change tactics, but they keep hunting. ADMADE was inflexible, and did not direct economic benefits through democratic institutions. It does not provide a strong model for

Plate 10.3 School in Zimbabwe funded by the CAMPFIRE Programme. This school is in Mahenye District in south-east Zimbabwe, where the CAMPFIRE Programme enjoyed considerable success in the 1990s (Murphree 2002). Revenues were derived from a successful safari hunting operation and a small hotel from which tourists could enter adjoining Gonorezhou National Park. Income was spent on classrooms and a communal grinding mill, and an all-weather road and water supply were built. More recently, there have been problems owing to the failure of local leadership and the withdrawal of outside supporting institutions (Balint and Mashinya 2005). Photo: W. M. Adams.

situations where human populations are denser and wildlife numbers less abundant than in the extended savannah woodlands of Zambia.

Early CBNRM programmes focused almost exclusively on safari hunting because this produced significantly higher and more visible benefits than other resources, although the maintenance of wildlife within a partly farmed landscape also produced more costs, and therefore conflict, between wildlife and people. All countries in the southern African region now have policies that allow for communities in communal areas to use and benefit from a range of natural resources on their lands, with varying degrees of success (Gibson and Marks 1995; Wainwright and Wehrmeyer 1998; Murombedzi 1999, 2001; Duffy 2000; Jones 2001; Murphree 2001; Fabricius *et al.* 2004).

CBNRM programmes were based on several hypotheses: first, that communities are more efficient managers of natural resources in their areas of jurisdiction than other agencies; second, that community management leads to improved incomes for communities, thus helping poverty reduction and providing economic incentives for conservation; third, that community management reduces conflicts with wild animals, and thus the cost they impose on people, leading to better tolerance of wildlife and better outcomes for biodiversity; fourth, that the community management of natural resources is more efficient and cheaper than state management (Hutton *et al.* 2005).

However, in southern Africa, there has been little real devolution of power and authority over resources, including land, from the state to local people (Murombedzi 2001). At best, power has been decentralized from central to local government. As a result, CBNRM has resulted in insufficient incentives for communities to internalize the costs of resources management (Jones 2001; Murphree 2001). Decentralization *per se* is not adequate to create the conditions required for significant community control over natural resources (Ribot and Larson 2004). The southern African programmes have stopped short of land-tenure reform (Murombedzi 2001). Communal tenure continues to function in ways that disadvantage its residents relative to those enjoying freehold and leasehold. CBNRM has also been weakened by its failure to engage with conventional rural development policy constituencies in either agriculture or land reform (Murombedzi 2001)

Oates (1999) is highly critical of attempts to integrate conservation and development, arguing that such projects, far from creating a 'win–win' outcome, end up satisfying neither human needs nor conservation objectives. In the case of the Okumu Forest Reserve in south-west Nigeria, he blames the sustainable development rhetoric in *Caring for the Earth* for new conservation programmes that have accelerated forest destruction by small farmers (Oates 1995).

The primary objective of conservationists is the preservation of biodiversity. Development, even if packaged as 'sustainable development', is attractive chiefly as a secondary strategy where it promotes their primary objective. From a narrow biodiversity conservation perspective, attempts to integrate conservation and development are valuable primarily where preservation is impossible. Thus, while selective logging of forests can be done in such a way as to minimize impacts on wildlife, this 'should not be regarded as an alternative to maintaining primary forest areas' (Johns 1985, p. 370). It is, therefore, where conservationists have little chance of commanding the kind of resources they need to guarantee preservation in well-managed strict reserves that they are willing to push those kinds of development (which they call sustainable development) least damaging to wildlife. In doing so, they may well temporarily align themselves with the needs and aspirations of indigenous groups and, rightly, use the language of win–win.

The politics of global conservation

While conservation costs are mostly borne locally, benefits accrue globally (Balmford and Whitten 2003). The existence value of species and habitats preserved through the creation of parks is mostly appropriated by remote and relatively wealthy wildlife-lovers in developed countries both virtually (for example, through wildlife television) and directly through tourism. The biologist E. O. Wilson (1992) observed the 'awful symmetry' of economic wealth and biodiversity, 'whereby the richest nations preside over the smallest and least interesting biotas, whilst the poorest nations, burdened by exploding populations and little scientific knowledge, are stewards of the largest' (p. 260). Of course, since the end of colonial rule, conservation in developing countries has been the responsibility of independent government agencies. Conservation can no longer be dismissed as an alien colonial ideology: like democracy and Coca-Cola, it has been incorporated into the activities of modern states in almost every country of the world. Indeed, some countries – for example, Tanzania – have adopted and expanded on the international model of conservation with enormous enthusiasm, with more than 20 per cent of the country in protected areas. The establishment of PAs has been part of its thrust for modernity and its apparatus of nation creation, as well as a vital element in its foreign-affairs strategy (Neumann 2004a).

However, it is somewhat disingenuous to argue that conservation in the Third World is not profoundly influenced by the interests of interest groups in industrialized countries. There is a standardized global conservation ideology, created and disseminated by visionaries, scientists and the media, that strongly reflects the interests of a global literate urban wildlife-loving elite predominantly based in industrialized countries. The ideologies that dominate media coverage of the environment in the North (indeed, the very concept of 'the environment' as a separate cognitive category) are very different from those in the South (Chapman *et al.* 1997). The strength and spread of global conservation ideology are reflected in the high membership of environmental organizations in industrialized countries, and the limited (and often purely elite membership) of such organizations in the South. It is articulated and turned into policy by international NGOs and aid donors. To the developing world, they offer ideas and values; they identify problems and suggest technical and bureaucratic structures to solve them. They seek partners in the South, including grass-roots environmental organizations, and by their support they recruit them to their worldview and preferred strategy.

Developed-country governments and NGOs push to establish international meetings (such as the Rio Conference in 1992 and the World Parks Congress in Durban in 2003) that developing-country governments attend and at which they are urged to sign agreements that will further conservation. Analysis drew attention to the influence of a small number of Northern (mostly US-based) environmental organizations at Rio. These NGOs (the Sierra Club, the National Audubon Society, the National Parks and Conservation Association, the Izaak Walton League, the Wilderness Society, the National Wildlife Federation, Defenders of Wildlife, the Environmental Defence Fund, Friends of the Earth, the Natural Resources Defence Council, IUCN, the WWF and the World Resources Institute (Chatterjee and Finger 1994)) have acquired a global corporate culture. Their leaders wear business suits and seek to talk the power language of bankers and world leaders. At Rio they had the resources, the experience and the expertise to lobby to some effect, and their voices were far louder than those of most of the other 4,200 accredited lobbyists, who were disorientated, confused and disorganized (see Chapter 4).

The international conservation movement also seeks more direct influences on policies within Third World countries. Among international conservation NGOs there is significant competition, for membership, for grant income from trusts and aid donors, and particularly for corporate funds (Chapin 2004). NGOs provide capital and recurrent funding for projects, and they run pilot projects to show how conservation can be done. They also influence First World donors, and push them to disburse money in particular ways, establishing environmental conditionality on aid. International NGOs also seek to secure effective environmental gains within particular countries through 'debt-for-nature' swaps – whereby part of a country's debt to a foreign bank is bought by an NGO at a discounted price in return for agreed expenditure on conservation in the country (Gullison and Losos 1993).

All these things may on balance be desirable, and the argument here is not that NGOs should not seek to have influence on development. Nor is it true that conservation necessarily involves actions against the interest of people (particularly poorer and less powerful people). It is simply that, at all scales from the local to the global, decisions about conservation are highly political. Actions aimed at the conservation of nature are also, by definition, actions that engage with society. However convincing the technical analysis of conservation biologists, and however pressing their conclusions about the need for drastic conservation action, conservation policy has significant political, social and economic impact. It is, therefore, within the framework of political ecology that conservation policy needs to be understood.

The power of international conservation is seeking to expand the scale in the way park systems are imagined to the landscape and international scales (e.g. Duffy 1997; Fonseca *et al.* 2005; Wolmer 2003, 2007). The rapid diversification of biodiversity mapping algorithms (Brooks *et al.* 2006) to an extent reflects the desire of each NGO to create its own classification (Redford *et al.* 2003). There is the renewed advocacy for traditional socially exclusive parks (e.g. Kramer *et al.* 1997; Brandon *et al.* 1998; Oates 1999; Struhsaker 1999;

Terborgh 1999), a 'resurgence of the protectionist paradigm' (Wilshusen *et al.* 2002) or a 'back-to-the-barriers' movement (Hutton *et al.* 2005).

At the same time, conservation is undergoing a process of self-criticism and reform as it seeks technical improvement and tighter self-regulation with respect to its social policies and procedures. This process is a form of ecological modernization (Hajer 1995), technically orientated and regulation-based responses to environmental problems (see Chapter 5). Conservation planners in governments and NGOs are urged to adopt established methods such as Social Impact Assessment in search of more socially equitable and effective conservation planning (Geisler 2003; see also Chapter 6). PAs share with other major projects imposed by the state in partnership with international actors (notably large dams (Scudder 2005)) the capacity to deliver significant public goals but also to impose significant local costs. Those who plan and manage PAs lag seriously behind in their response to these issues. A broad constituency supports an end to forced displacement for conservation. Planning for resettlement must involve a serious commitment to equity and finance for the complex and challenging task of reconstruction (Cernea 1997).

Traditional top-down conservation has been transformed in a number of ways by participatory approaches, and the attempt to integrate development and conservation aims. The idea of sustainable development has been fundamental to this transition, and conservation projects are now at the forefront of experiments in achieving sustainability in practice. Such projects reveal the confused diversity of thinking inherent to sustainable development, and may not create win–win solutions, but instead reveal divided interests and awaken latent controversy.

Summary

- Sustainable development has been an important strand in thinking about wildlife conservation since the *World Conservation Strategy* (*WCS*) (IUCN 1980). However, attempts to make wildlife 'pay its way', and particularly to contribute to poverty reduction, have often been problematic.
- The dominant approach to conservation has been the creation of protected areas (PAs) where nature is treated as wilderness, and people are excluded. The displacement of people from PAs has been an important feature of the political ecology of conservation.
- PAs create both benefits and costs. Most costs are borne locally by those displaced and protected area neighbours. Benefits of conservation in developing countries are more widely spread and include benefits enjoyed in developed countries.
- Since the 1970s numerous efforts have been made to develop conservation policies that generate benefits for local people. These include part outreach projects, integrated conservation and development projects and community-based natural resource management (CBNRM). Results have been mixed, but there is some local success. The wisdom and efficiency of such approaches are debated.

- International conservation organizations have become relatively large and powerful actors. Global science-based conservation priorities have become increasingly ambitious and effective.

Further reading

Adams, W. M. (2004) *Against Extinction: the story of conservation*, Earthscan, London.

Adams, W. M. and Mulligan, M. (2003) (eds) *Decolonizing Nature: strategies for conservation in a post-colonial era*, Earthscan, London.

Brandon, K., Redford, K. H. and Sanderson, S. E. (1998) *Parks in Peril: people, politics and protected areas*. Island Press, for the Nature Conservancy, Washington.

Brechin, S. R., Wilhusen, P. R., Fortwangler, C. L. and West, P. C. (2003) (eds) *Contested Nature: promoting international biodiversity with social justice in the twenty-first century*, State University of New York Press, New York.

Brockington, D. (2002) *Fortress Conservation: the preservation of the Mkomazi Game Reserve, Tanzania*. James Currey, Oxford.

Brosius, J. P., Tsing, A. L. and Zerner, C. (2005) *Communities and Conservation: histories and politics of community-based natural resource management*, Altamira Press, Walnut Creek, CA.

Duffy, R. (2000) *Killing for Conservation: wildlife policy in Zimbabwe*, James Currey, Oxford.

Ghimire, K. and Pimbert, M. (1997) (eds), *Social Change and Conservation: environmental politics and impacts of national parks and protected areas*, Earthscan, London.

Hulme, D. and Murphree, M. (2001) (eds) *African Wildlife and Livelihoods: the promise and performance of community conservation*, James Currey, Oxford.

Jacoby, K. (2001) *Crimes against Nature: squatters, poachers, thieves, and the hidden history of American conservation*, University of California Press, Berkeley and Los Angeles.

Neumann, R. P. (1998) *Imposing Wilderness: struggles over livelihood and nature preservation in Africa*, University of California Press, Berkeley and Los Angeles.

Oates, J. F. (1999) *Myth and Reality in the Rain Forest: how conservation strategies are failing in West Africa*, University of California Press, Berkeley and Los Angeles.

Western, D., White, R. M. and Strum, S. C. (eds) (1994) *Natural Connections: perspectives in community-based conservation*, Island Press, Washington.

Wolmer, W. (2007) *From Wilderness Vision to Farm Invasions: conservation and development in Zimbabwe's South-East Lowveld*, James Currey, Oxford.

Woodroffe, R., Thirgood, S. and Rabinowitz, A. (2005) (eds) *People and Wildlife: conflict or coexistence?* Cambridge University Press, Cambridge.

Web sources

http://www.iucn.org/themes/ceesp/ The World Conservation Union (IUCN) Commission on Environmental, Economic and Social Policy (CEESP).

http://www.iucn.org/themes/ssc/susg/ The World Conservation Union (IUCN) Sustainable Use Specialist Group (describes efforts to promote the sustainable use of wild renewable resources).

http://www.undp.org/equatorinitiative/ The United Nations Development Programme Equator Initiative, which promotes poverty reduction through the conservation and sustainable use of biodiversity.

http://www.refugeesinternational.org/ Refugees International with information on conservation displacements particularly in Ethiopia.

http:// www.africanparks-conservation.com The African Parks Network: 'a business approach to the management of protected areas'. They run a series of protected areas under contract for African governments in Ethiopia, Sudan, Zambia, Malawi and the Democratic Republic of Congo; including the Omo National Park in Ethiopia.

http://www.biodiv.org/cooperation/joint.shtml Joint website of the Convention on Biological Diversity (CBD), the Convention on International Trade in Endangered Species of Wild Fauna and Flora (CITES), the Convention on the Conservation of Migratory Species of Wild Animals (or the Bonn Convention), the Convention on Wetlands (the Ramsar Convention) and the World Heritage Convention.

http://www.peaceparks.org Peace Parks Foundation, 'Africa without fences'

http://www.fauna-flora.org/ Fauna & Flora International, UK-based conservation NGO, founded 1903.

http://www.wwf.org/ Worldwide Fund for Nature 'for a living planet'.

http://www.conservation.org Conservation International US-based conservation NGO, founded 1987.

http://www.povertyandconservation.info Poverty and Conservation Learning Group: 'a forum for facilitating mutual learning between key stakeholders, from a range of back-grounds, on conservation-poverty linkages'.

http://www.tradeoffs.org Advancing Conservation in a Social Context: 'working in a world of trade-offs'; project funded by the John D. and Catherine T. MacArthur Foundation.

11 Sustainability and river control

I am always bothered by the Western arrogance, by its assurance that it knows all the answers and can quite readily fix everything so that the tropical peoples can live happily ever after, if only they will listen.

(Marston Bates, *Where Winter Never Comes*, 1953)

Surely, if decades of failed international development efforts have taught anything, it is the folly of induced, uniform, top-down projects. Such schemes ignore and often destroy the local knowledge and social organization on which sound stewardship of ecosystems as well as equitable economic development depend.

(Bruce Rich, *Mortgaging the Earth*, 1994)

Rationality and environmental control

Environmental modification is an inherent part of the development process. Cowen and Shenton (1996) distinguish between immanent development (the changes to economy and society that take place) and intentional development (the 'active practice of the state' (p. 61)). Both dimensions involve environmental change, but it is intentional development that has produced the most dramatic environmental transformations. Formal development schemes can themselves be deeply unsustainable because of their environmental and social impacts. Nothing demonstrates this better than the development of water resources. Dams, in particular, demand intensive technical planning and massive environmental and socio-economic transformation. They flood large areas of often highly productive land. They affect the dynamic natural systems of rivers and floodplains on which large numbers of people depend and where human livelihoods are intricately linked to ecology. Many dams have had serious environmental impacts. This is the subject of this chapter.

Development was formed as an ideology and a practice in the ex-colonial world in the decades that followed the end of the Second World War (see Chapter 2). It was founded on an instrumental–rationalist approach to planning, to the notion that social, economic and environmental resources should be assessed, harnessed and brought to bear on the systematic improvement of human welfare. Rationalization has four dimensions (Murphy 1994): first, the development of science and technology, the 'calculated, systematic expansion of the means to understand and manipulate nature' as Murphy puts it (p. 28), and the related belief in the possibility (and desirability) of the mastery of nature through increased scientific and technical knowledge; second, the expanding capitalist economy (and the rationality of the market); third, formal hierarchical organization (to translate social action into rationally organized action); and, fourth, the elaboration of the legal system to manage social conflict and promote the predictability and calculability of the consequences of social action.

Rationalization is fundamental to the process of development. It underpins the formalized process of planning social, economic and environmental change, and the ideology of developmentalism that drives it: the view that environments and societies must be transformed in an all-out drive to modernize and achieve economic transformation. The spirit of post-war development was technocratic, optimistic, modernist and Promethean. This was the era of ecological managerialism, from which mainstream sustainable development (MSD) emerged (see Chapter 2). Across a wide range of disciplines (agriculture, health, veterinary science, fisheries, forestry and education, among others), colonial regimes adopted new ideas, employed staff with new skills, and invented new institutional forms for planning and delivering change.

The technocratic strategies and institutions of war were adopted and retooled to deliver development. Swords were beaten into ploughshares, sometimes almost literally: the Groundnut Scheme in Tanganyika tried to convert Sherman tanks for stump clearance (unsuccessfully as it turned out – a reflection of the disastrous performance of the scheme as a whole, Wood 1950). The industrial successes of Fordist factory organization in the 1930s, developed so successfully in the Second World War in the production of machines of war, were now applied conceptually to the delivery of development. Development projects adopted large-scale and mechanized farming, an attempt to mass-produce food (Jones 1938). Alas, the urge to rush development led in many instances to failure. Baldwin (1957), reviewing the Niger Agricultural Project in the Nigerian Middle Belt (another unsuccessful groundnut production scheme), commented that 'the removal of limitations on money led almost inevitably to removal of limitations on the size of project. Hence there grew up a fallacious notion that the bigger the scheme the better the results likely to be obtained' (p. 2).

Nowhere has what Herbert Frankel called 'the twin dangers in all development of grandiosity and arrogance' (in Baldwin 1957) been better demonstrated than in the control of river flows behind dams, and in the associated desire to make the desert bloom through irrigation. The juxtaposition of large rivers, un-urbanized and unindustrialized economies, drought-prone lands and large numbers of poor people has repeatedly led engineers and planners to dream of harnessing the waters and energy of rivers for irrigation or hydroelectric power generation. River water un-harnessed for power or other purposes is all too easily seen to be running to waste (Adams 1992; Rosenberg *et al.* 1995; Usher 1997a; World Commission on Dams 2000). In China, for example, rhetoric justifying the Three Gorges dam consciously echoed the language of the Great

Leap Forward: the communist party was literally fighting the Yangtze, with science as its weapon, justified by a pervasive language of rationality (Beattie 2002).

In semi-arid regions, the technology of irrigation has a powerful appeal (Moris 1987; Adams 1992): the provision of just the right amount of water at the right time to growing crops; the encouragement of skilled and market-orientated groups of farmers; the provision of new technologies (improved seeds, fertilizers, pesticides, machinery) and access to markets; the efficient location of people in settlements where health, education and clean water can be conveniently supplied; all the costs of investment paid for by double- and triple-cropped irrigated fields, and each investment gaining from and contributing to the next. Irrigation encapsulates the developers' determination to control the variability of nature and transform its productivity (Adams 1992).

Plate 11.1 Waterlogging on an irrigated field corner, Bakolori Project, Nigeria. Many established irrigation schemes suffer from poor management, or what are called 'operation and maintenance' ('O and M') problems. Physical problems include the blockage of canals and structures by accumulated sediment or vegetation, water losses through leakage from canals and evaporation from open water bodies, and poor drainage. This can lead to some parts of an irrigation scheme receiving too little water, while others receive so much that crops suffer from waterlogging and die. Poor water allocation can be made worse by farmers taking water out of turn, with those at the tail end of the canal system suffering disproportionately when water distribution is unfair. Waterlogging and associated salinization owing to poor drainage is a major global problem on irrigated land. This photograph shows maize suffering in a poorly drained field corner. Photo: W. M. Adams.

Dams and irrigation schemes hold a Promethean promise as dramatic strategies for disciplining nature and intensifying its exploitation to produce economic benefits (Kaika 2006). In many cases these benefits have proved persistently elusive. Irrigation projects are effectively exercises in large-scale applied hydraulics, rather than a way to provide farmers with an affordable and reliable water supply (Rydzewski 1990). Many irrigation projects struggle to cope with problems of high construction and maintenance costs, poor economic returns, inefficient water supply, salinity and disease (Chambers 1988a; Moris and Thom 1990). Dams are engineering miracles, but their costs are often far greater and their benefits far smaller and more narrowly shared than planned (Scudder 2005). Dams are technologies of control, built to discipline unruly nature, to bring a new order to the environment to speed wealth creation. Their magic is illusory: the very transformations they bring about create new and often serious problems in their wake.

Engineering nature

Grand river development schemes date back to the ancient Mesopotamian kingdoms, but in the modern era they were one of the hallmarks of colonial expertise. British Imperial India saw major experiments in irrigation development from the early nineteenth century, when existing irrigation systems – for example, in the kingdoms of Delhi and Tajore, and Madras – were found profitable, and were refurbished and replicated (Singh 1997), while the importance of the annual flood of the Nile to the prosperity of Egypt was a major factor in the imperial ambitions of the major European powers in the second half of the nineteenth century.

Canals and barrages were built in the Nile Delta early in the nineteenth century to allow perennial irrigation, and two major barrages were built north of Cairo between 1843 and 1861 (Waterbury 1979; Collins 1990). The first Aswan Dam was built on the Nile in 1902, and heightened in 1912 to double its storage. These works were simply the springboard for more extensive studies of the upper Nile in the Sudan, and in due course for more dramatic project proposals. Sir William Garstin, Under-Secretary of State for Public Works in the Anglo-Egyptian condominium of the Sudan, published a series of technical studies of the Nile in 1904 that discussed ideas for dams and water storage on the Blue Nile, on the River Atbara and in the East African Great Lakes. Once launched into government consciousness, these ideas were studied and debated repeatedly in successive attempts to bind the flows of both Blue and White Niles more efficiently to the task of irrigation. Egypt and Sudan had concluded the Nile Waters Agreement, which allocated 94 per cent of the available flows (48 billion out of an estimated 51 billion cubic metres) to Egypt. Dams were built at Sennar on the Blue Nile in 1925, and upstream at Jebel Aulia in 1937. The Aswan Dam itself was heightened again in 1934.

More dramatic plans were developed. In 1931 the Egyptian Ministry of Works began to publish a series of volumes, *The Nile Basin*, containing ideas and proposals for the development of the Nile. Volume 7 (1947) proposed the Equatorial Nile Project, involving storage in the Great Lakes and a vast canal to carry water past the extensive swamps and wetlands of the Sudd (where evapotranspiration significantly reduced river flow) to deliver the 'saved' water downstream to the northern Sudan and Egypt. The Jonglei Investigation Team undertook extensive hydrological and environmental research to assess the feasibility of the project (and its environmental impacts) between 1946 and 1954. However, largely for political reasons (not least the Suez crisis and a new relationship with the Soviet Union), Egyptian attention switched to the Aswan High Dam, built within Egypt's own borders with the services of Soviet engineers. This was begun in 1960 and finished in 1971 (Greener 1962; Waterbury 1979; Collins 1990). Over the same period the Sudan built the Roseires Dam on the Blue Nile (completed in 1966) and the Khashm el Girba Dam on the Atbara (completed in 1965), the latter solely to provide a refuge for evacuees from Nubia, whose land beside the Nile was lost below the waters of the upper end of Lake Nasser. Meanwhile, the idea of the Jonglei Canal lurked still in the official mind in Egypt, and it surfaced again in the 1970s (Waterbury 1979; Howell *et al.* 1988). Construction of the canal eventually began in 1976, although the project eventually became an issue in the ongoing civil war, and all construction work was brought to a halt in 1983 (Collins 1990).

The Nile was by no means a lone example of the growing ambitions and competence of river-basin planners and engineers. The US experience with development planning as part of Roosevelt's 'New Deal' in the 1930s, and particularly the work of the Tennessee Valley Authority (TVA) and state intervention in planning and creating development infrastructure that extended far beyond water resources, was influential across the globe. Planning in the Mekong Basin began in 1957 with the formation of the Committee for the Coordination of the Investigations of the Lower Mekong River (the Mekong Committee) by the UN Economic Commission for Asia and the Far East. A cascade of seven dams from northern Laos down to Cambodia's Tonle Sap was envisaged (Usher and Ryder 1997). Planning in the Mekong Basin continued to some extent throughout the Vietnam War. A basin plan was completed in 1972, triggering concern among environmentalists (Bardach 1973). The Mekong Committee's work reflected an 'almost evangelical faith in science and technical progress' (Sneddon and Binh 2001, p. 237), and major dams were built at Nam Pong in north-east Thailand and Nam Ngum in Laos in the 1960s. Despite decades of war in the region, international collaboration continued, and the Mekong River Commission was reformed in 1995, this time dedicated to achieving sustainable development of the basin.

The 'TVA model' for integrated river-basin planning (or, rather, a selective interpretation of what the TVA represented in terms of planning) was adopted by a United Nations Panel of Experts in 1958 and disseminated widely. In Africa, river basin authorities were created in many countries in the 1960s – for example, in Ghana (the Volta River Authority) and in Nigeria (the Niger Delta Development Board in 1960 and the Niger Dams Authority in 1961). The river basin model continued to expand – for example, in Nigeria (where eighteen multi-functional River Basin Development Authorities had been created by 1984) and in Kenya (Adams 1992). International river basins within Africa were also made subject to planning agencies, for example, the Central African Power Corporation, concerned with development on the Zambezi, particularly the dams at Kariba (between Zimbabwe and Zambia) and Cahora Bassa (Mozambique), and the Organisation pour la Mise en Valeur du Fleuve Sénégal (OMVS) established with international aid donor support by Mali, Mauritania and Senegal in 1972.

In addition to planning, the 1960s also saw the first rash of large dam projects in the developing world. In Mexico, rapid economic development from the 1940s to the 1960s was accompanied by the construction of numerous multi-purpose dams: for example, hydroelectric power capacity rose from 400MW to 5GW over the period (Castellán 2002). By 1986 the International Commission on Large Dams reported a global total of 40,000 large dams (over 15 metres high), with 18,820 in China alone (McCully 1996). In India, 1,554 large dams had been built by 1979, for both hydroelectric power and irrigation, absorbing almost 10 per cent of total public-sector investment (S. Singh 1997).

In Africa the 1960s was the decade of rapid decolonization, and as part of that process a series of major dams was built: the Aswan High Dam on the Nile, the Akosombo on the Volta in Ghana (the USA's riposte to Soviet support for Aswan, a concrete demonstration of the powers of democracy (Mabogunje 1973; D. Hart 1980)), the Kariba Dam on the Zambezi, and the Kainji on the Niger (Adams 1985b). The reservoirs created by these dams became the subject of considerable international research attention (Lowe-McConnell 1966; Rubin and Warren 1968; Obeng 1969; Oglesby *et al.* 1972; Ackerman *et al.* 1973). Furthermore, their ecological impacts became one of the focuses of growing environmental concern about development in the Third World – for example, in the volume *The Careless Technology* (Farvar and Milton 1973). Tropical dams have never quite lost the notoriety they acquired at that time (e.g. Goldsmith and Hildyard 1984; Pearce 1992; McCully 1996; Usher 1997a; Scudder 2005).

Concern about the environmental effects of dams grew rapidly and with an astonishing unison, as critics identified and explained failings in the way dam projects were conceived and designed. Lagler (1969) described the problems of economic loss and human suffering that could arise where planning failed to look far enough ahead. Sulton (1970) described the problems of 'myopia in the planning process, misinterpretation of ecological signs, poor timing and indifference to human suffering' (p. 128), which observers suggested accounted for many of the unfortunate yet often avoidable consequences of the construction of dams. The problem was one of planning failure or, to use Peter Hall's memorable phrase (1980), 'planning disaster'.

Protests against dams were important in the development of the environmental movement in industrialized countries. In the USA the protests against the Echo Park Dam in Utah, and the Glen Canyon Dam on the River Colorado in the 1950s and 1960s were seminal, and there were protests too in Scandinavia in the 1970s, in Tasmania and India in the 1980s (McCully 1996; see also Chapter 13).

Links between local and global anti-dam activists grew in the 1990s, particularly associated with dams on the Narmada River in India. In 1988 activists from round the world met in San Francisco and demanded a moratorium on all new large dams that failed to meet certain criteria: participation by those affected, access to project information and environmental, social health, safety and economic performance (McCully 1996).

The 1990s saw increasingly vocal and coherent protest against dam construction (including the 1994 Manibeli Declaration, calling for a moratorium on all World Bank funding for large dams, and the Bank's own independent review of the Sardar Sarovar Dam in 1992, the Morse Report (World Commission on Dams 2000; see also Chapter 13). A workshop organized in Switzerland in April 1997 by the International Union for the Conservation of Nature and Natural Resources (IUCN) – the World Conservation Union – and the World Bank brought together a wide range of parties, including governments, funding organizations, engineering companies and protest groups, to debate the benefits and costs of dam construction. From this meeting emerged the World Commission on Dams (WCD), an international commission comprising twelve people and a technical secretariat based in Cape Town. The work of the WCD is discussed later in this chapter. Its report was published in 2000, offering a detailed and constructive critique of dam planning, design and operation, and practical guidelines for dealing with environmental and socal costs (World Commission on Dams 2000; Scudder 2005). The report was rejeicted by India and China, and criticized by many industry parties. It was welcomed by many industrialized countries (particularly in the European Union), many non-governmental organizations (NGOs) and international organizations. Debate about the social and environmental costs of dams continued, better informed, but no less sharp.

The political ecology of resettlement

Development programmes and projects create both winners and losers, and dams create both in very stark ways. The potential benefits of dams (particularly hydro-electric power, irrigation and drinking water) have been so obvious for so long that dam construction can be seen as inevitable. But that does not mean that all dams should be built: the central argument of the WCD was precisely that dams should be built only when there is no better way to create benefits (whether other ways to generate power, or other forms of investment altogether), and where costs can be met effectively (Scudder 2005). Many groups of people stand to lose from dam construction, as this chapter will show. However, the biggest losers from dams are those displaced, forced to resettle because their homes are flooded or their livelihoods destroyed. These people often have little share of the benefits of development (Scudder 1991a, 2005).

The problem of resettlement was recognized by the 1960s (Chambers 1970; Colson 1971; Scudder 1975). However, despite extensive planning exercises, forced displacement of reservoir evacuees has rarely been satisfactory. As Gosling

(1979) commented in the context of the Mekong Basin, 'reality is harsher than dreams, and the opportunities for evacuees are more limited in reality than on paper' (p. 119).

Very substantial numbers of people have been moved to make way for reservoirs. McCully (1996) suggests a figure of 2.2 million people displaced by just 243 completed dams in a range of countries. However, this figure excludes China and India: Chinese government figures suggest that dams had displaced between one million and two million people before the end of the twentieth century. McCully (1996) suggests that a figure of thirty million may be more accurate. China's controversial Three Gorges Dam will displace a staggering 1.3 million people, although only 0.8 million of these will be directly flooded (Yuefang and Steil 2003). In India perhaps four million people have been displaced by reservoirs and irrigation schemes (McCully 1996). While many Indian schemes cause limited resettlement, several (for example, Polaran, Kangsabati, Kumari and Bansagar) have displaced over 100,000 people, while the Sardar Sarovar Dam and Narmada Sagar Dam on the Narmada River will eventually flood 265,000 and 170,000 respectively (Singh 1997). A disproportionate number of oustees in India have been people from scheduled tribes, or landless people (43 per cent of the oustees from the Narmada Sagar Dam, for example, were landless). The human cost of dam construction varies greatly between countries, but globally it is significant, and locally it can be devastating (World Commission on Dams 2000).

Plate 11.2 Village cleared for reservoir resettlement, Kiri Dam, Nigeria. Work on resettlement for dams or other development projects makes clear the multidimensional nature of the problems caused, their longevity, and the fact that they extend to communities receiving people as well as those moved (Scudder 2005). Photo: W. M. Adams.

Scudder (1991a) identifies a four-stage model of resettlement projects. Stage 1 is the stage of planning, infrastructural development and settler recruitment. Stage 2 is transition, a period of one to five years during which people actually move and seek to re-establish livelihoods in a new location, making use of whatever invest-ment has been made for them (for example, health facilities, roads, housing or employment). In stage 3, settlers ideally start to become more risk-taking, making investment strategies to increase productivity through diversification of family labour (investing in education, livestock, off-farm income). In stage 4, resettlement project activities are handed over to local organizations, and a generation of settlers takes over.

It is in the relocation stage, however, that evacuees face the greatest costs, particularly where it is rushed, as in the case of refugee relocation and the resettlement following the construction of dams (Scudder and Colson 1982). Compulsory resettlement is traumatic, causing 'multidimensional stress' (Scudder 1975, p. 455). This stress arises from the way in which people are uprooted from homes and occupations and brought to question their own values and behaviour, and the power of their leaders (Lumsden 1975).

Land flooded by reservoirs can have cultural or religious significance, as Colchester (1985) described in the case of the Gond tribal people losing land beneath hydroelectric reservoirs in Maharashtra and Madhya Pradesh in India. Evacuees stranded in isolated settlements on the edge of new reservoirs, or decanted into the urban squalor of dam construction towns, can face large cultural and socio-economic costs, and severe challenges in establishing new livelihoods. People who own land that is flooded, but who are not resettled, or whose lives are disrupted by the dam construction process, can also lose out severely as a result of new projects.

As Sutton (1977) points out, there is little difference between forced reset-tlement for 'developmental' purposes and anti-guerrilla resettlement such as the *regroupement* policy undertaken by the French army in Algeria between 1954 and 1961. Both are externally imposed and in the end both often involve coercion. State violence against evacuees, up to and including murder, is widely recognized – for example, in the case of Guatemala's Chixoy Dam in the 1970s (McCully 1996) and the Sardar Sarovar Project on the Narmada River in India in the 1990s (Baviskar 1995). Such violence usually represents an attempt to quench protest by citizens and civil disobedience and non-compliance with orders by evacuees. Protest at resettlement at Kariba in 1958 led to deaths (in Northern Rhodesia, present-day Zambia (Howarth 1961)), as did protest about compensation at the Bakolori Project in Nigeria in the early 1980s (Adams 1988a). These are by no means isolated incidents. Opposition to forced removal for dam construction has grown in a number of countries – for example, in Thailand, where farmers blockaded the Rasi Salai weir on the Mun River in 2000 to protest against slow payment of compensation (Sneddon 2003). Some protests have had high inter-national profile protests – notably the Narmada Bachao Andolan (the Movement to Save the Narmada) in western India (Baviskar 1995; McCully 1996). In the 1980s the World Bank recognized 'the hardship and human suffering caused by involuntary resettlement', and made a policy commitment to avoid or minimize it, and explore alternative solutions (Cernea 1988, p. 4).

The impacts of resettlement linger long after the relocation phase. The Gwembe Tonga, relocated from the Zambian portion of Lake Kariba in the late 1950s, suffered enormous initial dislocation, but the development of a gillnet fishery on Lake Kariba, the eradication of tsetse fly (and sleeping sickness in humans and cattle) and new roads into and out of the area meant that living standards rose between 1956 and 1974 (Scudder and Habarad 1991; Scudder 1993). However, they fell (with those in the rest of Zambia) with the collapse of copper prices in 1974, and remained depressed thereafter as infrastructure degraded and economic opportunities within and outside the area remained limited.

The key problem in reservoir resettlement is the unequal way in which project costs and benefits are allocated. The Bakun Hydroelectric Project in the rainforests of Sarawak involved a dam on the Balui River impounding 4 billion cubic metres of water in a reservoir covering 695 square kilometres, and demanded resettlement of as many as 4,300 Kenyah, Kayan and Kajang people (Mohun and Sattaur 1987). The Bakun Dam was one of a series of four dams planned in Sarawak to generate power for Peninsular Malaysia, supplied through an undersea cable. As so often, the benefits of development (hydroelectric power) were being enjoyed by one group of people, the costs (resettlement) by another. This is highly inequitable. In theory, development planning should be done in such a way that evacuees do not bear disproportionate amounts of the costs of dam projects (Gosling 1979). In practice, they often do bear these costs, and it is this failure of planning that has been the focus of attention by environmentalist critics of dam construction (e.g. Goldsmith and Hildyard 1984; Roggeri 1985; McCully 1996).

Very often, resettlement projects begin with high hopes of re-establishing evacuees in conditions no worse than those they have left. Thus the Resettlement Working Party convened by the Volta River Authority aimed to use resettlement to 'enhance the social, cultural and physical conditions of the people' (Chambers 1970), and, in the giant Three Gorges project in China, resettlement policy is intended to be 'development-oriented', improving the long-term living standard and production conditions of evacuees (Yuefang and Steil 2003). Such high hopes are often not realized (Scudder 2005). Lightfoot (1978) argues that the literature shows that 'most reservoir resettlements have been badly planned and inadequately financed, and that most evacuees have become at least temporarily and in many cases permanently worse off as a result, both economically and socially' (p. 63). To an extent, the litany of failure may be a function of a lack of published research, and the syndrome that social scientists get more kudos from researching failure than success (Chambers 1983). Nonetheless, there are well-documented case histories – for example, from Ghana (Chambers 1970) and Nigeria (Mabogunje 1973).

The reasons for the poor record of reservoir resettlement projects are manifold. The most fundamental is the disciplinary bias within dam-building organizations. Technical disciplines such as engineering, geology and hydrology dominate the dam project planning field, and the appraisal process concentrates on technical problems relevant to these disciplines. The characteristics of the population of the resettlement area, their economy and society, are neither recognized nor understood. Field investigations concentrate on the dam site and not the inundation area. Frequently the only research in the reservoir area itself concerns bedrock geology and a perfunctory check on topography to confirm the integrity of the reservoir's proposed storage level. The engineering companies called in at each successive stage of project appraisal lack the skills necessary to comprehend resettlement planning problems.

A second cause of the failure of resettlement projects is the fact that project appraisal rarely allows sufficient time for effective planning to be done. Socio-economic planning is both more difficult and more time-consuming than many technical experts assume. It is, therefore, rarely made part of the technical planning process. It is usually introduced too late to be effective. In the case of the Akosombo Dam in Ghana, for example, although the Preparatory Commission had made recommendations about surveys for resettlement, uncertainties over the future of the project meant that nothing was done until the construction contract for the dam was awarded in 1961. The first resettlement staff were appointed in that year, but effective resettlement work did not begin until nine months after the start of dam construction. The stress of resettlement is exacerbated when resettlement is rushed. The human costs of a crash resettlement programme are great, particularly on the elderly and infirm. Furthermore, speed brings errors, and problems of food shortage and poor water supplies are often increased (Scudder 1975).

A third problem with resettlement planning is its inherent complexity. To many observers, the hard part about building a dam would probably seem to be the geotechnical or engineering challenge, or the sheer logistics of building a massive artefact in an area remote from supplies of petrol, engineers and cement. In fact, resettlement planning has often proved the Achilles heel of reservoir projects, seemingly straightforward but in practice fiendishly complicated. Resettlement planning typically involves a series of tasks, including a population survey and an inventory of property and land within the reservoir area, and surveys to locate possible new settlement sites either near to or further away from the home area. Only at this stage can the scale of the resettlement problem be assessed. Knowledge of the costs of resettlement is, of course, a vital element in the assessment of the practicability and acceptability of the project as a whole. Without it, all the technical planning and design may prove useless. However, data on resettlement needs are rarely available in a suitable form in sufficient time to influence decision-making about dam construction. Often the surveys necessary to assess resettlement costs (or compensation payments) are simply not done (Sneddon 2003).

Resettlement is regarded as a secondary problem, to be addressed once the technical feasibility of the project is known. The sunk costs of technical engineering appraisal are such that it can be hard to stop a project once the full human (and environmental) costs are finally factored in. There are clearly quite specific skills associated with population resettlement planning that have been known for many years (e.g. Butcher 1967). Goodland (1978) outlined the measures necessary in the fields of social and cultural ecology with respect to the Tucuruí Dam in the Tocantins River basin in Brazil. He urged that mitigation of the project's impact on indigenous Amerindian people should be allocated time and resources commensurate with its importance, and that the associated costs should be considered an integral part of the costs of the whole project. Guidelines have been available for some years, not only outlining the essential elements of a resettlement project, but also discussing their integration into the operational procedures of the project development cycle (Cernea 1988).

A fourth problem is that of cost. Resettlement is expensive, even if done badly, and it is usually under-resourced. This is partly because costs can be calculated only once preliminary surveys have been carried out. But these surveys begin only after the decision to go ahead with the project has been made. Resettlement costs are therefore seen somehow as an added extra, additional to the costs of construction. It is as if they were in some way optional. An interesting example of this is the development of the Volta Scheme in Ghana. The Akosombo Dam was begun in 1961, but was preceded by an extensive study published in 1956 by a government Preparatory Commission. This recommended that extensive further studies be carried out into resettlement problems, but suggested that self-help resettlement with cash compensation might be the best solution. Partly because of its concern with resettlement and the resulting costs, the work of the Preparatory Commission made the economic prospects of the dam seem sufficiently unattractive for the work not to proceed. Within a few years, however, in the aftermath of the agreement between the USSR and Egypt over the Aswan High Dam, and in an independent Ghana, the scheme took on a new aspect. Revisions of the Preparatory Commission's figures were made that excluded a number of costs, and work began before further consideration of resettlement issues (Lanning and Mueller 1979; DsHart 1980).

Resettlement costs are often underestimated, owing to failure to specify the basis for resettlement, lack of data on the affected population and inadequate budgetary provision, with the lion's share of this being eaten up by the survey and planning elements of the resettlement process. This was the experience in the case of resettlement of 26,000 people in another Nigerian scheme, the Dadin Kowa Dam. Planning for resettlement began only once construction had started in 1980. It soon became clear that the cost of compensation at the rates set by the federal government would be about 60 million naira (then about $US50 million), more than an order of magnitude greater than the total resettlement budget. On top of this, further expenditure would be needed on infrastructural development (for example, roads and water supplies in new villages). Even then, all that would have been achieved would be relocation, since such a resettlement was far from re-creating a viable new economy for evacuees. Obviously resettlement was a very significant element in total project costs; if tackled seriously it would have altered the cost–benefit calculations of the dam irrevocably. In the event, attempts to seek additional funds from the federal government were overtaken by the fall in world oil prices and severe financial stringency at federal level. Resettlement planning effectively became a paper exercise (Adams 1985b).

A typical example of poorly organized and implemented resettlement planning is provided by the Bakolori Dam, completed on the River Sokoto in Nigeria in 1978, flooding the homes of 12,000 people. In this case there had been no attempt to involve evacuees in planning, and no specific provision at design stage to set out a specification for resettlement. Survey in the reservoir area was left to the river basin authority, which lacked people with technical skills. Most of it was in fact done by teams of students, and surveys were only narrowly completed in advance of the waters of the filling reservoir (Adams 1988a). Complaints about the inaccuracies and haste of the survey, the inadequacy of the compensation and the poor resettlement site (on a barren hilltop remote from the river valley) escalated, and, together with the similar grievances of those in the irrigation area served by the dam, led to blockades of the project area by protesters, and violent repressive action by federal police.

The lack of consultation at Bakolori (and the consequences of its ill-planned, hasty and coercive implementation) may be contrasted with the method eventually adopted in the Volta Resettlement Project in Ghana ten years before (Chambers 1970). The scale here was much larger, involving some 80,000 people in 739 villages, about 1 per cent of the population of Ghana. Nonetheless, the selection of new village sites attempted to take account of the views of evacuees. As part of an extensive social survey, enumerators recorded the positive and negative preferences of each village about relocation. Seventy-two sites were identified in this way; of these, twenty-seven were rejected by the Volta Resettlement Authority on technical grounds, although some were accepted after further discussion. Fifty-two village sites were eventually agreed (Chambers 1970). By the standards of many large dam projects, the Volta resettlement was exemplary; it is unfortunate that its lessons have so repeatedly been ignored in subsequent schemes elsewhere. However, even here the project fell short of its own targets, for example, in its attempt to clear land for mechanized farming. It was estimated that 42,000 hectares would be needed to support evacuees, but land clearance was slow, and by 1967 only 2,500 hectares were being cultivated, over half of them manually. In 1968 only 52 per cent of the adult males resettled could farm at all, and food relief had been necessary for three years (Hart 1980).

Perhaps the best measure of the 'success' of these resettlement projects is the loyalty of those resettled. The Kainji Dam in Nigeria was completed in 1968, creating a lake of 1,200 square kilometres, flooding 203 villages containing 44,000 people, the towns of Bussa and Yelwa, and 15,000 hectares of farmland (Adeniyi 1973). Resettlement planning began before independence in Nigeria, and surveys began in 1962 with a view to providing cash compensation to evacuees. By the end of 1963, 2,338 people in eighteen villages had been moved. However, resettlement was slow and inefficient. In November 1964 the policy on housing was reversed, and the authority began to build houses for evacuees using a design by British architects using sandcrete block walls and ferro-cement roofs (Atkinson 1973). Cash was paid for farmland and trees of economic importance, and agricultural advice was offered to evacuees. The main immediate focus of complaints by Kainji evacuees was poor village location and the design of the new houses, particularly the layout of rooms, and the houses' thermal properties and leaking roofs. By 1989 some houses had been abandoned, but most resettlement villages were still inhabited (Roder 1994).

Like the Volta project, Kainji might claim a modest success for its new settlement planning, but it is eloquent testimony that of the 67,500 evacuees on the Volta project only 25,900 were still present in resettlement villages in 1968. Other evacuees had moved elsewhere, while outsiders, including fishermen displaced by declining catches downstream, had arrived. Other resettlement schemes exhibit similarly high turnover rates. Evacuees from the Aswan Dam were resettled on the New Halfa Agricultural Scheme (originally called the Khashm el Girba Scheme in Sudan). Of 300,000 people in the area in 1977, about 24,000 had left by 1980. Both the productivity and cultivated area under irrigation began to fall, owing to a combination of technical and management problems (Khogali 1982).

To achieve parity between economic conditions before and after resettlement requires fair systems of compensation. Not only is it technically complex and expensive to organize effective surveys of land and household effects, but there is a need for probity, transparency and even-handedness in the bases for compensation and the payment process. These are often lacking. Satyajit Singh (1997) comments that, in many Indian projects, compensation is arbitrary, and depends on lawyers and middlemen, to whom only richer oustees have access, and the payment of bribes. Compensation is, therefore, often least available to those most in need. It is also usually available only to those who are registered land title-holders, excluding many tribal households, and those dependent on informal use of lands such as forests (SSingh 1997). The attempt to fit informal systems of resource access and the complex dynamics of human need into the bureaucratic rationality of a resettlement planning process is rarely wholly successful.

Both the Kainji and the Volta resettlements began with self-managed resettlement based on cash compensation but abandoned it. In the Kainji case, for example, self-managed settlement was too slow, and villages were sometimes located in sites unsuitable for water supply or other infrastructure. Perhaps more fundamentally, evacuees were using compensation money for purposes other than house-building. Presumably this reflected either the existence of economic opportunities other than re-establishment of a farming household and land clearance, or (more probably) hidden costs facing evacuees. However, it was interpreted as a shortcoming of the laissez-faire approach, and the centralized planning input was strengthened.

In most cases, resettlement planning is based on this kind of top-down centre-outwards approach to planning. Planners assume that their expertise allows them to 'understand and manage the interests of the farmers better than the farmers do for themselves' (Lightfoot 1979, p. 30). They tend to favour direct control of the resettlement process, despite the fact that, among evacuees from the Nam Pong Project (5,012 households resettled in 1964), those who had been resettled in planned schemes were worse off than those who had resettled themselves (Lightfoot 1978, 1979, 1981). Resettlement planning without public participation forms part of a development process that is imposed from outside, and meets an agenda that may be little influenced by local experience of past or present. There is increasing recognition of the need to revise and improve upon dam-planning procedures (World Commission on Dams 2000; Scudder 2005).

The environmental impacts of dams

Aquatic and riparian (floodplain) ecosystems are enormously complex (Ward and Stanford 1979; Petts 1984). Changes in river-flow regimes can have significant effects on river and floodplain environments (Malanson 1993; Hughes 1997). Freshwater habitats associated with river systems include both static water bodies (such as floodplain pools and meander cut-offs) and flowing water environments. Floodplain environments are ecotonal, ranging from dryland environments to low-lying wetland areas (Malanson 1993). Aquatic and floodplain ecosystems are each subject to the dynamic flow patterns of the river, in terms both of the annual discharge regime, and the size and longevity of shorter-term flood events. These ecosystems are also subject to the distribution of groundwater in space and time that these river flows support. The impacts of dams and other large water development projects on economy and society give them particular importance in debates about sustainability (World Commission on Dams 2000).

In many developing countries, people depend in a very direct way on the productivity of natural or semi-natural ecosystems for their livelihoods. The intensity of human use, and hence the potential severity of socio-economic impact, is particularly great in extensive floodplain wetlands – for example, those of arid or semi-arid Africa that are used for agriculture, hunting, fishing, grazing and gathering. Floodplain wetlands are among the most ecologically productive of global ecosystems and are acutely vulnerable to the impacts of narrowly conceived and poorly planned development projects (Hollis 1990).

The economic importance of tropical rivers and their associated wetlands can be very great. The Niger Inland Delta in Mali, for example, supports over half a million people, and in the dry season provides grazing for about 1–1.5 million cattle, 2 million sheep and goats and 0.7 million camels; there are some 80,000 fishermen (Moorhead 1988). These economic functions may overlap in time and space, or may be used by different communities in different ways through a year. Hunting, gathering and grazing and fishing activities are closely linked to the seasonal cycle of river discharge. Thus, for example, the seasonal grazing resources of the Niger Inland Delta are based on the perennial aquatic grass *bourgou* (*Echinochloa stagnina*), which can yield up to 25 tonnes per hectare of forage, and is accessible to livestock once seasonal floodwaters have retreated (Skinner 1992).

The economic values of rivers and wetlands are dependent on the interconnection of geo-morphological, hydrological and ecological processes. Thus floodplain agriculture in the West African Sahel may take the form of farming on the rising flood (planting before the flood arrives), or on the falling flood, using residual soil moisture left by retreating floods. Farmers usually have extensive knowledge

of crop ecological requirements and flooding patterns, and a detailed appreciation of the variation in land types in the floodplain. Such agriculture is ancient
(West African rice (*Oryza glaberrima*) was domesticated three thousand years ago
in the Niger Inland Delta), and is still widespread (Adams 1992). Indigenous use
of water resources also extends to irrigation, including in West Africa and the Nile
Valley the use of simple wells dug into floodplain sediments, sometimes with a
shadoof. Human-powered water-lifting for irrigation is being rapidly replaced by
small motor-powered pumps – for example, in northern Nigeria, where compact,
portable and relatively cheap petrol pumps began to be introduced in the early
1980s (Kimmage 1991; Adams 1992).

Wetlands are also important in other ways – for example, in sustaining
regional groundwater levels and as an ecological and economic resource for
very extensive surrounding drylands, particularly in times of drought (Scoones
1991). The wetlands of the Sahel, for example, such as the Inland Delta of the
River Niger in Mali, and Lake Chad, not only provide dry-season grazing for
huge numbers of livestock, but are important staging posts for Palaearctic birds
on migration to and from wintering grounds in tropical Africa. The 'functions'
of wetlands include flood control, food production and wildlife conservation
(Barbier 1998).

Dams and reservoirs modify and delay the peak of river floods (the so-called
flood-routing effect). The nature of hydrological effects varies with the purpose

Plate 11.3 Small petrol pump, Nigeria. Small-scale irrigation using shallow groundwater
and open water enabled some floodplain farmers to respond effectively to
the loss of floodwaters on their land owing to upstream dams and poor rainy
seasons. Photo: W. M. Adams.

of the dam (for flood control, power generation or the storage of irrigation or drinking water) and the seasonal regime of the river. Such impacts can be particularly significant where the river regime is seasonal – for example, in rivers in the semi-arid tropics.

Reservoirs behind dams tend to trap sediment in their relatively still waters, leading to considerable problems of loss of storage capacity. This is particularly a problem in river basins with high rates of erosion and sediment transport, for example in China (Yuqian and Qishun 1981), South Asia or Africa. In Pakistan, for example, the Tarbela Dam (3 kilometres long and 143 metres high), lost 12 per cent of its live storage after only eighteen years of operation (McCully 1996). The Khashm el Girba Reservoir on the Atbara River in Sudan, built to supply water for irrigation by Nubian evacuees from the Aswan Dam, had lost 59 per cent of its original storage capacity by 1977 (Khogali 1982).

Sediment deposition within reservoirs can lead to clear water releases below the dam and erosion (Rasid 1979). This can affect the composition of sediment in the river bed, affecting instream ecology, for example, the possibility of fish spawning. It can also affect the rate at which river channels erode, with impacts on riverside settlements and agriculture, and infrastructure such as bridges. Sediment-free water released from the Tarbela Dam in Pakistan began to erode the spillway soon after full operation began in 1976, requiring a massive and costly project to stabilize the plunge pool: by 1986, the dam had cost $1.5 billion, almost twice the 1968 estimate (McCully 1996).

Further downstream the geomorphological impacts of dams become more complex as patterns of erosion and sedimentation adjust to changed discharge and sediment load. Impacts stretch well beyond the river channel into the wider floodplain. Downstream of the Kariba Dam, low levels of sediment, and rapid fluctuations in discharge in the Zambezi, caused increased river bank erosion and affected regeneration of riparian *Acacia albida* woodland in the Mana Pools Game Reserve (Guy 1981; see also Plate 11.4). In Brazil, below the Tucuruí Dam on the Tocantins River, impacts on hydrology and sediment loads affected riverine and floodplain ecosystems, and the sustainability of riparian cultivation on *varzea* (Barrow 1987).

The physical impacts of dams can extend for many hundreds of kilometres downstream into deltaic and coastal environments. Dams on the Mekong River are predicted to affect flooding patterns in the Mekong Delta in Vietnam, with impacts on fisheries, soil acidity and biodiversity (Sneddon and Binh 2001). In Egypt, the Aswan High Dam (completed in 1969) allowed saline penetration of coastal aquifers (Biswas 1980), affected fishing villages and highly productive coastal lagoon fisheries (Kassas 1973; Sharaf el Din 1977) and offshore marine fisheries in the eastern Mediterranean (George 1973). The closure of the Cahora Bassa Dam on the Zambezi in Mozambique changed the seasonal flow regime of the river and the associated supply of nutrients to the shallow coastal waters. This has had a significant negative impact on the recruitment of shrimps on the Sofala Bank and the lucrative inshore shrimp fishery (Gammelsrød 1996).

Ecological succession takes place in new reservoirs as the organisms of flowing

Plate 11.4 Woodland along the Zambezi, downstream of the Kariba Dam in Zimbabwe.
Silt is trapped by the dam, and low levels of sediment, and rapid fluctuations in
discharge, have caused increased river-bank erosion and reduced regeneration
of riparian *Acacia albida* woodland in the Mana Pools Game Reserve. Photo:
W. M. Adams.

water ecosystems are replaced by those of still water, and planktonic and littoral
species arrive (Baxter 1977). Patterns of ecological change are affected by whether
standing vegetation is cleared (forests are often cut, although other measures are
used – for example, chemical defoliants were used in the Tucuruí reservoir in
Brazil). Blooms of algae or macrophytes may occur in nutrient-rich waters (as, for
example, in Cabora Bassa or Kariba on the Zambezi (Davies *et al.* 1972; Balon
and Coche 1974; see also Plate 11.5). Floating plants such as the Nile cabbage
(*Pistia stratiotes*) or the water hyacinth (*Eichornia crassipes*) can create problems
for hydroelectric turbines, and increase evapotranspiration losses from the reservoir surface. On the other hand, they can, to an extent, provide substrates where
fish can find food sources. Rotting vegetation and seasonally flooded drawdown
areas are significant sources of CO_2 and methane, with implications for anthropogenic climate change (Fearnside 2001).

Some reservoirs develop substantial fisheries, but this depends on the way in
which the reservoir ecosystem evolves (Lowe-McConnell 1975). There is often
an initial peak in fish population as nutrients from flooded areas feed into the
ecosystem, followed by a slump to a lower level (Jackson 1966; Petr 1975).
Reservoir fishing may require new methods such as trawling unfamiliar to (and
unaffordable by) previous river fishing communities.

The downstream impacts of dam construction on running-water ecosystems are
relatively well understood in temperate rivers, because of the drastic impacts on

Plate 11.5 Floating mats of water hyacinth (*Eichornia crassipes*) on Lake Kariba,
Zimbabwe, 1990s. The Kariba Dam on the Zambezi, between Zimbabwe
and Zambia, is an important source of hydroelectric power, but caused
forced resettlement of large numbers of Tonga people. Water hyacinth
restricts navigation on the lake (which has an important fishery), and can clog
hydroelectric turbines, as well as increase evapotranspiration losses from the
lake surface. Photo: W. M. Adams.

salmon and other sport fish. Research in the 1960s and 1970s yielded a relatively
detailed understanding of the migratory patterns and in-stream habitat requirements of migratory game fish, and the development of responses in dam design
(for example, fish ladders), dam operation (the release of artificial floods) and
downstream river management (for example, in-stream flow diversion structures)
to minimize adverse impacts of control on fish stocks. Knowledge of tropical
rivers is less complete (but see Payne 1986), although there are numerous data
on freshwater fisheries (Lowe-McConnell 1975; Welcomme 1979), and on the
limnology of certain rivers such as the Nile.

Downstream of dams, the capacity of rivers to support life may be limited by
water that is cold, deoxygenated and rich in hydrogen sulphide or mercury. The
simple barrier effect of a dam can severely curtail movement in active aquatic
species, and can be a serious threat to endangered species such as the Indus river
dolphin (Reeves and Chaudhry 1998) or the recently extinct *baji* of the Yangtze.
In tropical floodplain rivers, many fish exhibit fairly short 'lateral' migrations or
longitudinal migrations of greater length in response to seasonal fluctuations in
the river (Lowe-McConnell 1975; Welcomme 1979). Fish follow floodwaters out
of the river channel to breed in the warm, nutrient-rich, waters of the floodplain.
As the flood subsides, fish move back to the river channel, and in many cases

eventually to the small and deoxygenated pools of largely dry river beds. This
season sees high mortality of fish stranded in evaporating floodplain pools and
taken by predators and people (Lowe-McConnell 1975).

There have been many studies of the way impacts of dams on fish popula-
tions can affect the livelihoods of fishing people – for example, in the Mekong
(Sneddon and Binh 2001), the Tocantins River in Brazil (Fearnside 2001), the
Logone River above Lake Chad in Cameroon (Benech 1992; Mouafo *et al.*
2002), the Phongolo floodplain in South Africa (Jubb 1972), the Niger in West
Africa (Lowe-McConnell 1975) and the *Egeria* clam fishery of the lower Volta
(Chisholm and Grove 1985).

Floodplain vegetation communities, both those immediately bounding river
channels and those of larger and more extensive river-fed wetlands, are influenced
by flooding patterns in much the same way as aquatic ecosystems. Outside the
river channel, floodplain ecosystems maintained by high groundwater tables and
occasional inundations are also vulnerable to changing flood patterns. Floodplain forest regeneration depends on periodic high flood flows. In Kenya, a series
of dams on the Tana River in Kenya (Kindaruma built in 1968, Kamburu in

Plate 11.6 Floodplain forest, Tana River, Kenya. The Tana River flows from the slopes
of Mount Kenya through dry bush to the sea. The river's water supports a
narrow belt of riverine high forest, parts of which include these impressive
Sterculia appendiculata trees. The forest is important both economically to
Pokomo and Malekote people and ecologically for endemic primate species,
but it depends on high river flood flows for its regeneration, and these
have effectively been prevented by the construction of dams in the river's
headwaters, calling into question the ability of the forest to replace itself.
Photo: F. Hughes.

1975, Gitaru in 1978 and Masinga in 1982) threatens the long-term viability of the forest ecosystem (Hughes 1984). Here groundwater recharge by the river, particularly at high flow stages, supports a narrow belt of forest 1–2 kilometres wide (Plate 11.6). Some of these blocks support two endemic primates, the Tana River red colobus and the Tana mangabey, which have received protection in the Tana River Primate reserve (Hughes 1984, 1990). The Tana floods twice a year, with intermittent high flows, particularly in May, which inundate extensive areas of the floodplain. The record shows particularly high flood years every few decades (for example, 1961 with three times the average annual maximum flow). Studies of tree girths and growth rates suggest that past regeneration has been associated with extreme flows of this kind. Upstream dams reduce both the height and the frequency of high flows in the lower Tana, and are likely to bring forest regeneration to an end (Hughes 1984). Other pressures of development local to the forests, notably the cutting of construction timber and fuelwood for the irrigation scheme at Bura, have also had serious impacts on the forests of the lower Tana (Hughes 1984).

The costs of water control

Major water resource projects can have serious impacts on floodplain people and their economies. In the Sokoto Valley of northern Nigeria, as in the floodplains of many West African rivers, the farming of seasonally flooded land has long been integrated with dryland cultivation in the local economy. There is a single rainy season, from about the end of May to the end of September, and the River Sokoto has a strongly seasonal regime with high peak flows in July, August and September. Rice and flood-resistant sorghum are planted in up to 90 per cent of the floodplain in the rains, with rain-fed millet and sorghum followed by relay crops of cotton, cowpeas or groundnuts on upland areas. Different local varieties of rice and sorghum are grown, adapted to particular conditions of flooding, soil waterlogging and desiccation across the floodplain. Both rain-fed and floodplain crops are harvested from October onwards, and the floodplain then comes into its own because a second crop (for example, peppers of various kinds, onions and sweet potatoes) can be grown using residual soil moisture and, sometimes, shallow groundwater irrigation.

In 1978 the Bakolori Dam (see Plate 11.7) was completed on the Sokoto River, and in subsequent years it brought about a significant reduction in peak flows in the Sokoto (Adams 1985a). This in turn halved the average depth, duration and extent of flooding in survey villages downstream in the 120-kilometre stretch before the next major confluence, that with the River Rima. The area farmed fell, with a particularly marked decline in the area under rice and in the amount of dry-season farming. Of a total of 19,000 hectares of floodplain land, the dam caused the loss of 7,000 hectares of rice and 5,000 hectares of dry-season crops.

To an extent, these losses were compensated for by increases in the area under millet and sorghum, but the new uncertainty about flooding patterns and the

Plate 11.7 The Bakolori Dam, Nigeria. The Bakolori Dam was built in the late 1970s on
the Sokoto River in northern Nigeria, to store water to supply an irrigation
scheme on floodplain and terrace land downstream. The reservoir caused
the resettlement of 12,000 people, and disrupted floodplain agriculture and
fishing for several hundred kilometres downstream. Photo: W. M. Adams.

intolerance of millet to waterlogging following rain on heavy soils meant that
farmers were not able to adapt wholly to the new conditions. Furthermore,
although irrigation was a known and tested technology in the Sokoto Valley, the
costs of well-digging and the labour demands of water-lifting were both increased
because reduced floods were accompanied by increased depth to water table. As
a result, irrigation became a more specialized technique, accessible only to larger
producers. The magnitude of lost production in the Sokoto floodplain needs to
be seen against predictions that the flood-control effects of the dam would allow
increased production of rice from downstream areas. The value of lost downstream production can be estimated, and shown to have a significant effect on the
benefit–cost ratio of the Bakolori Project as a whole (Adams 1985a).

The impacts of dams on wetlands have been similar elsewhere in West Africa –
for example, on the rivers flowing into Lake Chad (Hollis *et al.* 1994; Mouafo
et al. 2002). In the north-east of Nigeria, the Hadejia and Jama'are rivers join,
forming the Komadugu Yobe, and drain towards Lake Chad. Around their
junction is an extensive floodplain complex of seasonal wetlands and pools, fed
by the seasonal flood flows. This area is of enormous economic and ecological
importance, supporting a large human community engaged in extensive rice
farming, grazing and fishing. The combination of upstream dams (to supply
rather unsuccessful large-scale irrigation schemes) and low rainfall through the

Plate 11.8 The Hadejia-Jama'are wetlands, Nigeria. These wetlands in north-east Nigeria
consist of an extensive complex of seasonal wetlands and pools, formed where
the rivers flow through an ancient dunefield. The area has many villages
engaged in rice farming, grazing and fishing. The wetlands have progressively
dried out owing to a combination of low-rainfall years and upstream dams,
with serious environmental and economic impacts. These have been offset in
some instances by the successful adoption of small petrol irrigation pumps and
shallow tubewells, but these cannot be used in all areas. Photo: W. M. Adams.

1970s and 1980s has caused significantly reduced flooding in the wetland. This,
in turn, has had considerable impact on flood recession farming, fishing and
grazing, and has had measurable economic costs (Hollis *et al.* 1994; Polet and
Thompson 1996; Barbier 1998).

Similar impacts have been recorded in every continent, wherever dams have
affected floodplain river people (McCully 1996; Singh 1997; Usher 1997a). The
socio-economic impacts of dam construction may be drastic, but it is important
to note that floodplain people are not always passive victims of those impacts.
They tend to be ingenious and industrious in their attempts to adapt to their new
circumstances. Environmental impacts themselves evolve over time, as physical
and ecological systems adapt to changing flood frequency and duration. People
also respond to changed environmental conditions by changing their patterns
of resource use to take account of reduced flooding. If external conditions are
favourable, they may be able to adapt successfully. Thus, for example, farmers in
both the Sokoto and Hadejia-Jama'are floodplains were able to take advantage of
a boom in small-scale irrigation in the 1980s and 1990s (Kimmage 1991). Using
pumps and shallow tubewells, some farmers (those with money to invest and in

areas with access to water) were able to maintain or even improve their economic position in the face of environmental degradation (Kimmage 1991; Thomas and Adams 1999; see also Chapter 13). Those who could not invest in this way lost out: the impacts of the upstream dams need to be understood in the context of complex chains of impacts and responses that can interact with various other kinds of positive or negative change.

Why projects fail

In theory, before construction begins, development projects like dams undergo a strict and technically sophisticated appraisal procedure involving project iden-tification, pre-feasibility and feasibility studies, and detailed design. Assessment should involve some form of environmental assessment, and cost–benefit anal-ysis (CBA) (Vanclay and Bronstein 1995; Barrow 1997; see also Chapter 6). In theory, this multi-stage process should allow problems to be identified and ironed out, successive consultants obtaining contracts on the basis of experience and competitive bidding, and checking their predecessors' findings. In practice, project planning and design are often far from perfect. First, badly framed terms of reference can constrain the range of options the consultant is prepared to investigate, perhaps meaning that viable alternative schemes are not considered. Thus, if asked to investigate the potential for large-scale irrigation in a river basin, a consultant will get small thanks for a study investigating the benefits of small-scale alternatives. Since success in bidding for the next job depends on contacts made during this, there are strong reasons for not rocking the boat by presenting unpalatable appraisals. Second, competition has the effect of paring bids for jobs to a minimum, with the result that 'extras' get cut out. Such extras can often include environmental or social impact assessments, or at least the resources necessary to make them effective. Third, client organizations in the host country can steer the assessment process – for example, setting a narrow scope for envi-ronmental appraisal, or calling for revision of studies that find too many problems with a project. Professional and commercial considerations in consultancy firms may push in very different directions.

Many major dam projects have been built without thorough environmental and social appraisal. The Tucuruí Dam in Brazil, for example, was designed with very limited environmental assessment, and impact studies never consid-ered the 'no-build' option (Fearnside 2001). In the case of the Chinese-funded Merowe Dam in Sudan, an environmental impact assessment was completed only in 2002, a year before construction began. It was not done by independent consultants and its findings were not made public. Ten thousand people will be displaced and downstream impacts have not been taken adequately into account (Giles 2006).

Technical processes of appraisal are also affected by the commitment of plan-ners to the idea of dams in particular places. Once proposed, projects often keep resurfacing in the work of successive planners until they are finally built. Thus the history of the Three Gorges Dam in China dates back to a proposal for a dam on the Yangtze in 1921, and subsequent support by Mao Zedong and (remarkably) the USA in the 1940s (Beattie 2002). Similarly, Sir Albert Kitson, Director of the Gold Coast Geological Survey, first proposed a hydroelectric dam on the River Volta to generate the power to smelt aluminium in the 1920s, and the idea kept resurfacing through the 1930s and 1940s before the colonial government commissioned a survey of the basin in the 1950s, and eventually took the decision to build the Akosombo Dam in the 1960s (Hart 1980). Development projects have a tendency to live as a blueprint in a limbo of experts' minds, buried in the depths of planning bureaucracies, until awakened to re-emerge in the light of a more auspicious dawn.

A suite of political factors underlie, and can override, technical project appraisal. Strategic factors can be important influences on development decisions, and so can the need for governments to be seen creating flagship projects: even a project with a sharply negative cost–benefit ratio may be attractive if it is 'a rural manifestation of the state's active presence' (Hart 1982, p. 89). Thus, in semi-arid areas prone to drought, the lure of irrigation is very strong (Moris 1987; Adams 1992). Ironically, of course, irrigation has an unfortunate reputation for economic inefficiency, particularly in sub-Saharan Africa (Moris 1987; Moris and Thom 1990; Adams 1991, 1992).

Corruption is a significant factor in decision-making by Third World bureaucracies, although it is hard to trace and substantiate (Usher 1997c). Wade (1982) has made it clear that corruption in canal irrigation in India is not some kind of curious aberration, but an integral part of the structure of decision-making and performance. Large projects mean large profits. International construction firms compete for shares of lucrative construction projects like irrigation schemes whose farmers and land can never possibly generate a surplus large enough to pay for all these overheads (Hart 1982). Commercial competition between consultants, and commercial independence between consultants appraising projects and contractors building them, may be more apparent than real (Usher 1997c). Commercial pressures also influence aid donor decision-makers, sensitive to the need for domestic companies to win contracts on aid projects: the dam construction and turbine industry of Norway and Sweden, for example, is a major influence on Nordic donor decisions (Usher 1997c).

The process of conception, design and approval of major development projects is in practice highly complex and often affected by political considerations. Usher (1997b) describes the anatomy of decision-making in the approval of Swedish and Norwegian aid to allow the Nordic multinational Kvaerner to supply turbines for the Pangue Dam on the Bíobío River in Chile in the early 1990s. This was a complex case, not least because it was being built by a private company (the privatized electricity utility ENDESA), and not eligible for aid under the rules of the Organization for Economic Cooperation and Development (OECD). Moreover, although Pangue was one of an integrated suite of dams for the Bíobío River, its environmental impacts were assessed as if the other dams, with which it formed an integrated package, were not being planned. Reports were prepared under tight deadlines, and decision-making by the Swedish and

Norwegian governments, and other donors, was highly political (Usher 1997b). Chile's environmental assessment (EA) process was weak, and dependent on excessively close relationships between consultants and the dam construction industry (Silva 1997). In the case of Pangue, the EA did not consider down-stream impacts and water-release patterns of Pangue, and failed to provide a full picture of the impact of the Pangue and Ralco dams on the environment of the Bíobío valley (Silva 1997).

Usher and Ryder (1997) argue that such failures of environmental assess-ment are characteristic of dam projects. The impact of the Theun Hinboun Dam in Laos (on a tributary of the Mekong) on subsistence fisheries and the food security of floodplain communities was systematically ignored or underplayed in 'expert' environmental impact assessments. In the case of the Pak Mun Dam on the Mekong, an environmental assessment was carried out, but not released. When opponents saw a leaked copy in the USA in the early 1990s, they attacked it fiercely for its inaccuracy. The shortcomings of the environmental assessment of this dam (which lies within the Koeng Tana National Park) were a significant factor in criticisms of the World Bank's environmental record (Rich 1994). Usher and Ryder (1997) conclude that 'the failure to identify such crucial issues raises questions about both the experts and about the aid agencies that take these claims at face value' (p. 99). Usher (1997c) suggests that the process for reviewing the environmental impacts of Third World dams is 'rigged' (p. 59).

Barnett (1980) suggests that development should not be understood as a kind of black-box process, where known inputs create entirely predictable outputs. It is more like Pandora's box: investment produces change, but it is not possible to predict the extent and direction of that change. In other words, development is a probabilistic process, full of uncertainties and full of risks. If development projects are not to fail, they must be allowed to evolve as circumstances change. Environmental appraisals need to evolve with them as they change. Development is a continuing process, and development planning is a form of permanent crisis and risk management. Both development planning and environmental appraisal need to aim for the same flexible and interactive relationship with change.

Beyond river control

The adverse environmental impacts of river control on downstream environments and people are now reasonably widely recognized. While the development of more sophisticated techniques for assessing environmental and social impacts allows economists to calculate net benefits and costs more accurately, and allows dam designers to assess the relative merits of different dam sites on a wider range of criteria, conventional approaches to river-basin planning are still locked into a rather sterile techno-centrist 'control and transform' mindset in response to the natural functions of rivers and floodplains and the existing uses people make of them. It is possible to start thinking outside this straitjacket. The challenge is to link sophisticated understandings of ecosystem dynamics and human impacts with innovative institutional models for planning change. There have been some interesting experiments in addressing this challenge in the field of river-basin planning.

Thayer Scudder (1980, 1988, 1991b) calls for a quite different approach to dam design and operation. In a number of African river basins, artificial floods have been proposed and in some cases released to re-establish ecological function and sustain economic activity in huge downstream floodplains (e.g. Horowitz and Salem-Murdock 1991; Scudder 1991a; Acreman and Howard 1996). This reflects developments in restoration ecology (Hughes and Rood 2001; Perrow and Day 2002). By the end of the 1990s, the need to integrate dam releases and downstream environments had become widely recognized, and adopted by the World Commission on Dams (2000; see also Scudder 2005).

Scudder suggests that the simulation of the seasonal flood peak would make downstream production possible and also allow cultivation in the drawdown zone of the reservoir. In the case of a hydroelectric dam, management in this way would offset the many costs to downstream producers that need to be taken into account, and also open up a new resource for reservoir evacuees. In the case of dams built for flood control and irrigation, this form of management would reduce downstream flooding losses and allow further development to be piece-meal and locally instigated and managed, thus avoiding the high costs of centrally planned large-scale irrigation. Gross benefits might be smaller – for example, through reduced power generation or slower expansion of irrigation, but the cost–benefit ratio would improve considerably. There would, of course, be problems, for example, in management. However, as Scudder points out, it is time that attention was paid to the management of tropical river basins rather than simply their 'development', seen as a one-off process.

There have now been experiments with controlled flood releases in several places in Africa – for example, on the Phongolo River in Natal in South Africa (Scudder 1991b; Bruwer *et al.* 1996), on the Senegal River between 1988 and 1990 (Salem-Murdock and Horowitz 1991; Hollis 1996), on the Waza-Logone floodplain in Cameroon (Ngantou 1994; Wesseling *et al.* 1996) and at the Itezhitezhi Dam above the Kafue Flats in Zambia (Scudder and Acreman 1996). In each of these places, dam construction or other engineering works have created serious downstream ecological and economic impacts. On the River Senegal, flood cropping is practised on floodplain *waalo* land (Lericollais and Schmitz 1984). The area cultivated has varied from about 150,000–200,000 hectares in the 1960s (when rainfall was good) to about 20,000 hectares in the drought years of the 1970s. Studies showed that the value of lost downstream production following construction of the Manantali Dam outweighed marginal benefits of hydroelectric power generation. However, between 1904 and 1984 there was sufficient water both to generate 74 MW of power and to release an artificial flood large enough to inundate some 50,000 hectares of land downstream in every year except the most severe drought years (1913, 1977 and 1979–84). An experimental release of an artificial flood was carried out, although with mixed success.

The use of controlled floods to convert single-purpose dams (for example, for hydroelectric power generation) into a tool for multi-purpose multi-environment management – indeed, the wider notion of using dams to work with the natural patterns in the rivers of Africa – is attractive, but presents problems for river managers. In particular, the idea is extremely demanding technically, since it requires knowledge of a number of complex variables (see Table 11.1), and planning and project development that is based on effective real-time monitoring and decision-making. African river-basin planning agencies attempting such integrated planning will require extensive training, technical support and institutional strengthening.

The task of integrating the releases of water from upstream dams and the needs of people and ecosystems in downstream floodplains cannot simply be seen as a technical one. The diversity of downstream needs makes it effectively impossible to devise a single solution that automatically takes account of all interests. One approach to the complex planning required is to involve floodplain communities in the planning and management of releases. This has been done in the Phongolo floodplain in South Africa (Bruwer *et al.* 1996). Here some 70,000 people depend on wetland resources sustained by the flooding of the river. The decision to build the Pongolapoort Dam was taken in the 1950s for political reasons, but it was filled only to 30 per cent of capacity to avoid inundation of part of Swaziland. Surplus water was released from the dam to serve downstream communities, but the restructured floods up to 1984 were smaller and unpredictable in timing, and created a risky environment for floodplain resource use. In 1984 the reservoir was filled to capacity by floodwaters from cyclone Dominoa, and larger releases began to be possible. This enabled ecological conditions in the floodplain to be restored, but it did nothing to reduce uncertainty for floodplain people. From 1983 downstream villages began to organize themselves to present their needs and interests, and gradually 'combined water committees' were set up. In 1988 a 'liaison committee' met to hear the views of all stakeholders. There are now carefully agreed procedures for ward water committees to communicate their needs for floods to the Department of Water Affairs. The result is reported to be positive

Table 11.1 Knowledge required to integrate dams and downstream environments

- The topography of the downstream floodplain
- Predictive models of probable flood volumes and durations at all points downstream of the release site
- The implications of floodplain morphology for flood depth and duration
- The depth and duration of flooding in past years
- The nature of aquatic and riparian vegetation in past years
- Changes in aquatic and riparian ecology since changes in flooding, and the implications of reversing (or further changing) those flooding patterns on ecology
- The social and economic impacts of changes in past flooding patterns and assessment of the implications of reversing (or further changing) those flooding patterns
- Cost–benefit analyses of present and future management regimes
- Monitoring of the hydrology of released floods
- Monitoring of the ecological impacts of the released floods
- Monitoring of the social and economic impacts of the released floods

local attitudes to the possibility of managing floodplain water effectively, and a move towards sustainable utilization of the floodplain (Bruwer *et al.* 1996).

The idea of integrating the management of upstream dams and downstream environments through controlled flood releases essentially seeks to transform river-basin development from the conventional closely directed and externally imposed blueprint of future development based on large-scale projects to a more open-ended, flexible and diverse picture of locally initiated smaller-scale projects. Scudder (1980, 1991b, 2005) seeks to offer dam-builders and river-basin planners a practical alternative development model that can be implemented using existing planning frameworks. However, the implications of his suggestions are more fundamental. They start to challenge the whole established 'development from above' model of development planning (Stöhr 1981). They also challenge the Promethean arrogance of conventional approaches to development, and start to demonstrate how development planning can 'design with nature' (McHarg 1969), tackling development problems holistically and intelligently.

The new approaches discussed here are among those debated in the work of the WCD. That commission set out to transform the way dams were planned and designed. The WCD's terms of reference were to review the development effectiveness of large dams, to assess alternatives for water resource and energy development and to develop internationally recognized criteria, guidelines and standards for the planning, implementation, operation and decommissioning of large dams (Biswas 2004; Scudder 2005).

The WCD began work in May 1998, first to review the development effectiveness of large dams and assess alternatives for water resources and energy development, and second to develop internationally accepted criteria, guidelines and standards for the planning, design, appraisal, construction, operation, monitoring and decommissioning of dams (World Commission on Dams 2000). It took evidence for two years, holding four regional consultations and taking 900 submissions from individuals and organizations, commissioning seventeen 'thematic reviews' and eight detailed case studies of individual dams, as well as carrying out a cross-check survey of dams in fifty-two countries.

The WCD's report, launched in London in November 2000 with a speech from Nelson Mandela, was ambitious. It offered a clear and new basis for planning water and energy resources that embraced participatory decision-making and an explicit engagement with both rights and risks; it emphasized the centrality of social and environmental issues in dam planning (World Commission on Dams 2000). The WCD report was MSD at its best: carefully professional, yet challenging to the status quo, building on best practice and yet pushing (and pushing hard) at the outside of the envelope of normal planning practice.

The WCD dissolved itself on the publication of its report, although dissemination of its proposals and debate were carried on by the United Nations Environment Programme (UNEP) Dams and Development Project (www.unep-dams. org). Inevitably, perhaps, there was a rearguard action from those who imagined their interests threatened by its recommendations. Key dam-building nations (China and India) rejected its findings, and a chorus of criticism came from dam construction interests. It remains to be seen to what extent the proposals will prove effective in changing the way dams are planned, designed and operated (Scudder 2005).

Development from below

The conventional model of development, development 'from above', became increasingly battered in the last two decades of the twentieth century. The presumption of a monolithic value system and a uniform basis for human happiness (which 'automatically or by policy intervention will spread over the entire world' (Stöhr 1981, p. 41)) began to be quite widely challenged in development studies. Criticism of individual projects and development outcomes grew into (and in a cycle of criticism and affirmation fed upon) a profound shift in the dominant discourses of development during the 1970s. 'Top-down', 'technocratic', 'blueprint' approaches to development came under increasing scrutiny as they failed to deliver the economic growth and social benefits that had been promised. An alternative agenda emerged, associated in particular with the work of Robert Chambers (1983, 1988b, 1997). It began to be widely argued that development goals could be achieved only by 'bottom-up planning', 'decentralization' and 'participation' and 'community development' (Agrawal and Gibson 1999). By the early 1990s, aid donors and development-planners were heavily committed to participatory approaches.

This 'development from below' demanded a reversal of conventional development thinking, working from the 'bottom up' and the 'periphery inwards' (Stöhr 1981, p. 39). It was an approach, not a package; an idea, not a set of rules. It suggested that, for success, developments must be not only innovative and research based, but locally conceived and initiated, flexible, participatory and based on a clear understanding of local economics and politics. There was an alternative to large-scale centralized development, one 'characterised by small-scale activities, improved technology, local control of resources, widespread economic and social participation and environmental conservation' (Ghai and Vivian 1992a, p. 15). Development had to begin to 'put people first' (Cernea 1991).

Development projects are the product of a planning process called into being as a result of a search for betterment of the human condition. On the face of it, development projects should not cause significant persistent adverse environmental and socio-economic costs, and certainly not costs that are not properly compensated from the benefits elsewhere. It is an unavoidable fact, however, that they often do. National or regional interests may conflict with the interests of those people immediately and adversely affected by the immediate context of development (Cernea 1988, 1991). It is commonplace that impacts on these people are not predicted, not recognized and not compensated. It is still standard practice that those bearing environmental and social costs are not consulted about whether large-scale developments should go ahead, even if increasingly they are consulted about options consequently available to them. Development is still imposed from above, the balance of costs and benefits weighed only by ranks of technical experts, dispassionately viewed by planners and decision-makers who are themselves insulated from both the consequences of their actions and the consequences of any failure to plan well.

In the final analysis, development is a political and not a planning issue. Questions of equity and justice are fundamental to sustainability. Whatever the planning process, environmental impacts are experienced by particular people in particular places. The nature of those impacts, and any steps taken to compensate or reduce them, reflects patterns of power, wealth and influence as well as the geography of the environment.

There are few studies of development written from the perspective of the developed. Researchers, particularly anthropologists, sometimes claim to speak for those with faint voices in writing the academic literature of development, but most researchers and almost all development professionals are guilty of some form of 'rural development tourism' (Chambers 1983). They turn up in a village on a tight timetable with a limited budget, seeking something. Whether that something is information or agreement to a development plan, village people might well be slow to trust the stranger in the four-wheel drive, with city clothes, a bagful of papers and a computer.

Researchers are particularly suspect. In her 'letter to a young researcher', Adrian Adams (1979) tried to explain the experience of the village of Jamaane on the River Senegal of Europeans. The researcher had come from the development agency (OMVS) filled with its presumptions and philosophy. He was but one in a long chain of visitors, mostly white, mostly staying only short periods, who transformed *un développement paysan* into *un développement administratif*. A peasant-run project based on communal work groups was gradually taken over by the Société d'État with the introduction of mechanized irrigation exclusively for rice, funded by the United States Agency for International Development (USAID).

To the technician through whom the alienation of indigenous change began, 'develop' was 'an intransitive verb, and "development" was one and indivisible' (A. Adams 1979, p. 473). The changes he and his successors introduced were 'presented as neutral steps, objectively required on technical grounds as part of the development process'. The technician was an 'expert', someone with 'a halo of impartial prestige' lent by his skills, able to neutralize conflicts and package political issues as technical ones. Such an expert embodies 'modernity, progress, efficiency', and, as Adrian Adams (1979) pointed out, 'none may be an expert in his own country: it's an expatriates' title' (p. 474). Ulrich Beck (1994) recognizes the problematic role of the 'expert' in risk society, and calls for a 'demonopolization of expertise'. He says that 'people must say farewell to the notion that administrations and experts always know exactly, or at least better, what is right and good for everyone' (p. 29).

The technicians visiting Jamaane village were such 'experts', full of knowledge and ideas, but they could not communicate and they would not listen. They failed to see that Jamaane village was the home of thinking people, with memories of past visitors, problems of the present and ideas for the future (Adams 1979). Croll and Parkin (1992) suggest that development itself can be thought of as 'a form of self-conscious or planned construction, mapping and charting both landscapes and mindscapes' (p. 31). The control of rivers for irrigation by dam construction is pre-eminently part of that mapping and charting process. A search for development that is equitable and environmentally benign must ask who should have influence over the making of those charts.

Summary

- Development itself can be a significant source of *un*sustainability. Projects such as dams and irrigation schemes can generate a chain of environmental and socio-economic impacts that are serious and complex. These result, in particular, from the urge to regulate the environment, and rationalize and modernize society and the human use of nature.
- Upstream of dams, serious cultural, social and economic costs are imposed on those forced to evacuate from reservoir areas. Despite decades of experience of reservoir resettlement planning, these costs remain serious, and are often underestimated and inadequately compensated.
- Dams and reservoirs alter the pattern of river flows, typically lowering and extending flood peaks; they affect aquatic ecosystems (for example, fish and hence the livelihoods of fishing people) and downstream floodplain wetlands (for example, floodplain forests and farmlands). Socio-economic impacts on people fishing, farming and grazing can be significant.
- Irrigation development is also capable of delivering fewer benefits and more costs than planners anticipate. The problems of new large-scale irrigation schemes in areas such as sub-Saharan Africa have been particularly acute, but even in areas of established irrigation, such as South Asia, problems of disease, inefficient and inequitable water distribution, poor yields, farmer debt and poor economic performance can be significant.
- An alternative approach to the management of dams involves the controlled release of water to maintain and enhance production and ecosystems in downstream environments.

Further reading

Adams, W. M. (1992) *Wasting the Rain: rivers, people and planning in Africa*, Earthscan, London.

Andrae, G. and Beckman, B. (1985) *The Wheat Trap: bread and underdevelopment in Nigeria*, Zed Press, London.

Chambers, R. (1988) *Managing Canal Irrigation: practical analysis from South Asia*, Cambridge University Press, Cambridge.

Collins, R. O. (1990) *The Waters of the Nile: hydropolitics and the Jonglei Canal, 1900–1988*, Clarendon Press, Oxford.

McCully, P. (1996) *Silenced Rivers: the ecology and politics of large dams*, Zed Press, London.

Moris, J. R. and Thom, D. J. (1990) *Irrigation Development in Africa: lessons from experience*, Westview Press, Boulder, CO.

Murphy, R. (1994) *Rationality and Nature: a sociological inquiry into a changing relationship*, Westview Press, Boulder, CO.

Pearce, F. (1992) *The Dammed: rivers, dams and the coming world water crisis*, Bodley Head, London.

Scudder, T. (2005) *The Future of Large Dams: dealing with the social, environmental and political costs*, Earthscan, London.

Usher, A. D. (ed.) (1997) *Dams as Aid: a political anatomy of Nordic development thinking*, Routledge, London.

World Commission on Dams (2000) *Dams and Development: a new framework for decision-making*, Earthscan, London.

Web sources

http://www.dams.org The World Commission on Dams: site contains the texts of the reports that were prepared for the commission, and information about and reactions to the final report in November 2000.

http://www.unep.org/dams/ UNEP Dams and Development Project.

http://internationalrivers.org/ International Rivers, leading international non-governmental organization campaigning on dams: 'works to protect rivers and rights, and promote real solutions for meeting water, energy and flood management needs'.

http://www.narmada.org/ Friends of the River Narmada. A support organization for Barchao Narmada Andolan (Save the Narmada Movement), campaigning against the Indian government's Narmada Valley development project.

http://www.icid.org The International Commission on Irrigation and Drainage, established in 1950 as an international technical organization devoted to improving water and land management and the productivity of irrigated and drained land through irrigation, drainage and flood management.

http://www.icold-cigb.org The International Commission on Large Dams (ICOLD), the major global advocate of large dams, with interesting insights into its efforts to support 'environmentally and socially responsible development and management of the world's water resources'.

http://www.cgiar.org/iwmi/ The International Water Management Institute (formerly International Irrigation Management Institute) is a CGIAR international research organization on water and irrigation. Its work has developed from having a specific focus on irrigation (including farmer-managed irrigation) to a broader concern for water management. The website reports research reports and other information, tools and software models.

http://www.worldwatercouncil.org The World Water Council is an international water policy think-tank, founded in 1996, with members from public institutions, UN organizations, private firms and non-governmental organizations, and it tries to take an independent view of water issues.

12 Industrial and urban hazard

> Not only do the rich occupy privileged niches in the habitat while the poor tend to
> work and live in the more toxic or hazardous zones ... but the very design of the
> transformed ecosystem is redolent of its social relations.
>
> (David Harvey, *Justice, Nature and the Geography of Difference*, 1996)

Development's environmental cost

On the night of 2 December 1984, an explosion in a pesticide factory in Bhopal
in Madhya Pradesh, India, spread a cloud of pollution over the city. The plant,
owned by Union Carbide India Limited (in turn 50.9 per cent owned by the
Union Carbide Corporation of the USA), was built on the outskirts of Bhopal
city, 1 kilometre from the railway station and 3 kilometres from two hospitals.
Over the next hour, about 30 metric tons of methyl isocyanate gas escaped, and
sank over the surrounding streets. Within two days, five thousand people had
died from inhaling the gas, and half the population of the city had fled in terror.
The eventual total death toll was probably about 20,000 people, with another
200,000 poisoned (Low and Gleeson 1998; Varma and Varma 2005).

The accident in Bhopal happened when a tank was being cleaned, water coming
into contact with liquid methyl isocyanate and generating gas that vented to the
atmosphere. Responsibility for the tragedy remains a moot point: the explosion
might have been averted had the automatic controls standard in such Union
Carbide plants in the USA been installed, had the Indian government not insisted
that manual controls be fitted, or had the Indian subsidiary of Union Carbide
followed plans more closely. The human cost of the disaster would have been
much less if the plant had not been built so close to the city. The Indian government made an out-of-court settlement with Union Carbide in 1989 for $470
million, although legal actions continued, both to make the state pay out this
compensation to victims, and to make the plant's new owners (Dow Chemicals)
clean up pollution (Varma and Varma 2005).

The Bhopal disaster has come to epitomize the hazards of industrialization in
the poorly regulated urban environments of the developing world. The hazards
of dangerous, poorly understood and unregulated industrial process have long
offered a threat to the welfare of workers – as, for example, in the notorious case

of the employees of the US Radium Corporation in the 1920s who contracted mouth cancer licking the brushes with which they put luminescent dots of radio-active paint on watch faces (Caufield 1989). As industrial processes are scaled up, so too is the magnitude of accidents. There had been extensive industrial disasters before in industrialized countries over many decades – for example, the fire at the first nuclear reactor in the UK at Sellafield (then Windscale) in the UK in October 1957, the explosion at Flixburgh in the UK in 1974, and the release of dioxin at Seveso in Italy in 1976. But Bhopal was in a developing country. Now it was clear that the hazards of industrial processes could be experienced anywhere.

Fifteen months after the Bhopal disaster, the German sociologist Ulrich Beck published his book *Risikogesellschaft* (published in English in 1992 as *Risk Society*). Beck (1992) described new forms of techno-scientific risk, inherent to modernity yet offering a global threat. He defined risk as 'the probabilities of physical harm due to given technological or other processes' (p. 4). The risks of pesticide production at Bhopal were, to Beck, part of 'a bargain struck by the Indian state on behalf of its people', a deal that balanced risk of poisoning and death against the benefits of work and useful agricultural fertilizers, pesticides and herbicides. The plant itself, and its belching pipes and steaming tanks, were an expensive symbol of developmental success, their products signifying 'emancipa-tion from material need'. By contrast, the 'death threat they contain' remained largely invisible (p. 42).

The pesticides being produced in Bhopal were part of the package of technical inputs of the 'Green Revolution', intended to revolutionize agricultural produc-tivity in India as elsewhere, and push back hunger and poverty; as Beck (1992) comments, 'in the competition between the visible threat from hunger and the invisible threat of death from toxic chemicals, the evident fight against misery is victorious' (p. 42).

This apparent choice between these unlike abstractions, of promise and risk, is inherent to developmentalist thinking. Economic growth is regarded as necessary to reduce human poverty, and industrialization and urbanization are conven-tionally taken to be a prerequisite of economic growth. That is how the earth's currently industrialized (and wealthy) core countries got where they are, and copying them is the one big idea of the developmentalist paradigm. But does the argument that only economic growth can allow escape from the downward spiral of poverty and environmental degradation (Broad 1994) take account of the increased environmental hazards of industrialization?

There is an argument that these hazards, and specifically the problem of pollution, are transitory. The so-called environmental Kuznets curve (Neumayer [1999] 2003) suggests that, while environmental quality does tend to fall as poor countries start to grow economically (with effects such as forest being felled and rising urban pollution), it rises again when they become wealthy (see Chapter 5). The experience of industrializing countries in South-East Asia suggests a shift over time as industry develops, from water-borne organic pollution through solid waste and airborne pollution towards toxic pollution (Auty 1997). The World

Bank argues that industrialized countries have begun to manage to delink economic growth and pollution through control of emissions of sulphur dioxide, nitrogen dioxide and particulates and lead, and argue from this that industrialization and environmental improvement do not need to be incompatible (World Bank 1992; Munasinge 1999).

The usual argument that the history of industrialized countries provides a useful forecast of what might happen in developing countries suggests this might be the case: cities in nineteenth-century Britain were profoundly polluted, and today they are clean. The information and service economies of developed countries, which have lost much of their heavy industry, do tend to have reduced pollution levels (Neumayer [1999] 2003). However, historical data on already industrialized countries are a poor guide to contemporary industrialization, because market, corporate and regulatory environments have changed hugely since industrialization began during an era of naked European imperialism. The ability of a wealthy country such as the UK to reduce air pollution, for example, is a function of the rapid decline of its manufacturing industry, and may not be a good guide to the future in a rapidly industrializing country like India or China. One reason why environmental conditions improve in richer countries is that they simply export their dirty industries to poorer countries and import clean processed products (Nahman and Antrobus 2005). Growth in population and consumption increases overall global industrial pollution, whatever happens locally. Huq (1994) raises the question of whether the transition of all countries to full industrialization can be sustained, even assuming it could be achieved.

Gadgil and Guha (1995) analyse the pattern of industrialization in India following independence in 1947. Government investment subsidized water, power, transport and communication facilities in larger centres. Gadgil and Guha divide India's people into three categories: omnivores, ecosystem people and ecological refugees. A sixth of Indians are the omnivores: entrepreneurs, larger landowners, professionals and formal-sector workers. Four decades of planned development have created islands of relative prosperity in a sea of poverty, where the omnivores thrive. Half of India's population they describe as ecosystem people, locked in poverty and dependent on fields, forests and rivers for their subsistence. A third of Indian people live as environmental refugees, displaced by poverty, landlessness or development projects such as dams, and living on the margins of rural or urban life. Most of India's ecosystem people are submerged in the sea of poverty, while the omnivores inhabit islands securely on firm ground. The ecological refugees are 'hangers on at the end of the islands of prosperity'

> somewhat like mud-skipper fishes hopping around on the muddy beaches fringing mangrove islands. From time to time the tide swallows them; they can manage to clamber back on to the mud, but can never make it to dry land.
>
> (Gadgil and Guha 1995, p. 34)

An 'iron triangle' has been created by India's omnivores, at the expense of the subsistence sector and the environment (see Figure 12.1). This triangle is

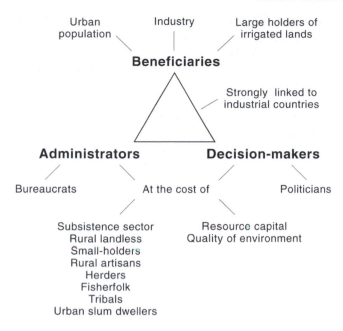

Figure 12.1 The 'iron triangle' governing resource use in India (after Gadgil and Guha 1995)

an alliance between the omnivores and those who decide the size and scale of the favours of the state (the politicians) and those who implement their delivery (technocrats and bureaucrats).

Urban political ecology

Manufactured hazard seems an inescapable part of the development process, part of the price paid for the opportunity to enter the development race. It is a price disproportionately borne by the poor. Techno-scientifically produced risk arises from processes of wealth creation in the South as it does in the global industrial core. Indeed, its power and scope are less confined and its proliferation less moderated because technologies and systems of industrial organization are applied without the institutions of understanding, regulation and governance developed in the North to control them. The global industrial frontier is a wild and open place – sometimes a profitable place to invest capital, but a dangerous place to live and work.

Work on 'natural hazards' in the 1980s (like drought, flood or earthquake) demonstrated the centrality of political economy in determining who suffers their impacts (Hewitt 1983; Blaikie *et al.* 1994). Risk is therefore the combined product of physical hazard and human vulnerability. Geological or meteorological events may occur independently of human action (although decreasingly so as anthropogenic climate change bites), but their effects are very directly mediated through human institutions, and societies that are structurally unequal.

The distribution of hazard and risk among people is therefore the outcome of those processes that determine poverty and powerlessness.

This is just as true of the hazards people face as a result of human activities in industrial and urban contexts. There are 'natural' physical hazards in urban areas, such as landslide or floods (Mustafa 2005), as well as uniquely human-made hazards, such as fire, lack of drinking water and sanitation, and air and water pollution. The question of who is at risk and who is not depends on their ability to evade or adapt to the hazards threatening welfare. Above all, this is affected by where they live. It is the poorest people who live on the steepest and most failure-prone slopes, the areas liable to flood, and who live in the most polluted urban environments. These are basic parameters of urban political ecology (Pelling 2003).

Around the start of the first decade of the twenty-first century, urban people became the majority of the world's population. In *Planet of Slums*, Mike Davis (2006) estimates that 95 per cent of the extra four billion people on earth in the next generation will live in cities. In 2004, there were eighteen cities with over ten million inhabitants in the developing world. Numerous cities are expected to top twenty million people, especially in East Asia. By 2025, three-quarters of the world's urban people will be in developing counties (Pelling 2003; Swyngedouw and Heynen 2003).

The majority of them will live in poverty, in makeshift informal settlements clustered around the peripheries of huge urban agglomerations. Marginalized in space and outside the reach of the formal state in terms of the provision of services (water, waste, transport), these slums are exposed to the sharp economic disciplines of an unfettered informal economy. In some countries huge numbers of people live in this way: over 90 per cent of the urban population in Ethiopia, Chad, Afghanistan and Nepal. The slums of cities like Mumbai, Mexico City and Dhaka hold more than ten million people (M. Davis 2006). Meanwhile, in mega-cities such as Jakarta and its surrounding municipalities, affluent people have 'managed to build a city through which they can drive in their cars without having to experience the poverty of the majority': elevated highways lined with glass tower blocks or landscapes barriers shield the affluent in their cars from the poor, just as private water supplies, waste disposal and sewage facilities protect the inhabitants of gated elite residential communities from the perils of urban life (Atkinson 1993).

And it is here, at the margins of cities and the margins of state developmental action, that human living conditions are at their worst. Earthquakes, landslides, floods, infectious disease, air and water pollution, traffic, exert a huge death toll of the urban poor. The most extreme differences in life expectancy in the developing world are between urban and rural areas, but between rich and poor in urban areas, between the gated community of the prosperous suburb and the slurried canyons of the slum (M. Davis 2006). Slum-dwellers carry a triple health burden, debilitated by malnutrition, afflicted by chronic and epidemic infectious disease, and exposed to the hazards of industrial pollution.

Although, in theory, relatively dense settlement makes it cheaper to supply services such as health or water to the urban poor, rates of mortality and sickness are high. Hazards exist in the residential environment (due to water-borne disease,

Plate 12.1 Rubbish-blocked urban stream channel, Nigeria. Problems of municipal waste disposal, poor-quality water supplies and flooding are connected and form a major challenge to raising living standards in developing-world cities.

limited water availability and poor water quality, and to lack of sewage and waste disposal), in the workplace (due to unsafe industrial practices, pollution, low wages and lack of sickness or disability provision) and in the general environment. In Metro Manila, for example, only 15 per cent of the population were served by sewers or septic tanks at the time of the Rio Conference, and 1.8 million people lacked adequate water supply or sanitation (Hardoy *et al.* 1992). Children are particularly exposed to hazards, including the hazards of workplaces.

Urban economies in developing countries are highly commodified, with many people paying directly for services such as drinking water, which cannot be accessed through common rights or household labour. In Guayaquil in Ecuador, almost half of urban residents (one million people) lack reliable sources of potable water. The distribution of water reflects the distribution of power. Those without piped supply pay high prices to water vendors for water of dubious quality (Swyngedouw 1997). The supply of water has been the site of intense social struggle. The distribution of people across a city is the result of the working of the urban economy and the circulation of capital (Harvey 1973). The location of slums and industrial plants, the occurrence of pollution and the lack of water supply reflect the outworking on the urban landscape of political, economic and social forces. Richer and more powerful residents move to safer, cleaner and more spacious areas; classically in the Victorian city, factory-owners lived upwind, their workers lived clustered near the factory gates. The same safe and salubrious locations attract wealthy urban residents today.

Plate 12.2 Village well, Nigeria. The supply of clean water is a critical development
    challenge. Millennium Development Goal 7 aims to reduce by half the
    proportion of people without sustainable access to safe drinking water. Much
    effort is being focused on urban water supply, but the problem is also great in
    rural areas, particularly in sub-Saharan Africa.

Serious and intractable pollution problems exist in urban areas in developing
countries (Hardoy *et al.* 1992). The Millennium Development Goals set specific
targets for cities, but the challenges are huge (UN-HABITAT 2006). Megacities
such as São Paulo in Brazil, Mexico City or Mumbai contain some of the most
heavily polluted environments in the world. Problems include the disposal of
solid wastes (Myers 2005; Tuan and Maclaren 2005), lack of provision for the
safe disposal of sewage and consequential pollution of watercourses and aquifers
(Showers 2002).

The disposal of the by-products of industrial processes is a particular problem
(Tuntamiroon 1985; Zhao and Sun 1986). Castleman (1981), for example,
described piles of asbestos-cement waste and the discharge of untreated waste
water outside a factory in Ahmedabad in India making building materials. There
are many such factories, innumerable instances of hazardous production processes
and inadequate regulation or control of pollution. Heavy metals (for example,
lead, mercury and cadmium) in various forms, polychlorinated biphenyls (PCBs),
hydrocarbons, organic solvents, asbestos, cyanide and arsenic are all recognized
problems. The threat to health posed by many of these chemicals is invisible,
their effects cumulative – and exacerbated in the bodies of children, and people
weakened by hunger and disease.

Air pollution is a major issue, causing both acute and chronic (low-level) health effects (Hardoy *et al.* 1992). The burning of coal and woodfuel produces smoke or suspended particulates, sulphuric acid and polycyclic aromatic carbons (the classic 'London smog' complex, still a problem in many Third World cities) (Schwela *et al.* 2006). Vehicle traffic and other hydrocarbon combustion produce photochemical pollutants (including hydrocarbons from evaporating petrol or other sources, nitric oxide, nitrogen dioxide, ozone and aldehydes and other oxidation products: the classic 'Los Angeles smog' complex; see Plate 12.4). Other common air pollutants associated with vehicle use include carbon monoxide and lead. Respiratory disease is a major killer in Third World cities: in Bangkok there are estimated to be 1,400 deaths a year due to airborne particulates (Hardoy *et al.* 1992). Industrial emissions can be very significant: Boon *et al.* (2001) describe how the release of sulphur dioxide from a single copper smelter in Peru exceeded the total emissions of Norway, Sweden, Denmark, Finland and the Netherlands, with serious health effects on the 70,000 residents of the nearby town of Ilo (respiratory problems, effects on the skin, vomiting and headaches). In 1999, before action by both city and corporation to reduce pollution, peak SO_2 concentrations at night reached 10,000 $\mu g/m^3$, twenty times World Health Organization (WHO) ten-minute guidelines (Boon *et al.* 2001).

The very speed of economic growth in some developing countries threatens the future sustainability of city life for the poor. In Vietnam, for example, GDP

Plate 12.3 Cement factory, Nigeria. Cement is critical to building projects in the developing world, and often has to be imported. Yet, like so many other products, its manufacture can bring significant environmental problems, generating large amounts of CO_2 as well as particulate matter, nitrogen and sulphur oxides, volatile organic compounds and heavy metals.

rose by 8.9 per cent per year in the early 1990s, primarily because of industrial and construction sectors (Drakakis-Smith and Dixon 1997). Problems included acute slum crowding, increasing social stratification, lack of water, sanitation and rising air pollution.

The rapid industrialization of China attracts particular attention among analysts of the relationship between development and environment (Liu and Diamond 2005). By the 1990s, China was the world's biggest coal producer (Nolan 2005), and the small-scale mining sector in particular caused extensive environmental externalities (Andrews-Speed *et al.* 2003). Heavy particulate pollution characterized many Chinese cities in the 1970s; acid rain was a problem in the 1980s. Rapid rises in the number of motor vehicles – still only one-tenth of US levels of 'hyperautomobility' (Martin 2007) – led to classic photochemical smogs owing to nitrous oxides and carbon monoxide in the 1990s (He *et al.* 2002). By 2000 China had the highest SO_2 emissions in the world (Liu and Diamond 2005). Rapidly industrializing regions share with the world's rustbelt's industrial heartlands susceptibility to problems such as acidification (see Figure 12.2).

Carbon dioxide output also rose sharply in China, as its industrial economy and income levels grew, although per capita emissions in 2005 were still at 10 per cent of those in the USA (Nolan 2005). There have been major top-down government efforts to control pollution – for example, of sulphur dioxide and vehicle

Plate 12.4 Traffic in Mexico City. Mexico City, built on a dry lake bed surrounded by hills, has legendary air pollution problems. Vehicle traffic is a major culprit, with hydrocarbons, nitrogen dioxide, ozone and other compounds creating a heavy smog. Traffic levels continue to rise, despite the creation of an underground metro system. Photo: W. M. Adams.

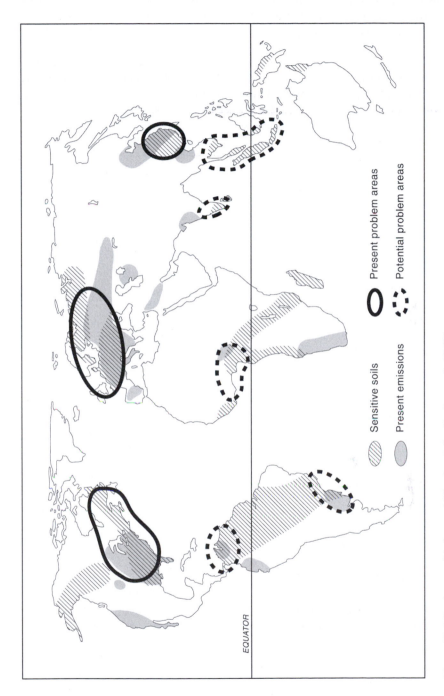

Sensitive soils
Present emissions

Present problem areas
Potential problem areas

EQUATOR

Figure 12.2 Regions at risk from acidification (after Rodhe *et al.*, 1988)

emissions. China's overall performance in reducing pollution while increasing gross domestic product is variable (de Groot *et al.* 2004).

Urban air quality can also be affected by processes outside the city. The forest fires of 1997–8 in Sumatra and Kalimantan (associated with low rainfall because of an El Niño event) caused massive palls of smoke and haze across Indonesia, Malaysia and Singapore (Byron and Shepherd 1998). The health of twenty million people was adversely affected (especially that of the old and the young) through asthma and upper respiratory tract infections, and there were serious economic impacts through business shutdowns, airport delays, accidents and depressed tourist revenues. These were not 'natural' fires, but were often started deliberately, not least as a cheap (and government-sanctioned) way to clear forest for oil-palm plantations. The hazard of forest fires in Indonesia was as much a product of modernity, its system of governance and regulation and its capitalist economy, as any more obviously technological pollution hazard.

The distribution of risk

Ulrich Beck's *Risk Society* (1992) has been much discussed by both sociologists and environmentalists. For sociologists, the work was important for the way in which he addressed the nature of modernity, and offered an alternative to the fragmentation and relativism of postmodernism. Environmentalists identified with his vision: 'a new twilight of opportunities and hazards comes into existence – the contours of the Risk Society' (p. 15). Beck took seriously environmentalist concern that the scientific-technical complex was producing toxic materials and life-threatening processes that were novel in their longevity, the scale of their impact, their invisibility and their complexity.

Beck's work, both in his book and in subsequent discussions of modernity and reflexivity (e.g. Beck 1995; Beck *et al.* 2004) has been widely reviewed, and criticized (Elliot 2002; Mythen 2005). The very popularity of Beck's ideas did not endear him to academic sociologists, and, in retrospect, it is probably better to see his work as deliberately provocative and conceptual rather than empirical (Matten 2004a). Beck (1994) suggested that there is a break within modernity, a transition from classical industrial society towards a new (but still industrial) form of Risk Society. He suggested that the end of the twentieth century was seeing a developmental phase of modern society characterized by the fact that 'social, political, economic and individual risks increasingly tend to escape the institutions for monitoring and protection' (p. 5). Global ecological crisis could, therefore, no longer be thought of as 'environmental problems' in the world outside, but needed instead to be understood as 'a profound institutional crisis of industrial society itself' (p. 8).

In these arguments, Beck reflected the thrust of 1980s environmentalism, and made an argument that seemed singularly prescient, in that his book was published just a month after the nuclear accident at Chernobyl in the Ukraine, which trailed a swath of radioactive caesium across the Soviet Union, eastern Europe, Scandinavia and the UK, as far as the eastern seaboard of the USA.

Beck suggested that the universal principles of modernity (civil rights, equality, functional differentiation, methods of argumentation and scepticism) were now in conflict with existing industrial society, and as a result industrial modernity would be forced to change by a process of 'reflexive modernization'. Far from a retreat from modernity, Beck believed that modernity itself was being 'radicalized *against* the paths and categories of the classical industrial setting' (Beck 1992, p. 14; emphasis in the original).

Beck's concept of Risk Society is generally seen in the context of the idea of 'reflexive modernization' proposed by Anthony Giddens (Beck *et al.* 2004). Modern societies have to find rational solutions to the environmental hazards that they have brought into being (Matten 2004b). The task of eliminating risk reveals 'a vacuum of institutionalized political competence' in existing governance institutions (Beck 1992, p. 48). To tackle risk, those institutions have to go beyond the established scientific, legal, economic and political strategies of industrial society, placing reliance on experts and on the bureaucracy to regulation and control risk. Risk Society demands 'institutional innovations' (Beck 1992). Risk Society challenges industrialism, the social hegemony of science and the legal system: Beck (1995) asks what good is a legal system 'that prosecutes technically manageable small risks, but legalizes large-scale hazards on the strength of its authority, foisting them on everyone, including even those multitudes who resist them?' (p. 69).

Beck is clearly right that the modern techno-economic system produces risks, at the same time as it produces wealth. The risks of modernization are 'a *whole-sale product* of industrialization and are systematically intensified as it becomes global' (Beck 1992, p. 21; emphasis in the original). He is surely also right that the novel form and scale of industrial risks (for example, the long-term health effects of complex organic compounds or the trans-boundary pollution risks of nuclear industrialization) make it hard for conventional governance institutions to control them. Techno-scientific risks are not readily calculable, and are not limited by traditional boundaries of location or time span or social groups (Matten 2004b).

What is more controversial is Beck's further suggestion that risks are no longer limited to particular social groups. He argues that risk is individualized: each and all are exposed to techno-scientific risk. He maintains that the risks of modernization will eventually rebound, even on those who profit from them (the risks of a major nuclear accident being hard for even the rich and powerful to avoid, for example), and are therefore, potentially at least, universal. The threat of late modern risks is so enormous that they cut across class lines. Yet both perception and experience of risk are translated through socio-economic status, geographical location and cultural values and belief (Mythen 2005). Hazards reinforce existing class divisions rather than transcending them (Marshall 1999). Not everybody experiences the risks of modernization in the same way, or to the same extent. Some classes of people are more exposed to risk than others. The impact of risk therefore remains fundamentally affected by social stratification. The idea of techno-scientific risk as a boomerang that comes back to afflict the societies that create it may be true, but some societies (in the developed world) and some class of people (those who can move, get treatment and invest to protect themselves) are much less vulnerable than others: as Mythen (2005) puts it, 'the boomerang effect may bruise some, but it will administer knock-out blows to others' (p. 141). The Risk Society thesis fails to make the case that risk effects can be separated from class: political ecology still matters.

There is a distributional logic to risk, both between rich and poor in industrialized countries, and internationally between developed and developing countries. There is a danger that the pursuit of development leads to neglect of environmental justice (Merryfield and Swyngedouw 1996; Low and Gleeson 1998). In economic terms, it is logical to locate hazardous pollution where costs are low, where the poor are willing to accept low levels of compensation; thus community leaders are faced with the dilemma of 'trading their people's environmental health in return for basic material security' (Low and Gleeson 1998, p. 119). Debate in the USA about 'environmental racism' drew attention to the differential exposure of minorities, native people and people of colour to environmental risks, particularly in the location of toxic waste dumps and other polluting facilities. The manifest unfairness of the outcome of corporate and municipal decisions about noxious facilities gave rise to the environmental justice movement in the USA (Harvey 1996b).

Environmental justice was an important element in wider concern about environmental health, a key element in post-war environmentalism (Hays 1987). The movement has been built around particular cases, many of them highly disturbing and high profile. One such was the protest about Love Canal in Buffalo, New York, in 1977, where, between 1942 and 1952, the Hooker Chemical and Plastics Corporation buried about 22,000 tons of toxic waste. Subsequently a school and houses were built on the landfill. By the late 1970s, severe health problems were being experienced in the area, and children were being born with severe handicaps. Following protest, a Federal State of Emergency was declared, and residents were rehoused. The Environmental Protection Agency sued Occidental Petroleum (which had taken over the Hooker company), which paid a substantial sum in restitution in 1995.

Love Canal was followed by many other protests in the 1980s and 1990s about the location of hazardous facilities and threats to human health and well-being – for example the opposition to the proposed Brooklyn Navy Yard incinerator in New York (Gandy 2002). By the 1990s the environmental justice movement was well established in the USA, with strong offshoots elsewhere (Harvey 1996b). In response, the US Environmental Protection Agency established an Office of Environmental Equity in 1992, and in 1996 an Executive Order required every federal agency to consider the effects of its programmes on the health and well-being of minority communities (Low and Gleeson 1998).

Trading hazard

There is a wider geographical scale at which the locational logic of industrial hazard and pollution works. Just as the logic of capitalism at the urban scale locates noxious facilities in poor districts (and ensures that they remain poor by degrading the environment), so the same logic at the global scale dictates that the most efficient logical place to locate noxious industrial processes is not in the industrial world's increasingly clean and self-proclaimed 'sustainable' cities, but at the feet of the poor in developing countries. There, determination to over-come poverty and lax environmental regulations attract hazardous industries 'like magnets', and create an explosive mixture: 'The devil of hunger is fought with the Beelzebub of multiplying risks', and to poverty is added the 'destructive powers of the developed risk industry' (Beck 1992, p. 43). In the face of globalized corporate power, the nation state (particularly in the developing world) is not necessarily a particularly powerful player in regulating environmental risk.

It can be argued that Third World countries without strict pollution legisla-tion can achieve a competitive advantage over industrialized countries (Walter and Ugelow 1979). There is some evidence for this 'pollution haven hypothesis' – for example, in southern Africa (Nahman and Antrobus 2005). Lack of pollution controls (like cheap labour) cuts the cost of production, or rather transfers costs to the host environment and community. In a sense, therefore, pollution is a hidden subsidy for developing-world industry. The hazards and costs of industrialization without pollution control can, therefore, seem to have a certain attractiveness. When this is combined with the inertia, inefficiency and corruption with which Third World government bureaucracies are plagued, the result can be – as in the case of Bhopal – disastrous.

Low pollution control standards in the developing world offer attractions to transnational companies faced with rising production costs in industrial coun-tries. Simple economics is fundamental to many incidents of chronic industrial pollution. The costs of pollution control tend to be passed either forwards to product prices (thus making manufactured goods more expensive), or backwards (for example, through development and adoption of new technology), reducing returns on capital. Pollution control is, therefore, often unattractive to the individual corporation, unless the efficiency gains (for example, through recovery of saleable waste products or greatly improved process efficiency) are large, or unless the regulatory regime makes change essential. Yet, ever since the early days of British industrialization, industrial corporations were aware of the potential impacts of environmental damage on their earning power, and yet sought to off-load liability and disputed nascent attempts by the state to regulate their activities (Enzens-berger 1996).

Of course, there is an argument that it is in the interests of a corporation to improve its efficiency, and pollution may demonstrate inefficient processes. It is a basic tenet of ecological modernization that there is money in it for business to become cleaner, because thereby it becomes more profitable. There may in particular be an early-mover advantage. Thus one can imagine a benign circle, where individual corporations compete to clean up their act (Dryzek 1995; Vlachou 2004). As discussed in Chapter 5, some 'greening' of industry has gone on as a result of processes like this, and there has been some measure of self-reg-ulation – for example, by the mining industry (see below). However, observers note that forms of regulation adopted tend to be restricted to those that favour corporate profit and growth, and that competition takes place between firms to bend government regulation in directions that favour their particular situation (Vlachou 2004). Matten (2004b) argues that corporations all too often engage in self-regulation to avoid stricter imposed regulation (indeed, governments sometimes explicitly threaten enforced regulation in the hope that self-regulation will be offered instead). State environmental regulation can strengthen the competitive advantage of capital-rich and technologically strong companies within industrialized economies, and can provide a means to establish *de facto* trade barriers against corporate competitors in the developing world. In all these cases, the capacity of the state to enforce environmental regulations is a critical factor in the control of pollution.

There are also corporate interests in the definition of risk. The underestimation of risk (and attempts to estimate risks where no estimate is possible) is a strategy useful to corporate interests, since it serves 'to legitimate the imposition of risks on the population' (Murphy 1994, p. 142). Wynne (1992) demonstrates that the common use of the word 'risk' actually refers to several quite distinct things. He distinguishes between risk (where the chances of something happening are known); uncertainty (where the chances are not known, at least in detail); ignorance (where it is not even known what the problems are, let alone the chances of their happening); and indeterminacy (where the outcomes are inherently not predictable). Formal scientific techniques of 'risk assessment' alone cannot address all these. Yet it is risk analysis carried out by scientific-technical experts, either within industry or government, that is given weight in setting regulatory targets.

Developing countries often lack the environmental and anti-pollution safeguards that have become standard in the industrialized world. Walter and Ugelow (1979) showed an inverse relationship between level of development and rigour of environmental policies in 145 countries in 1976. There is, therefore, an incentive for transnational companies to move industrial plants that are highly polluting to locations in the Third World as a direct result of environmental protection policies in industrialized countries. Thus Suckcharoen *et al.* (1978) discussed the location of a caustic soda factory in Thailand by a Japanese company in 1966 in response to weak local environmental protection legislation. The risk was of pollution by methyl mercury in aquatic ecosystems, and especially in fish, which formed half the animal protein intake in the average Thai diet. The location of this factory, and its pollution, outside Japan was in part a response to tight Japanese regulations following appalling organic mercury poisoning among the fishing community at Mumacoto and Niigata in Japan from the Chisso Corporation's fertilizer factory (Vlachou 2004). This brought about 'Minamata disease', extensive birth defects in children, and had been a major factor in raising consciousness of the dangers of industrial pollution in industrialized countries (D'Itri and D'Itri 1977).

Many developing countries have weak or poorly resourced regimes of environmental regulation, and hence effectively trade off hopes of future economic security against present risk. The movement of polluting industries to the developing world is seen as important industrial investment on all sides: by governments, international banks and financial houses, and global corporations. The 'traffic in risk' (Low and Gleeson 1998) may be attractive to developing-country governments eager for investment, and to industrialized-country governments, which are able to externalize industrial risks by moving hazardous elements in the production process outside their borders. 'Sustainability' within one jurisdiction is being achieved by exporting risk and hazard elsewhere. It is also attractive to corporate strategic planners seeing low-cost locations with restricted environmental regulation. Lack of facilities for identifying or measuring environmental hazards and costs means they are often underestimated or ignored. Long trans-global production chains tend to restrict transparency about the conditions of production. It can seem profitable to 'manipulate and bribe politicians and bureaucrats rather than worry about technological innovation, efficient resource use or pollution control' (Gadgil and Guha 1995, p. 31).

There is evidence to support theoretical arguments about the attractions of lax pollution controls in the developing world. Walter and Ugelow (1979) presented data on the percentage of capital expenditure spent on pollution control by US-based multinationals in the USA and overseas. In primary metals industries, pollution control constituted 21 per cent of domestic capital outlays, but less than 10 per cent of overseas capital, and proportions were similar for other industries – for example chemicals (9 per cent in the USA, 5 per cent abroad). Industrial plants converting raw materials into more complex products can be particularly polluting, because the costs of pollution control can be high compared to the value of the product. The manufacture of pulp and paper is an important case in point. Christiansson and Ashuvud (1985) discussed the potential environmental effects of a paper mill in Mufundi District in Tanzania. Standards do vary between the pollution risk from plants in the First and Third Worlds. Walter and Ugelow (1979) found that US-based multinationals in the USA spent 22 per cent of capital expenditure on pollution control in US plants, and 12 per cent abroad.

Raymond Murphy (1994) argues that manufacturers have a good idea of the dangers implicit in their operations, although they seek to shield this knowledge from environmental movements and the public, and sometimes from the state: 'Transnational companies, often in complicity with state organizations, have overestimated the safety of their factories in order to convince the public to allow them to pursue their search for profit' (p. 137). Accidents, therefore, are in a sense not unexpected, but events whose likelihood of occurrence is carefully calculated by corporate planners. Techno-scientific risk emerges from modernity, an integral element in the evolution of industrial society.

A particular example of this process is the trade in toxic waste. There is a growing industry breaking ships and reprocessing shipped electrical waste in Asia. Mobile phones, computers, fridges and other consumer devices, many of them rendered obsolete by technical advance and the relentless dictates of fashion, are dispatched by the container load to the developing world. In 2002 the USA exported over ten million obsolete computers (Iles 2004). Waste is either diverted from landfill around developed-world cities, simply because it is cheaper to send it to India or

China, or sent for recycling (often as a result of a search for 'sustainability' on the part of their former owner). Working conditions in these industries are very poor, with workers exposed to hazardous compounds and heavy metals without training, protective clothing or any form of social insurance.

Economists have made the case that it makes sense to locate toxic and hazardous waste dumps in poor countries, not least from within the World Bank in 1992 (Rich 1994; Harvey 1996b). Because of the rapid and unregulated growth in the trade in toxic and hazardous wastes, it has become an increasingly urgent international environmental issue since the 1980s. In 1986 the British company Thor Chemicals began importing waste from the USA and Europe to its reprocessing plant in South Africa. By 1988 mercury levels 1,000 times the WHO standards were discovered in a river 50 kilometres away, and by 1992 workers had started to die in the plant, and others had become disabled. The plant closed in 1994, but the pollution remained (Lipman 2002).

There are many such examples. In 1988 Guinea-Bissau was offered a contract for $600 million (four times its GNP) to dispose of 15 million tons of toxic waste over five years (Lipman 2002). That deal did not go though, partly because of opposition warned by a story from nearby Nigeria. In 1988 Greenpeace revealed that two Italian corporations had paid a businessman in the town of Koko in the Niger Delta in Nigeria about $100 per month to store 18,000 drums of hazardous wastes disguised as building materials (including asbestos, dimethyl formaldehyde and PCBs). Workers clearing the site suffered illness, and the town was virtually abandoned (Adeola 2001; Lipman 2002). Meanwhile the ship the *Karin B* toured the ports of Europe in search of a place to unload the toxic waste where it could be properly processed.

While the evasion of proper practices was particularly blatant in the Koko case, and the 1991 Bamako Convention subsequently banned waste imports to Africa, international hazardous waste transfers remained standard. Greenpeace estimates that more than 2.5 million tons of hazardous waste were exported to developing countries between 1989 and 1994 (Lipman 2002). The trade is covered by the 1989 Basel Convention on the Control of Transboundary Movements of Hazardous Wastes, and was addressed in *Agenda 21*. In 1994 the parties to the Basel Convention agreed to ban all export of hazardous waste from countries in the Organization for Economic Cooperation and Development (OECD) (that is, developed countries) to non-OECD countries. The door to international traffic in waste has, therefore, closed somewhat further. The World Trade Organization's determination to remove restrictions on free trade offers little hope for tighter regulation on the global market in pollution and environmental hazard.

Metals, oil and environment

The impact of the extraction and processing of mineral resources in terms of environmental degradation and social impact has long been of particular concern to environmentalists. Many standard processes create substantial risks: the treatment of bauxite with caustic soda to produce alumina creates caustic alkaline slurry (in Jamaica, Bell (1986) reported that the aluminium industry produced 13 million tonnes every year). The common way to extract gold involves the use of cyanide, lethal to fish in even small quantities. Tailings from such processes are expensive to process so they are stored in slurry ponds, at risk of reaching watercourses or aquifers. Many mines create huge quantities of overburden, often dumped in streams and rivers. Such practices are common even in industrialized countries. Orr (2007) describes the massive scale of open-cast coal mining in the Appalachians of the USA by corporations such as Massey Energy, Inc. and Arch Coal. By 2007 they had literally removed the forested mountains of West Virginia and Kentucky (456 mountains across 6,000 square kilometres) and dumped the overburden into valleys with 2,400 kilometres of streams over the preceeding twenty-five years). Coal washing fills containment ponds with 'billions of gallons of a dilute asphalt-like gruel laced with toxic flocculants and heavy metals' (p. 289); ponds are abandoned and unmanaged when the coal seams are worked out. The impacts of mines and associated facilities can extend over long distances (through water or air pollution). Because mines are often in remote locations, effects on people can be both unrecognized and uncompensated.

In the mid-1980s, an Australian corporation, Broken Hill Proprietary (BHP), developed an opencast gold- and copper-mining operation in the western mountains of Papua New Guinea (Low and Gleeson 1998). The government of Papua New Guinea owned 30 per cent of the shares of the operating company, Ok Tedi Mining Ltd. The mine employed 1,700 people, and supplied over 16 per cent of national export earnings. However, it also released about 80,000 tonnes of mine tailings daily into the Tedi River (which drains into the Fly River system). The tailings contain a suite of heavy metals. The original plan was to hold these within a tailings dam, but this failed in 1984 in a massive landslide, and they have since flowed directly into the river (Hyndman 1994). Over the life of the mine, some 250 billion tonnes of waste will have been dumped, and, despite sanguine predictions of the potential for containing the downstream effects (Petr 1979), floodplain forest and river fisheries have been drastically affected; the bed of the upper river has been raised by several metres.

What neither BHP nor the government appears to have realized (or thought to check) was that substantial numbers of people lived downstream of the mine. They had been excluded from benefits in the original mining agreement. Between 1994 and 1996, lawyers sued the company in the Australian courts on behalf of 30,000 people living downstream along the Ok Tedi and Fly rivers (Banks and Ballard 1997). An out-of-court settlement was eventually reached, involving resettlement, worth $350 million placed in a trust fund for compensation. However, the mine continued to operate, and in 1999 it was announced that none of the technical options on the table (including closure) would significantly improve downstream conditions. BHP withdrew in 2002, transferring its 52 per cent stake to Papua New Guinea Sustainable Development Programme Ltd. The mine is due to close when the ore body is exhausted, probably around 2010.

The environmental impacts of the Ok Tedi mine are not the result of some unfortunate accident. They are integral to the ore extraction process, and are inevitable unless pollution control facilities are built. Such facilities cost money, and their construction cuts profits, and may not (as in the case of the Ok Tedi tailings dam) be technically feasible. A major attraction of mining in the Third World is that capital and operating costs can be kept low. Limited statutory protection for workers and weak requirements for the control of pollution (whether in the environment or in workers' bodies) often mean that costs are lower than for an equivalent First World mine, and profits correspondingly greater.

There are many mining projects with similar impacts on downstream communities. Banks (2002) describes disputes, elsewhere in Melanesia, at Porgera in Papua New Guinea, and at the Grasberg mine in Irian Jaya (operated by PT Freeport Indonesia) and the Gold Ridge mine in the Solomon Islands (operated by Ross Mining). He notes that, while problems are generally portrayed as purely environmental, the range of issues involved includes economic, social, political and cultural concerns of local indigenous communities – indeed, disputes are as much about control of resources and people's futures as about the specifics of environmental damage. Yet such damage can certainly be extensive. Baluyut (1985) describes the impacts of three gold and copper mines on the Agno River in the Philippines that used cyanidation and flotation to obtain ores. Large volumes of tailings have been impounded, but failure of retaining dams released fine sediments and toxic materials into the Agno River. Downstream ecosystems, particularly coastal mangroves, have been adversely affected by sedimentation, and heavy metals (mercury, cadmium and lead) have accumulated in fish and shellfish downstream.

Oil and gas exploration and extraction bring with them well-known environmental problems. These have been prominently reported in a number of areas, particularly in the Niger Delta in Nigeria. Aluko (2004) describes this region as 'in turmoil, restive, poor, backward and neglected' (p. 67). Crude oil pollution has been a critical problem in the Niger Delta for decades (Ikporukpo 1983). Environmental degradation has taken place against the background of Nigeria's turbulent politics (back to and including the Biafran War), and the complex policies of fiscal federalism: oil constituted over 50 per cent of the revenue to Nigeria's Federal Government for the first time in 1972, and has since mostly been between 70 and 80 per cent (Ikporukpo 2004). State governments, politicians and businessmen have all fed at the one pot.

The impacts of oil and natural gas flaring on the delta region include loss of crops, sterilization of soils, the death of fish and air pollution. In the decade 1970–80 Nigeria experienced eighteen major oil spillages involving over one million barrels of oil. The problem was not addressed by either the Nigerian government or the oil corporations (of which the most prominent was Shell). Both state and corporation won a reputation for being high-handed, arbitrary and periodically violent in securing access to land for facilities and in ignoring protest. Environmental degradation associated with oil extraction and gas burning persisted through the 1980s and 1990s. There were 300 spills per year in Delta and Rivers States in 1991 and 1993, in addition to massive gas flaring, and repeated disputes over land rights with local communities (Frynas 2000).

Protests against pollution, and infringements of human and civil rights, particularly by the Ogoni people, met with heavy-handed treatment by government forces. Demands for a separate Ogoni state (and the share of oil revenue this would bring) were met with violence (Simonsen 1995; Rowell 1996). The judicial murder of the Nigerian playwright Ken Saro-Wiwa and other activists in 1995 brought the environmental and political performance of oil companies, and particularly Shell, into the international spotlight. Shell subsequently extended its approach to corporate social responsibility and sustainability, and changed the way it dealt with communities in the Niger Delta (Ite 2004). However, social and political unrest and environmental degradation remain very significant problems in the area.

Environmentalist and human-rights NGOs were vociferous in their opposition to the more destructive activities of mineral companies in the 1980s and 1990s: for example, Friends of the Earth ran a major campaign against Rio Tinto's plans to exploit mineral sands in the forests of south-east Madagascar in 1995, their campaigner Andrew Lees losing his life there. In common with other large corporations facing pressure from shareholders and public opinion in the North, most such companies took steps to become more 'green', or at least to present themselves as doing so (as discussed in Chapter 5).

Biodiversity conservation NGOs began a policy of engagement (e.g. Rosenfeld *et al.* 2000). Some companies in the minerals and energy sector, such as Rio Tinto, Placer Dome and BP, adopted a policy of seeking out and engaging their critics in the search for legitimacy (e.g. Mulligan 1999). There was a major collaboration between a series of NGOs (Conservation International, Fauna & Flora International, IUCN, The Nature Conservancy, Smithsonian Institution) and several leading oil corporations, including BP, ChevronTexaco, Shell and Statoil (Energy and Biodiversity Initiative 2003). There is increasing interest by both corporate and NGO strategists about the best ways to engage and the degree to which common aims can be defined. Most analysts welcome the new initiatives, but wait to make judgement on the significance of their benefits in terms of sustainability (Mulligan 1999).

The problem of pesticides

Pesticides (the word embraces insecticides, herbicides, fungicides and other biocidal agents) bring the world of chemical hazard from factory to farm. They are central elements in the modernist strategy of increasing food and commodity production in developing countries to deal with hunger and poverty. They have also been enormously effective in reducing the incidence of disease spread by insects or other vectors. The effectiveness of pesticides in controlling crop pests and disease vectors is one of the most commonly cited pieces of evidence in support of a technology-driven strategy of development: do they not allow the hungry world to be fed, and the sick to be protected from disease?

The scale of pesticide use, and the speed with which that use developed following the Second World War, are staggering. In Africa, sleeping sickness, and an equivalent disease of livestock, *nagana*, was a major scourge over ten million square kilometres of Africa, in thirty-four countries. It is spread by the hefty biting tsetse fly (*Glossina*). From the 1940s tsetse began to be controlled in Central Africa by aerial dusting with the organochlorine pesticides DDT and gamma-BHC (lindane). In due course this was replaced with spraying of these pesticides and dieldrin, from the ground using manual or vehicle sprayers on tsetse resting sites, and more recently with fine aerosol spraying of synthetic pyrethroids from aircraft over larger areas (Ford 1971; Matthiessen and Douthwaite 1985; Ormerod 1986). Fortunately, these methods have now given way to a much more targeted technique using traps coated with pesticides baited with chemicals attractive to tsetse (cheaper, more effective and less risky).

Crop losses through pest attack are a major problem in developing countries. In the 1970s in-field crop losses were estimated to be 42 per cent in Africa: 13 per cent from insects, 13 per cent from disease and 16 per cent from weeds (Ghatak and Turner 1978). In Asia in-field losses were 43 per cent, in Latin America 33 per cent. Food losses in storage are also considered a major factor in hunger and poverty (see Plate 12.5). The larger grain borer, a beetle pest of stored maize in particular, was accidentally introduced to Tanzania in the late 1970s and has spread rapidly through both East and West Africa despite international control efforts. Standard control involves fumigation with methyl bromide (a serious ozone-depleting chemical) or phosphine (Taylor and Harris 1994).

Pesticides are one of the linchpins of the Green Revolution, which depended on the spread of high-yielding hybrid crops capable of much greater yields than local crops. Such crops demand inorganic fertilizers and often irrigation, and are particularly vulnerable to insect and fungal attack and weed competition. The Green Revolution depended on the availability of cheap and effective pesticides. The distribution and sale of pesticides is eagerly promoted by transnational agribusiness companies.

The Green Revolution has been the subject of endless academic debate. Work on crop breeding was primarily done within the public sector in the International Agricultural Research Centres (IARCs) – for example, the International Institute for Tropical Agriculture (IITA) at Ibadan in Nigeria, the International Rice Research Institute (IRRI) at Los Baños in the Philippines and ICRISAT in Pakistan, although increasingly new seed development (now drawing on genetic manipulation technologies) has been privatized (Lappé and Bailey 1999). The Green Revolution required farmers to purchase manufactured chemicals and other inputs to guarantee their crop, 'devaluing the "reproductive power" of nature by substituting the "productive power" of industrial inputs' (Yapa 1996, p. 82).

Higher-yielding and more fertilizer-responsive varieties of several major crops (particularly rice, wheat and maize) have transformed food production in parts of the Third World, particularly in East and South Asia, although much less has been achieved in Africa (Richards 1985; Lipton and Longhurst 1989; Evenson and

Plate 12.5 Millet attacked by pests in storage, Nigeria. Food losses in storage are a major factor in hunger and poverty. Insects such the larger grain borer, a beetle pest of stored maize in particular, has spread rapidly through both East and West Africa. Control usually involves intensive use of pesticides. Photo: W. M. Adams.

Gollin 2003). The impact of the package has been fiercely criticized, for example, because it helps rich farmers more than poor farmers or landless labourers, although others point out that even the poor can gain from increases in demands for labour on richer neighbours' fields and crops (Lipton and Longhurst 1989).

The economic costs and benefits of pesticides are complex to calculate, particularly the benefits from crop damage avoided, and particularly by prophylactic use before pest outbreaks occur. Ghatak and Turner (1978) note the foreign-exchange costs of importing pesticides (for few countries in the Third World have indigenous production capacity). They point out that pesticides often serve as substitutes for labour inputs (for example, in weeding), and labour is neither expensive nor in short supply in most developing countries. It is difficult to obtain data on the economic significance of pollution effects, including problems of the impact of pesticides on non-target organisms and the development of resistance of target organisms. Cox (1985) studied the economics of DDT and inorganic copper fungicides in Tanzania. These are both persistent in the environment, but the implications of this – for example, on the use of drainage water for drinking downstream – are not known, and long-term social costs cannot, therefore, be calculated.

The risks of pesticide use are, however, clear: first, the development of resistance; second, wider environmental impacts of application; third, persistence of pesticides in food, and the problem of acute poisoning, particularly of people using them (Conway and Pretty 1991; Pretty 1995; Ecobichon 2001). The development of resistance to pesticides is well documented. The problem of the rapid development of resistance to pesticides is combined with the impact of broad-spectrum insecticides and herbicides on non-target species, particularly natural predators. Classic examples are the Gezira Scheme in the Sudan, where cotton losses to whitefly kept pace with massive increases in pesticide use in the 1960s and 1970s, and the resurgence of the brown plant hopper, a pest of rice, in the late 1970s in the face of growing and intensifying pesticide use (Bull 1982). Pest resurgence following pesticide use is an acknowledged problem, and can affect adjacent unsprayed fields because of the spatial dynamics of recolonization. Decisions about the benefits and costs of pesticide use in terms of pest numbers are, therefore, more complex than is often assumed (Trumper and Holt 1998).

The problem of pesticide resistance, and the escalating cost of multiple pesticide applications, have led to new strategies to try to reduce levels of pesticide use and control crop losses instead by integrated pest management (IPM). In the Philippines, for example, a national IPM programme was launched in 1993 to train farmers in pest-management strategies that minimize chemical pesticide use (Bolido 1998). It is now clear that the brown plant hopper is controlled by a wide guild of predators, and rapid resurgence follows accidental control by pesticides of these organisms. Strategies against the brown plant hopper include breeding rice varieties with pest-resistant attributes. There is increasing interest in the possibility of IPM for cotton, which is one of the most heavily treated with pesticides (FAO 1994a). Other IPM strategies include the release of natural enemies of pests, the use of a bio-pesticide such as the bacterium *Bacillus thuringiensis* and

Plate 12.6 Transplanting rice, Bangladesh. While local varieties of rice may have some pest resistance, high-yielding varieties tend to be genetically uniform and are vulnerable without intensive pesticide use. Resistance to pesticides is a major problem in rice as in other crops. Photo: W. M. Adams.

the integrated management of fields and surrounding habitats to minimize pest damage and maximize natural predation (Pretty 1995, 2005).

IPM, and related approaches such as integrated plant nutrition, are part of the increasingly important field of sustainable agriculture (Pretty 1995). While some new technologies may be beyond the reach of poor Third World farmers, many of the modifications of agribusiness practice called for by environmentalists in industrialized economies (including organic cropping – without inorganic pesticides and fertilizers – or low-input farming) are perforce normal practice in the Third World. Indeed, ironically, they are among the kinds of 'backward' practices that agricultural modernizers from governments and agribusiness companies spent a great deal of time and money in the twentieth century trying to persuade farmers to abandon.

There have also been problems of pesticide resistance in public health applications. The Onchocerciasis Control Programme in West Africa began spraying the organophosphorus insecticide temephos on rivers in Upper Volta, Mali, the Ivory Coast and Ghana to control the biting fly *Simulium damnosum*, the vector for river blindness, in 1974. Resistance was first recorded on the lower Bandama River in the Ivory Coast in March 1980, but spread rapidly. The project switched to chlophoxim (another organophosphorus compound), but resistance to this appeared by mid-1981, and this too was withdrawn. It appears that resistance to one organophosphorus pesticide promoted resistance to others (Walsh 1985).

The second kind of issue associated with the use of pesticides is the externalities associated with their impacts on the wider environment (Conway and Pretty 1991; Wilson and Tisdell 2001). The organochlorines, such as DDT and dieldrin, were the subject of much controversy in Britain and other industrialized countries in the 1950s and 1960s because they do not metabolize, but become stored in fatty tissue. There were a number of direct poisoning incidents on farmland birds in the 1950s, and subsequent research showed that, in top predators such as birds of prey, concentrations were reached that caused physiological problems such as eggshell thinning. There were serious population declines in some species (Sheail 1985). In Nigeria, Perfect (1980) found that DDT accumulated in the soil over the four-year experimental period. There were changes in numbers and kinds of soil organisms and in the pathways and rates of litter breakdown. Perfect suggested that DDT may impair the system's capacity to regenerate fertility in fallow periods.

The misuse of pesticides can contribute to wider environmental impacts. In Guyana, the rodenticide thallium sulphate was imported in 1981 to kill rats in sugar cane plantations. It was misused as a fertilizer and in other ways, and contaminated milk and grain caused a number of deaths (MacKenzie 1987). Pesticide use rose in the Pacific, as elsewhere, in the 1970s, and there have been a number of reported cases of the death of fish from pesticide spills – for example, of lindane and DDT in the lagoon of Tokelau, or the leakage of endrin and sodium arsenate into streams feeding the lagoon on Yap (Brodie and Morrison 1984).

The third aspect of pesticide use is its persistence in food. Guatemala is one of a number of countries that have become suppliers of out-of-season vegetables and fruit for Northern markets. The US Food and Drug Administration regu-larly detains shipments of fruit and vegetables from Guatemala because of excessive pesticide residues (costing the country $18 million between 1984 and 1994), but there are no equivalent controls on food eaten in Guatemala itself (Arbona 1998).

The final, and in many ways most serious, problem of pesticide use is the threat it poses to the health of those applying them. Accidental poisoning by pesticide use is a major problem (Conway and Pretty 1991). In the 1970s, developing countries accounted for only 15 per cent of global pesticide use, but over half the cases of pesticide poisoning (Bull 1982). In the 1980s it was estimated that there were 375,000 cases of poisoning by pesticides in the Third World each year, 10,000 of them fatal (Caufield 1984). WHO data for 1990 suggest that between three million and twenty-five million agricultural workers may be poisoned by pesticides annually, with perhaps 20,000 deaths (Pretty 1995).

These problems are now ubiquitous through the developing world. In Vavu-niya in north-east Sri Lanka, for example, there were 938 deaths from pesticide poisoning in 1977, more than from malaria, tetanus, diphtheria, whooping cough and polio put together (Bull 1982, p. 44). In six countries in Central America, estimates suggest that there are over 400,000 pesticide poisonings per year, 1.9 per cent of the total population, and 4.9 per cent of those who use or are otherwise exposed to pesticides (Murray *et al.* 2002). In the high-lands of western Guatemala most farmers apply pesticides in greater doses and more frequently than is recommended, they rarely purchase protective clothing and they regularly wash out containers in irrigation canals. There are about 1,200 cases of acute reaction to pesticides a year, in addition to a greater (but unrecorded) incidence of chronic low-level exposure (particularly problematic for children, and those with immune systems suppressed by malnourishment (Arbona 1998)). In the Philippines, sales of pesticides increased by 70 per cent between 1988 and 1992, and between 1980 and 1987 there were over 4,000 cases of pesticide poisoning and over 600 deaths (Pretty 1995). A rise in pesticide purchases in the early 1970s was accompanied by a 27 per cent rise in the death rates of men of working age (Pearce 1987). Deaths up until 1976 peaked in August, the height of the spraying season. In that year double cropping began on the local irrigation scheme, and spraying in February was matched by a second peak in deaths in that month. Protective clothing was not worn, and the backpack sprayers that were used put 40 mg of active ingredients of pesticide onto operators per hour.

The danger of accidental pesticide poisoning has been made worse by the transition from organochlorine pesticides to organophosphorus compounds. The organochlorine insecticides (for example, DDT, dieldrin and lindane) are cheap and relatively safe to human users. Their replacements, organophosphorus pesticides such as pirimphos-ethyl, endosulfan or disulfoton, are hugely toxic to humans and (unlike organochlorines) can poison through skin contact. Disulfoton, for example, inhibits the production of cholinesterase, an enzyme important in nerve function, lack of which causes convulsions. Product labelling is often poor and inappropriate, often not in a relevant language and often without any provision for use by illiterate people, who are, of course, the majority. Farmers are therefore unable to read warnings about toxicity, and lack the knowledge to interpret dangers, or the equipment to apply pesticides safely. It is quite unrealistic to expect rules of application devised in developed countries – for example, using clothing proof against sprays and protective masks, or washing before eating – to be practicable in the rural Third World. Medical support in the case of poisoning is rarely obtainable.

Ulrich Beck (1992) notes the way in which the risk of pesticides is embedded within corporate decision-making:

> The 'industrial naiveté' of the rural population, which often can neither read nor write, much less afford protective clothing, provides management with un-imagined opportunities to legitimize the ways of dealing with risks that would be unthinkable in the more risk-conscious milieus of the industrial states. Management can issue strict safety instructions, knowing they will be unenforceable, and insist that they be obeyed. This way they keep their hands clean, and can shift responsibility for accidents and death to the people's cultural blindness to hazards, cheaply and in good conscience.
>
> (p. 42)

As controls on pesticide use tighten in the industrialized world, Third World markets are increasingly attractive to producers, particularly when pesticides banned at home can be exported to countries where environmental controls are more lax. In 1979 25 per cent of the pesticides exported by US companies were either banned or unregistered in the USA (Caufield 1984). However, there has been growing international concern about the spread and longevity in organisms and the environment of persistent organic pollutants (POPs), including a number of synthesized pesticides (aldrin, dieldrin, DDT, chlordane and endrin), as well as dioxins and PCBs. There is now a process led by UNEP to negotiate an international agreement banning such chemicals (McGinn 2000).

The expansion of pesticide use in the Third World has been an integral part of the development process. It is sanctioned and promoted by development agencies, and financed by First World loans. The industry is run from the industrialized world, and the expansion of pesticide use in the name of development is good business. There are both ecological and economic costs to pesticide use, although these are difficult to identify, and it is clear that at certain times and in certain places these costs will outweigh the benefits of increased yields and disease control. There are also human costs, particularly from those who apply poisons unprotected and in ignorance of their toxicity, and these are not always borne by those who stand to benefit. Higher productivity may mean cheaper food for urban consumers, but maybe also illness and perhaps an early death for a peasant farmer. Pesticides may raise yields, but if they also raise the cost of production, and take the farmer onto a treadmill where pest resistance demands new and larger pesticide applications, their benefits may be a cruel illusion.

Pesticides are a brilliant product of the modern techno-scientific system, long regarded by conventional development thinkers as an essential element in strategies to promote economic growth and freedom from hunger in the Third World. They epitomize developmentalism's disconnection of people from nature (Pretty 2002). They offer serious technological hazards that are inseparable from their intended function. Like industrial pollution, they are just one element in the complex equation linking poverty and environmental quality.

The hazard of development

Development is a two-edged sword, promising to hack away at the choking creepers of poverty, but at the same time bringing with it unrecognized, unregulated and often deeply hazardous change. Furthermore, the risks development creates are not distributed uniformly, but are concentrated in space and time. In cities, hazard and risk are concentrated in the zone occupied by the poor. On top of the burdens of ill-health and poverty they bear the hazards of technology's products in their workplace (if they can find work) and in the degraded, crowded and unhealthy environments where they live. Low governance capacity conspires with the naked urge of capital to treat the environment and the bodies of the poor as externalities, absorbing the hard edge of techno-scientific risk.

Internationally, the logic of capitalism that drives industry to relocate in search of cheap labour and resources and the freedom to externalize the costs of production is inherent to capitalism, but the profitability of relocation to developing countries is a relatively recent phenomenon (Marshall 1999). It is a function of increasing globalization, a revolution in communications and the mobility of capital. The global capitalist system 'thrives on passing on the costs of environmental degradation to the ecosystem people of the Third World' (Gadgil and Guha 1995, p. 122).

By the end of the twentieth century industrialized countries had begun to evade the risks they had conjured up with their industrial technology by moving them elsewhere. One effect of the search for sustainable development in industrialized countries has been the export of unsustainability offshore. Even as sustainable development became part of the language of developed-country governments, their citizens were importing the products of new factories in new industrial zones on the global periphery. The gleaming products that power the West's insatiable consumer demand show little or nothing of how they were made. Child labour, sweatshops, industrial pollution and urban poverty are far away down long chains of trade that are only slowly being made visible.

Summary

- Ulrich Beck's *Risk Society* offers a challenging analysis of modernity and environmental hazard that is directly relevant to debates about sustainable development. Risk Society is the product of change forced on modernity by global ecological crisis. Techno-scientific risk is ubiquitous, arising from processes of wealth creation, yet hidden from straightforward perception.
- Environmental risk is unevenly distributed and unequally shared. Environmental

justice, environmental racism and environmental class are critical issues for sustainable development, and relevant at both the global scale (for example, in the trade in hazardous wastes) and the more local scale (for example, the location of polluting industrial plants).

- Industrial and urban pollution are major problems in many parts of the developing world. While in theory it can be argued that industrial pollution declines over time with economic growth, environmental hazards of development are both real and persistent.

- Mineral extraction and manufacturing are both forms of industrial investment where pollution is endemic, and where poor environmental and employment regulation effectively offer an incentive to pollution and hazard creation.

- The integration of hazard and potential development benefit is well demonstrated by the case of pesticides. Accidental pesticide poisoning, environmental impacts and pest resistance to biocides all represent environmental or social hazards that can balance or outweigh the planned benefits from reduced incidence of food losses to pests.

- The existence of environmental hazards from pesticides or industrial pollution is integral to the modern project of development through industrialization. Sustainable development challenges the assumptions and methods of conventional developmentalism, but there are important questions as to strategy between the confusion of mainstream sustainable development (MSD) and the radicalism of countercurrents to that mainstream.

Further reading

Beck, U. (1992) *Risk Society: towards a new modernity*, Sage, London.

Beck, U. (1995) *Ecological Politics in an Age of Risk*, Polity Press, Cambridge.

Conway, G. R and Pretty, J. N. (1991) *Unwelcome Harvest: agriculture and pollution*, Earthscan, London.

Cutter, S. L. (2006) *Hazards, Vulnerability and Environmental Justice*, Earthscan, London.

Davis, M. (2006) *Planet of Slums*, Verso, London.

Gadgil, M. and Guha, R. (1995) *Ecology and Equity: the use and abuse of nature in contemporary India*, Routledge, London.

Harvey, D. (1996) *Justice, Nature and the Geography of Difference*, Blackwell, Oxford.

Mol, A.P.J. (2001) *Globalization and Environmental Reform: the ecological modernization of the global economy*, The MIT Press, Cambridge, MA.

Murphy, R. (1994) *Rationality and Nature: a sociological inquiry into a changing relationship*, Westview Press, Boulder, CO.

Myers, G. A. (2005) *Disposable Cities: garbage, governance and sustainable development in urban Africa*, Ashgate, Burlington, VT.

Pretty, J. N. (1995) *Sustainable Agriculture: policies and practice for sustainability and self-reliance*, Earthscan, London.

Pretty, J. (2002) *Agri-culture: reconnecting people, food and nature*, Earthscan, London.

Web sources

http://www.bhopal.org/ The Bhopal Medical Appeal and Sambhavna Trust.

http://www.epa.gov/history/topics/lovecanal/index.htm US Environmental Protection Agency site on Love Canal.

http://www.mosop.net/ The Movement for Survival of the Ogoni People (MOSOP), formed in 1990, an umbrella organization for a number of Ogoni organizations.

http://www.shell.com/home/content/nigeria/ Shell Nigeria: follow community and environment.

http://www.chem.unep.ch/pops/ United Nations Environment Programme on persistent organic pollutants (POPs).

http://www.natural-resources.org/minerals/CD/twb.htm United Nations Environment programme data on mining, environment and development.

http://www.whirledbank.org/environment/mining.htm The spoof campaigning website.

http://www.oktedi.com/ Ok Tedi Mining Limited: entries on 'sustainable development'.

http://www.barrick.com/CorporateResponsibility/default.aspx The Barrick Gold Corporation, Canada (which merged with Placer Dome, the fifth largest gold-mining company in the world, in 2006): 'responsible mining'.

http://www.riotinto.com/ourapproach/sustainabledevelopment.asp Rio Tinto: major mining company on sustainability: 'We believe that our contribution to sustainable development is not just the right thing to do. We also understand that it gives us business reputational benefits that result in greater access to land, human and financial resources.'

http://www.pan-uk.org/ Pesticide Action Network UK: international campaigns against pesticides.

http://www.iied.org/NR/agbioliv/index.html The International Institute for Environment and Development Sustainable Agriculture Biodiversity and Livelihoods Programme: pesticides in context.

http://www.fao.org/organicag/ The Food and Agriculture Organization: organic agriculture.

http://www.fao.org/WAICENT/FAOINFO/AGRICULT/agp/agpp/Pesticid/ Default.HTM The Food and Agriculture Organization on pesticide management. This site contains information on recommendations for the distribution and use of pesticides, and maximum pesticide residual levels in food.

13 Green development: reformism or radicalism?

> But be ye doers of the word, and not hearers only, deceiving your own selves.
> (James 1: 22, Holy Bible, Authorized Version)

Claiming sustainability

There is a tension that has run through this book, and that also runs through a great deal of writing about the environment and development, both South and North. The argument set out here has drawn a distinction between 'mainstream' approaches to sustainability, and various more radical countercurrents (Chapter 5–7). This distinction is not an absolute one, as even a cursory closer analysis of the kaleidoscope of ideas about sustainability soon demonstrates. Analysis of the political ecology of sustainability in drylands, forests, conservation, water resources and urbanization and industrialization in Chapter 8–12 shows the complexity of the technical and political issues. Much sustainable-development thinking is pragmatic, seeking technical and implementable steps towards the reform of development practice. Some analyses, however, are more radical in their explanation of the impacts of development on environment and people. They suggest that the 'greening' of development demands a response that goes beyond tinkering with the technologies and bureaucratic detail of development.

It has not been my intention in this book to argue that one end of this reformist–radical continuum is right and the other wrong. Both demand action, and there is a great deal that needs to be done to change locked-in 'business-as-usual' mindsets that sustainability thinking counters. I am not trying to argue that there is a comfortable synthesis of different approaches, some pure essence of thinking within which 'real' sustainability is to be found, or a secret policy formula with which the stew of muddled and well-meaning talk of sustainability of recent decades can be clarified and made effective. There is no simple and single recipe for sustainability, and no easy answers for those who address the legacy of global 'development', in the classic normative sense as the idea of progress towards the goal of universal human improvement (Cowen and Shenton 1996).

The challenge of sustainability is precisely that there are hard decisions to be made. These are not simple, but contingent on endlessly repeated dilemmas in different places and at different times. Sustainability is not something that can be delivered formulaically through the adoption of new and improved analytical structures (even those as effective as ecological economics), or new planning procedures (even those that seek to bring the environment to the heart of economic decision-making) or some new technology (however much we need new innovations to allow dematerialization and power-down strategies). Nor is it enough to promote 'development from below', and hope that somehow decentralization or more participation in planning will promote better decisions. Pitfalls await those who expect a quick and unproblematic switch from 'unsustainable' to 'sustainable' development simply by altering the style of development planning, the nature of consultation with affected people and the scale of projects (Conroy and Litvinoff 1988; Redclift 2005).

The idea of sustainable development flowered at the end of the twentieth century at a time that saw the collapse of confidence in the myth of development as continuous human improvement, not least the rise of post-development theory (Sachs 1992; Escobar 1995). There are indeed limits to what technocentric strategies can achieve, and what ecological modernization can mould into sustainable paths; as Escobar (2004) notes, 'modernity's capacity to provide solutions to modernity's problems has been increasingly compromised' (p. 209).

The conventional development debate has been transformed by a renewed focus on poverty, notably in the Millennium Development Goals. However, the problems of chronic poverty are relentless and persistent (Chen and Ravallion 2007). Development in practice too often holds little comfort for the poor. The environmental and human costs of rapid industrialization and economic change, of grandiose development projects, of supposedly Promethean technologies such as dams or pesticides are real. Even development success involves substantial costs. These are not shared equally. The rising tide of global prosperity does not raise all ships, and the poor too often drown on the mud (Gadgil and Guha 1995).

Furthermore, development everywhere tends to transform, homogenize and degrade biological diversity (Imhoff *et al.* 2004; Sanderson 2005; Hails *et al.* 2006; Secretary of the Convention on Biological Diversity 2006). Belatedly, the extent of anthropogenic climate change, and awareness of its disastrous implications for the biosphere and human well-being, have been recognized (Parry *et al.* 2007). As David Orr (2007) comments, with atmospheric carbon dioxide at its highest concentration for over 650,000 years and coal, oil and gas extraction continuing unchecked, 'our mismanagement of carbon threatens the human future' (p. 292). Development offers mixed prospects to environmentalists, whether wealthy nature lovers in industrialized countries or poor 'ecosystem people' dependent on ecosystem functions and resources for survival (Gadgil and Guha 1995).

To an extent, sustainability can be sought through the classic strategies of the mainstream, improving governance and regulation and planning, revising economic assessments to internalize costs, reforming industrial processes to minimize risk and seeking to mitigate and compensate for environmental and socio-economic costs. Much, indeed, can be achieved by a reformist approach to sustainability. Yet this leaves much that is undone. Arguments about the costs and benefits of development that aggregate across place and time may well not satisfy those whose livelihoods and lives are shattered by development, or who suffer environmental degradation today. For them, a reformist approach may not yield sustainability or environmental justice on a meaningful timescale. Therefore, sustainability must not only be planned for, but must also be claimed.

However completely the sustainable development mainstream is adopted into orthodox development thinking, it has been necessary for individual people and communities to lay claim to the prerequisites for a sustainable livelihood. Beyond a certain point, sustainability is not something that can be administered from above or driven by scientific and technical innovation; it has to be seized from below. The poor, in particular, are environmental activists, both against their own degradation of the environment on which they depend, and against the environmental impacts of development. Three kinds of responses to environmental risk and degradation, and to the unwanted impacts of development, can be distinguished: adaptation, resistance and protest.

Adapting for sustainability

The most basic response to environmental degradation or risk is for people to adapt their lives and systems of production to cope with it. It is now commonplace to celebrate the skills and adaptability of small farmers in the developing world. Thus Huijsman and Savenije (1991) argue that building strong community-based environmental management systems and decision-making structures demands that we 'respect and make use of native wisdom and indigenous knowledge and experience, and to accept local decision making' (p. 25). Ideas of this kind became popular in development thinking in the 1970s, part of a populist reaction against the unsuccessful technological triumphalism of rural development practice, for example, in Belshaw's call (1974) to 'take indigenous technology seriously', and in the book *Indigenous Systems of Knowledge and Development* (Brokensha *et al.* 1980). There has, for example, been extensive research on the skills of local agricultural management in wet tropical areas (e.g. Rambo 1982; Denevan *et al.* 1984; Gliessman 1984; Richards 1985, 1986). Thus Barker and Spence (1988) described the 'rich and functional environmental knowledge' of Maroon farmers in Jamaica, which underpins their 'food forests'. These are multi-tier crop complexes involving tree crops (such as coconut), shrubs (coffee or cocoa) and smaller plants, sometimes of up to fifty species.

A. J. Bebbington (1996) warns of dangers in an over-romantic view of local knowledge, and the proposal that agricultural programmes should build primarily, or only, on farmers' own existing techniques and innovations. He finds local organizations in the Andes in Ecuador pursuing agrarian development not by standing against modernization, but by reforming, managing and adapting it. Specifically, Indian federations in the Andes incorporate Green Revolution technologies in programmes that seek to promote development while reinforcing Indian culture and society. Bebbington argues that 'what gives a strategy its alternative, indigenous orientation is not its *content* (i.e. that it uses indigenous technologies) but rather its *goal* (i.e. that it aims to increase local control of social change)' (p. 88; emphasis in the original). The importance of indigenous knowledge therefore lies less in the technical superiority of existing over new ideas (although this may hold true) than in issues of ownership of ideas and control of change. Ghai and Vivian (1992b) comment that 'sustainable development requires that local communities enjoy genuine autonomy, have control over adequate resources, and, in some cases, that they be provided with financial and technical assistance to restore their resource base and re-establish their control over resources' (p. 19).

The livelihoods of people in high-risk or highly variable environments tend to exhibit considerable self-reliance and flexibility, as well as a high degree of careful adaptation to local environmental resources and environmental change. Thus, for example, pastoralist groups in drought-prone savannah environments are remarkably flexible in their use of space and resources in the face of seasonal and inter-annual rainfall variability. Flexibility is built into the composition of their herds, their relations with adjacent agricultural groups and their awareness of rich kinship networks (Behnke *et al.* 1993; see also Chapter 8).

Farmers in drought-prone dryland environments such as the Sahel have also developed livelihood systems that allow them to adapt to environmental and economic conditions between and within years (Mortimore and Adams 1999). In agriculture, these include diverse crop varieties, diverse cropping systems and integrated management of crops and livestock. Farm households choose crops, particular varieties of crops and cropping mixtures to suit the soils of their fields, their observations and expectations of rainfall and the availability of labour to manage them. However, agricultural activity is also balanced against off-farm income from buying and selling food or petty products, from craft activities (making palm-leaf mats, for example) and seasonal and even longer-term migration in search of work. The effort invested at the household level by different people into these different activities is a direct response to the opportunities each offers, and particularly to the amount and timing of rain. For any one household, the portfolio of human and technical resources available varies through each season and from one season to the next, and so too does their decision about what they do. There is, of course, considerable variation between one household and another in both their endowment of social and environmental resources and their decisions about how they should be allocated (Mortimore 1998; Mortimore and Adams 1999).

The same organizing principles of livelihood adaptation prove useful when environmental or socio-economic change is exogenous (from outside local society), driven by the development process. In the face of deforestation, peasant farmers respond to shrinking forest and land resources defensively, trying to maintain traditional systems of resource management to intensify crop, livestock and forest production, and squeezing consumption (Barraclough and Ghimire 1995). Thus, in Lusotho District in the Usambara Mountains of Tanzania, growing population pressure was contained with minimal deforestation by the adaptation of consumption and production systems. Poorer families used less fuelwood and other forest products and more bricks and other non-forest materials. Livestock were fewer in number but of better quality, and were stall fed and not open grazed; higher-yielding crops such as cassava, maize and Irish potatoes were grown, even supplanting coffee; vegetables were grown for urban markets, and farmers invested in irrigation, contour-bunding and tree-planting. This kind of productive adaptation of agricultural husbandry is echoed elsewhere – for example, in Machakos in Kenya (Tiffen *et al.* 1994), or Kigezi in Uganda (Carswell 2003, 2007). However, structural factors constrain opportunities for such adaptation. In Tanzania, Barraclough and Ghimire (1995) found their relative success related to the supportive institutional environment created by the state. Similar pressures in different circumstances in Totonicapan in Guatemala (particularly with regard to land tenure) prevented Mayan farmers from making a success of such adaptations (Barraclough and Ghimire 1995).

Sometimes production systems can be adapted to cope with even traumatic development-induced environmental change. The example of the adaptation of farmers in the wetlands of the Hadejia-Jama'are Valley of north-eastern Nigeria to desiccation caused by upstream dam construction and several years of low rainfall was described in Chapter 11. Production of wet-season flood rice and flood-recession crops such as cowpeas declined through much of the area, as did the fishery (Thomas and Adams 1999). However, some communities were able to adapt their agriculture and re-establish or even enhance their livelihoods, because of the availability of new agricultural technology in the form of imported petrol pumps and shallow tubewells. It would be a mistake to expect this kind of luck to hold more generally, but the principle that people can and do adapt successfully to regain and retain sustainable livelihoods, and that they will work to make and keep those livelihoods sustainable, is fairly universally valid. The institutional context (of state, market and civil society) is enormously important to their capacity to adapt successfully, but in favourable circumstances they may be able to do so.

Of course, in this Nigerian example by no means everybody was able to profit by the newly available technology and regain their lost livelihoods. The patterns of impact and response were socially, spatially and temporally quite complex. Some parts of the floodplain remained dry, and not all areas were suitable for irrigation. Desiccation of the floodplain and changes in water flows caused by the invasive growth of bulrushes (*Typha*) led to disagreements about fishing rights between long-established and recently arrived fish-catchers, while the new technology of irrigation pumps allowed dry-season farming in places previously available for Fulani cattle to graze, resulting in a series of violent local conflicts between farmers and graziers (Penrose *et al.* 2005). Even local adaptation is a highly political process.

Wealth is a near-universal discriminator of ability to adapt successfully. Barraclough and Ghimire describe how, in the hill districts of Rasuwa and Nuwakot in Nepal, poorer households used less fuelwood and no longer kept fires burning overnight, while richer households with sufficient land planted trees for fodder and fuelwood; poor households cut even mango trees for wood and to make room for crops (Barraclough and Ghimire 1995). Not everybody can adapt to achieve sustainability, even when circumstances are kind.

Resistance to development

A second response strategy to environmental risk and the environmental and social impacts of development is resistance. The transformation of rural economies by colonial and postcolonial states has often involved direct state coercion of rural producers in the name of development (Williams 1981; Crummey 1986). The state's ability to 'capture' the peasantry politically and economically (or its failure to do so) has been widely discussed (Hyden 1980). Many Third World governments expected peasant farmers to contribute to development by providing the resources for others to develop the urban industrial economy (Williams 1976). With an agricultural policy aimed at taxing producers to pay for urban infrastructure, and price controls to keep urban food prices down, the construction of projects such as hydroelectric dams in rural areas may well make the state and its 'development' a highly unattractive actor on the rural scene (Good 1986).

The myth of the conservative peasant ran deep in the colonial mind. When colonial administrators came to tropical Africa, for example, they mostly failed to see order or skill in rural production systems. Practices such as mixed cropping or intercropping presented an image of confusion and poor husbandry, and the cautious risk-avoidance strategies of peasant farmers were dismissed as the result of a stultified conservatism. Yet colonial development projects, and their successors, were often marked by unrealistic ambitions, limited knowledge and technical incompetence. In Sierra Leone, Richards (1985, 1986) sharply contrasts the high degree of ecological adaptation in Mende swamp rice-production systems and the grim comedy of repeated attempts by the colonial and postcolonial developers to transform them. As described in Chapter 11, colonial agriculturalists tried to grow groundnuts on the Niger Agricultural Project at Mokwa in Nigeria for several years in the late 1940s, with conspicuous lack of success. Their plans for mechanized production by smallholders simply did not work. Nigerian farmers were reluctant to come to the scheme as settlers; they did not think that they were being offered 'development', but that they were being called to the rescue (Baldwin 1957). The inadequacy of such blundering projects in post-war Africa, most notoriously the Groundnut Scheme at Kongwa in Tanganyika, makes the reluctance of rural people to become involved seem entirely understandable.

There were contemporary commentators with greater vision. In West Africa, for example, Howard Jones (1936) celebrated the diversity of native farming, and Stamp (1938) praised the soil conservation practices of Nigerian farmers. In time, official respect began to grow for the 'African husbandman' (Allan 1965). It began to be appreciated that Western science might not have the monopoly of answers: in describing the shifting cultivation of the Zande in Sudan, de Schlippe (1956) wrote, 'the teacher of a culture is its environment, and agriculture is its classroom' (p. xiv). Faulkner and Mackie pointed out in 1933 that 'the prevalent idea that the native farmer is excessively conservative is largely due to the mistakes of Europeans in the past' (p. 7).

Yet, then as now, the failure of development projects was often blamed on the refusal of local people to do as they were bid, and that was explained in terms of the 'intransigence and primitiveness of peasants' (Hill 1978). Developers saw – and to some extent still see – a problem in the recalcitrance of peasants to outsiders' conceptions of progress, and their reluctance to join in as required (G. Williams 1981). Such attitudes persist to a greater extent than might be expected given the almost universal protestations about indigenous knowledge and participatory development – for example, in official attitudes to pastoralists, people living in areas designated as national parks, or those unhappy about being forced to move to make way for a lake behind a hydroelectric dam.

Reluctant participation in development projects needs to be understood in the context of wider resistance to modernity, to commodification of social interactions and nature and to the demands of development's champions. Commonly, subordinate classes resist impositions and demands made upon them (whether by the state or by richer neighbours) silently, subtly, passively and without overt organization. They meet the demands for food, labour, taxes, rents and interest on loans with what Scott (1985) calls 'everyday forms of peasant *resistance*' (p. xvi; emphasis in the original): 'the ordinary weapons of relatively powerless groups: foot dragging, dissimulation, desertion, false compliance, pilfering, feigned ignorance, slander, arson, sabotage and so on' (p. xvi).

Spittler (1979) describes the defensive strategies of peasants faced with the unreasonable demands of the colonial state in Niger. One strategy was evasion: lookouts outside villages reported the arrival of soldiers and policemen, and people hid in the bush to evade tax collection, requisitioning or labour demands, and hid cattle and children during censuses. Another strategy was silent disobedience – for example, ignoring orders about groundnut smuggling or voting; another to misuse material or money or livestock given for a specific purpose; another to end difficult interviews by agreeing with everything. The state in turn responded with a paradoxical mix of 'laissez-faire and force' (p. 33), often alternating the two – for example, rounding up people who refused to participate in a development project, but eventually dropping it.

Scott's own study *Weapons of the Weak* concerns conflict within a village on the Muda Irrigation Scheme in Peninsular Malaysia, where mechanization of the rice harvest in the second half of the 1970s had drastically reduced field labour opportunities for poor households. He is concerned to understand both the acts of resistance and their symbolic context. In the theatrical metaphor he uses, he is interested in both the 'onstage' behaviour and the 'offstage' language that people use to contextualize it. The rich are effectively immune to material sanctions, but not to what Scott calls 'symbolic sanctions', such as slander, gossip and character assassination. He comments that

> those with power in the village are not, however, in total control of the stage. They may write the basic script of the play but, within its confines, truculent

and disaffected actors find sufficient room for manoeuvre to suggest subtly their disdain for the proceedings. The necessary lines may be spoken, the gesture made, but it is clear that many of the actors are just going through the motions and do not have their hearts in the performance.

(Scott 1985, p. 26)

Protest for sustainability

The third form of response to environmental degradation and imposed development is open protest. Resistance to coercion may be typified by 'silence and stealth' (Crummey 1986, p. 10), but many protests are more vocal or active. In the context of colonial Africa, for example, Bates (1983, p. 104) links the commercialization of agriculture with the rise of political protest in the rural areas of colonial Africa, and the importance of forced terracing campaigns in focusing political mobilization in countries such as Kenya and South Africa is now widely recognized (Beinart 1984; Throup 1987; Beinart and Coates 1995; Mackenzie 2000). However, peasant movements are typically limited in aims and achievements, and deficient in organization and execution, as Beinart and Bundy (1980) describe in the context of the Transkei.

While peasant revolts fascinated Western scholars excited by Marxism or the Vietnam War, rural people rarely engage in open rebellion. When they do, they are rarely successful. Hildermeier (1979) distinguishes between agrarian revolts, often short-lived and violent, and longer-term and more peaceful agrarian movements. The latter are less dangerous and more common. The land occupancy protests that followed the alienation of common lands in Latin America by haciendados in the 1920s were 'more or less spontaneous and localized affairs, involving peaceable squatting in the first place, but almost inevitably developing later, as the troops moved in, into violent confrontations, at great cost in peasant lives' (Rudé 1980, p. 69). Crummey (1986) refers to this as the waking of 'the other beast' of state violence (p. 21).

Protest is a fundamental aspect of the Western environmental movement. It is easy to dismiss Western environmentalism as selfish in its concern about obscure rare species and the money spent on their protection. Its obsessions with marginal issues of environmental quality (access to the countryside, scenic beauty, organic food) can seem like the special pleading of an effete and selfish class of global hedonists. Western environmental concern does reflect class-based interests in lifestyle quality, yet behind this lies a challenge to the emerging pattern of economy and society of the late twentieth century. Environmentalism works because of shared conviction that 'things don't have to be like this'. Most successful environmental protests of the past three decades have been built on such a call for mainstream practices to change, whether the issue be the removal of lead from petrol, the disposal of oil platforms or of non-returnable bottles, the thinning of the ozone layer or the clearance of old-growth forests (Rowell 1996; Rawcliffe 1998). From the Monkey Wrench Gang to the Sierra Club, Friends of the Earth to Earth First!, environmentalism has long had a radical stream within it that knew that sustainability (or whatever they called it) was not something that could safely be left to industry and government, to expert technical planners and politicians (Abbey 1975; Devall 1988; Bookchin and Foreman 1991).

In the South, too, sustainability is something that has to be claimed. Work on environmental movements in developing countries has emphasized the importance of actions by the poor and marginalized to protect the environmental basis of sustenance and livelihood (Gadgil and Guha 1995; Chapman *et al.* 1997; Guha 2000; Dwivedi 2005). Increasingly, however, such local actions have been linked to larger transnational networks and global actors (Dwivedi 2005), and environmental organizations have developed to express a similar range of concerns to those in the North. Thus, in Chile, David Carruthers (2001) identifies three kinds of environmentalist: conservationists (their organizations strongly science based and often linked to biodiversity organizations in North America); environmentalists (addressing a broad array of environmental problems and issues and engaging in lobbying, advocacy and education); and ecologists, or '*duros*' ('hard-liners'), with a strong tradition of political critique and a focus on social and environmental justice.

There are many celebrated examples of organized grass-roots protests against development. One is the Chipko movement in the Garhwal Himalaya in India. The historical roots of Chipko are long and complex (Guha 1989). The adoption of 'scientific' forestry, in both the traditionally governed Tehri Garhwal and the colonially governed Kumaon, and the reservation of forests for timber production towards the end of the nineteenth century caused considerable disruption of production systems, and considerable hardship (Guha 1989). In both areas, in different ways and with different effects, peasants protested about their exclusion from the forests throughout the twentieth century. Tactics included strikes and go-slows in providing porters for visiting officials, refusals to pay fines and incendiarism (the standard technique of burning litter on the forest floor to open it up for grazing was banned in order to improve tree regeneration). By 1921 Guha (1989) reports 'a near total rupture between the colonial state and its subject population' (p. 137).

Chipko therefore built on a long history of challenges to the demands of commercial forestry. Such protests had been openly rebellious, and had been met with state violence (notably in Tehri Garhwal in the 1940s). Protests, and government attempts to dissipate or appease them, continued. Conflict over access to forests and allocation of felling rights in trees, as well as problems of landslides and floods, led to a series of actions to prevent contractors from logging forests in the 1970s. The action of women to prevent the cutting of trees in the village of Reni in 1974 (Guha 1989) has become the centre of the global environmentalist myth of Chipko. Guha (1989) concludes *The Unquiet Woods* by suggesting that 'peasant movements like Chipko are not merely a defence of a little community and its values, but also an affirmation of a way of life more harmoniously adjusted with natural processes' (p. 196).

Chipko was in fact a very diverse movement politically. It was also remarkably effective, achieving an effective ban on green felling in the Himalaya above

1,000 metres in 1981. However, there are wider ramifications of Chipko both within India and globally, and more particularly of beliefs about it. Bandyopadhyay (1992) describes the way in which Chipko metamorphosed from a peasant movement to a global campaign for the sustainable management of forests in general, and those of the Himalaya in particular. He suggests that Chipko was 'no longer a hill people's movement against forest fellings', but has become a philosophy, an extension of Gandhian thought. This may well be true, but other observers are critical of the romanticism of some of those enthralled by Chipko. Rangan (1996) criticizes environmentalists inside and outside India for being 'rapt and slavish in their adoration and assiduous pursuit of romance with Chipko's ecological reincarnation' (p. 222). He suggests that Chipko's leaders were in fact reactionary, their allegiance to the myth of Chipko preventing the protest and arguments of village leaders and activists from being heard outside the region. There were militant local calls for more development, for tree-felling rather than tree-hugging. Chipko was one strand in political calls for the establishment of the new state of Uttaranchal (Mawdsley 1998, 1999). However, Rangan argues that, although the Uttaranchal movement's message of secular development and social justice commanded widespread political support locally, Chipko's very success in winning the admiration of international environmentalists actually helped stifle the wider movement's voice.

As Chapter 11 showed, controversies over the construction of large dams have been a trigger for active opposition, both internationally within the environmental movement, and locally among reservoir evacuees – not least in the Tehri Garhwal, against construction of the Tehri Dam (Bandyopadhyay 1992). Anti-dam protest within the environmental movement before the 1970s involved essentially preservationist opposition to the inundation of wilderness areas. Gradually, however, protest extended to include directly threatened communities, and to address the plight of evacuees and the wider issues of unsustainable development (McCully 1996). One of the formative moments in the early history of the environmental movement in the USA was the battle between John Muir (and the Sierra Club) and the city of San Francisco over the proposal to flood the Hetch Hetchy Valley in Yosemite National Park in the early years of the twentieth century. That campaign was self-consciously re-echoed in the 1950s in the ultimately successful campaign against the US Bureau of Reclamation's proposal to build the Echo Park Dam on the Green River in Utah in the 1950s, and the unsuccessful campaign against Glen Canyon Dam on the Colorado a few years later (McCully 1996). That dam was closed in 1963, but protests by conservationists continued, and by the end of the 1980s the Bureau of Reclamation had more or less abandoned large dams as a formal and official policy. From 1976 a Dam Fighters' Conference was held annually in the USA.

Elsewhere in the industrialized world, similar protest campaigns took place. Indeed, Usher (1997d) argues that opposition to dams exists in almost every country where there is democratic space to express dissent. In Sweden, for example, there was fierce debate about dam construction on the Vindel River in the 1960s, and opposition to dams was strong through the 1970s. An inter-basin River Savers

Association was established in 1974, and in 1987 a Natural Resources Act halted all dam construction on the last four large free-flowing rivers (Lövgren 1997). In Arctic Norway a proposal to dam the Alta River in the 1970s led to intense protest (including civil disobedience, protest camps on-site and hunger strikes by Sami protesters). Protest was stopped by the police (with the army in reserve), and the dam was finally finished in 1987 (Dalland 1997). In Australia the Tasmanian Wilderness Society led protest against the proposal to dam the Gordon River in the early 1980s (McCully 1996). In India a campaign by the Kerala Shastra Sahitya Parishad against the flooding of the Silent Valley in Kerala was successful: the land threatened was tropical forest, rich in wildlife, although it contained few human evacuees (Singh *et al.* 1984; S. Singh 1997).

Protest against dams elsewhere in the world is less easily classified within the confines of the conservation movement. Protest can take the form of legal challenges (as, for example, in the case of the Japanese-funded Motapanjang Dam in Sumatra (Karimi *et al.* 2005)), or more overt political action. Protest about the construction of dams formed the focus for broad social movements in a number of countries in Eastern Europe during the 1980s, before the collapse of the Iron Curtain, for example, in Bulgaria and Latvia, and in Hungary against dams on the Danube at Nagymaros and Gabcíkovo (McCully 1996). In the Third World, dams have triggered powerful social movements of social and environmental protest, for example in Brazil in the 1980s.

The outstanding example of protest against large-scale dams is probably that of Narmada Bachao Andolan against the Sardar Sarovar Dam on the Narmada River (Baviskar 1995). The notion of damming the Narmada was first considered in 1946, although planning waited until the resolution of disputes between Maharashtra, Gujarat and Madhya Pradesh on water sharing in 1978. The Sardar Sarovar Project forms only one element in a vast programme of river engineering, the Narmada Valley Project, which includes 135 medium and 3,000 minor dams. The Sardar Sarovar is the second largest, and construction began in earnest in 1985 with World Bank funding. The dam is 139 m high and is intended to supply drinking and irrigation water, and to generate hydroelectricity. It will flood 37,000 hectares of land and displace 152,000 people in 245 villages in the three states. These include hill *adivasis* and other groups. In addition, substantially greater numbers of people (up to one million) will be affected who live outside the reservoir area – for example, in linked forestry schemes or in irrigation infrastructure (Baviskar 1995).

There was a short-lived movement against the project in the 1970s, which died when the politician who began it was elected and dropped the issue. Anti-dam mobilization began again in 1985; its initial focus was a demand for adequate rehabilitation for evacuees, but the Andolan subsequently developed its position to outright opposition. The campaign has been based on popular mobilization within the Narmada valley. Since 1988, people threatened by the reservoir have removed survey markers, held demonstrations both in the reservoir area and outside, and staged hunger strikes (Baviskar 1995). In 1989, for example, the Andolan organized a National Rally against Destructive Development, and in

1990 they marched on the dam site and declared a programme of non-cooperation with the state. Many of these protests were met with violence by the police and Indian Administrative Service.

However, the work of the Narmada Bachao Andolan has not been confined to localized protest. It has been extensively supported by other groups, both within India and internationally. Urban environmental NGOs within India raised funds and lobbied, and rural mass organizations from elsewhere in the country joined protests. The movement has carried out research and launched legal challenges within India, and has proposed alternative development strategies (Baviskar 1995). Internationally, lobbying of the US Congress by NGOs put pressure on the World Bank to stop funding the dam, and an independent review was carried out. Its report, in 1992, was highly critical of the project, and the Bank pulled out of the project in 1993. At the same time, the Japanese government suspended aid to Sardar Sarovar in response to lobbying by Friends of the Earth Japan.

The response of the Indian government to the World Bank's withdrawal of funding was an escalation of violence against protesters, and an attempt to complete the project using Indian finance. Flooding began as a result of the part-completed works in 1993. In 1994 the Narmada Bachao Andolan filed a case against the project in the Supreme Court in Delhi, which called for a detailed review of the project; further hunger strikes were begun (McCully 1996). The Supreme Court of India halted construction of the Sardar Sarovar dam in 1995 at a height of 80.3 m. In 1999, it gave the go-ahead for the dam's height to be raised to a height of 88 m. On 18 October 2000 the Supreme Court delivered a 2 to 1 majority judgment, allowing immediate construction on the dam up to a height of 90 m. The judgment also authorized construction up to the original planned height of 138 m in 5 m increments subject to receiving approval from the Relief and Rehabilitation Subgroup of the Narmada Control Authority.

Protests against dams have become increasingly internationalized, and increasingly linked with wider social movements. For example, the Narmada Bachao Andolan helped establish the National Alliance of People's Movements in 1996 (McCully 1996). Internationally, the *Ecologist* magazine began to campaign on large dams in the 1980s (Goldsmith and Hildyard 1984), while, in the USA, publication of the *International Dams Newsletter* began in 1984. From this grew the International Rivers Network and its newsletter, *World Rivers Review* (McCully 1996). In 1988 activists from round the world met in San Francisco and signed a 'San Francisco Declaration', which demanded a moratorium on all new large dams that failed to meet criteria for participation by those affected, access to project information and environmental, social health, safety and economic performance. In 1994 the Manibeli declaration, presented to the World Bank's president on its fiftieth anniversary, by 326 environmental groups and coalitions in 44 countries, called for a moratorium on all loans for large dams until certain conditions were met (McCully 1996).

The work of the World Commission on Dams (WCD) (1998–2000) was intended to move beyond the oppositional politics of protest. Its report (2000) was the result of extensive research and consultation, involving both the dam-building industry and its opponents. Its recommendations were clear, practical and robust. It acknowledged that dams were necessary but problematic, and proposed a series of strategic priorities that would help ensure that their benefits were maximized and their costs minimized (Scudder 2005). UN agencies and European governments endorsed the report. However, to many in the dam-construction community, the recommendations went too far, and a substantial rebuttal machine kicked into gear. Two leading dam-building nations, India and China, quickly rejected the findings of the Commission, and others (for example, Turkey) were highly critical. The debate about dams continued around questions of the technical feasibility of the WCD's proposals and the Commission's standing. Thus Thatte (2001) found 'obvious imbalances' in the report and Biswas (2004) found the process flawed. Gagnon *et al.* (2002) contrasted the WCD's unworkable guidelines that sought consensus on project development with the more robust approach of the International Energy Agency, which emphasized efficient government decision-making to avoid endless negotiations. Ted Scudder, a Commissioner on the WCD and a lifelong researcher on the costs of resettlement, points to a lack of political will on the part of governments and project authorities to change the ways dams are planned. There are technical guidelines that would allow this to happen, but the minds of dam-planners are closed. Unless they open, Scudder (2005) believes nothing will change. Protest will continue to be the only option in the face of the unsustainabilty of imposed development.

Social movements and sustainability

Adaptation, resistance and protest often grade into one another when people oppose development. Sometimes resistance to development is informal, small scale and hidden. At other times it is organized and overt, sometimes even involving illegal acts, sometimes provoking violence by the state or its allies. Collective social action over environmental issues can be remarkably effective. In the Mexican city of Monterrey, housewives meeting together organized effectively to force an improvement in water supply (Bennett 1998). By 1980 water supplies to over half of Monterrey were rationed. Meeting at the collective tap or at a street or neighbourhood meeting, housewives began organizing direct meetings with government officials, and meetings and protest rallies. They bypassed the water authority and contacted the state governor or the mayor. They organized direct action, either blocking roads or kidnapping water service vehicles. In the short term protests usually persuaded hard-pressed engineers to find water for the protesting neighbourhood. In the longer term they helped demonstrate the scale of the water supply problem and forced the centralized government planning process to move towards investment in improved water supplies.

Successful environmental action is commonly firmly rooted in locality. Thus, opposition to the Ilisu dam on River Tigris in south-east Turkey (which will flood 621,000 people and the historic Kurdish town of Hasankef, and have major impacts downstream) eventually led to the withdrawal of one of the main contractors, Balfour Beatty. Morvaridi (2004) comments that, despite growing global visibility of issues such as this, it is still local people who have to take up the challenge of making the political and legal and institutional changes necessary to resolve contentious development issues and obtain their rights.

Under what circumstances can local people organize, making their resistance directed and active? Broad (1994) identifies three necessary conditions for activism. First, the natural resource base on which people depend has to be threatened; second, they have to have lived in the area for some time or have some sense of permanence there; third, civil society must already be somewhat politicized and organized. On Mindanao in the Philippines, for example, people acted 'once environmental degradation began to transform poor people who lived in a stable ecosystem into marginal people living in vulnerable and fragile ecosystems' (p. 814).

James Scott (1985) points out that changes that are sudden, sharp and unexpected (for example, the introduction of combine harvesters in the village of Sedaka on the Muda Irrigation Scheme, with all its implications for rural employment) may stimulate people to organize, but many changes are gradual and piecemeal. People cope with them progressively, and often alone. Changes that create complex (and especially not visible) categories of winner and loser are also less likely to lead to collective responses than those that create a large and self-aware group of losers. Other obstacles to collective action lie in class structure, and in the cleavages and alliances that cut across it such as kinship, friendship, faction, patronage and ritual ties (Scott 1985).

It might seem obvious from the perspective of Western environmentalism that it is a good idea to organize against developments that challenge sustainability and in favour of those that foster it. However, there are strong reasons why people do not organize in this way. Very often, people fear the consequences of collective action, whether from the state or from other actors. In the language of institutional economics, the anticipated costs of deviation from established (and often imposed) norms are not balanced by anticipated benefits.

In some countries, like China, freedom to dissent or protest against projects such as the Three Gorges Dam is severely restricted by the state (Beattie 2002). In China, environmentalism has evolved slowly and in a piecemeal way: there is no environmental movement capable of organizing national demonstrations. Indeed, many government organizations have set up their own 'non-governmental organizations', to devolve certain functions, or to attract foreign funding: independent grass-roots activism in China is muted, fragmented and highly localized (Ho 2001).

Local actors also often lack capacity to make protests effective. Without education to make a case in the formal terms required by a court of law, without the money to pay lawyers or transport to reach a politician to petition, without the knowledge of the development planned or of the bureaucratic process of project decision-making, without enough time to do anything significant, how can people organize themselves to protest, or to suggest what should be done?

A further factor limiting prospects for grass-roots demands for sustainable management of the environment, and for sustainable livelihoods, is the nature of civil society. One way of looking at this is to use the slightly confused term 'social capital' (Harriss and de Renzio 1997; Putzel 1997). Putnam (1993) argued that networks of trust and shared norms in society facilitate cooperation for mutual benefit, and hence provide a secure base for effective government and economic development. With weak social capital, communities are fragmented, and shared action is difficult to organize and unlikely to be effective. Places with recent immigration or rapid turnover of population, where there are unresolved tensions of ethnicity or class, where political leaders are corrupt or ineffectual, where people are chronically sick or hungry, where skills like literacy are weak: all these would typically have weak social capital. There are many such places in the developing world whose citizens will not find claiming the conditions for a decent environment, life and livelihood easy.

Green theorists emphasize the importance of social movements to bring about sustainability (Friberg and Hettne 1985; see also Chapter 7). In an increasingly networked global society, protesters are able to plug into international networks of activists to try to exert pressure not only locally but also at different points through the structures that drive or regulate change (Castells 2000; Dwivedi 2005). Calls for environmental justice are not confined to the USA, but are characteristic of environmental claims in developing countries (Gadgil and Guha 1995; Harvey 1996b). People can, and do, organize to address the problems of poverty and environmental degradation, and the organizations and structures they see as responsible. Sometimes it works.

Green development: reformism or radicalism?

There is no magic formula for sustainable development. Despite the enthusiastic rhetoric, the technical guidelines and the celebrated greening of development agencies, corporations and governments, there is no easy reformist solution to the dilemmas and tragedies of poverty and environmental degradation, whether at the local or global scale. There is no 'magic bullet' to defeat these threats to human well-being. Behind the slogans about environment and development lies the hard process of development itself, wherein choices 'are indeed cruel' (Goulet 1971, p. 326). One early message from the twenty-first century must be that the state of future conditions cannot be assured, even for the wealthy.

Development ought to be what human communities do to themselves. In practice, it is usually what is done to them by others, whether governments or their bankers or 'expert' agents, in the name of modernity, national integration, economic growth or a thousand other slogans. Fundamentally, it is this reality of development – imposed, centralizing and often unwelcome – that the greening of development challenges. It throws attention back on the ethical questions that underlie the idea of development itself. It recognizes that societies are 'developing' whether or not they are the targets of some specific government 'development' scheme. In practice, in the developing world (as elsewhere) ideas, culture and the nature of society are in flux. Farming practice, production system, economy, are all sucked into the whirlpool of the world economy to some extent, moving in response to the pull of capital. There is no 'real development' to be reached for that escapes this pull, and sustainable development is no magic bridge by which it can be attained.

Part of the limitation of the sustainable-development thinking and the reformist technical approaches discussed in this book is their failure to address the politics of environment and development. Without a theory of how the world economy works, and without theories about the relations between people, capital and state power, sustainable-development thinking – and most conservation action – is profoundly limited. In practice, plans for both development planning and environmental conservation tend to be formed by technocratic elites and imposed, although both kinds of planners seek to involve (and co-opt) local interests and both believe they are operating in the interests of some notional wider constituency. Development initiatives and the context of aid-giving and project formulation have to be understood in terms of the way the world economy functions. Pollution and environmental degradation reflect economic and political structures, and have to be understood in terms of their relations to the urban, industrial cores of the world economy (Chambers 1988b).

'Development' is not necessarily good; it depends on who you are, it depends on how the structures in society expose you to its hazards or open to you its fruits. It depends on how you value the changes created around you by others, and whether your own voice can gain purchase on the behemoths of state planning and business profit-seeking. Development planning involves choices, and tough decisions. Very often in the past those decisions have been taken by 'experts', trained to see the world through clever but reductionist lenses, and insulated by wealth, culture and place of residence from the consequences of their decisions. However, even where planning is brought down to earth, dragged out of the tangle of government bureaucracy and politics, extracted from the spreadsheets of experts and freed from the stranglehold of consultancy contracts, the hard decisions do not go away. Sometimes improved development planning is sufficient to move towards sustainability, and 'win–win' solutions are possible. At others the hard choices inherent in development still have to be made, and, when they are made, the decision comes down against the poor, the marginal, the uneducated and the powerless.

We know the limitations and failures of development, of course. They have been key elements in the litany of sustainable development since the Stockholm Conference in 1972. Indeed, 'sustainable' development has been one of the ideas through which we have sought to recapture a sense of moral trajectory, and a means of measuring our success in driving economies and societies forwards. Since the 1980s, more and more have been added to the concept, until it groans under the weight of ideas not only about the environment, but also about equity, democracy, openness and freedom. As the economic system has become increasingly globalized, with power leaking from nation states towards transnational corporations linked in a highly interconnected global order (Lash and Urry 1994), the attraction of the moral agenda apparently offered by sustainable development has grown. However, 'sustainable development' offers no escape from the dilemmas of development. The huge achievement of the debate about sustainability has been that it has expanded the horizons of development thinking to embrace the environment. Yet, it offers no resolution of the moral ambiguities inherent in development. It offers no route around development's hard choices.

Development does create victims, such as those displaced by dams, those whose subsistence is taken and sold by logging companies, or those whose livelihoods or health are destroyed by factory effluent. Even if these development projects generate benefits – generating cheap power for small enterprises, creating employment, paying taxes for government to recycle in improved health services or water-supply systems – these benefits are often reaped at a larger spatial scale by others, elsewhere, later. If politicians prefer soldiers to teachers and limousines to schools, these benefits are dissipated. If environmental regulation is poor and the costs of pollution are successfully externalized by industry, the benefits may never outweigh costs, and the victims of development may keep appearing for generations to come.

The 'green' challenge in development is not therefore simply about reforming environmental policy; it also issues a challenge to the very structures and assumptions of development. It is, first and foremost, about poverty and human need, about sustainable livelihood security (Chambers 1988b). It is about the state of the environment, and the rights of people to enjoy its benefits. Debates about the mechanisms and dynamics of development have tended to obscure its ethical basis, but the concept of sustainable development is inherently and inevitably ethical (Jacobs 1995). There are strong moral as well as practical reasons for putting poor people first in development planning. Goulet (1971) suggests that the 'shock of underdevelopment' can be overcome only by creating 'conditions favourable to reciprocity', in which 'stronger partners ... offset the structural vulnerability of weaker interlocutors by being themselves rendered politically, economically and culturally vulnerable' (p. 328). The feasibility of such a vision as a political project can be debated, but the extent of its challenge to reformist tinkering with environmental aspects of development policy is clear.

Green development focuses on the rights of the individual to choose and control his or her own course for change, rather than having it imposed. The green agenda is therefore necessarily radical, but it is also open-ended, flexible, and diverse. Green development is almost a contradiction in terms, not something for which blueprints can be drawn, not something easily absorbed into structures of financial planning, or readily co-opted by the state. It shares the very real tensions between techno-centric and ecocentric environmentalism (O'Riordan 1988; Turner 1988b). It requires the state of nature and the state of society to be considered together. It demands an interdisciplinary approach to analysis, training and policy. Green development is something that very often emerges in spite of, rather than as a direct result of, the actions of development bureaucracies. Green development programmes must start from the needs, understanding and aspirations of individual people, and must work to build and enhance their capacity to help themselves. As Robert Chambers (1988b) comments: 'The poor are not the problem, they are the solution' (p. 3).

Culture, society, economy and environment interact in complex and dynamic ways, changing continually, sometimes slowly and in subtle ways and sometimes dramatically and fast. 'Development' based on programmes and policies that are conceived within institutions distanced from those they affect is unlikely to be able to cope with these changes effectively, or to meet a wide range of human needs. Better environmental and developmental planning is both needed and possible, and is at the core of mainstream sustainable development (MSD). But this is just the beginning of the challenge of greening development.

Green development is not about the way the environment is managed, but about who has the power to decide how it is managed. Its focus must be the capacity of the poor to exist on their own terms. At its heart, therefore, greening development involves not just a pursuit of new forms of economic accounting or ecological guidelines or new planning structures, but an attempt to redirect environmental and developmental change so as to maintain or enhance people's capacity to sustain their livelihoods and to direct their own engagements with nature. Escobar (2004) calls for 'dissenting imaginations' that can think beyond modernity and the regimes of the globalized economy and the exploitation of marginalized people and nature.

'Sustainable development' is a way of talking about the future shape of the world. To conceive of the future in these terms marks the beginning of a process of political reflection and action, not the end. To call for sustainable development is not to set out a blueprint for the future but to issue a statement of intent and a challenge to action.

References

Abbey, E. (1975) *The Monkey Wrench Gang*, J. B. Lippincott, New York.

Ackerman, W. C., White, G. F. and Worthington, E. B. (1973) (eds) *Man-Made Lakes: their problems and environmental effects*, Geophysical Monograph 17, American Geophysical Union, Washington.

Acreman, M. and Howard, G. (1996) 'The use of artificial floods for floodplain restoration and management in sub-Saharan Africa', *IUCN Wetlands Programme Newsletter* 12: 20–5.

Adams, A. (1979) 'An open letter to a young researcher', *African Affairs* 38(313): 451–79.

Adams, J. (1992) 'Horse and rabbit stew', pp. 65–73 in A. Coker and C. Richards (eds) *Valuing the Environment: economic approaches to environmental valuation*, Belhaven, London.

Adams, W. M. (1985a) 'The downstream impacts of dam construction: a case study from Nigeria', *Transactions of the Institute of British Geographers* NS 10: 292–302.

Adams, W. M. (1985b) 'River basin planning in Nigeria', *Applied Geography* 5: 292–308.

Adams, W. M. (1987) 'Approaches to water resource development, Sokoto Valley, Nigeria: the problem of sustainability', pp. 307–25 in D. M. Anderson and R. H. Grove (eds), *Conservation in Africa: people, policies and practice*, Cambridge University Press, Cambridge.

Adams, W. M. (1988) 'Rural protest, land policy and the planning process on the Bakolori Project, Nigeria', *Africa* 58: 315–36.

Adams, W. M. (1991) 'Large scale irrigation in northern Nigeria: performance and ideology', *Transactions of the Institute of British Geographers* NS 16: 287–300.

Adams, W. M. (1992) *Wasting the Rain: rivers, people and planning in Africa*, Earthscan, London.

Adams, W. M. (1996) *Future Nature: a vision for conservation*, Earthscan, London.

Adams, W. M. (2003) 'Nature and the colonial mind', pp. 16–50 in W. M. Adams and M. Mulligan (eds), *Decolonizing Nature: strategies for conservation in a post-colonial era*, Earthscan, London.

Adams, W. M. (2004) *Against Extinction: the story of conservation*, Earthscan, London.

Adams, W. M. and Hulme, D. (2001a) 'Conservation and communities: changing narratives, policies and practices in African conservation', pp. 9–23 in D. Hulme and M. Murphree (eds), *African Wildlife and Livelihoods: the promise and performance of community conservation*, James Currey, London.

Adams, W. M. and Hulme, D. (2001b) 'If community conservation is the answer, what is the question?' *Oryx* 35: 193–200.

Adams, W. M. and Infield, M. (2003) 'Who is on the gorilla's payroll? Claims on tourist revenue from a Ugandan National Park', *World Development* 31: 177–90.

Adams, W. M. and Mortimore, M. J. (1997) 'Agricultural intensification and flexibility in the Nigerian Sahel', *Geographical Journal* 163: 150–60.

Adams, W. M. and Mulligan, M. (2003) (eds) *Decolonising Nature: strategies for conservation in a postcolonial era*, Earthscan, London.

Adams, W. M., Aveling, R., Brockington, D., Dickson, B., Elliott, J., Hutton, J., Roe, D., Vira, B. and Wolmer, W. (2004) 'Biodiversity conservation and the eradication of poverty', *Science* 306: 1146–9.

Adams, W. M., Brockington, D., Dyson, J. and Vira, B. (2003) 'Managing tragedies: understanding conflict over common pool resources', *Science* 302: 1915–16.

Adamson, R. S. (1938) *The Vegetation of South Africa*, British Empire Vegetation Committee, London.

Addo, H., Amin, S., Aseniero, G., Frank, A. G., Friberg, M., Frobel, F., Heinrichs, J., Hettne, B., Kreye, O. and Seki, H. (1985) *Development as Social Transformation: reflections on the global problématique*, Hodder and Stoughton, London, for the United Nations University.

Adeniyi, E. O. (1973) 'The impact of change in river regime on economic activities below Kainji Dam', pp. 169–77 in A. L. Mabogunje (ed.), *Kainji: a Nigerian man-made lake; Kainji Lake Studies vol. 2: Socio-economic conditions*, Nigerian Institute for Social and Economic Research, Ibadan.

Adeola, F. O. (2001) 'Environmental injustice and human rights abuse: the states, MNCs, and repression of minority groups in the world system', *Human Ecology Review* 8: 39–59.

Agrawal, A. (2001) 'Common property institutions and sustainable governance of resources', *World Development* 29(10): 1649–72.

Agrawal, A. and Gibson, C. (1999) 'Enchantment and disenchantment: the role of community in natural resource management', *World Development* 27: 629–49.

Agrawal, A. and Ostrom, E. (2006) 'Political science and conservation biology: a dialogue of the deaf?', *Conservation Biology* 20: 681–2.

Ahmed, Y. J. and Sammy, G. K. (1985) *Guidelines to environmental impact assessment in developing countries*, Hodder and Stoughton, Sevenoaks.

Åkerman, M. (2003) 'What does "Natural Capital" do? The role of metaphor in economic understanding of the environment' *Environmental Values* 12: 431–48.

Alcorn, J. B. (1993) 'Indigenous peoples and conservation', *Conservation Biology* 7: 424–6.

Allan, J. A. (1983) 'Natural resources as natural fantasies', *Geoforum* 14: 243–7.

Allan, W. (1965) *The African Husbandman*, Oliver and Boyd, Edinburgh.

Allen, D. E. (1976) *The Naturalist in Britain*, Penguin, Harmondsworth.

Allen, J. C. and Barnes, D. F. (1985) 'The causes of deforestation in developing countries', *Annals of the Association of American Geographers* 75: 163–84.

Allen, R. (1980) *How to Save the World: strategy for world conservation*, Kogan Page, London.

Alley, P. (1999) 'From the dragon's tail to suburbia', *Earth Matters* 41: 24–6.

Aluko, M. A. O. (2004) 'Sustainable development, environmental degradation and the entrenchment of poverty in the Niger Delta of Nigeria', *Journal of Human Ecology* 15: 63–8.

Ambio (1979) 'Review of environmental development', *Ambio* 8: 114–15.

Amin, S. (1985) 'Apropos the "green" movements', in H. Addo, S. Amin, G. Aseniero, A. G. Frank, M. Friberg, F. Frobel, J. Heinrichs, B. Hettne, O. Kreye and H. Seki (eds),

Development as Social Transformation: reflections on the global problématique, Hodder and Stoughton, Sevenoaks, for the United Nations University.

Anderson, C. W. (1912) *Forests of British Guiana* (2 volumes), Department of Lands and Mines, Georgetown, British Guiana.

Anderson, D. M. (1984) 'Depression, dust bowl, demography and drought: the colonial state and soil conservation in East Africa during the 1930s', *African Affairs* 83: 321–44.

Anderson, D. and Grove, R. (1987) 'The scramble for Eden: past, present and future in African conservation', pp. 1–12 in D. Anderson and R. Grove (eds), *Conservation in Africa: people, policies and practice*, Cambridge University Press, Cambridge.

Andrae, G. and Beckman, B. (1985) *The Wheat Trap: bread and underdevelopment in Nigeria*, Zed Press, London.

Andrews-Speed, P., Yang, M., Shen, L. and Cao, S. (2003) 'The regulation of China's township and village coal mines: a study of complexity and ineffectiveness', *Journal of Cleaner Production* 11: 185–96.

Annan, K. A. (2002) 'Toward a sustainable future', *Environment* 44(7): 10–15.

Arbona, S. I. (1998) 'Commercial agriculture and agrochemicals in Almolonga, Guatemala', *Geographical Review* 88: 47–63.

Armitage, D. (2002) 'Socio-institutional dynamics and the political ecology of mangrove forest conservation in Central Sulawesi, Indonesia', *Global Environmental Change* 12: 203–17.

Armsworth, P. R. and Roughgarden, J. E. (2001) 'An invitation to ecological economics', *Trends in Ecology & Evolution* 16: 229–34.

Arvelo-Jiménez, N. (1984) 'The politics of cultural survival in Venezuela: beyond indigismo', pp. 105–26 in M. Schmick and C. H. Wood (eds), *Frontier Expansion in Amazonia*, University of Florida Press, Gainesville.

Arts, B. (2002) '"Green alliances" of businesses and NGOs: new styles of self regulation or "dead-end roads"?', *Corporate Social Responsibility and Environmental Management* 9: 26–36.

Aseniero, G. (1985) 'A reflection on developmentalism: from development to transformation', pp. 48–85 in H. Addo, S. Amin, G. Aseniero, A. G. Frank, M. Friberg, F. Frobel, J. Heinrichs, B. Heltne, O. Kreye and H. Seki, *Development as Social Transformation: reflections on the global problématique*, Hodder and Stoughton, Sevenoaks, for the United Nations University.

Atkinson, A. (1993) 'Are Third World megacities sustainable? Jabotabek as an example', *Journal of International Development* 5: 605–22.

Atkinson, G., Dietz, S. and Neumayer, E. (2007) (eds) *Handbook of Sustainable Development*, Edward Elgar, Cheltenham.

Atkinson, G., Dubourg, R., Hamilton, K., Munasinghe, M., Pearce, D. and Young, C. (1997) *Measuring Sustainable Development: macroeconomics and the environment*, Edward Elgar, Aldershot.

Atkinson, T. R. (1973) 'Resettlement programmes in the Kainji Lake Region', in A. L. Mabogunje (ed.), *Kainji: a Nigerian man-made lake; Kainji Lake Studies vol. 2: Socioeconomic conditions*, Nigerian Institute for Social and Economic Research, Ibadan.

Attaran, A. (2005) 'An immeasurable crisis? A criticism of the Millennium Development Goals and why they cannot be measured', *PLoS Medicine* 2: 955–1.

Aubréville, A. (1949) *Climats, forêts et désertification de l'Afrique tropicale*, Société d'Édition Géographiques, Maritimes et Coloniales, Paris.

Auty, R. M. (1997) 'Pollution patterns during the industrial transition', *Geographical Journal* 163: 216–24.

B&Q (1995) *How Green is my Front Door? B&Q's second environmental review, February 1993–July 1995*, B&Q plc, Eastleigh, Hants.

Bahro, R. (1978) *The Alternative in Eastern Europe*, New Left Books, London (republished Verso, London, 1981).

Bahro, R. (1982) *Socialism and Survival*, Heretic, London.

Bahro, R. (1984) *From Red to Green: interviews with 'New Left Review'*, Verso, London.

Bailey, I. (2007) 'Market environmentalism, new environmental policy instruments, and climate policy in the United Kingdom and Germany', *Annals of the Association of American Geographers* 97: 530–50.

Baker, K. M. (1995) 'Drought, agriculture and environment: a case study from the Gambia, West Africa', *African Affairs* 94: 67–86.

Baldwin, K. D. S. (1957) *The Niger Agricultural Project: an experiment in African development*, Blackwell, Oxford.

Balint, P. J. and Mashinya, J. (2005) 'The decline of a model community-based conservation project: governance, capacity, and devolution in Mahenye, Zimbabwe', *Geoforum* 37: 805–15.

Balmford, A. and Whitten, T. (2003) 'Who should pay for tropical conservation, and how could the costs be met?' *Oryx* 37: 238–50.

Balmford, A., Bruner, A., Cooper, P., Costanza, R., Farber, S., Green, R. E., Jenkins, M., Jefferiss, P., Jessamy, V., Madden, J., Munro, K., Myers, N., Naeem, S., Paavola, J., Rayment, M., Rosendo, S., Roughgarden, J., Trumper, K. and Turner, R. K. (2002) 'Why conserving wild nature makes economic sense', *Science* 297: 950–3.

Balmford, A., Bennun, L., ten Brink, B., Cooper, D., Côté, I. M. Crane, P. Dobson, A., Dudley, N., Dutton, I., Green, R. E., Gregory, R., Harrison, J., Kennedy, E. T., Kremen, C., Leader-Williams, N., Lovejoy, T., Mace, G., May, R., Mayaux, P., Phillips, J., Redford, K., Ricketts, T. H., Rodriguez, J. P., Sanjayan, M., Schei, P., van Jaarsveld, A. and Walther, B. A. (2005) 'Science and the Convention on Biological Diversity's 2010 target', *Science* 307: 212–13.

Balon, E. K. and Coche, A. G. (eds) (1974) *Lake Kariba: a man-made tropical ecosystem in Central Africa*, Monographiae Biologicae 24, Junk, The Hague.

Baluyut, E. A. (1985) 'The Agno River Basin (The Philippines)', pp. 15–54 in T. Petr (ed.), *Inland Fisheries in Multi-Purpose River Basin Planning and Development in Tropical Asian Countries: three case studies*, FAO Fisheries Technical Paper 265, Rome.

Bandarage, A. (1984) 'Women in development: liberalism, Marxism and Marxist-feminism', *Development and Change* 15: 495–515.

Bandyopadhyay, J. (1992) 'Sustainability and survival in the mountain context', *Ambio* 21: 297–302.

Banks, G. (2002). 'Mining and environment in Melanesia: contemporary debates reviewed', *Contemporary Pacific* 14(1): 39–67.

Banks, G. and Ballard, C. (1997) (eds) *The Ok Tedi Settlement: issues, outcomes and implications*, Pacific Policy Paper No. 27, Resource Management in Asia-Pacific and National Centre for Development Studies, Canberra.

Barbier, E. B. (1994) 'Natural capital and the economics of environment and development', pp. 290–322 in A. Jansson, M. Hammer, C. Folke and R. Costanza (eds), *Investing in Natural Capital: the ecological economics approach to sustainability*, Island Press, Washington.

Barbier, E. B. (1998) 'Valuing environmental functions: tropical wetlands', pp. 344–69 in E. B. Barbier, *The Economics of Environment and Development: selected essays*, Edward Elgar, Cheltenham.

Barbier, E. B., Burgess, J. C., Bishop, J. and Aylward, B. (1995) *The Economics of the Tropical Timber Trade*, Earthscan, London.

Barbier, E., Markandya, A. and Pearce, D. (1990) 'Environmental sustainability and cost–benefit analysis', *Environment and Planning A* 22: 1259–66, reprinted as pp. 54–64 in E. B. Barbier, *The Economics of Environment and Development: selected essays*, Edward Elgar, Cheltenham.

Bardach, J. F. (1973) 'Some ecological consequences of Mekong River development plans', in M. T. Farvar and J. P. Milton (eds), *The Careless Technology: ecology and international development*, Stacey, London.

Barker, D. and Spence, B. (1988) 'Afro-Caribbean agriculture: a Jamaican Maroon community in transition', *Geographical Journal* 154: 198–208.

Barnett, T. (1980) 'Development: black box or Pandora's box?', pp. 306–24 in J. Heyer, P. Roberts and G. Williams (eds), *Rural Development in Tropical Africa*, Macmillan, London.

Barraclough, S. L. and Ghimire, K. B. (1995) *Forests and Livelihoods: the social dynamics of deforestation in developing countries*, Macmillan, London.

Barrett, C. B. and Arcese, P. (1995) 'Are integrated conservation–development projects (ICDPs) sustainable? On the conservation of large mammals in sub-Saharan Africa', *World Development* 23: 1073–84.

Barrow, C. J. (1987) 'The environmental effects of the Tucuruí Dam on the middle and lower Tocatins River Basin, Brazil', *Regulated Rivers* 1: 44–60.

Barrow, C. J. (1997) *Environmental and Social Impact Assessment: an introduction*, Arnold, London.

Barrow, C. J. (2000) *Social Impact Assessment: an introduction*, Arnold, London.

Barrow, E. and Murphree, M. (2001) 'Community conservation: from concept to practice', pp. 24–37 in D. Hulme and M. Murphree (eds), *African Wildlife and Livelihoods: the promise and performance of community conservation*, James Currey, London.

Bartelmus, P. (1994) *Environment, Growth and Development: the concepts and strategies of sustainability*, Routledge, London.

Bartelmus, P. (2007) 'SEEA-2003: accounting for sustainable development?', *Ecological Economics* 61: 613–16.

Barton, G. A. (2002) *Empire Forestry and the Origins of Environmentalism*, Cambridge University Press, Cambridge.

Bass, S., Reid, H., Satterthwaite, D. and Steele, P. (2005) *Reducing Poverty and Sustaining the Environment: the politics of local engagement*, Earthscan, London.

Bass, S. M. J. (1988) 'National conservation strategy, Zambia', pp. 186–91 in C. Conroy and M. Litvinoff (eds), *The Greening of Aid: sustainable livelihoods in practice*, Earthscan, London.

Bassett, T. J. (1993) 'Introduction: the land question and agrarian transformation in sub-Saharan Africa', pp. 3–31 in T. J. Bassett and D. E. Crummey (eds), *Land in African Agrarian Systems*, University of Wisconsin Press, Madison.

Bate, J. (1991) *Romantic Ecology: Wordsworth and the romantic tradition*, Routledge, London.

Bates, M. (1953) *Where Winter Never Comes: a study of man and nature in the tropics*, Victor Gollancz, London.

Batisse, M. (1975) 'Man and the biosphere', *Nature* 256: 156–8.

Batisse, M. (1982) 'The biosphere reserve: a tool for environmental conservation and management', *Environmental Conservation* 9: 101–11.

Baviskar, A. (1995) *In the Belly of the River: tribal conflicts over development in the Narmada Valley*, Oxford University Press, Delhi.

Baxter, R. M. (1977) 'Environmental effects of dams and impoundments', *Annual Review of Ecology and Systematics* 8: 255–84.

Bayliss-Smith, T. P., Bedford, R., Brookfield, H. and Latham, M. (1988) *Islands, Islanders and the World: the colonial and post-colonial experience of eastern Fiji*, Cambridge University Press, Cambridge.

Beattie, J. (2002) 'Dam building, dissent and development: the emergence of the Three Gorges Project', *New Zealand Journal of Asian Studies* 4(1): 138–58.

Bebbington, A. (2004) 'Social capital and development studies 1: critique, debate, progress?' *Progress in Development Studies* 4: 343–9.

Bebbington, A. J., Guggenheim, S., Olson, E. and Woolcock, M. (2004) 'Exploring Social Capital Debates at the World Bank', *Journal of Development Studies* 40: 33–64.

Bebbington, A. J. (1996) 'Movements, modernisations and markets: indigenous organisations and agrarian strategies in Ecuador', pp. 86–109 in R. Peet and M. Watts (eds), *Liberation Ecologies: environment, development, social movements*, Routledge, London (1st edn).

Beck, U. (1992) *Risk Society: towards a new modernity*, Sage, London (originally published in German, 1986).

Beck, U. (1994) 'The reinvention of politics: towards a theory of reflexive modernization', pp. 1–55 in U. Beck, A. Giddens and S. Lash, S. (eds), *Reflexive Modernization: politics, traditions and aesthetics in the modern social order*, Polity Press, Cambridge.

Beck, U. (1995) *Ecological Politics in an Age of Risk*, Polity Press, Cambridge (originally published in German, Suhrkamp Verlag, Frankfurt am Main, 1988).

Beck, U., Giddens, A. and Lash, S. (1994) (eds) *Reflexive Modernization: politics, traditions and aesthetics in the modern social order*, Polity Press, Cambridge.

Becker, C. D. (2003) 'Grassroots to grassroots: why forest preservation was rapid at Loma Alta, Ecuador', *World Development* 31: 163–76.

Becker, H. A. and Vanclay, F. (2003) (eds) *The International Handbook of Social Impact Assessment: conceptual and methodological advances*, Edward Elgar, Cheltenham.

Beckerman, W. (1974) *In Defence of Economic Growth*, Jonathan Cape, London.

Beckerman, W. (1994) 'Sustainable development: is it a useful concept?', *Environmental Values* 3: 191–209.

Beckerman, W. (1995a) 'How would you like your "sustainability" sir? Weak or strong? A reply to my critics', *Environmental Values* 4: 169–79.

Beckerman, W. (1995b) *Small is Stupid: blowing the whistle on the Greens*, Duckworth, London.

Behnke, R. H. and Scoones, I. (1991) *Rethinking Range Ecology: implications for range management in Africa*, ODI/IIED, London.

Behnke, R. H., Scoones, I. and Kerven, C. (1993) *Range Ecology at Disequilibrium: new models of natural variability and pastoral adaptation in African savannas*, Overseas Development Institute, London.

Beinart, W. (1984) 'Soil erosion, conservation and ideas about development: a Southern African exploration 1900–1960', *Journal of Southern African Studies* 11: 52–84.

Beinart, W. and Bundy, C. (1980) 'State intervention and rural resistance: the Transkei 1900–1965', pp. 272–315 in M. Klein (ed.) *Peasants in Africa: historical and contemporary perspectives,* Sage Publishing, Beverly Hills, CA, and London.

Beinart, W. and Coates, P. (1995) *Environment and History: the taming of nature in the USA and South Africa*, Routledge, London.

Bell, J. (1986) 'Caustic waste menaces Jamaica', *New Scientist* 3 April: 33–7.

Belshaw, D. (1974) 'Taking indigenous technology seriously: the case of intercropping techniques in East Africa', *Institute of Development Studies Bulletin* 10: 24–7.

Benech, V. (1992) 'The northern Cameroon floodplain: influence of hydrology on fish production', pp. 155–64 in E. Maltby, P. J. Dugan and J. C. Lefeuve (eds), *Conservation and Development: the sustainable use of wetland resources*, IUCN, Gland, Switzerland.

Bennett, V. (1998) 'Housewives, urban protest and water policy in Monterrey, Mexico', *Water Resources Development* 14: 481–97.

Benton, T. (1996) (ed) *The Greening of Marxism*, Guilford Press, New York.

Berkes, F. and Folke, C. (1994) 'Investing in natural capital for the sustainable use of natural capital', pp. 128–49 in A. Jansson, M. Hammer, C. Folke and R. Costanza (eds), *Investing in Natural Capital: the ecological economics approach to sustainability*, Island Press, Washington.

Berkes, F. and Folke, C. (1998) (eds) *Linking Social and Ecological Systems: management practices and social mechanisms for building resilience*, Cambridge University Press, Cambridge.

Bews, J. W. (1916) 'An account of the chief types of vegetation in South Africa, and notes on the plant succession', *Journal of Ecology* 4: 129–59.

Bideleux, R. (1987) *Communism and Development*, Methuen, London.

Bigg, T. (1995) 'The UN Commission on sustainable development: a non-governmental perspective', *Global Environmental Change* 5: 251–3.

Bigg, T. (2004) (ed.). *Survival for a Small Planet: the sustainable development agenda*, Earthscan, London.

Biswas, A. K. (1980) (ed.) 'The Nile and its environment', *Water Supply and Management* 4: 1–113.

Biswas, A. K. (2004) 'Dams: Cornucopia or Disaster?', *Water Resources Development* 20: 3–14.

Biswas, A. K. and Biswas, M. R. (1976) 'Hydropower and the environment', *Water Power and Dam Construction* 28: 40–3.

Biswas, M. R. and Biswas, A. K. (1984) 'Complementarity between environment and development processes', *Environmental Conservation* 11: 35–44.

Blaikie, P. (1981) 'Class, land use and soil erosion', *ODI Review* 2: 57–77.

Blaikie, P. (1985) *The Political Economy of Soil Erosion in Developing Countries*, Longman, London.

Blaikie, P. (1988) 'The explanation of land degradation in Nepal', pp. 132–58 in J. Ives and D. C. Pitt (eds), *Deforestation: social dynamics in watersheds and montane ecosystems*, Routledge, London.

Blaikie, P. (1995) 'Understanding environmental issues', pp. 1–30 in S. Morse and M. Stocking (eds), *People and the Environment*, UCL Press, London.

Blaikie, P. and Brookfield, H. (1987) *Land Degradation and Society*, Methuen, London.

Blaikie, P., Cannon, T., Davis, I. and Wisner, B. (1994) *At Risk: natural hazards, people's vulnerability and disasters*, Routledge, London.

Bledsoe, C. (1994) '"Children are like young bamboo trees": potentiality and reproduction in sub-Saharan Africa', pp. 105–38 in K. L. Kiessling and H. Landberg (eds), *Population, Economic Development and the Environment: the making of our common future*, Clarendon Press, Oxford.

Bliss-Guest, P. A. and Keckes, S. (1982) 'The regional seas programme of UNEP', *Environmental Conservation* 9: 43–9.

Boardman, R. (1981) *International Organizations and the Conservation of Nature*, Indiana University Press, Bloomington.

Boffey, P. M. (1976) 'International Biological Programme: was it worth the cost and effort?', *Science* 193: 866–8.

Bohringer, C. (2003) 'The Kyoto Protocol: A Review and Perspectives', *Oxford Review of Economic Policy* 19(3): 451–66.

Bolido, L. (1998) 'Helping farmers fight pests', *People and the Planet* 7(1): 18–19.

Bond, R., Curran, J., Kirkpatrick, C. and Lee, N. (2001) 'Integrated impact assessment for sustainable development: a case study approach', *World Development* 29: 1011–24.

Bonnie, R. and Schwartzman, S. (2003) 'Tropical reforestation and deforestation and the Kyoto Protocol', *Conservation Biology* 17: 4–5.

Bookchin, M. (1971) *Post-Scarcity Anarchism*, Ramparts, Berkeley, CA.

Bookchin, M. (1979) 'Ecology and revolutionary thought', *Antipode* 10(3)/11(1): 21–32.

Bookchin, M. (1982) *The Ecology of Freedom: the emergence and dissolution of hierarchy*, Cheshire, Palo Alto, CA.

Bookchin, M. and Foreman, D. (1991) *Defending the Earth: a dialogue between Murray Bookchin and Dave Foreman*, ed. S. Chase, South End Press, Boston.

Boon, R. G. J., Alexaki, A. and Becerra, E. H. (2001) 'The Ilo Clean Air Project: a local response to industrial pollution control in Peru', *Environment and Urbanization* 13: 215–32.

Boserup, E. (1965) *The Conditions of Agricultural Growth: the economics of agrarian change under population pressure*, Allen and Unwin, London.

Botkin, D. B. (1990) *Discordant Harmonies: a new ecology for the twenty-first century*, Oxford University Press, New York.

Boulding, K. E. (1966) 'The economics of the coming spaceship earth', pp. 3–14 in H. Jarrett (ed.), *Environmental Quality in a Growing Economy*, Resources for the Future, Johns Hopkins University Press, Baltimore, MD; reprinted as pp. 121–32 in H. E. Daly (ed.), *Towards a Steady-State Economy*, W. H. Freeman, New York, 1973.

Bourdieu, P. (1986) 'The forms of capital', pp. 141–258 in J. Richardson (ed.), *Handbook of Theory and Research for the Sociology of Education*, New York: Greenwood Press.

Bovill, E. W. (1921) 'The encroachment of the Sahara on the Sudan', *Journal of the African Society* 20: 174–85.

Boyd-Orr, Sir J. (1953) *The White Man's Dilemma: food and the future*, Allen and Unwin, London.

Bradnock, R. W. and Saunders, P. L. (2000) 'Sea-level rise, subsidence and emergence: the political ecology of environmental change in the Bengal delta', pp. 66–90 in P. Stott and S. Sullivan (eds), *Political Ecology: science, myth and power*, Arnold, London.

Bragdon, S. (1996) 'The Convention on Biological Diversity', *Global Environmental Change* 6: 177–9.

Brahic, C. (2007) 'As polluters quibble, the poor learn their fate', *New Scientist* 2599 (14 April): 11.

Brandon, K. E. and Wells, M. (1992) 'Planning for people and parks: design dilemmas', *World Development* 20: 557–70.

Brandon, K., Redford, K. H. and Sanderson, S.E. (1998) *Parks in Peril: people, politics and protected areas*. Island Press, for the Nature Conservancy, Washington.

Brandt, W. (1980) *North–South: a programme for survival*, Pan, London.

Brandt, W. (1983) *Common Crisis North–South: cooperation for world recovery*, Pan, London.

Bray, D. B. and Klepeis, P. (2005) 'Deforestation, forest transitions, and institutions for sustainability in southeastern Mexico, 1900–2000', *Environment and History* 11: 195–223.

Bray, D. B., Merino-Perez, L., Negeros-Castillo, P., Segura-Warnholtz, G., Torres-Rojo, J. M. and Vester, H. F. M. (2003) 'Mexico's community-managed forests as a global model for sustainable landscapes', *Conservation Biology* 17: 672–7.

Brechin S. R., Wilhusen, P. R., Fortwangler, C. L. and West, P. C. (2003) (eds), *Contested Nature: promoting international biodiversity with social justice in the twenty-first century*, State University of New York Press, New York.

Breitbart, M. (1981) 'Peter Kropotkin: the anarchist geographer', pp. 134–53 in D. R. Stoddart (ed.), *Geography, Ideology and Social Concern*, Blackwell, Oxford.

Breman, H. and de Wit, C. T. (1983) 'Rangeland productivity and exploitation in the Sahel', *Science* 221: 1341–7.

Brent, R. J. (1990) *Project Appraisal for Developing Countries*, Harvester Wheatsheaf, New York.

Bridger, G. A. and Winpenny, J. T. (1987) *Planning Development Projects: a practical guide to the choice and appraisal of public sector developments*, HMSO, London.

Broad, R. (1994) 'The poor and the environment: friends or foes?', *World Development* 22: 811–23.

Brockington, D. (2002) *Fortress Conservation: the preservation of the Mkomazi Game Reserve, Tanzania*, James Currey, Oxford.

Brockington, D. and Homewood, K. (1996) 'Wildlife, pastoralists and science: debates concerning Mkomazi Game Reserve, Tanzania', pp. 91–104 in M. Leach and R. Mearns (eds), *The Lie of the Land*, James Currey, Oxford.

Brockington, D. and Schmidt-Soltau, K. (2004) 'The social and environmental impacts of wilderness and development', *Oryx* 38: 140–2.

Brodie, J. and Morrison, J. (1984) 'The management and disposal of hazardous wastes in the Pacific Islands', *Ambio* 13: 331–3.

Brokensha, D. W., Warren, D. M. and Warner, O. (1980) (eds) *Indigenous Systems of Knowledge and Development*, University Press of America, Washington.

Bromley, D. W. (1989) *Economic Interests and Institutions: the conceptual foundations of public policy*, Blackwell, Oxford.

Bromley, D. W. (1992) (ed.) *Making the Commons Work: theory, practice and policy*, ICS Press, San Francisco.

Brookfield, H. (1975) *Interdependent Development*, Methuen, London.

Brooks, F. T. (1925) (ed.) *Imperial Botanical Conference, London 1924: report of proceedings*, Cambridge University Press, Cambridge.

Brooks, S. (2005) 'Images of "Wild Africa": nature tourism and the (re)creation of Hluhluwe Game Reserve, 1930–1945', *Journal of Historical Geography* 31: 220–49.

Brooks, T. M., Mittermeier, R. A., Fonseca, G. A. B. da, Gerlach, J., Hoffmann, M. Lamoreux, J. F., Mittermeier, C. G., Pilgrim, J. D. and Rodrigues, A. S. L. (2006) 'Global biodiversity conservation priorities', *Science* 313: 58–60.

Brosius, J. P., Tsing, A. L. and Zerner, C. (1998) 'Representing communities: histories and politics of community-based natural resource management', *Society and Natural Resources* 11: 157–68.

Brosius, J. P., Tsing, A. L. and Zerner, C. (2005) *Communities and Conservation: histories and politics of community-based natural resource management*, Altamira Press, Walnut Creek, CA.

Brosius, P. (2004) 'Indigenous peoples and protected areas at the World Parks Congress', *Conservation Biology* 18: 609–12.

Brouwer, R. (2005) (ed.) *Cost–Benefit Analysis and Water Resources Management*, Edward Elgar, Cheltenham.

Brown, K. (1997) 'The road from Rio', *Journal of International Development* 9: 383–9.

Brown, K. and Pearce, D. (1994) (eds) *The Causes of Tropical Deforestation: the economic and statistical analysis of factors giving rise to the loss of the tropical forests*, UCL Press, London.

Brown, K., Adger, N. W. and Turner, R. K. (1993) 'Global environmental change and mechanisms for North–South resource transfers', *Journal of International Development* 5: 571–89.

Brown, L. R. (2006) *Plan B. 2.0: rescuing a planet under stress and a civilization in trouble*, W. W. Norton, New York, for the Earth Policy Institute.

Brown, L. and Wolf, E. C. (1984) *Soil Erosion: quiet crisis in the world economy*, World-watch Paper 60, Washington.

Brundtland, H. (1987 *Our Common Future*, Oxford University Press, Oxford, for the World Commission on Environment and Development.

Bruwer, C., Poultney, C. and Nyathi, Z. (1996) 'Community-based hydrological management of the Phongolo floodplain', pp. 199–211 in M. C. Acreman and G. E. Hollis (eds), *Water Management and Wetlands in Sub-Saharan Africa*, IUCN, Gland, Switzerland.

Bryant, R. (1998) 'Power, knowledge and political ecology in the Third World', *Progress in Physical Geography* 22: 79–94.

Bryant, R. and Bailey, S. (1997) *Third World Political Ecology*, Routledge, London.

Buckley, G. P. (ed.) (1989) *Biological Habitat Reconstruction*, Belhaven, London.

Buckley, P. (1995) 'Critical natural capital: operational flaws in a valid concept', *Ecos: A Review of Conservation*, 16(3/4): 13–18.

Budiardjo, C. (1986) 'The politics of transmigration', *Ecologist* 16: 57–116.

Bull, D. (1982) *A Growing Problem: pesticides and the Third World*, Oxfam Books, Oxford.

Bunce, M. (1994) *The Countryside Ideal: Anglo-American images of landscape*, Routledge, London.

Bunker, S. G. (1980) 'The impact of deforestation on peasant communities in the Medio Amazonas of Brazil', pp. 45–60 in V. H. Sutlive, N. Altshuler and M. D. Zamora (eds), *Where Have All the Flowers Gone? Deforestation in the Third World*, Studies in Third World Societies No. 13, College of William and Mary, Williamsburg, VA.

Burgess, R. (1978) 'The concept of nature in geography and Marxism', *Antipode* 10: 1–11.

Bush, E. J. and Harvey, L. D. D. (1997) 'Joint implementation and the ultimate objective of the United Nations Framework Convention on Climate Change', *Global Environmental Change* 7: 265–85.

Butcher, D.A. (1967) *An Operational Manual for Resettlement: a systematic approach to the resettlement problem created by man-made lakes, with special relevance for West Africa*, FAO, Rome.

Buxton, P. A. (1935) 'Seasonal changes in vegetation in the north of Nigeria,' *Journal of Ecology* 23: 134–9.

Byres, T. (1979) 'Of neo-populist pipe-dreams', *Journal of Peasant Studies* 4: 210–44.

Byron, N. and Arnold, M. (1999) 'What future for the people of the tropical forests?', *World Development* 27: 789–805.

Byron, N. and Shepherd, G. (1998) *Indonesia and the 1997–8 El Niño: fire problems and long-term solutions*, ODI Natural Resource Perspectives 28, Overseas Development Institute, London.

Cairns, J., Jr (1991) 'The status of the theoretical and applied science of restoration ecology', *Environmental Professional* 13: 186–94.

Caldecott, J. (1996) *Designing Conservation Projects*, Cambridge University Press, Cambridge.

Caldwell, L. K. (1984) 'Political aspects of ecologically sustainable development', *Enviromental Conservation* 11: 299–308.

Campbell, C. in collaboration with the Women's group of Xapuri (1996) 'Out on the front line but still struggling for voice: women in the rubber tappers' defense of the forest in Xapuri, Acre, Brazil', pp. 27–61 in D. Rocheleau, B. Thomas-Slayter and E. Wangari (eds), *Feminist Political Ecology: global issues and local experiences*, Routledge, London.

Campbell, L. M. (2002a) 'Conservation narratives and the received wisdom of ecotourism: case studies from Costa Rica', *International Journal of Sustainable Development*, 5: 300–25.

Campbell L. M. (2002b) 'Science and sustainable use: views of conservation experts', *Ecological Applications* 12: 122–46.

Campbell, L. M. (2005) 'Overcoming obstacles to interdisciplinary research', *Conservation Biology* 19: 574–7.

Campbell, M., Sahin-Hodoglugil, N. N. and Potts, M. (2006) 'Barriers to fertility regulation: a review of the literature', *Studies in Family Planning* 37: 87–98.

Capra, F. and Spretnak, C. (1984) *Green Politics*, E. P. Dutton, New York.

Carney, D. (1998) (ed.) *Sustainable Rural Livelihoods: what contribution can we make?*, Department for International Development, London.

Carney, D. (2002) *Sustainable Livelihoods Approaches: progress and possibilities for change*, Department for International Development, London.

Carney, J. (1993) 'Converting the wetlands, engendering the environment: the intersection of gender with agrarian change in the Gambia', *Economic Geography* 69: 329–48.

Carr, S. and Mpande, R. (1996) 'Does the definition of the issue matter? NGO influence and the International Convention to Combat Desertification in Africa', pp. 143–66 in D. Potter (ed.), *NGOs in Africa and Asia*, Frank Cass, London.

Carr-Saunders, A. M. (1936) *World Population: past growth and present trends*, Clarendon Press, Oxford.

Carruthers, D. (2001) 'Environmental politics in Chile: legacies of dictatorship and democracy', *Third World Quarterly* 22: 343–58.

Carruthers, J. (1995) *The Kruger National Park: a social and political history*, Natal University Press, Durban.

Carswell, G. (2003) 'Soil conservation policies in colonial Kigezi, Uganda: successful implementation and an absence of resistance', pp. 131–54 in W. Beinart and J. McGregor (eds), *Social History and African Environments*, James Currey, Oxford.

Carswell, G. (2007) *Cultivating Success in Uganda: Kigezi farmers and colonial policies*, James Currey, Oxford.

Carter, A. (1993) 'Towards a green political theory', pp. 39–62 in A. Dobson and P. Lucardie (eds), *The Politics of Nature: explorations in green political theory*, Routledge, London.

Castelán, E. (2002) 'Role of large dams in the socio-economic development of Mexico', *Water Resources Development* 18: 163–77.

Castells, M. ([1996] 2000) *The Information Age: economy, society,and culture*, vol. 1, *The Rise of the Network Society*, Blackwell, Oxford (revised edn).

Castleman, B. (1981) 'Double standards: asbestos in India', *New Scientist* 26 February: 522–3.

Caufield, C. (1982) *Tropical Moist Forests: the resource, the people, the threat*, Earthscan/ IIED, London.

Caufield, C. (1984) 'Pesticides: exporting death', *New Scientist* 16 August: 15–17.

Caufield, C. (1985) *In the Rainforest*, Pan Books, London.

Caufield, C. (1987) 'Conservationists scorn plans to save tropical forests', *New Scientist* 25 June: 33.

Caufield, C. (1989) *Innocent Exposures: chronicles of the radiation age*, Secker and Warburg, London.

Cernea, M. M. (1988) *Involuntary Resettlement in Development Projects: policy guidelines in World Bank-financed projects*, Technical Paper 80, World Bank, Washington.

Cernea, M. M. (1991) (ed.) *Putting People First: sociological variables in rural development*, Oxford University Press, Oxford, for the World Bank.

Cernea, M. M. (1997) 'The risks and reconstruction model for resettling displaced populations', *World Development* 25: 1569–89.

Cernea, M. M. (2006) 'Population displacement inside protected areas: a redefinition of concepts in conservation politics', *Policy Matters* 14: 8–26.

Cernea, M. M. and McDowell, C. (2000) (eds) *Risks and Reconstruction Experiences of Resettlers and Refugees*, World Bank, Washington.

Chambers, R. (1970) (ed.) *The Volta Resettlement Experience*, Pall Mall, London.

Chambers, R. (1983) *Rural Development: putting the last first*, Longman, London.

Chambers, R. (1988a) *Managing Canal Irrigation: practical analysis from South Asia*, Cambridge University Press, Cambridge.

Chambers, R. (1988b) 'Sustainable rural livelihoods: a key strategy for people, environment and development', pp. 1–17 in C. Conroy and M. Litvinoff (eds), *The Greening of Aid: sustainable livelihoods in practice*, Earthscan, London.

Chambers, R. (1997) *Whose Reality Counts? putting the first last*, Intermediate Technology Publications, London.

Chambers, R. (2001) 'The World Development Report: concepts, content and a chapter 12', *Journal of International Development* 13: 299–306.

Chambers, R. (2005) *Ideas for Development*, Earthscan, London.

Chape, S., Harrison, J., Spalding, M. and Lysenko, I. (2005) 'Measuring the extent and effectiveness of protected areas as an indicator for meeting global biodiversity targets', *Philosophical Transactions of the Royal Society B* 360: 443–55.

Chapin, M. (2004) 'A challenge to conservationists', *World Watch* 17(6): 17–31.

Chapman, G. P., Kumar, K., Fraser, C. and Gaber, I. (1997) *Environmentalism and the Mass Media: the North–South divide*, Routledge, London.

Charney, J. (1975) 'Dynamics of deserts and drought in the Sahara', *Quarterly Journal of the Royal Meteorological Society* 101: 193–202.

Charney, J., Stone, P. H. and Quirk, W. J. (1975) 'Drought in the Sahara: a biogeophysical feedback mechanism', *Science* 187: 434–5.

Chatterjee, P. and Finger, M. (1994) *The Earth Brokers: power, politics and world development*, Routledge, London.

Chen, S. and Ravallion, M. (2007) 'Absolute poverty measures for the developing world, 1981–2004', *Proceedings of the National Academy of Sciences* 104: 16757–62

Chisholm, A. (1972) *Philosophers of the Earth: conversations with ecologists*, Sidgwick and Jackson, London.

Chisholm, N. and Grove, J. M. (1985) 'The lower Volta', pp. 229–50 in A. T. Grove (ed.), *The Niger and its Neighbours: environmental history and hydrobiology, human use and health hazards of the major West African rivers*, Balkema, Rotterdam.

Christiansson, C. and Ashuvud, J. (1985) 'Heavy industry in a rural tropical ecosystem', *Ambio* 14: 122–33.

Christoff, P. (1996) 'Ecological modernisation, ecological modernities', *Environmental Politics* 5: 476–99.

Chronic Poverty Research Centre (2005) *Chronic Poverty Report 2004–5*, Chronic Poverty Research Centre, University of Manchester.

Ciriacy-Wantrup, S. V. (1952) *Resource Conservation, Economics and Policies*, University of California Press, Berkeley.

Clark, W. C. and Munn, R. E. (1986) (eds) *Sustainable Development of the Biosphere*, Cambridge University Press, Cambridge.

Clarke, R. and Timberlake, L. (1982) *Stockholm Plus Ten: promises promises? the decade since the 1972 UN Environment Conference*, Earthscan, London.

Clay, E. J. and Schaffer, B. B. (1984) (eds) *Room for Manoeuvre: an exploration of public policy in agriculture and rural development*, Heinemann, London.

Cleaver, K. (1994) 'Deforestation in the Western and Central African forest: the agricultural and demographic causes, and some solutions', pp. 65–78 in K. Cleaver (ed.) *Conservation of West and Central African Rainforests*, World Bank Environmental Paper No. 1, World Bank, Washington.

Clements, F. E. (1905) *Research Methods in Ecology*, University Publishing, Lincoln, Neb.

Cleveland, C. J., Stern, D. I. and Costanza, R. (2001) (eds) *The Economics of Nature and the Nature of Economics*, Edward Elgar, London.

Cline-Cole, R. and Madge, C. (2000) (eds) *Contesting Forestry in West Africa*, Ashgate, Aldershot.

Clynes, T. (2002) 'They shoot poachers, don't they?' *Observer Magazine*, 24 November.

Coe, M. J., Cummings, D. H. and Phillipson, J. (1976) 'Biomass and production of large African herbivores in relation to rainfall and primary production', *Oecologia* 22: 341–54.

Cohen, J. E. (2007) 'Human population: the next half century', pp. 13–21 in D. Kennedy and the editors of *Science* magazine (eds), *Science Magazine's State of the Planet 2006–2007*, Island Press, Washington.

Cohen, S., Demeritt, D., Robinson, J. and Rothman, D. (1998) 'Climate change and sustainable development: towards dialogue', *Global Environmental Change* 8: 341–71.

Colchester, M. (1985) 'An end to laughter? Hydropower projects in central India', in *An End to Laughter?*, Review 44, Survival International, London.

Colchester, M. (1994) 'Sustaining the forests: the community-based approach in South and South East Asia', *Development and Change* 25: 69–93.

Colchester, M. (1997) 'Salvaging nature: indigenous peoples and protected areas', pp. 97–130 in K. Ghimire and M. Pimbert (eds), *Social Change and Conservation: environmental politics and impacts of national parks and protected areas*, Earthscan, London.

Colchester, M. (2002) *Salvaging Nature: indigenous peoples, protected areas and biodiversity conservation*, World Rainforest Movement, Montevideo.

Colchester, M. (2004) 'Conservation policy and indigenous peoples', *Cultural Survival Quarterly* 28(1): 17–22.

Cole, S. (1978) 'The global futures debate 1965–1976', in C. Freeman and M. Jahoda (eds), *World Futures: the great debate*, Martin Robinson, London.

Coleman, J. S. (1990) *Foundations of Social Theory*, Harvard University Press, Cambridge, MA.

Collett, D. (1987) 'Pastoralists and wildlife: image and reality in Kenyan Masailand', pp. 129–48 in D. M. Anderson and R. H. Grove (eds), *Conservation in Africa: people, policies and practice*, Cambridge University Press, Cambridge.

Collins, R. O. (1990) *The Waters of the Nile: hydropolitics and the Jonglei Canal 1900–1988*, Clarendon Press, Oxford.

Colson, E. (1971) *The Social Consequences of Resettlement*, Kariba Studies IV, University of Manchester Press, Manchester.

Colston, A. (2003) 'Beyond preservation: the challenge of ecological restoration', pp. 247–67 in W. M. Adams and M. Mulligan (eds), *Decolonizing Nature: strategies for conservation in a post-colonial era*, Earthscan, London.

Common, M. and Stagl, S. (2005) *Ecological Economics: an introduction*, Cambridge University Press, Cambridge.

Commoner, B. (1972) *The Closing Circle: confronting the environmental crisis*, Jonathan Cape, London.

Conklin, B. A. and Graham, L. R. (1995) 'The shifting middle ground: Amazonian Indians and eco-politics', *American Anthropologist* 97: 695–710.

Conroy, C. (1988) 'Introduction', pp. xi–xiv in C. Conroy and M. Litvinoff (eds), *The Greening of Aid: sustainable livelihoods in practice*, Earthscan, London.

Conroy, C. and Litvinoff, M. (1988) (eds), *The Greening of Aid: sustainable livelihoods in practice*, Earthscan, London.

Conway, G. R. (1985) 'Agroecosystem analysis', *Agricultural Administration* 20: 31–55.

Conway, G. R. and Pretty, J. N. (1991) *Unwelcome Harvest: agriculture and pollution*, Earthscan, London.

Conwentz, H. (1914) 'On national and international protection of nature', *Journal of Ecology* 2: 109–22

Cooke, B. and Kothari, U. (2001) (eds), *Participation: the new tyranny?* Zed Press, London.

Coomes, O. T. and Barham, B. L. (1994) 'The Amazon rubber boom: labor control, resistance, and failed plantation development revisited', *Hispanic American History Review* 74: 231–57.

Coomes, O. T. and Barham, B. L. (1997) 'Rain forest extraction and conservation in Amazonia', *Geographical Journal* 163: 180–8.

Copans, J. (1983) 'The Sahelian drought: social sciences and the political economy of underdevelopment', pp. 83–97 in K. Hewitt (ed.) *Interpretations of Calamity: from the viewpoint of human ecology*, Allen and Unwin, Hemel Hempstead.

Coppock, D. L., Ellis, J. E. and Swift, D. M. (1986) 'Livestock feeding ecology and resource utilisation in a nomadic pastoral ecosystem', *Journal of Applied Ecology* 23: 573–85.

Corbridge, S. E. (1982) 'Interdependent development? Problems of aggregation and implementation in the Brandt Report', *Applied Geography* 2: 253–65.

Corbridge, S. E. (1993) 'Marxisms, modernities, and moralities: development praxis and the claims of distant strangers', *Environment and Planning D: Society and Space* 11: 449–72.

Cornwall, A. (2007) 'Buzzwords and fuzzwords: deconstructing development discourse', *Development in Practice* 17: 471–84.

Costanza, R. (1989) 'What is ecological economics?', *Ecological Economics* 1: 1–7.

Costanza, R. (1991) (ed.) *Ecological Economics: the science and management of sustainability*, Columbia University Press, New York.

Costanza, R. (1998) 'The value of ecosystem services', *Ecological Economics* 25: 1–2 (special section on valuation of ecosystem services).

Costanza, R. and Daly, H. E. (1992) 'Natural capital and sustainable development', *Conservation Biology* 6: 37–46.

Costanza, R., Cumberland, J. H., Daly, H., Goodland, R. and Norgarard, R. B. (1997a) *An Introduction to Ecological Economics*, CRC Press, London.

Costanza, R., d'Arge, R., de Groot, R., Farber, S., Grasso, M., Hanon, B., Limburg, K., Naeem, S., O'Neill, R. V., Paruelo, J., Raskin, R. G., Sutton, P. and van den Belt, M. (1997b), 'The value of the world's ecosystem services and natural capital', *Nature* 387: 253–60.

Costanza, R., d'Arge, R., de Groot, R., Farber, S., Grasso, M., Hanon, B., Limburg, K., Naeem, S., O'Neill, R. V., Paruelo, J., Raskin, R. G., Sutton, P. and van den Belt, M. (1998) 'The value of the world's ecosystem services: putting the issues in perspective', *Ecological Economics* 25: 67–72 (special section on valuation of ecosystem services)

Cotgrove, S. (1982) *Catastrophe or Cornucopia: the environment, politics and the future*, Wiley, Chichester.

Cotgrove, S. and Duff, A. (1980) 'Environmentalism, middle class radicalism and politics', *Sociology Review* 28: 235–351.

Coughenour, M. B., Ellis, J. E., Swift, D. M., Coppock, D. L., Galvin, K., McCabe, J. T. and Hart, T. C. (1985) 'Energy extraction and use in a nomadic pastoral ecosystem', *Science* 230(4726): 619–25.

Courel, M. F., Kandel, R. S. and Rasool, S. I. (1984) 'Surface albedo and the Sahel drought', *Nature* 307: 528–31.

Cowen, M. P. and Shenton, R. W. (1995) 'The invention of development', pp. 27–43 in J. Crush (ed.), *Power of Development*, Routledge, London.

Cowen, M. P. and Shenton, R. W. (1996) *Doctrines of Development*, Routledge, London.

Cowlishaw, G., Mendelson, S. and Rowcliffe, J. M. (2005) 'Evidence for post-depletion sustainability in a mature bushmeat market', *Journal of Applied Ecology* 42: 460–8.

Cox, P. (1985) *Pesticide Use in Tanzania*, ODI Economic Research Bureau, Dar es Salaam.

Croll, E. and Parkin, D. (1992) 'Cultural understandings of the environment', pp. 11–36 in E. Croll and D. Parkin (eds), *Bush Base: Forest Farm: culture, environment and development*, Routledge, London.

Cronon, W. (1995) 'The trouble with wilderness, or, getting back to the wrong nature', pp. 69–90 in W. Cronon (ed.), *Uncommon Ground: toward reinventing nature*, New York: W. W. Norton, New York.

Crosby, A. W. (1986) *Ecological Imperialism: the ecological expansion of Europe, 1600–1900*, Cambridge University Press, Cambridge.

Crummey, D. (1986) (ed.) *Banditry, Rebellion and Social Protest in Africa*, James Currey/ Heinemann, London.

Crush, J. C. (1995) 'Imagining development', pp. 1–23 in J. C. Crush (ed.), *Power of Development*, Routledge, London.

Culwick, A.T. (1943) 'New beginning', *Tanganyika Notes and Records* 15: 1–6.

Curry-Lindahl, K. (1986) 'The conflict between development and nature conservation with special reference to desertification', pp. 106–130 in N. Polunin (ed.), *Ecosystem Theory and Application*, Wiley, Chichester.

Cushing, D. H. (1988) *The Provident Sea*, Cambridge University Press, Cambridge.

Cutter, S. L. (2006) *Hazards, Vulnerability and Environmental Justice*, Earthscan, London.

D'Itri, P. A. and D'Itri, F. M. D. (1977) *Mercury Contamination: a human tragedy*, Wiley, Chichester.

Dahlberg, K. A., Soroos, M. S., Ferau, A. T., Harf, J. E. and Trout, B. T. (1985) *Environment and the Global Arena: actors, values, policies and futures*, Duke University Press, Durham, NC.

Daily, G. C. (1997) *Nature's Services: societal dependence on natural systems*, Island Press, Washington.

Dalland, Ø. (1997) 'The last big dam in Norway: whose victory?', pp. 41–56 in A. D. Usher (ed.), *Dams as Aid: a political anatomy of Nordic development thinking*, Routledge, London.

Dalton, R. J., Recchia, S. and Rohrschneider, R. (2003) 'The environmental movement and the modes of political action', *Comparative Political Studies* 36: 743–71.

Daly, H.E. (1973) (ed.) *Towards a Steady-State Economy*, W. H. Freeman, New York.

Daly, H. E. (1977) *Steady-State Economics: the economics of biophysical equilibrium and moral growth*, W. H. Freeman, New York.

Daly, H. E. (1990) 'Toward some operational principles of sustainable development', *Ecological Economics* 2: 1–6.

Daly, H. E. (1994) 'Operationalising sustainable development by investing in natural capital', pp. 22–37 in A. Jansson, M. Hammer, C. Folke and R. Costanza (eds), *Investing in Natural Capital: the ecological economics approach to sustainability*, Island Press, Washington.

Daly, H. E. (2007) *Ecological Economics and Sustainable Development: selected essays of Herman Daly*, Edward Elgar, London.

Daly, H. E. and Cobb, J. R. (1989) *For the Common Good: redirecting the economy towards community, the environment and a sustainable future*, Beacon Press, Boston.

Dasmann, R. F. (1980) 'Ecodevelopment: an ecological perspective', pp. 1331–5 in J. I. Furtado (ed.), *Tropical Ecology and Development*, International Society of Tropical Ecology, Kuala Lumpur.

Dasmann, R. F., Milton, J. P. and Freeman, P. H. (1973) *Ecological Principles for Economic Development*, Wiley, Chichester.

Davies, B. R., Hall, A. and Jackson, P. B. N. (1972) 'Some ecological aspects of the Cabora Bassa Dam', *Biological Conservation* 8: 189–201.

Davis, D. K. (2005) 'Indigenous knowledge and the desertification debate: problematising expert knowledge in North Africa', *Geoforum* 36: 509–24.

Davis, M. (2004) 'The political ecology of famine: the origins of the Third World', pp. 48–62 in R. Peet and M. Watts (eds), *Liberation Ecologies: environment, development, social movements*, Routledge, London (2nd edn).

Davis, M. (2006) *Planet of Slums*, Verso, London.

Davy, J. B. (1925) 'Correlation of taxonomic work in the Dominions and Colonies with work at home', pp. 214–34 in F. T. Brooks (ed.), *Imperial Botanical Conference, London 1924, Report of Proceedings*, Cambridge University Press, Cambridge.

de Groot, H. L. F., Withagen, C. A. and Minliang, Z. (2004) 'Dynamics of China's regional development and pollution: an investigation into the Environmental Kuznets Curve', *Environment and Development Economics* 9: 507–37.

de Schlippe, P. (1956) *Shifting Cultivation in Africa: the Zande system of agriculture*, Routledge and Kegan Paul, London.

de Waal, A. (1989) *Famine that Kills: Darfur, Sudan, 1984–1985*, Clarendon Press, Oxford.

Delang, C. O. (2005) 'The political ecology of deforestation in Thailand'. *Geography* 90: 225–37.

Demeritt, D. (2001) 'Scientific forest conservation and the statistical picturing of nature's limits in the Progressive Era United States', *Environment and Planning D: Society and Space* 19: 431–59.

Denevan, W. M. (1973) 'Development and the imminent demise of the Amazonian forest', *Professional Geographer* 25: 130–5.

Denevan, W. M. (1992) 'The pristine myth: the landscape of North America in 1492', *Annals of the Association of American Geographers* 82: 269–85.

Denevan, W. M., Treacy, J. M., Alcorn, J. B., Padock, C., Denslow, J. and Paitan, S. F. (1984) 'Indigenous agroforestry in the Peruvian Amazon: Bora Indian management of swidden fallows', *Interciencia* 9: 346–57.

Derman, W. (1984) 'USAID in the Sahel: development and poverty', in J. Barker (ed.), *The Politics of Agriculture in Tropical Africa*, Sage, Beverly Hills, CA.

Deshmukh, I. (1986) *Ecology and Tropical Biology*, Blackwell, Oxford.

Devall, B. (1988) *Simple in Means, Rich in Ends: practicing deep ecology*, Gibbs Smith, Layton, UT.

Devall, B. (2001) 'The deep, long range ecology movement 1960–2000 – a review', *Ethics and the Environment* 6: 18–41.

Devall, B. and Sessions, G. (1985) *Deep Ecology: living as if nature mattered*, Peregrine Smith Books, Salt Lake City.

Devereux, S. (2006) *The New Famines: why famines persist in an era of globalization*, Routledge, London.

di Castri, F. (1986) 'Interdisciplinary research for the ecological development of mountain and island areas', pp. 301–16 in N. Polunin (ed.), *Ecosystem Theory and Application*, Wiley, Chichester.

Diamond, J. (2006) *Collapse: how societies choose to fail or survive*, Penguin, Harmondsworth.

Dietz, S. and Neumayer, E. (2007) 'Weak and strong sustainability in the SEEA: concepts and measurement', *Ecological Economics* 61: 617–26.

Dietz, T. (1996) *Entitlements to Natural Resources: contours of political environmental geography*, International Books, Utrecht.

Dixon, J. A. and Fallon, L. A. (1989) 'The concept of sustainability: origins, extensions, and usefulness for policy', *Society and Natural Resources* 2: 73–84.

Dobson, A. (1990) *Green Political Thought*, HarperCollins, London (1st edn).

Dobson, A. ([1990] 2007) *Green Political Thought*, HarperCollins, London (4th edn).

Dove, M. I. and Noguiera, J. M. (1994) 'The Amazon rain forest, sustainable development and the biodiversity convention: a political economy perspective', *Ambio* 23: 491–5.

Dovers, S. R. and Handmer, J. W. (1992) 'Uncertainty, sustainability and change', *Global Environmental Change* December: 262–76.

Drakakis-Smith, D. and Dixon, C. (1997) 'Sustainable urbanization in Vietnam', *Geoforum* 28: 21–38.

Drayton, R. (2000) *Nature's Government: science, imperial Britain and the 'improvement' of the world*, Yale University Press, New Haven.

Dregne, H. E. (1984) 'Combating desertification: evaluation of progress', *Environmental Conservation* 11: 115–21.

Dresner, S. (2002) *The Principles of Sustainability*, Earthscan, London.

Drèze, J. and Sen, A. (1989) *Hunger and Public Action*, Clarendon Press, Oxford.

Dryzek, J. S. (1995) 'Toward an ecological modernity', *Policy Sciences* 28: 231–41.

Duffy, R. (1997) 'The environmental challenge to the Nation-State: superparks and National Parks policy in Zimbabwe', *Journal of Southern African Studies* 23: 441–51.

Duffy, R. (2000) *Killing for Conservation: wildlife policy in Zimbabwe*, James Currey, Oxford.

Dunlap, T. R. (1999) *Nature and the English Diaspora: environment and history in the United States, Canada, Australia and New Zealand*, Cambridge University Press, Cambridge.

Dwivedi, R. (2005) 'Environmental movements in the global South: issues of livelihood and beyond', *International Sociology* 16: 11–31.

Dykstra, D. P. and Heinrich, R. (1996) *The FAO Model Code of Forest Harvesting Practice*, Food and Agriculture Organization, Rome.

Eckersley, R. (1992) *Environmentalism and Political Theory: toward an ecocentric approach*, UCL Books, London.

Ecobichon, D. J. (2001) 'Pesticide use in developing countries', *Toxicology*, 160: 27–33.

Eden, S., Tunstall, S. M., Tapsell, S. M. (1999) 'Environmental restoration: environmental management or environmental threat?, *Area* 31: 151–9.

Ehrlich, P. R. (1972) *The Population Bomb*, Ballantine, London.

Ehrlich, P. R. and Ehrlich, A. H. (1970) *Population, Resources and Environment: issues in human ecology*, W. H. Freeman, New York.

Ekins, P. (1992) *A New World Order: grassroots movements and global change*, Routledge, London.

Ekins, P. and Jacobs, M. (1995) 'Environmental sustainability and the growth of GDP: conditions for compatibility', pp. 9–46 in V. Bhaskar and A. Glyn (eds), *The North, the South and the Environment: ecological constraints and the global economy*, United Nations University Press and Earthscan, London.

El Serafy, S. (1996) 'In defence of weak sustainability: a response to Beckerman', *Environmental Values* 5: 75–81.

Elliot, R. (1997) *Faking Nature: the ethics of environmental restoration*, Routledge, London.

Elliott, A. (2002) 'Beck's sociology of risk: a critical assessment', *Sociology* 36: 293–315.

Elliott, J. A. (2005) *An Introduction to Sustainable Development*, Routledge, London (3rd edn).

Elmhirst, R. (1998) 'Reconciling feminist theory and gendered resource management in Indonesia', *Area* 30: 225–35.

Elton, C. (1927) *Animal Ecology*, Sidgwick and Jackson, London.

Eltringham, S. K. (1994) 'Can wildlife pay its way?', *Oryx* 28: 163–8.

Emerton, L. (2001) 'The nature of benefits and the benefits of nature: why wildlife conservation has not economically benefited communities in Africa', pp 208–26 in D. Hulme and M. Murphree (eds), *African Wildlife and Livelihoods: the promise and performance of community conservation*, James Currey, Oxford.

Energy and Biodiversity Initiative (2003) *Integrating Biodiversity Conservation into Oil and Gas Development*, Conservation International, Washington for BP, ChevronTexaco, Conservation International, Fauna & Flora International, IUCN, The Nature Conservancy, Shell, Smithsonian Institution, Statoil.

Enzensberger, H. M. (1974) 'A critique of political ecology', *New Left Review* 8: 3–32.

Enzensberger, H. M. (1996) 'A critique of political ecology', pp. 117–49 in T. Benton (ed.), *The Greening of Marxism*, Guilford Press, New York.

Escobar, A. (1995) *Encountering Development: the making and unmaking of the Third World*, Princeton University Press, Princeton.

Escobar, A. (1996) 'Constructing nature: elements of a poststructural political ecology', pp. 46–68 in R. Peet and M. Watts (eds), *Liberation Ecologies: environment, development, social movements*, Routledge, London (1st edn).

Escobar, A. (1999) 'After nature: steps to an antiessentialist political ecology', *Current Anthropology* 40: 1–30.

Escobar, A. (2000) 'Beyond the search for a paradigm? Post-development and beyond', *Development* 43: 11–14.

Escobar, A. (2004) 'Beyond the Third World: imperial globality, global coloniality and anti-globalisation social movements', *Third World Quarterly* 25: 207–30.

Esteva, G. (1992) 'Development', pp. 6–25 in W. Sachs (ed.), *The Development Dictionary: a guide to knowledge as power*, Witwatersrand University Press, Johannesburg, and Zed Books, London.

European Commission (1997) *Addressing Desertification: a review of EC policies, programmes, financial instruments and projects*, European Commission, Brussels.

Evans, D. (1992) *A History of Nature Conservation in Great Britain*, Routledge, London.

Evans, G. C. (1939) 'Ecological studies on the rainforest of southern Nigeria: II. The atmospheric environmental conditions', *Journal of Ecology* 26: 436–82.

Evenson, R. E. and Gollin, D. (2003) 'Assessing the Impact of the Green Revolution, 1960 to 2000', *Science* 300: 758–62.

Fa, J. E., Seymour, S., Dupain, J. Amin, R., Albrechtsen, L. and Macdonald, D. (2006) 'Getting to grips with the magnitude of exploitation: bushmeat in the Cross–Sanaga rivers region, Nigeria and Cameroon', *Biological Conservation* 129: 497–510.

Fabricius, C., Koch, E., Magome, H and Turner, S. (2004) (eds), *Rights, Resources and Rural Development: community-based natural resource management in southern Africa*, Earthscan, London.

Fairhead, J. and Leach, M. (1995a) 'Local agro-ecological management and forest–savanna transitions: the case of Kissidougou, Guinea', pp. 163–70 in T. Binns (ed.), *People and Environment in Africa*, Wiley, Chichester.

Fairhead, J. and Leach, M. (1995b) 'False forest history, complicit social analysis: rethinking some West Africa environmental narratives', *World Development* 23: 1023–35.

Fairhead, J. and Leach, M. (1996) *Misreading the African Landscape: society and ecology in a forest savanna land*, Cambridge University Press, Cambridge.

Fairhead, J. and Leach, M. (1998) *Reframing Deforestation: global analysis and local realities*, Routledge, London.

Fairhead, J. and Leach, M. (2000) 'The nature lords', *Times Literary Supplement* 5 May: 3–4.

FAO (1994a) 'Expert consultation on cotton pest problems and their control in the Near East', *FAO Plant Protection Bulletin* 42: 139–48.

FAO (1994b) *Mangrove Forest Management Guidelines*, FAO Forestry Paper 117, Food and Agriculture Organization, Rome.

Farley, K. A. (2007) 'Grasslands to tree plantations: forest transition in the Andes of Ecuador', *Annals of the Association of American Geographers* 97: 755–71.

Farvar, M. T. and Milton, J. P. (1973) (eds), *The Careless Technology: ecology and international development*, Stacey, London.

Faulkner, O. T. and Mackie, J. R. (1933) *West African Agriculture*, Cambridge University Press, Cambridge.

Fearnside, P. M. (1980) 'The effects of cattle pasture on soil fertility in the Brazilian Amazon: consequences for beef production sustainability', *Tropical Ecology* 21: 125–37.

Fearnside, P. M. (1986) 'Spatial concentration of deforestation in the Brazilian Amazon', *Ambio* 15: 74–81.

Fearnside, P. M. (2001) 'Environmental impacts of Brazil's Tucuruí Dam: unlearned lessons for hydroelectric development in Amazonia', *Environmental Management* 27: 377–96.

Ferau, A. T. (1985) 'Environmental actors', pp. 43–67 in K. A. Dahlberg *et al.*, *Environment and the Global Arena: actors, values, policies and futures*, Duke University Press, Durham, NC.

Fisher, M. (2006) 'Don't desert drylands', *Oryx* 40: 1–2.

Fitsimmons, M. (1989) 'The matter of nature', *Antipode* 21(2): 106–20.

Fitter, R. S. R. and Scott, P. (1978) *The Penitent Butchers: the Fauna Preservation Society, 1903–1978*, Collins, London.

Flanders, L. (1997) 'The United Nations' Department for Policy Coordination and Sustainable Development (DPCSD)', *Global Environmental Change* 7: 391–4.

Fold, N. (2002) 'Lead firms and competition in "bi-polar" commodity chains: grinders and branders in the global cocoa-chocolate industry', *Journal of Agrarian Change* 2: 228–47.

Folke, C., Hammer, M., Costanza, R. and Jansson, A. (1994) 'Investing in natural capital – why what and how?', pp. 1–20 in A. Jansson, M. Hammer, C. Folke and R. Costanza (eds), *Investing in Natural Capital: the ecological economics approach to sustainability*, Island Press, Washington.

Folland, C. K., Palmer, T. N. and Parker, D. E. (1986) 'Sahel rainfall and worldwide sea temperatures 1901–1985', *Nature* 320: 602–7.

Fonseca, G. A. B. da, Bruner, A., Mittermeier, R. A., Alger, K., Gascon, C. and Rice, R. E. (2005) 'On defying nature's end: the case for landscape-scale conservation', *George Wright Forum* 22(1): 46–60.

Ford, J. (1971) *The Role of Trypanosomiasis in African Ecology: a study of the tsetse fly problem*, Clarendon Press, Oxford.

Forrester, J. W. (1971) *World Dynamics*, Wright-Allen Press, Cambridge, MA.

Forsyth, T. (2003) *Critical Political Ecology: the politics of environmental science*, Routledge, London.

Forsyth, T. (2005) *Encyclopedia of International Development*, Routledge, London.

Fosberg, F. R. (1963) *Man's Place in the Island Ecosystem*, Bishop Museum Press, Honolulu.

Foucault, M. (1975) *Discipline and Punish: the birth of the prison*, Gallimard, Paris; in translation, Allen Lane, London, 1997.

Fox, J. A. (1998) 'When does reform policy influence practice?', pp. 303–44 in J. A. Fox and L. D. Brown (eds), *The Struggle for Accountability: the World Bank, NGOs and grassroots movements*, MIT Press, Cambridge, MA.

Fox, J. A. and Brown, L. D. (1998a) 'Introduction', pp. 1–47 in J. A. Fox and L. D. Brown (eds), *The Struggle for Accountability: the World Bank, NGOs and grassroots movements*, MIT Press, Cambridge, MA.

Fox, J. A. and Brown, L. D. (1998b) (eds), *The Struggle for Accountability: the World Bank, NGOs and grassroots movements*, MIT Press, Cambridge, MA.

Fox, W. (1984) 'Deep ecology: a new philosophy of our time?', *Ecologist* 14: 194–200.

Fox, W. (1990) *Toward a Transpersonal Ecology: developing new foundations for environmentalism*, Shambhala, Boston.

Frank, A. G. (1980) 'North–South and East–West: Keynesian paradoxes in the Brandt Report', *Third World Quarterly* 2: 669–80.

Frank, A. G. (1981) *Crisis in the Third World*, Heinemann, London.

Frank, L. (1987) 'The development game', *Granta* 22: 231–43.

Franke, R. W. and Chasin, B. H. (1979) 'Peanuts, peasants and pastoralists: the social and economic background to ecological deterioration in Niger', *Journal of Peasant Studies* 8: 1–30.

Franke, R. W. and Chasin, B. H. (1980) *Seeds of Famine: ecological destruction and the development dilemma in the West African Sahel*, Allenheld and Osman, Montclair, NJ.

Fraser Darling, F. (1955) *West Highland Survey: an essay in human ecology*, Oxford University Press, Oxford.

Fresco, L. O. (1997) 'The 1996 World Food Summit', *Global Environmental Change* 7: 1–3.

Fresco, L. O. and Kroonenberg, S. B. (1992) 'Time and spatial scales in ecological sustainability', *Land Use Policy* July: 155–68.

Friberg, M. and Hettne, B. (1985) 'The greening of the world: towards a non-deterministic model of global processes', pp. 204–70 in H. Addo, S. Amin, G. Aseniero, A. G. Frank, M. Friberg, F. Frobel, J. Heinrichs, B. Heltne, O. Kreye and H. Seki, *Development as Social Transformation: reflections on the global problématique*, Hodder and Stoughton, Sevenoaks, for the United Nations University.

Frobel, F., Heinrichs, J. and Kreye, O. (1985) 'The global crisis and developing countries', pp. 111–24 in H. Addo, S. Amin, G. Aseniero, A. G. Frank, M. Friberg, F. Frobel, J. Heinrichs, B. Heltne, O. Kreye and H. Seki, *Development as Social Transformation: reflections on the global problématique*, Hodder and Stoughton, Sevenoaks, for the United Nations University.

Frynas, J. G. (2000) *Oil in Nigeria: conflict and litigation between oil companies and village communities*, LIT, Münster/Hamburg.

Fryzuk, M. D. (2004) 'Ammonia transformed', *Nature* 427: 498–9.

Fullbrook, E. (2004) (ed.) *A Guide to What's Wrong with Economics*, Anthem Press, London.

Furley, P. A. (1994) *The Forest Frontier: settlement and change*, Routledge, London.

Furon, R. (1947) *L'Érosion du sol*, Payot, Paris.

Gadgil, M. and Guha, R. (1995) *Ecology and Equity: the use and abuse of nature in contemporary India*, Routledge, London.

Gagnon, L., Klimpt, J. and Seelos, K. (2002) 'Comparing recommendations from the World Commission on Dams and the IEA initiative on hydropower', *Energy Policy* 30: 1299–304.

Galois, B. (1976) 'Ideology and the idea of nature: the case of Peter Kropotkin', *Antipode* 8: 1–16.

Galtung, J. (1984) 'Perspectives on environmental politics in overdeveloped and under-developed countries', pp. 9–21 in B. Glaeser (ed.), *Ecodevelopment: concepts, projects, strategies*, Pergamon Press, Oxford.

Gammelsrød, T. (1996) 'Effect of Zambezi management on the prawn fishery of the Sofala Bank', pp. 119–23 in M. C. Acreman and G. E. Hollis (eds), *Water Management and Wetlands in Sub-Saharan Africa*, IUCN, Gland, Switzerland.

Gandy, M. (2002) *Concrete and Clay: reworking nature in New York City*, MIT Press, Cambridge, MA.

Gash, J. H. C., Nobre, C. A., Roberts, J. M. and Victoria, R. L. (1996) (eds), *Amazonian Deforestation and Climate*, Wiley, Chichester.

Geisler, C. (2003) 'Your park, my poverty: using impact assessment to counter the displacement effects of environmental greenlining', pp. 217–29 in S. R. Brechin, P. R. Wilhusen, C. L. Fortwangler and P. C. West (eds), *Contested Nature: promoting international biodiversity with social justice in the twenty-first century*, State University of New York Press, New York.

Geist, H. J. and Lambin, E. F. (2002) 'Proximate causes and underlying driving forces of tropical deforestation', *BioScience*, 52: 143–50.

Geldof, B. (1986) *Is that It?*, Sidgwick and Jackson, London.

George, C. J. (1973) 'The role of the Aswan High Dam in changing the fisheries of the southeastern Mediterranean', pp. 159–78 in M. T. Farvar and J. P. Milton (eds), *The Careless Technology: ecology and international development*, Stacey, London.

Gerlagh, R., Dellink, R., Hofkes, M. and Verbruggen, H. (2002) 'A measure of sustainable national income for the Netherlands', *Ecological Economics* 41: 157–74.

Ghai, D. and Vivian, J. M. (1992a) 'Introduction', pp. 1–19 in D. Ghai and J. M. Vivian, *Grassroots Environmental Action: people's participation in sustainable development*, Routledge, London.

Ghai, D. and Vivian, J. M. (1992b) (eds) *Grassroots Environmental Action: people's participation in sustainable development*, Routledge, London.

Ghatak, S. and Turner, R. K. (1978) 'Pesticide use in less developed countries: economic and environmental considerations', *Food Policy* 3: 136–46.

Ghimire, K. and Pimbert M. (1997) (eds), *Social Change and Conservation: environmental politics and impacts of national parks and protected areas*, Earthscan, London.

Gibbs, D. (2000) 'Ecological modernisation, regional economic development and regional development agencies', *Geoforum* 31: 9–19.

Gibson, C. C. and Marks, S. A. (1995) 'Transforming rural hunters into conservationists: an assessment of community-based wildlife management programs in Africa', *World Development* 23: 941–57.

Gilbert, V. C. and Christy, E. J. (1981) 'The UNESCO Program on Man and the Biosphere (MAB)', pp. 701–20 in E. J. Kormondy and J. F. McCormick (eds), *Handbook of Contemporary Developments in World Ecology*, Greenwood Press, Westport, CT.

Giles J. (2006) 'Tide of censure for African dams', *Nature* 440: 393–4.

Gimenez, M. E. (2000) 'Does ecology need Marx?', *Organization and Environment* 13: 292–304.

Glacken, C. J. (1967) *Traces on the Rhodian Shore: nature and culture in Western thought from ancient times to the end of the Eighteenth Century*, University of California Press, Berkeley and Los Angeles.

Glaeser, B. (1984) (ed.) *Ecodevelopment: concepts, projects, strategies*, Pergamon Press, Oxford.

Gliessman, S. R. (1984) 'Resource management in traditional agroecosystems: southeast Mexico', pp. 191–201 in G. K. Douglass (ed.), *Agricultural Sustainability in a Changing World Order*, Westview, Boulder, CO.

Goldman, M. (2005) *Imperial Nature: the World Bank and struggles for social justice in the age of globalization*, Yale University Press, New Haven.

Goldsmith, E. (1987) 'Open letter to Mr Conable, President of the World Bank', *Ecologist* 17: 58–61.

Goldsmith, E. and Hildyard, N. (1984) *Social and Environmental Impacts of Large Dams*, vol. 1, Ecologist Magazine, Wadebridge, Cornwall.

Goldsmith, E., Allen, R., Allaby, M., Davoll, J. and Lawrence, S. (1972) *Blueprint for Survival: Ecologist* 2: 1–50 (paperback edn, Penguin, Harmondsworth, 1972).

Golub, R. and Townsend, J. (1977) 'Malthus, multinationals and the Club of Rome', *Social Studies of Science* 7: 202–22.

Gómez-Pompa, A., Whitmore, T. C. and Hadley, M. (1991) (eds), *Rain Forest Regeneration and Management*, UNESCO, Paris.

Good, K. (1986) 'The reproduction of weakness in the state and agriculture: the Zambian experience', *African Affairs* 85: 239–65.

Goodland, R. J. (1978) *Environmental Assessment of the Tucurui Hydroproject, Rio Tocatins, Amazonia, Brazil*, Eletronorte SA, Brasilia.

Goodland, R. J. (1980) 'Environmental rankings of Amazonian development projects in Brazil', *Environmental Conservation* 7: 9–26.

Goodland, R. J. (1984) *Environmental Requirements of the World Bank*, Environment and Science Unit, Projects Policy Department, World Bank, Washington.

Goodland, R. J. (1990) 'Environment and development: progress of the World Bank', *Geographical Journal* 156: 149–57.

Goodland, R. J. and Ledec, G. (1984) *Neoclassical Economics and Principles of Sustainable Development*, World Bank Office of Environmental Affairs, Washington.

Goodland, R. J. A., Daly, H. E. and El Serafy, S. (1993) 'The urgent need for rapid transition to global environmental sustainability', *Environmental Conservation* 20: 297–309.

Gore, A. (2006) *An Inconvenient Truth: the planetary emergency of global warming and what we can do about it*, Bloomsbury Publishing, London.

Gosling, L. (1979) 'Resettlement losses and compensation', pp. 119–31 in L. Gosling (ed.), *Population Resettlement in the Mekong River Basin*, Studies in Geography No. 10, University of Northern Carolina, Chapel Hill, NC.

Gottfried, R. R., Brockett, C. D. and Davis, W. C. (1994) 'Models of sustainable development and forest resource management in Costa Rica', *Ecological Economics* 9: 107–20.

Gouldner, L. H. and Kennedy, D. (1997) 'Valuing ecosystem services: philosophical bases and empirical methods', pp. 23–47 in G. C. Daily (ed.), *Nature's Services: societal dependence on natural ecosystems*, Island Press, Washington.

Goulet, D. (1971) *The Cruel Choice: a new concept in the theory of development*, Athenaeum, London.

Goulet, D. (1992) 'Development: creator and destroyer of values', *World Development* 20: 467–75.

Goulet, D. (1995) *Development Ethics: a guide to theory and practice*, Zed Books, London.

Gradwohl, J. and Greenberg, R. (1988) *Saving the Tropical Forests*, Earthscan, London.

Graham, A. (1973) *The Gardeners of Eden*, Allen and Unwin, Hemel Hempstead.

Grainger, A. (1982) *Desertification: how people make deserts, how people can stop and why they don't*, Earthscan/IIED, London.

Grainger, A. (1996) 'An evaluation of the FAO *Tropical Forest Resource Assessment 1990*', *Geographical Journal* 162: 73–9.

Gray, L. C. and Moseley, W. G. (2005) 'A geographical perspective on poverty-environment interactions', *Geographical Journal* 171: 9–23.

Greener, L. (1962) *High Dam over Nubia*, Cassell, London.

Griffiths, T. and Robin, L. (1997) (eds), *Ecology and Empire: environmental history of settler societies*, Keele University Press, Keele.

Grojean, R. (1991) *Sand Encroachment Control in Mauritania*, United Nations Sudan-Sahelian Office Technical Publication Series No. 5, New York.

Groombridge, B. (1992) *Global Biodiversity: status of the earth's living resources*, Chapman and Hall, London.

Grove, A. T. (1958) 'The ancient erg of Hausaland, and similar formations on the south side of the Sahara', *Geographical Journal* 124: 526–33.

Grove, A. T. (1973) 'A note on the remarkably low rainfall of the Sudan Zone in 1913', *Savanna* 2: 133–8.

Grove, A. T. (1977) 'Desertification', *Progress in Physical Geography* 1: 296–310.

Grove, A. T. (1981) 'The climate of the Sahara in the period of meteorological records', in J. A. Allan (ed.), *Sahara: ecological change and early economic history*, Menas Press, London.

Grove, A. T. and Warren, A. (1968) 'Quaternary landforms and climate on the south side of the Sahara', *Geographical Journal* 134: 194–208.

Grove, R.H. (1987) 'Early themes in African conservation: the Cape in the nineteenth century', pp. 21–40 in D. M. Anderson and R. H. Grove (eds) *Conservation in Africa: people, policies and practice*, Cambridge University Press, Cambridge.

Grove, R. H. (1990a) 'The origins of environmentalism', *Nature* 345(6270): 11–14.

Grove, R. H. (1990b) 'Colonial conservation, ecological hegemony, and popular resistance: towards a global synthesis', pp. 15–50 in J. M. McKenzie (ed.), *Imperialism and the Natural World*, Manchester University Press, Manchester.

Grove, R. H. (1995) *Green Imperialism: colonial expansion, tropical island Edens and the origins of environmentalism, 1600–1800*, Cambridge University Press, Cambridge.

Grove R. H. (1998a) 'The East India Company, the Raj and the El Niño: the critical role played by colonial scientists in establishing the mechanisms of global climate teleconnections 1770–1930', pp. 301–23 in R. H. Grove, V. Damodaran and S. Sangwan (eds), *Nature and the Orient: the environmental history of South and Southeast Asia*, Oxford University Press, Delhi.

Grove, R. H. (1998b) *Ecology, Climate Change and Empire: the Indian legacy in global environmental history 1400–1940*, Oxford University Press, New Delhi.

Grove, R. H. and Damodaran, V. (2006) 'Imperialism, intellectual networks, and environmental change: origins and evolution of global environmental history, 1670–2000: Part I', *Economic and Political Weekly* 41: 4345–54.

Grove, R. H., Damodaran, V. and Sangwan, S. (1998) (eds) *Nature and the Orient: the environmental history of South and South East Asia*, Oxford University Press, Delhi.

Grubb, M. (1998) 'International emissions trading under the Kyoto Protocol: core issues in implementation', *Review of European Community and International Environmental Law* 7: 140–6.

Grubb, M., Koch, M., Thompson, K., Munson, A. and Sullivan, F. (1993) (eds) *The 'Earth Summit' Agreements: a guide and assessment*, Earthscan, London (for the Royal Institute of International Affairs, London).

Guha, R. (1989) *The Unquiet Woods: ecological change and peasant resistance in the Himalaya*, Oxford University Press, Delhi.

Guha, R. (2000) *Environmentalism: a global history*, Oxford University Press, Delhi.

Guha, R. and Martinez-Alier, J. (1997) *Varieties of Environmentalism: essays North and South*, Earthscan, London.

Gulbrandsen, L. H. (2006) 'Creating markets for eco-labelling: are consumers insignificant?' *International Journal of Consumer Studies*, 30: 477–89.

Gullison, R. E. (2003) 'Does forest certification conserve biodiversity?' *Oryx* 37: 153–65.

Gullison, R. E. and Losos, E. C. (1993) 'The role of foreign debt in deforestation in Latin America', *Conservation Biology* 7: 140–7.

Guy, P. R. (1981) 'River bank erosion in the mid-Zambezi Valley downstream of Lake Kariba', *Biological Conservation* 19: 199–212.

Gwynne, M. D. (1982) 'The Global Environment Monitoring System (GEMS) of UNEP', *Environmental Conservation* 9: 35–42.

Haila, Y. and Levins, R. (1992) *Humanity and Nature: ecology, science and society*, Pluto Press, London.

Hailey, Lord (1938) *An African Survey*, Royal Institute for African Affairs and Oxford University Press, London.

Hails, C., Loh, J. and Goldfinger, S. (2006) (eds) *Living Planet Report 2006*, WWF International, Zoological Society of London and Global Footprint Network, Gland, Switzerland.

Hajer, M. A. (1995) *The Politics of Environmental Discourse: ecological modernization and the policy process*, Oxford University Press, Oxford.

Hajer, M. A. (1996) 'Ecological modernisation as cultural politics', pp. 246–68 in S. Lash, B. Szerzynski and B. Wynne (eds), *Risk, Environment and Modernity: towards a new ecology*, Sage, London.

Hall, P. (1980) *Great Planning Disasters*, Weidenfeld and Nicolson, London.

Hamilton, K. and Johnson, I. (2004) 'Responsible growth to 2050', *World Economics* 5: 33–51.

Hanley, N. and Spash, C. L. (1994) *Cost–Benefit Analysis and the Environment*, Edward Elgar, London.

Hanlon, J. (1996) *Peace without Profit: how the IMF blocks rebuilding in Mozambique*, James Currey/Heinemann for the International African Institute, London.

Hannah, L., Lohse, D., Hutchinson, C., Carr, J. L. and Lankerani, A. (1994) 'A preliminary inventory of human disturbance of world ecosystems', *Ambio* 23: 246–50.

Haraway, D. J. (1991) *Simians, Cyborgs and Women: the reinvention of nature*, Free Association Books, London.

Hardin, G. (1968) 'The tragedy of the commons', *Science* 1628: 1243–8.

Hardin, G. (1974) 'Living on a lifeboat', *Bioscience* 24: 561–8.

Hardjono, J. M. (1977) *Transmigration in Indonesia*, Oxford University Press, Selangor, Malaysia.

Hardjono, J. M. (1983) 'Rural development in Indonesia: the "top down" approach', pp. 38–66 in D. A. M. Lea and D. P. Chaudhri (eds), *Rural Development and the State: contradictions and dilemmas in developing countries*, Methuen, London.

Hardoy, J. E., Mitlin, D. and Satterthwaite, D. (1992) *Environmental Problems in Third World Cities*, Earthscan, London.

Harris, F. M. A. (1998) 'Farm-level assessment of the nutrient balance in northern Nigeria', *Agriculture, Ecosystems and Environment* 71: 201–14.

Harris, F. M. A. (1999) 'Nutrient management of smallholder farmers in a short-fallow farming system in north-east Nigeria', *Geographical Journal* 165: 275–85.

Harrison, P. (1987) *The Greening of Africa: breaking through in the battle for land and food*, Paladin, London.

Harrison, P. (1992) *The Third Revolution: population, environment and a sustainable world*, Penguin, Harmondsworth.

Harriss, J. and de Renzio, P. (1997) '"Missing link" or analytically missing? The concept of social capital: an introductory bibliographic essay', *Journal of International Development* 9: 919–37.

Harrop, S. R. (2003) 'From cartel to conservation and on to compassion: animal welfare and the International Whaling Commission', *Journal of International Wildlife Law and Policy*, 6: 79–104.

Harroy, J.-P. (1949) *Afrique: terre qui meurt: la dégradation des sols africains sous l'influence de la colonisation*, Marcel Hayez, Brussels.

Hart, D. (1980) *The Volta River Project: a case study in politics and technology*, Edinburgh University Press, Edinburgh.

Hart, K. (1982) *The Political Economy of West African Agriculture*, Cambridge University Press, Cambridge.

Hartwick, E. and Peet, R. (2003) 'Neoliberalism and nature: the case of the WTO', *Annals of the American Academy of Political Economy* 590: 188–211.

Harvey, D. (1973) *Social Justice and the City*, Johns Hopkins University Press, Baltimore.

Harvey, D. (1974) 'Population, resources and the ideology of science', *Economic Geography* 50: 256–77.

Harvey, D. (1990) *The Condition of Postmodernity: an enquiry into the origins of cultural change*, Blackwell, Oxford.

Harvey, D. (1996a) *Justice, Nature and the Geography of Difference*, Blackwell, Oxford.

Harvey, D. (1996b) 'The environment of justice', pp. 63–99 in A. Merryfield and E. Swyngedouw (eds), *The Urbanization of Injustice*, Lawrence and Wishart, London.

Haub, C. (1999) 'Six billion – and counting', *People and the Planet* 8(1): 6–9.

Hawken, P., Lovins, A. and Lovins, H. (1999) *Natural Capitalism: creating the next industrial revolution*, Earthscan, London.

Hayden, C. (2003) *When Nature Goes Public: the making and unmaking of bioprospecting in Mexico*, Princeton University Press, Princeton.

Hays, S. P. (1959) *Conservation and the Gospel of Efficiency: the progressive conservation movement, 1890–1920*, Harvard University Press, Cambridge, MA.

Hays, S. P. (1987) *Beauty, Health and Permanence: environmental politics in the United States, 1955–1985*, Cambridge University Press, Cambridge.

He, K., Huo, H. and Zhang, Q. (2002) 'Urban air pollution in China: current status, characteristics, and progress', *Annual Review of Energy Environment*, 27: 397–431.

Hecht, S. B. (1980) 'Deforestation in the Amazon basin: magnitude, dynamics and soil resource effects', pp. 61–108 in V. H. Sutlive, N. Altshuler and M. D. Zamora (eds), *Where Have All The Flowers Gone? Deforestation in the Third World*, Studies in Third World Societies No. 13, College of William and Mary, Williamsburg, VA.

Hecht, S. B. (1984) 'Cattle ranching in Amazonia: political and ecological considerations', pp. 366–400 in M. Schmick and C. H. Wood (eds), *Frontier Expansion in Amazonia*, University of Florida Press, Gainesville.

Hecht, S. B. (1985) 'Environment, development and politics: capital accumulation and the livestock sector in Eastern Amazonia', *World Development* 13: 663–84.

Hecht, S. B. (2004) 'Invisible forests: the political ecology of forest resurgence in El Salvador', pp. 64–103 in R. Peet and M. Watts (eds), *Liberation Ecologies: environment, development, social movements*, Routledge, London (2nd edn).

Hecht, S. and Cockburn, A. (1989) *The Fate of the Forest: developers and defenders of the Amazon*, Verso, London.

Hecht, S. B., Kandel, S., Gomes, I., Cuellar, N. and Rosa, H. (2006) 'Globalisation, forest resurgence and environmental politics in El Salvador', *World Development* 34: 308–23.

Heine, B. (1985) 'The Mountain People: the Ik of north-eastern Uganda', *Africa* 55: 3–16.

Hellden, U. (1988) 'Desertification monitoring: is the desert encroaching?', *Desertification Control Bulletin* 17: 8–12.

Hellerman (2007) 'Things Fall Apart? Management, Environment and Taungya Farming in Edo State, Southern Nigeria', *Africa* 77: 371–92

Henderson-Sellers, A. and Gornitz, V. (1984) 'Possible climatic impacts of land cover transformations, with particular emphasis on tropical deforestation', *Climatic Change* 6: 231–58.

Hewitt, K (1983) (ed.) *Interpretations of Calamity*, Allen and Unwin, London.

Hildermeier, M. (1979) 'Agrarian social protest, populism and economic development: some problems and results from recent studies', *Social History* 4: 319–32.

Hill, A. W. (1925) 'The best means of promoting a complete botanical survey of different parts of the empire', pp. 196–204 in F. T. Brooks (ed.), *Imperial Botanical Conference, London 1924: report of proceedings*, Cambridge University Press, Cambridge.

Hill, F. (1978) 'Experiments with a public sector peasantry: agricultural schemes and class formation in Africa', pp. 25–41 in A. K. Smith and C. E. Welch (eds), *Peasants in Africa*, African Studies Association, Boston.

Hill, K. A. (1995) 'Conflicts over development and environmental values: the international ivory trade in Zimbabwe's historical context', *Environment and History* 1: 335–49.

Hill, P. (1986) *Development Economics on Trial: the anthropological case for the prosecution*, Cambridge University Press, Cambridge.

Hingston, R. W. G. (1931) 'Proposed British national parks for Africa', *Geographical Journal* 77: 401–28.

Hinterberger, F., Luks, F. and Schmidt-Bleek, F. (1997) 'Material flows vs. natural capital: what makes an economy sustainable?' *Ecological Economics* 23: 1–14

Hiraoka, M. and Yamamoto, S. (1980) 'Agricultural development in the Upper Amazon of Ecuador', *Geographical Review* 70: 423–45.

Ho, P. (2001) 'Greening without conflict? Environmentalism, NGOs and civil society in China', *Development and Change* 32: 893–901.

Hoben, A. (1995) 'Paradigms and politics: the cultural construction of environmental politics in Ethiopia', *World Development* 23: 1007–22.

Hobley, C. W. (1914) 'The alleged desiccation of East Africa', *Geographical Journal* 44: 467–77.

Hofer, T., and Messerli, B. (2006) *Floods in Bangladesh; history, dynamics and rethinking the role for the Himalayas.* United Nations University Press, New York and Tokyo.

Hogg, R. (1983) 'Irrigation agriculture and pastoral development: a lesson from Kenya', *Development and Change* 14: 577–91.

Hogg, R. (1987a) 'Development in Kenya: drought, desertification and food security', *African Affairs* 86: 47–58.

Hogg, R. (1987b) 'Settlement, pastoralism and the commons: the ideology and practice of irrigation development in northern Kenya', pp. 293–306 in D. M. Anderson and R. H. Grove (eds), *Conservation in Africa: people, policies and practice*, Cambridge University Press, Cambridge.

Holden, C. (1987) 'World Bank launches new environmental policy', *Science* 236: 769.

Holden, C. (1988) 'The greening of the World Bank', *Science* 240: 1610–11.

Holdgate, M. (1996) *From Care to Action: making a sustainable world*, Earthscan, London.

Holdgate, M. (1999) *The Green Web: a union for world conservation*, Earthscan, London.

Holdgate, M. W., Kassas, M. and White, G. F. (1982) 'World environmental trends between 1972 and 1982', *Environmental Conservation* 9: 11–29.

Holland, A. and Roxbee Cox, J. (1992) 'The valuing of environmental goods: a modest proposal', pp. 12–24 in A. Coker and C. Richards (eds), *Valuing the Environment: economic approaches to environmental valuation*, Belhaven, London.

Hollis, G. E. (1990) 'Environmental impacts of development on wetlands in arid and semi-arid lands', *Hydrological Sciences Journal* 35: 411–28.

Hollis, G. E. (1996) 'Hydrological inputs to management policy for the Senegal River and its floodplain', pp. 155–184 in M. C. Acreman and G. E. Hollis, *Water Management and Wetlands in Sub-Saharan Africa*, IUCN, Gland, Switzerland.

Hollis, G. E., Adams, W. M. and Aminu-Kano, M. (1994) (eds) *The Hadejia–Nguru Wetlands: environment, economy and sustainable use of a Sahelian floodplain wetland*, IUCN Wetlands Programme, Gland, Switzerland.

Holmberg, J., Thomson, K. and Timberlake, L. (1993) *Facing the Future: beyond the Earth Summit*, Earthscan/International Institute for Environment and Development, London.

Homewood, K. and Rodgers, W. A. (1984) 'Pastoralism and conservation', *Human Ecology* 12: 431–41.

Homewood, K. and Rodgers, W. A. (1987) 'Pastoralism, conservation and the overgrazing controversy', pp. 111–28 in D. M. Anderson and R. H. Grove (eds), *Conservation in*

Africa: people, policies and practice, Cambridge University Press, Cambridge.

Homewood, K. and Rodgers, W. A. (1991) *Maasailand Ecology*, Cambridge University Press, Cambridge.

Hopper, W. (1988) *The World Bank's Challenge: balancing economic need with environmental protection*, Seventh Annual World Conservation Lecture, 3 March, World Wide Fund for Nature UK, Godalming.

Horowitz, M. M. (1987) 'Destructive development', *Institute of Development Anthropology Network* 5: 1–2.

Horowitz, M. M. and Little, P. D. (1987) 'African pastoralism and poverty: some implications for drought and famine', pp. 59–82 in M. Glantz (ed.), *Drought and Hunger in Africa: denying famine a future*, Cambridge University Press, Cambridge.

Horowitz, M. M. and Salem-Murdock, M. (1991) 'Management of an African floodplain: a contribution to the anthropology of public policy', *Landscape & Urban Planning* 20: 215–21.

Houghton, J. ([1994] 1997) *Global Warming: the complete briefing*, Cambridge University Press, Cambridge (2nd edn).

Houghton, J. T., Meira Filho, L. G., Callander, B. A., Harris, N., Kattenberg, A. and Maskell, K. (1995) (eds) *Climate Change 1995: the science of climate change (contribution of Working Group I to Second Assessment Report of the Intergovernmental Panel on Climate Change)*, Cambridge University Press, Cambridge.

Hovi, J., Skodvin, T. and Andresen, S. (2003) 'The persistence of the Kyoto Protocol: why other Annex 1 countries move on without the United States', *Global Environmental Politics* 3(4): 1–23.

Howard, P. (1978) *Weasel Words*, Hamish Hamilton, London.

Howarth, D. (1961) *The Shadow of the Dam*, Collins, London.

Howell, P., Lock, M. and Cobb, S. (1988) *The Jonglei Canal: impact and opportunity*, Cambridge University Press, Cambridge.

Hughes, A. (2001) 'Global commodity networks, ethical trade and governmentality: organizing business in the Kenyan cut flower industry', *Transactions of the Institute of British Geographers* NS 26: 390–406.

Hughes, F. M. R. (1984) 'A comment on the impact of development schemes on the floodplain forests of the Tana River of Kenya', *Geographical Journal* 150: 230–44.

Hughes, F. M. R. (1990) 'The influence of flooding regimes on forest distribution and composition in The Tana River Floodplain, Kenya', *Journal of Applied Ecology* 27: 475–91.

Hughes, F. M. R. (1997) 'Floodplain biogeomorphology', *Progress in Physical Geography* 21: 501–29.

Hughes, F. M. R. and Rood, S. (2001) 'Floodplains', pp. 105–201 in A. Warren and J. R. French (eds), *Habitat Conservation: managing the physical environment*, Wiley, Chichester.

Hughes, F. M. R., Colston, A. and Owen Mountford. J. (2005) 'Restoring riparian ecosystems: the challenge of accommodating variability and designing restoration trajectories', *Ecology and Society* 10(1): 12; http://www.ecologyandsociety.org/vol10/iss1/art12/

Huijsman, B. and Savenije, H. (1991) 'Making haste slowly', pp. 13–34 in H. Savenije and B. Huijsman (eds), *Making Haste Slowly: strengthening local environmental management in agricultural development*, Royal Tropical Institute, Amsterdam.

Hulme, D. and Murphree, M. (1999) 'Communities, wildlife and the "new conservation" in Africa', *Journal of International Development* 11: 277–86.

Hulme, D. and Murphree, M. (2001) (eds) *African Wildlife and Livelihoods: the promise and performance of community conservation*, James Currey, Oxford.

Hulme, M. (1995) (ed.) *Climate Change and Southern Africa: an exploration of some potential impacts and implications in the SADC region*, Climate Research Unit and World Wide Fund for Nature, Norwich.

Hulme, M. (1996) 'Climate change within the period of meteorological records', pp. 88–102 in W. M. Adams, A. S. Goudie and A. R. Orme (eds), *The Physical Geography of Africa*, Oxford University Press, Oxford.

Humphreys, D. (2006) *Logjam: deforestation and the crisis of global governance*, Earthscan, London.

Huq, S. (1994) 'Global industrialization: a developing country perspective', pp. 107–13 in R. Socolow, C. Andrews, F. Berkhout and V. Thomas (eds), *Industrial Ecology and Global Change*, Cambridge University Press, Cambridge.

Hutton, J. M. and Leader-Williams, N. (2003) 'Sustainable use and incentive-driven conservation: realigning human and conservation interests', *Oryx* 37: 215–26.

Hutton, J., Adams, W. M. and Murombedzi, J. C. (2005) 'Back to the barriers? Changing narratives in biodiversity conservation', *Forum for Development Studies* 32: 341–70.

Huxley, E. (1960) *A New Earth: an experiment in colonialism*, Chatto and Windus, London.

Huxley, J. L. (1930) *African View*, Chatto and Windus, London.

Huxley, J. L. (1977) *Memories II*, Harper and Row, New York.

Hviding, E. and Bayliss-Smith, T. (2000) *Islands of Rainforest: agroforestry, logging and ecotourism in Solomon Islands*, Ashgate, Aldershot.

Hyden, G. (1980) *Beyond Ujamaa in Tanzania: underdevelopment and an uncaptured peasantry*, Heinemann, London.

Hyndman, D. (1994) *Ancestral Rain Forests and the Mountain of Gold: indigenous people and mining in New Guinea*, Westview Press, Boulder, CO.

Ickowitz, A. (2006) 'Shifting cultivation and deforestation in tropical Africa: critical reflections', *Development and Change*, 37: 599–626.

ICOLD (1980) *Dams and their Environment*, International Commission on Large Dams, Paris.

ICOLD (1981) *Dam Projects and Environmental Success*, International Commission on Large Dams, Paris.

Idso, S. B. (1977) 'A note on some recently proposed mechanisms of genesis of deserts', *Quarterly Journal of the Royal Meteorological Society* 103: 369–70.

Ikporukpo, C. O. (1983) 'Environmental deterioration and public policy in Nigeria', *Applied Geography* 3: 303–16.

Ikporukpo, C. O. (2004) 'Petroleum, fiscal federalism and environmental justice in Nigeria', *Space and Polity* 8: 321–54.

Iles, A. (2004) 'Mapping environmental justice in technology flows: computer waste impacts in Asia', *Global Environmental Politics* 4: 76–107.

Iliffe, J. (1995) *Africans: the history of a continent*, Cambridge University Press, Cambridge.

Illich, I. (1973) 'Outwitting the "developed" countries', pp. 357–68 in H. Bernstein (ed.), *Underdevelopment and Development: the Third World today*, Penguin, Harmondsworth.

Imhoff, M. L., Bounoua, L., Ricketts, T., Loucks, C., Harriss, R. and Lawrence, W. T. (2004) 'Global patterns in human consumption of net primary production', *Nature* 429: 870–3.

International Institute for Sustainable Development (1997) 'Summary of the First Conference of the Parties to the Convention to Combat Desertification', *Desertification Control Bulletin* 31: 1–5.

IPCC (2007a) *Synthesis of the Fourth Assessment Report of the Intergovernmental Panel on Climate Change*, Intergovernmental Panel on Climate Change, Geneva, www.ipcc.ch/.

IPCC (2007b) *Climate Change 2007: the physical science basis.* Contribution of Working Group I to the *Fourth Assessment Report* of the Intergovernmental Panel on Climate Change, IPCC Secretariat, Geneva.

Ite, U. E. (1996) 'Small farmers and forest loss in Cross River National Park, Nigeria', *Geographical Journal* 163: 47–56.

Ite, U. E. (2001) *Global Thinking and Local Action: agriculture, tropical forest loss and conservation in South-East Nigeria*, Ashgate, Guildford.

Ite, U. E. (2004) 'Multinationals and corporate social responsibility in developing countries: a case study of Nigeria', *Corporate Social Responsibility and Environmental Management* 11: 1–11.

IUCN (1980) *The World Conservation Strategy*, International Union for Conservation of Nature and Natural Resources, United Nations Environment Programme, World Wildlife Fund, Geneva.

IUCN (1984) *National Conservation Strategies: a framework for sustainable development*, IUCN, Geneva.

IUCN (1991) *Caring for the Earth: a strategy for sustainable living*, IUCN, Gland, Switzerland.

Ivanova, M. (2007) 'Moving forward by looking back: learning from UNEP's history', pp. 262–47 in L. Swart and E. Perry (eds), *Global Environmental Governance: perspectives on the current debate*, Center for UN Reform, New York, see http://www.centerforunreform.org/node/251.

Ives, J. and Messerli, B. (1989) *The Himalayan Dilemma: reconciling development and conservation*, Routledge, London.

Jacks, G. V. and Whyte, R. O. (1938) *The Rape of the Earth: a world survey of soil erosion*, Faber and Faber, London.

Jackson, C. (1993) 'Questioning synergism: win–win with women in population and environment policies', *Journal of International Development* 5: 651–68.

Jackson, C. (1994) 'Gender analysis and feminisms', pp. 113–49 in M. R. Redclift and T. Benton (eds), *Social Theory and the Global Environment*, Routledge, London.

Jackson, P. B. N. (1966) 'The establishment of fisheries in man-made lakes in the tropics', in R. H. Lowe-McConnell (ed.), *Man-Made Lakes*, Academic Press, London.

Jackson, T. (2006) *The Earthscan Reader on Sustainable Consumption*, Earthscan, London.

Jacobs, M. (1991) *The Green Economy: environment, sustainable development and the politics of the future*, Pluto Press, London.

Jacobs, M. (1995) 'Sustainable development, capital substitution and economic humility: a response to Beckerman', *Environmental Values* 4: 57–68.

Jacoby, K. (2001) *Crimes against Nature: squatters, poachers, thieves, and the hidden history of American conservation*, University of California Press, Berkeley and Los Angeles.

Jäger, J. and O'Riordan, T. (1996) 'The history of climate change science and politics', pp. 1–31 in T. O'Riordan and J. Jäger (eds), *Politics of Climate Change: a European perspective*, Routledge, London.

Jansson, A., Hammer, M., Folke, C. and Costanza, R. (1994) (eds) *Investing in Natural Capital: the ecological economics approach to sustainability*, Island Press, Washington.

Jarosz, L. (1996) 'Defining deforestation in Madagascar', pp. 148–64 in R. Peet and M. Watts (eds), *Liberation Ecologies: environment, development, social movements*, Routledge, London (1st edn).

Jeffers, H. P. (2003) *Roosevelt the Explorer: Teddy Roosevelt's amazing adventures as a naturalist, conservationist, and explorer*, Taylor Trade Publishing, Lanham, NY.

Jenkins, M., Green, R. E. and Madden, J. (2003) 'The challenge of measuring global change in wild nature: are things getting better or worse?' *Conservation Biology* 17: 65–86.

Jewell, P. A. (1980) 'Ecology and management of game animals and domestic livestock in African savannas', pp. 353–81 in D. R. Harris (ed.), *The Human Ecology of Savanna Environments*, Academic Press, London.

Jewitt, S. (1995) 'Europe's "others"? Forestry policy and practices in colonial and post-colonial India', *Environment and Planning D: Society and Space* 13: 67–90.

Johns, A. D. (1985) 'Selective logging and wildlife conservation in tropical rain-forest: problems and recommendations', *Biological Conservation* 31: 355–76.

Johns, A. G. and Johns, B. G. (1995) 'Tropical forest primates and logging: long-term coexistence?', *Oryx* 29: 205–11.

Johnson, B. (1985) 'Chimera or opportunity? An environmental appraisal of the recently concluded International Tropical Timber Agreement', *Ambio* 14: 42–4.

Johnson, B. and Blake, R. O. (1980) *The Environment and Bilateral Aid Agencies*, IIED, Washington.

Johnston, R. J. (1989) *Environmental Problems: nature, economy and state*, Belhaven, New York.

Jones, B. (1938) 'Desiccation and the West African colonies', *Geographical Journal* 41: 401–23.

Jones, B. (2001) 'The evolution of community-based approaches to wildlife management in Kunene, Namibia', pp. 160–76 in D. Hulme and M. Murphree (eds), *African Wildlife and Livelihoods: the promise and performance of community conservation*, James Currey, Oxford.

Jones, G. H. (1936) *The Earth Goddess: a study of native farming in the West African context*, Longman, Green, London.

Jones, S. J. (1996) 'Farming systems and nutrient flows: a case of degradation? *Geography* 81: 289–300.

Jordan, A. and Voisey, H. (1998) 'The "Rio Process": the politics and substantive outcomes of "Earth Summit II"', *Global Environmental Change* 8: 93–7.

Jubb, R. A. (1972) 'The J. G. Strydon Dam, Pongolo River, northern Zululand: the importance of floodplain pans below it', *Piscator* 86: 104–9.

Justice, C. O., Townshend, J. R. G., Holben, B. N. and Tucker, C. J. (1985) 'Analysis of the phenology of global vegetation using meteorological satellite data', *International Journal of Remote Sensing* 6: 1271–318.

Kahn, H. (1979) *World Economic Development: 1979 and beyond*, Croom Helm, London.

Kahn, H. and Wiener, A. J. (1967) *The Year 2000: a framework for speculation on the next 33 years*, Macmillan, London.

Kaika M. (2006) 'Dams as symbols of modernization: the urbanization of nature between geographical imagination and modernity', *Annals of the Association of American Geographers* 96: 276–301.

Kanninen, M., Murdiyarso, D., Seymour, F., Angelsen, A., Wunder, S. and German, L. (2007) *Do Trees Grow on Money? The implications of deforestation research for policies to promote REDD*, Center for International Forestry Research (CIFOR), Bogor, Indonesia.

Karimi, S., Nakayama, M., Fujikura, R., Katsurai, T., Iwata, M., Mori, T. and Mizutani, K. (2005) 'Post-project review on a resettlement programme of the Kotapanjang Dam Project in Indonesia', *Water Resources Development* 21: 371–84.

Karrar, G. (1984) 'The UN Plan of Action to Combat Desertification and the concomitant UNEP campaign', *Environmental Conservation* 11: 99–102.

Kartawinata, K., Adisoemarto, S., Riswa, S. and Vayda, A. P. (1981) 'The impact of man on a tropical forest in Indonesia', *Ambio* 10: 115–19.

Kassas, M. (1973) 'Impact of river control schemes on the shoreline of the Nile Delta', pp. 179–88 in M. T. Farvar and J. P. Milton (eds), *The Careless Technology: ecology and international development*, Stacey, London.

Kates, C. A. (2004) 'Reproductive liberty and overpopulation', *Environmental Values* 13: 51–79.

Kates, R. W., Parris, T. M. and Leiserowitz, A. A. (2005). 'What is sustainable development? Goals, indicators, values, and practice', *Environment: Science and Policy for Sustainable Development* 47(3): 8–21.

Katz, C. and Kirby, A. (1991) 'In the nature of things: the environment and everyday life', *Transactions of the Institute of British Geographers* 16: 259–71.

Katz, E., Light, A. and Rothenberg, D. (2000) (eds) *Beneath the Surface: critical essays in the philosophy of deep ecology*, MIT Press, Cambridge, MA.

Kauppi, P. E., Ausubel, J. H, Fang, J., Mather, A. S., Sedjo, R. A. and Waggoner, P. E. (2006) 'Returning forests analyzed with the forest identity', *Proceedings of the National Academy of Sciences of the United States of America* 103: 17574–9.

Kennedy, D. and the Editors of *Science* (2006) (eds) *Science Magazine's State of the Planet 2006–7*, Island Press, for the American Association for the Advancement of Science, Washington.

Kennedy, V. W. (1988) 'Environmental impact assessment and bilateral development aid: an overview', pp. 272–82 in P. Wathern (ed.), *Environmental Impact Assessment: theory and practice*, Unwyn Hyman, London.

Kepe, T., Saruchera, M. and Whande, W. (2004) 'Poverty alleviation and biodiversity conservation: a South African perspective', *Oryx* 38: 143–5.

Khogali, M. M. (1982) 'The problem of siltation in Khasm el Girba Reservoir: its implications and suggested solutions', pp. 96–106 in H. G. Mensching (ed.), *Problems of the Management of Irrigated Land in Areas of Traditional and Modern Cultivation*, International Geographical Union Working Group on Resource Management in Drylands, Hamburg.

Kiessling, K. L. and Landberg, H. (1994) (eds) *Population, Economic Development and the Environment: the making of our common future*, Clarendon Press, Oxford.

Kimmage, K. (1991) 'Small scale irrigation initiatives in Nigeria: the problems of equity and sustainability', *Applied Geography* 11: 5–20.

Kirchner, J. W., Ledec, G., Goodland, R. J. A. and Drake, J. M. (1985) 'Carrying capacity, population growth and sustainable development', pp. 42–89 in D. J. Mahar (ed.), *Rapid Population Growth and Human Carrying Capacity*, World Bank Staff Working Paper 690, Washington.

Kitching, G. (1982) *Development and Underdevelopment in Historical Perspective: populism, nationalism and industrialism*, Methuen, London.

Kjellén, B. (2003) 'The saga of the Convention to Combat Desertification: the Rio/ Johannesburg Process and the global responsibility for the drylands', *Review of European Community and International Environmental Law* 12: 127–32.

Koh, T. T.-B. (1993) 'The Earth Summit's negotiating process: some comments on the art and science of negotiation', pp. v–xiii in N. Robinson (ed.) *Agenda 21: Earth's action plan*, IUCN Environmental Policy and Law Paper 27, Oceana Publications, New York.

Kowal, J. M. and Adeoye, K. B. (1973) 'An assessment of aridity and the severity of the 1972 drought in northern Nigeria and neighbouring countries', *Savanna* 2: 145–58.

Kramer, R. A., Schaik, C. P. van, and Johnson, J. (1997) (eds) *The Last Stand: protected areas and the defense of tropical biodiversity.* Oxford University Press, New York.

Kropotkin, P. ([1906] 1972) *The Conquest of Bread*, ed. P. Avrich, Allen Lane, London (first published in English as articles in *Freedom* 1892–4, and as a book in London, 1906).

Kropotkin, P. ([1899] 1974) *Fields, Factories and Workshops Tomorrow*, ed. C. Ward, Allen and Unwin, London.

Kull, C. A. (2000) 'Deforestation, erosion, and fire: degradation myths in the environmental history of Madagascar', *Environment and History* 6: 423–50.

Kull, C. A. (2004) *Isle of Fire: the political ecology of landscape burning in Madagascar*, University of Chicago Press, Chicago.

Kurlansky, M. (1999) *Cod: a biography of the fish that changed the world*, Vintage, London.

La Viña, A. G. M., Hoff, G. and DeRose, A. M. (2003) 'The outcomes of Johannesburg: assessing the World Summit on Sustainable Development', *SAIS Review* 23(1): 53–70.

Lagler, K. F. (1969) (ed.) *Man-Made Lakes: planning and development*, Food and Agriculture Organization, Rome.

Lamb, D. (1980) 'Some ecological consequences of logging rainforests on Papua New Guinea', pp. 55–64 in J. I. Furtado (ed.), *Tropical Ecology and Development*, International Society for Tropical Ecology, Kuala Lumpur (Proceedings of the Fifth International Symposium on Tropical Ecology).

Lamb, P. J. (1979) 'Some perspectives on climate and climatic dynamics', *Progress in Physical Geography* 3: 215–35.

Lamprey, H. (1988) 'Report on the desert encroachment reconnaissance in northern Sudan, October 21–November 10, 1875', *Desertification Control Bulletin* 17: 1–7 (reprinted).

Lane, C. (1992) 'The Barabaig pastoralists of Tanzania: sustainable land use in jeopardy', pp. 81–105 in D. Ghai and J. M. Vivian (eds), *Grassroots Environmental Action: people's participation in sustainable development*, Routledge, London.

Lane, C. R. (1998) (ed.) *Custodians of the Commons: pastoral land tenure in East and West Africa*, Earthscan, London.

Langton, M. (2003) 'The "wild", the market and the native: indigenous people face new forms of global colonization', pp. 79–107 in W. M. Adams and M. Mulligan (eds), *Decolonizing Nature: strategies for conservation in a post-colonial era*, Earthscan, London.

Lanly, J.-P. (1982) *Tropical Forest Resources*, Forestry Paper No. 30, Food and Agriculture Organization, Rome.

Lanning, G. and Mueller, M. (1979) *Africa Undermined: mining companies and the underdevelopment of Africa*, Penguin, Harmondsworth.

Lappé, M. and Bailey, B. (1999) *Against the Grain: genetic transformation of global agriculture*, Earthscan, London.

Lash, S. and Urry, J. (1994) *Economies of Signs and Space*, Sage, London.

Leach, M. and Mearns, R. (1996) (eds) *The Lie of the Land: challenging received wisdom on the African environment*, James Currey/International African Institute, London.

Leakey, R. and Morell, V. (2001) *Wildlife Wars: my battle to save Kenya's elephants*, Macmillan, London.

Lebel, L. (2005) 'Transitions to sustainability in production-consumption systems', *Journal of Industrial Ecology* 9(1–2): 11–13.

Lee, E. (1981) 'Basic needs strategies: a frustrated response to development from below?', pp. 107–27 in W. B. Stöhr and D. R. F. Taylor (eds), *Development: from above or below?*, Wiley, Chichester.

Lélé, S. M. (1991) 'Sustainable development: a critical review', *World Development* 19: 607–21.

Leopold, L. B., Clarke, F. E., Nanshaw, B. B. and Balsley, J. R. (1971) *A Procedure for Evaluating Environmental Impact*, US Geological Survey Circular 645, Washington.

Lericollais, A. and Schmitz, J. (1984) '"La calebasse et la houe": Techniques et outils de décrue dans la vallée du Sénégal', *Cahiers ORSTOM série Sciences Humaines* 20: 127–59.

Lesorogol, C. K. (2005) 'Privatising pastoral lands: economic and normative outcomes in Kenya', *World Development* 11: 1959–78.

Lewis, M. W. (1992) *Green Delusions: an environmentalist critique of radical environmentalism*, Duke University Press, Durham, NC.

Lightfoot, R. P. (1978) 'The costs of resettling reservoir evacuees in NE Thailand', *Journal of Tropical Geography* 47: 63–74.

Lightfoot, R. P. (1979) 'Alternative resettlement strategies in Thailand: lessons from experience', pp. 28–38 in L. Gosling (ed.), *Population Resettlement in the Mekong River Basin*, Studies in Geography 10, University of North Carolina, Chapel Hill.

Lightfoot, R. P. (1981) 'Problems of resettlement in the development of river basins in Thailand', pp. 93–114 in S. K. Saha and C. J. Barrow (eds), *River Basin Planning: theory and practice*, Wiley, Chichester.

Lindblade, K. A., Carswell, G. and Tumuhairwe, J. K. (1998) 'Mitigating the relationship between population growth and land degradation: land use change and farm management in southwestern Uganda', *Ambio* 27: 565–71.

Lindemann, R. L. (1942) 'The trophic-dynamic aspect of ecology' *Ecology* 23: 399–418.

Lipman, Z. (2002) 'A dirty dilemma: the hazardous waste trade', *Harvard International Review*, Winter: 67–71.

Lipton, M. (1991) 'A note on poverty and sustainability', *IDS Bulletin* 22(4): 12–16

Lipton, M. and Longhurst, R. (1989) *New Seeds and Poor People*, Johns Hopkins University Press, Baltimore.

List, J. A. and de Zeeuw, A. (2002) (eds) *Recent Advances in Environmental Economics*, Edward Elgar, London.

Liu, J. and Diamond, J. (2005) 'China's environment in a globalizing world: how China and the rest of the world affect each other', *Nature* 453: 1179–86.

Livingstone, D. N. (1995) 'The polity of nature: representation, virtue, strategy', *Ecumene* 2(4): 353–77.

Loh, J., Randers, J., MacGillivray, A., Kapos, V., Jenkins, M., Groombridge, B., Cox, N. and Warren, B. (1999) *Living Planet Report*, World Wildlife Fund, Gland, Switzerland.

Lomborg, B. (2001) *The Sceptical Environmentalist: measuring the real state of the world*, Cambridge University Press, Cambridge.

Lovejoy, T. E., Bierregaard, R. O., Rankin, J. and Schubart, H. O. R. (1983) 'Dynamics of forest fragments', pp. 377–84 in S. L. Sutton, T. C. Whitmore and A. C. Chadwick (eds), *Tropical Rain Forest: ecology and management*, Blackwell, Oxford.

Lövgren, L. (1997) 'Moratorium in Sweden: an account of the dams debate', pp. 21–30 in A. D. Usher (ed.), *Dams as Aid: a political anatomy of Nordic development thinking*, Routledge, London.

Low, D. A. and Lonsdale, J. M. (1976) 'Introduction: towards a new order 1945–1963', pp. 1–63 in D. A. Low and A. Smith (eds), *History of East Africa*, vol. 3, Clarendon Press, Oxford.

Low, N. and Gleeson, B. (1998) *Justice, Society and Nature: an exploration of political ecology*, Routledge, London.

Lowe, P. D. (1976) 'Amateurs and professionals: the institutional emergence of British plant ecology', *Journal of the Society for the Bibliography of Natural History* 7: 517–35.

Lowe-McConnell, R. H. (1966) (ed.) *Man-Made Lakes*, Institute of Biology and Academic Press, London.

Lowe-McConnell, R. H. (1975) *Fish Communities of Tropical Freshwaters*, Longman, London.

Löwy, M. (2005) 'What is ecosocialism?' *Capitalism, Nature, Socialism* 16(2): 15–24.

Lugo, A. E., Parrotta, J. A. and Brown, S. (1993) 'Loss of species caused by tropical deforestation and their recovery through management', *Ambio* 22: 106–69.

Luke, T. W. (2005) 'Neither sustainable nor development: reconsidering sustainability in development', *Sustainable Development* 13: 228–38.

Lummis, D. (1992) 'Equality', pp. 38–52 in W. Sachs (ed.), *The Development Dictionary*, Zed, London.

Lumsden, D. P. (1975) 'Towards a systems model of stress: feedback from an anthropological study of the impact of Ghana's Volta River Project', *Stress and Anxiety* 2: 191–227.

Lyons, M. (1985) 'From "death camps" to *cordon sanitaire*: the development of sleeping sickness policy in the Uele District of the Belgian Congo, 1903–1914', *Journal of African History* 26: 69–91.

Mabbutt, J. A. (1984) 'A new global assessment of the status and trends of desertification', *Environmental Conservation* 11: 103–13.

Mabogunje, A. L. (1973) (ed.) *Kainji: a Nigerian man-made lake, Kainji Lake Studies vol. 2: Socio-economic conditions*, Nigerian Institute for Social and Economic Research, Ibadan.

McCabe, J. T. (2004) *Cattle Bring us to our Enemies: Turkana ecology, politics, and raiding in a disequilibrium system*, University of Michigan Press, Ann Arbor.

McCormick, J. (1997) *Acid Earth*, Earthscan, London.

McCormick, J. S. (1986) 'The origins of the *World Conservation Strategy*', *Environmental Review* 10(2): 177–87.

McCormick, J. S. (1989) *The Global Environmental Movement: reclaiming paradise*, Belhaven, London.

McCully, P. (1996) *Silenced Rivers: the ecology and politics of large dams*, Zed Press, London.

Mace, R. (1991) 'Overgrazing overstated', *Nature* 349 (24 January): 280–1.

McGinn, A. P. (2000) 'POPs culture', *World Watch* 13(2): 26–36.

McHarg, I. (1969) *Design with Nature*, Natural History Press, Garden City, NY.

McIntosh, R. P. (1985) *The Background of Ecology: concept and theory*, Cambridge University Press, Cambridge.

Mackenzie, A. F. D. (1998) *Land, Ecology and Resistance in Kenya, 1990–1952*, Edinburgh University Press, Edinburgh.

Mackenzie, A. F. D. (2000) 'Contested ground: colonial narratives and the Kenyan environment, 1920–1945', *Journal of Southern African Studies* 26: 697–718.

MacKenzie, D. (1987) '"Thousands poisoned" by pesticide in Guyana', *New Scientist* 19 March: 18.

MacKenzie, J. M. (1988) *The Empire of Nature: hunting, conservation and British imperialism*, Manchester University Press, Manchester.

McKibben, B. (1990) *The End of Nature*, Penguin, Harmondsworth.

McNamee, K. (1993) 'From wild places to endangered spaces: a history of Canadian national parks', pp. 17–44 in P. Dearden and R. Rollins (eds), *Parks and Protected Areas in Canada: planning and management*, Oxford University Press, Toronto.

McNeely, J. A. (1993) 'Economic incentives for conserving biodiversity: lessons for Africa', *Ambio* 22: 144–50.

McNeely, J. A. (1996) 'Partnerships for conservation: an introduction', in J. A. McNeely (ed.), *Expanding Partnerships in Conservation*, Island Press, Washington.

McNeely, J. A. and Miller, K. R. (1984) (eds) *National Parks, Conservation and Development: the role of protected areas in sustaining society*, Smithsonian Institute Press, Washington.

McNeely, J. and Pitt, D. (1987) (eds) *Culture and Conservation: the human dimension in environmental planning*, Croom Helm, London.

McRobie, G. (1981) *Small is Possible*, Jonathan Cape, London.

McSweeney, K. (2004) 'Indigenous population growth in the lowland Neotropics: social science insights for biodiversity conservation', *Conservation Biology* 19: 1357–84.

Maddox, J. (1972) *The Doomsday Syndrome*, Macmillan, London.

Madely, J. (1995) 'Feeding 8 billion', *People and the Planet* 4(4): 7–9.

Malanson, G. P. (1993) *Riparian Landscapes*, Cambridge University Press, Cambridge.

Margalef, R. (1968) *Perspectives in Ecological Theory*, University of Chicago Press, Chicago.

Markandya, A., Harou, P., Bellù, L. G. and Cistulli, V. (2002) *Environmental Economics for Sustainable Growth: handbook for practitioners*, Edward Elgar, London.

Marsh, G. P. ([1864] 1965) *Man and Nature; or, physical geography as modified by human action*, Harvard University Press, Cambridge, MA, 1965) (first published Scribners, New York, and Sampson Low, London).

Marshall, B. K. (1999) 'Globalisation, environmental degradation and Ulrich Beck's Risk Society', *Environmental Values* 8: 253–75.

Martin, G. (2007) 'Global motorization, social ecology and China', *Area* 39: 66–73.

Maser, C. (1990) *The Redesigned Forest*, Stoddart, Toronto.

Mason, M. (1999) *Environmental Democracy*, Earthscan, London.

Mason, M. (2005) *The New Accountability: environmental responsibility across borders*, Earthscan, London.

Mather, A. (1992) 'The forest transition', *Area* 24: 367–79.

Matten, D. (2004a) 'Editorial: The Risk Society thesis in environmental politics and management – a global perspective', *Journal of Risk Research* 7: 371–6.

Matten, D. (2004b) 'The impact of the risk society thesis on environmental politics and management in a globalizing economy – principles, proficiency, perspectives', *Journal of Risk Research* 7: 377–98.

Matthews, S. (2004) 'Post-development theory and the question of alternatives: a view from Africa', *Third World Quarterly* 25: 373–84.

Matthiessen, P. and Douthwaite, B. (1985) 'The impact of tsetse control campaigns on African wildlife', *Oryx* 19: 202–9.

Mawdsley, E. (1998) 'After Chipko: from environment to region in Uttaranchal', *Journal of Peasant Studies* 25: 36–54.

Mawdsley, E. (1999) 'A new Himalayan state in India: popular perceptions of regionalism, politics and development', *Mountain Research and Development* 19: 101–12.

Maybury-Lewis, D. (1984) 'Demystifying the second conquest', pp. 127–34 in M. Schmick and C. H. Wood (eds), *Frontier Expansion in Amazonia*, University of Florida Press, Gainesville.

McLean, R. C. (1919) 'Studies in the ecology of tropical rain forest: with special reference to the forests of south Brazil', *Journal of Ecology* 7: 121–72.

Meadows, D. H., Meadows, D. K., Randers, J. and Behrens, W. W., III (1972) *The Limits to Growth*, Universe Books, New York.

Meine, C., Soulé, M. and Noss, R. F. (2006) '"A mission-driven discipline": the growth of conservation biology', *Conservation Biology* 20: 631–51.

Melillo, J. M., Palm, C. A. and Myers, N. (1985) 'A comparison of two recent estimates of forest disturbance in tropical forests', *Environmental Conservation* 12: 37–40.

Merchant, C. (1980) *The Death of Nature: women, ecology and the scientific revolution*, Harper and Row, New York.

Merryfield, A. and Swyngedouw, E. (1996) (eds) *The Urbanization of Injustice*, Lawrence and Wishart, London.

Mertens, B., and Lambin, E. F. (2000) 'Land-cover-change trajectories in southern Cameroon', *Annals of the Association of American Geographers* 90: 467–94.

Metcalfe, S. (1994) 'The Zimbabwe Communal Areas Management Programme for Indigenous Resources (CAMPFIRE)', pp. 161–92 in D. Western, R. M. White and S. C. Strumm (eds), *Natural Connections: perspectives in community-based conservation*, Island Press, Washington.

Middleton, N. J. (1985) 'Effect of drought on dust production in the Sahel', *Nature* 316: 431–4.

Middleton, N. (2003) *The Global Casino: an introduction to environmental issues*, Arnold, London (3rd edn).

Middleton, T. and Thomas, D. S. G. ([1992] 1997) (eds) *World Atlas of Desertification*, United Nations Environment Programme, Nairobi (2nd edn).

Midgley, J., Hall, A., Hardiman, M. and Narine, D. (1986) *Community Development, Social Participation and the State*, Methuen, London.

Mies, M. (1986) *Patriarchy and Accumulation on a World Scale: women in the international division of labour*, Zed Books, London.

Millennium Ecosystem Assessment (2005) *Ecosystems and Human Wellbeing: synthesis*, Island Press, Washington.

Miller, R. B. (1994) 'Interactions and collaboration in global change across the social and natural sciences', *Ambio* 23: 19–24.

Milner-Gulland, E. J., Bennett, E. L. and the SCB 2002 Annual Meeting Wild Meat Group (2003) 'Wild meat: the bigger picture', *Trends in Ecology and Evolution*, 18: 351–7.

Mishan, E. ([1967] 1969) *The Costs of Economic Growth*, Penguin, Harmondsworth (first published Staples Press).

Mishan, E. J. (1977) *The Economic Growth Debate: an assessment*, Allen and Unwin, Hemel Hempstead.

Mohun, J. and Sattaur, O. (1987) 'The drowning of a culture', *New Scientist* 15 January: 37–42.

Mol, A. P. (1996) 'Ecological modernisation and institutional reflexivity: environmental reform in a late modern age', *Environmental Politics* 5: 302–23.

Mol, A. P. J. (2001) *Globalization and Environmental Reform: the ecological modernization of the global economy*, MIT Press, Cambridge, MA.

Mol, A. P. J. (2006) 'Environment and modernity in transitional China: frontiers of ecological modernization', *Development and Change* 37: 29–56.

Monbiot, G. (2007) *Heat: how we can stop the planet burning*, Penguin, Harmondsworth.

Monfreda, C., Wackernagel, M. and Deumling, D. (2004) 'Establishing national natural capital accounts based on detailed Ecological Footprint and biological capacity assessments', *Land Use Policy* 21: 231–46.

Mooney, H. A. and Ehrlich, P. R. (1997) 'Ecosystem services: a fragmentary history', pp. 11–19 in G. C. Daily (ed.), *Nature's Services: societal dependence on natural ecosystems*, Island Press, Washington.

Moore, D. and Sklar, L. (1998) 'Reforming the World Bank's lending for water: the process and outcome of developing a water resources management policy', pp. 345–90 in J. A. Fox and L. D. Brown (eds), *The Struggle for Accountability: the World Bank, NGOs and grassroots movements*, MIT Press, Cambridge, MA.

Moore, H. and Vaughan, M. (1994) *Cutting down Trees: gender, nutrition and agricultural change in Zambia*, James Currey, London.

Moore, M. (1993) 'Good government? Introduction', *IDS Bulletin* 24(1): 1–6.

Moorhead, R. (1988) 'Access to resources in the Niger Inland Delta, Mali', pp. 27–39 in J. Seeley and W. M. Adams (eds), *Environmental Issues in African Development Planning*, Cambridge African Monographs No. 9, African Studies Centre, Cambridge.

Moran, E. F. (1983) (ed.) *The Dilemma of Amazonian Development*, Westview, Boulder, CO.

Moris, J. (1987) 'Irrigation as a privileged solution in African development', *Development Policy Review* 5: 99–123.

Moris, J. R. and Thom, D. J. (1990) *Irrigation Development in Africa: lessons from experience*, Westview Press, Boulder, CO.

Morrison, J. (1997) 'Protected areas, conservationists and aboriginal interests in Canada', pp. 270–96 in K. Ghimire and M. Pimbert (eds), *Social Change and Conservation: environmental politics and impacts of national parks and protected area*, Earthscan, London.

Mortimore, M. (1989) *Adapting to Drought: farmers, famines and desertification in West Africa*, Cambridge University Press, Cambridge.

Mortimore, M. (1993) 'Population growth and land degradation', *GeoJournal* 31: 15–21.

Mortimore, M. (1998) *Roots in the African Dust: sustaining the sub-Saharan drylands*, Cambridge University Press, Cambridge.

Mortimore, M. and Adams, W. M. (1999) *Working the Sahel: environment and society in northern Nigeria*, Routledge, London.

Mortimore, M. J. and Tiffen, M. (1995) 'Population and environment in time perspective: the Machakos story', pp. 69–89 in T. Binns (ed.), *People and Environment in Africa*, Wiley, Chichester.

Mouafo, D., Fotsing, E. R., Sighomnou, D. and Sigha, L. (2002) 'Dam, environment and regional development: case study of the Logone floodplain in Northern Cameroon', *Water Resources Development* 18: 209–19.

Mulligan, P. (1999) 'Greenwash or blueprint? Rio Tinto in Madagascar', *IDS Bulletin* 30(3): 50–7.

Munasinghe, M. (1993a) 'Environmental issues and economic decisions in developing countries', *World Development* 21: 1729–48.

Munasinghe, M. (1993b) *Environmental Economics and Sustainable Development*, World Bank Environmental Paper 3, World Bank, Washington.

Munasinghe, M. (1993c) 'Environmental economics and biodiversity management', *Ambio* 22: 126–35.

Munasinghe, M. (1999) 'Is environmental degradation an inevitable consequence of economic growth?: tunneling through the environmental Kuznets curve', *Ecological Economics* 29: 89–109.

Munn, R. E. (1979) (ed.) *Environmental Impact Assessment: principles and procedures*, SCOPE Report 5, Wiley, Chichester.

Munro, D. A. (1978) 'The thirty years of IUCN', *Nature and Resources* 14(2): 14–18.

Munslow, B., Katere, V., Ferf, A. and O'Keefe, P. (1988) *The Fuelwood Trap: a study of the SADCC region*, Earthscan, London.

Murdia, R. (1982) 'Forest development and tribal welfare: analysis of some policy issues', pp. 31–41 in E. G. Hallsworth (ed.), *Socio-Economic Effects and Constraints in Tropical Forest Management*, Wiley, Chichester.

Murombedzi, J. S. (1999) 'Devolution and stewardship in Zimbabwe's CAMPFIRE Programme', *Journal of International Development* 11: 287–93.

Murombedzi, J. S. (2001) 'Natural resource stewardship and community benefits in Zimbabwe's CAMPFIRE Programme', pp. 244–56 in D. Hulme and M. Murphree (eds), *African Wildlife and Livelihood*, James Currey, Oxford.

Murphree, M. W. (1994) 'The role of institutions in community-based conservation', pp. 403–27 in D. Western, R. M. White and S. C. Strumm (eds), *Natural Connections: perspectives in community-based conservation*, Island Press, Washington.

Murphree, M. (2001) 'A case study of ecotourism development from Mahenye, Zimbabwe', pp. 177–94 in D. Hulme and M. Murphree (eds), *African Wildlife and Livelihoods: the promise and performance of community conservation*, James Currey, Oxford.

Murphy, D. F. and Bendell, J. (1997) *In the Company of Partners: business, environmental groups and sustainable development post-Rio*, Policy Press, Bristol.

Murphy, D. F. and Bendell, J. (2005) 'New partnerships for sustainable development: the changing nature of business–NGO relations', pp. 216–44 in P. Utting (ed.), *The Greening of Business in Developing Countries: rhetoric, reality and prospects*, Zed Books, London.

Murphy, R. (1994) *Rationality and Nature: a sociological inquiry into a changing relationship*, Westview Press, Boulder, CO.

Murray, D., Wesseling, C., Keifer, M., Corriolis, M., and Henao, S. (2002) 'Surveillance of pesticide-related illness in the developing world: putting the data to work', *International Journal of Occupational Environmental Health* 8: 243–8.

Murton, J. (1999) 'Population growth and poverty in Machakos District, Kenya', *Geographical Journal* 165: 37–46.

Mustafa, D. (2005) 'The production of an urban hazardscape in Pakistan: modernity, vulnerability and the range of choice', *Annals of the Association of American Geographers* 95: 566–86.

Myers, G. A. (2005) *Disposable Cities: garbage, governance and sustainable development in urban Africa*, Ashgate, Burlington, VT.

Myers, N. (1980) *Conversion of Tropical Moist Forests*, National Academy of Sciences, Washington.

Myers, N. (1981) 'The hamburger connection: how Central America's forests become North America's hamburgers', *Ambio* 10: 3–8.

Myers, N. (1984) *The Primary Source: tropical forests and our future*, W. W. Norton, New York.

Myers, N. and Myers, D. (1982) 'From "duck pond" to the global commons: increasing awareness of the supranational nature of emerging environmental issues', *Ambio* 11: 195–201.

Myers, R. A. and Worm, B. (2003) 'Rapid worldwide depletion of predatory fish communities', *Nature* 423: 280–3.

Mythen, G. (2005) 'Employment, individualization and insecurity: rethinking the risk society perspective', *Sociological Review*, 53: 129–49.

Naess, A. (1973) 'The shallow and the deep, long-range ecology movement: a summary', *Inquiry* 16: 95–100.

Nahman, A. and Antrobus, G. (2005) 'Trade and the environmental Kuznets curve: is Southern Africa a pollution haven?', *South African Journal of Economics* 73: 803–14.

Najam A., Poling, J. M., Yamagishi, N., Straub, D. G., Sarno, J., de Ritter, S. M. and Kim, E. M. (2002) 'From Rio to Johannesburg: progress and prospects', *Environment* 44(7): 26–38.

Najam A., Huq, S. and Sokona, Y. (2003) 'Climate negotiations beyond Kyoto: developing countries, concerns and interests', *Climate Policy* 3: 221–31.

Nash, R. (1973) *Wilderness and the American Mind*, Yale University Press, New Haven, CT.

Nature (1948) 'Aspects of colonial development', *Nature* 162: 547–50.

Naughton-Treves, L. (1997) 'Farming the forest edge: vulnerable places and people around Kibale National Park, Uganda', *Geographical Review* 87: 27–46.

Nelson, J. G. (1987) 'National parks and protected areas, national conservation strategies and sustainable development', *Geoforum* 18: 291–320.

Nesmith, C. and Radcliffe, S. A. (1993) '(Re)mapping Mother Earth: a geographical perspective on environmental feminisms', *Environment and Planning D: Society and Space* 11: 379–94.

Neumann, R. P. (1996) 'Dukes, earls and ersatz Edens: aristocratic nature preservationists in colonial Africa', *Environment and Planning D: Society and Space* 14: 79–98.

Neumann, R. P. (1998) *Imposing Wilderness: struggles over livelihood and nature preservation in Africa*, University of California Press, Berkeley and Los Angeles.

Neumann, R. P. (2001) 'Africa's "last wilderness": reordering space for political and economic control in colonial Tanzania', *Africa* 71: 641–65.

Neumann, R. P. (2002) 'The postwar conservation boom in British colonial Africa', *Environmental History* 7: 22–47.

Neumann, R. P. (2004a) 'Nature–state-territory: towards a critical theorization of conservation enclosures', pp. 195–217 in R. Peet and M. Watts (eds), *Liberation Ecologies: environment, development, social movements*, Routledge, London (2nd edn).

Neumann, R. P. (2004b) 'Moral and discursive geographies in the war for biodiversity in Africa', *Political Geography* 23: 813–37.

Neumann, R. P. (2004c) *Making Political Ecology*, Hodder Arnold, London.

Neumayer, E. ([1999] 2003) *Weak versus Strong Sustainability: exploring the limits of two opposing paradigms*, Edward Elgar, Cheltenham (2nd edn).

Neumayer, E. (2004) 'The WTO and the environment: its past record is better than critics believe, but the future outlook is bleak', *Global Environmental Politics* 4(3): 1–8.

Nicholson, E. M. (1970) *The Environmental Revolution: a guide for the new masters of the world*, Hodder and Stoughton, London.

Nicholson, E. M. (1975) 'Conservation', pp. 12–14 in E. B. Worthington (ed.), *The Evolution of the IBP*, Cambridge University Press, Cambridge.

Nicholson, S. E. (1978) 'Climatic variation in the Sahel and other African regions during the past five centuries', *Journal of Arid Environments* 1: 3–24.

Nicholson, S. E. (1996) 'Environmental change within the historical period', pp. 60–88 in W. M. Adams, A. S. Goudie and A. R. Orme (eds), *The Physical Geography of Africa*, Oxford University Press, Oxford.

Nolan, P. H. (2005) 'China at the Crossroads', *Journal of Chinese Economic and Business Studies* 3: 1–22.

Norregaard, N. and Reppelin-Hill, V. (2000) *Controlling Pollution Using Taxes and Tradable Permits*, International Monetary Fund, Washington.

North, D. C. (1990) *Institutions, Institutional Change and Economic Performance*, Cambridge University Press, Cambridge.

Norton, B. (1991) *Toward Unity among Environmentalists*, Oxford University Press, London.

Norton-Griffiths, M. (1979) 'The influence of grazing, browsing and fire on the vegetation dynamics of the Serengeti', pp. 310–52 in A. R. E. Sinclair and M. Norton-Griffiths (eds), *Serengeti: dynamics of an ecosystem*, University of Chicago Press, Chicago.

Norton-Griffiths, M. and Southey, C. (1995) 'The opportunity costs of biodiversity conservation in Kenya', *Ecological Economics* 12: 125–39.

Nye, P. and Greenland, D. (1960) *The Soil under Shifting Cultivation*, Technical Report 51, Commonwealth Agricultural Bureau, London.

Nyerere, J. K. (1985) 'Africa and the debt crisis', *African Affairs* 84: 489–98.

Nyssen, J., Haile, M., Moeyersons, J., Poesen, J. and Deckers, J. (2004) 'Environmental policy in Ethiopia: a rejoinder to Keeley and Scoones', *Journal of Modern African Studies* 42: 137–47.

O'Connor, M. and Spash, C. L. (1999) (eds) *Valuation and the Environment: theory, method and practice*, Edward Elgar, London.

O'Riordan, T. ([1976] 1981) *Environmentalism*, Pion, London (2nd edn).

O'Riordan, T. (1988) 'The politics of sustainability', pp. 29–50 in R. K. Turner (ed.), *Sustainable Environmental Management: principles and practice*, Westview Press, Boulder, CO.

O'Riordan, T. and Jäger, J. (1996) 'The history of climate change science and policies', pp. 11–31 in T. O'Riordan and J. Jäger (eds), *Politics of Climatic Change in European Perspective*, Routledge, London.

O'Riordan, T. and Turner, R. K. (1983) *An Annotated Reader in Environmental Planning and Management*, Pergamon Press, Oxford.

Oates, J. F. (1995) 'The dangers of conservation by rural development: a case study from the forests of Nigeria', *Oryx* 29: 115–22.

Oates, J. F. (1999) *Myth and Reality in the Rain Forest: how conservation strategies are failing in West Africa*, University of California Press, Berkeley and Los Angeles.

Obeng, L. E. (1969) (ed.) *International Symposium on Man-Made Lakes, Accra*, Ghana University Press, Accra.

OECD (1995) *The Economic Appraisal of Environmental Projects and Policies: a practical guide*, Organization for Economic Cooperation and Development, Paris.

OECD (1996) *The Global Environmental Goods and Services Industry*, Organization for Economic Cooperation and Development, Paris.

OECD (2006) *Cost–Benefit Analysis and the Environment: recent developments*, Organization for Economic Cooperation and Development, Paris.

Oglesby, R. T., Carlson, C. A. and McCann, J. A. (1972) *River Ecology and Man*, Academic Press, New York.

Oldfield, S. (1988) 'Rare tropical timbers', IUCN, Gland, Switzerland.

Olson, D. M. and Dinerstein, E. (1998) 'The Global 200: a representation approach to conserving the earth's most biologically valuable ecoregions', *Conservation Biology* 12: 502–15.

Olsson, L. (1993) 'On the causes of famine: drought, desertification and market failure in the Sudan', *Ambio* 22: 395–403.

Olthof, W. (1994) 'Wildlife resources and local development: experiences from Zimbabwe's CAMPFIRE Programme', pp. 111–28 in J. P. M. van den Breemer, C. A. Drijver and L. B. Venema (eds), *Local Resource Management in Africa*, Wiley, Chichester.

Oman, C. P. and Wignarajah, G. (1991) *The Postwar Evolution of Development Thinking*, Macmillan, London, in association with the OECD Development Centre.

Ormerod, W. E. (1986) 'A critical study of the policy of tsetse eradication', *Land Use Policy* 3: 85–99.

Orr, D. (2007) 'The carbon connection', *Conservation Biology* 21: 289–92.

Osborn, D. and Bigg, T. (1998) *Earth Summit II: outcomes and analysis*, Earthscan, London.

Osborn, F. (1948) *Our Plundered Planet*, Faber and Faber, London.

Osborn, F. (1954) *The Limits of the Earth*, Faber and Faber, London.

Osmaston, A. E. (1922) 'Notes on the forest communities of the Garhwal Himalaya', *Journal of Ecology* 10: 129–67.

Ostrom, E. (1990) *Governing the Commons: the evolution of institutions for collective action*, Cambridge University Press, Cambridge.

Otten, M. (1986) 'Transmigrasi: from poverty to bare subsistence', *Ecologist* 16: 57–67.

Otterman, J. (1974) 'Baring high-albedo soils by overgrazing: a hypothesised desertification mechanism', *Science* 166: 531–3.

Pahl-Wostl, C. (1995) *The Dynamic Nature of Ecosystems: chaos and order intertwined*, Wiley, Chichester.

Palumbi, S. R. (2001) 'Humans as the World's Greatest Evolutionary Force', *Science* 293(5536): 1786–90.

Parayil, G. and Tong, F. (1998) 'Pasture-led to logging-led deforestation in the Brazilian Amazon', *Global Environmental Change* 8: 63–79.

Parikh, J., Babu, P. G. and Kavi Kumar, K. S. (1997) 'Climate change, North–South cooperation and collective decision-making post-Rio', *Journal of International Development* 9: 403–13.

Parnwell, M. J. G. and Bryant, R. L. (1996) 'Introduction', pp. 1–20 in M. J. G. Parnwell and R. L. Bryant (eds), *Environmental Change in South-East Asia: people, politics and sustainable development*, Routledge, London.

Parris, T. M. and Kates, R. W. (2003) 'Characterizing and measuring sustainable development', *Annual Review of Environment and Resources* 28: 559–86.

Parry, M., Canziani, O., Palutikof, J., van der Linden, P. and Hanson, C. (2007) (eds) *Climate Change 2007: Impacts, Adaptation and Vulnerability, Contribution of Working Group II to the Fourth Assessment Report of the Intergovernmental Panel on Climate Change*, Published for the Intergovernmental Panel on Climate Change, Cambridge University Press, Cambridge.

Parsons, H. L. (1979) (ed.) *Marx and Engels on Ecology*, Greenwood Press, Westport, CT.

Paudel, N. S. (2006) 'Protected areas and the reproduction of social inequality', *Policy Matters* 14: 155–69.

Pauley, D., Alder, J., Bennett, E., Christensen, V., Tyedmers, P. and Watson, R. (2003) 'The future for fisheries', *Science* 302: 1359–61.

Payne, A. J. (1986) *The Ecology of Tropical Lakes and Rivers*, Wiley, Chichester.

Pearce, D. (1988) 'The sustainable use of natural resources in developing countries', pp. 102–17 in R. K. Turner (ed.), *Sustainable Environmental Management*, Belhaven, London.

Pearce, D. (1995) *Blueprint 4: capturing global environmental value*, Earthscan, London.

Pearce, D. and Brown, K. (1994) 'Saving the world's tropical forests', pp. 2–26 in K. Brown and D. Pearce (eds), *The Causes of Tropical Deforestation: the economic and statistical analysis of factors giving rise to the loss of the tropical forests*, UCL Press, London.

Pearce, D., Markandya, A. and Barbier, E. (1989) *Blueprint for a Green Economy*, Earthscan, London.

Pearce, F. (1987) 'Pesticide deaths: the price of the Green Revolution', *New Scientist* 18 June: 30.

Pearce, F. (1991) 'North–South rift bars path to summit', *New Scientist* 22 November: 20–1.

Pearce, F. (1992) *The Dammed: rivers, dams and the coming world water crisis*, Bodley Head, London.

Pearce, F. (1994) 'Are Sarawak's forests sustainable?', *New Scientist* 26 November: 28–32.

Pearl, R. (1927) 'The growth of populations', *Quarterly Review of Biology* 2: 537–43.

Peet, R. (1991) *Global Capitalism: theories of societal development*, Routledge, London.

Peet, R. and Watts, M. (1996) 'Liberation ecology: development, sustainability, and environment in an age of market triumphalism', pp. 1–45 in R. Peet and M. Watts (eds), *Liberation Ecologies: environment, development, social movements*, Routledge, London (1st edn).

Peet, R. and Watts, M. (2004) (eds) *Liberation Ecologies: environment, development, social movements*, London: Routledge (2nd edn).

Pelling, M. (2003) 'Toward a political ecology of urban environmental risk: the case of Guyana', pp. 73–93 in K. S. Zimmerer and T. J. Bassett (eds), *Political Ecology: an integrative approach to geography and environment-development studies*, Guilford Press, New York.

Peluso, N. (1993) 'Coercing conservation: the politics of state resource control', *Global Environmental Change* 3: 199–217.

Penrose, J.-P., Bdliya, H. and Chettleborough, J. (2005) 'Stories on the environment and conflict from northern Nigeria', pp. 131–51 in S. Bass, H. Reid, D. Satterthwaite and P. Steele (eds), *Reducing Poverty and Sustaining the Environment: the politics of local engagement*, Earthscan, London.

Pepper, D. (1984) *The Roots of Modern Environmentalism*, Croom Helm, London.

Pepper, D. (1993) *Eco-Socialism: from deep ecology to social justice*, Routledge, London.

Pepper, D. (1996) *Modern Environmentalism: an introduction*, Routledge, London.

Perfect, J. (1980) 'The environmental impacts of DDT in a tropical agro-ecosystem', *Ambio* 9: 16–22.

Perrow, M. R. and Day, A. J. (2002) (eds) *Handbook of Ecological Restoration* (2 volumes) Cambridge University Press, Cambridge.

Petr, T. (1975) 'On some factors associated with high fish catches in new African man-made lakes', *Archiv für Hydrobiologie* 75: 32–49.

Petts, G. E. (1984) *Impounded Rivers: perspectives for ecological management*, Wiley, Chichester.

Phillips, J. (1931) 'Ecological investigations in South, Central and East Africa: outline of a progressive scheme', *Journal of Ecology* 14: 474–82.

Pieterse, J. N. (1998) 'My paradigm or yours? Alternative development, post-development and reflexive development', *Development and Change* 29: 343–73.

Pimbert, M. (1997) 'Issues emerging in implementing the Convention on Biological Diversity', *Journal of International Development* 9: 415–25.

Pimm, S. (1997) 'The value of everything', *Nature* 387 (15 May): 231–2.

Plumwood, V. (1993) *Feminism and the Mastery of Nature*, Routledge, London.

Plumwood, V. (2000) 'Deep ecology, deep pockets, and deep problems: a feminist ecosocialist analysis', pp. 59–84 in E. Katz, A. Light and D. Rothenberg (eds), *Beneath the Surface: critical essays in the philosophy of deep ecology*, MIT Press, Cambridge, MA.

Plumwood, V. (2003) 'Decolonizing relationships with nature', pp. 51–78 in W. M. Adams and M. Mulligan (eds), *Decolonizing Nature: strategies for conservation in a post-colonial era*, Earthscan, London.

Poirier, R., and Ostergren, D. (2002) 'Evicting people from nature: indigenous land rights and national parks in Australia, Russia and the United States', *Natural Resources Journal* 42: 331–51.

Polet, G. and Thompson, J. R. (1996) 'Maintaining the floods: hydrological and institutional aspects of managing the Komadugu–Yobe River basin and its floodplain wetlands', pp. 73–90 in M. C. Acreman and G. E. Hollis (eds), *Water Management and Wetlands in Sub-Saharan Africa*, IUCN, Gland, Switzerland.

Polunin, N. (1984) 'Genesis and progress of the World Campaign and Council for the Biosphere', *Environmental Conservation* 11: 293–8.

Poore, D. (1976) *Ecological Guidelines for Development in Tropical Rainforests*, IUCN, Gland, Switzerland.

Poore, D. and Sayer, J. (1987) *The Management of Tropical Moist Forest Lands: ecological guidelines*, IUCN, Gland, Switzerland.

Porritt, J. (2005) *Capitalism as if the World Matters*, Earthscan, London.

Potts, M. (2005) 'Why can't a man be more like a woman?', *Obstetrics and Gynaecology* 106: 1065–70.

Power, M. (2003) *Rethinking Development Geographies*, Routledge, London.

Prance, G. T. (1991) 'Rates of loss of biological diversity', pp. 27–44 in I. F. Spellerberg, F. B. Goldsmith and M. G. Morris (eds), *The Scientific Management of Temperate Communities for Conservation*, Blackwell, Oxford.

Pratt, M. L. (1992) *Imperial Eyes: travel writing and transculturation*, Routledge, London.

Prendergast, D. K. and Adams, W. M. (2003) 'Colonial wildlife conservation and the origins of the Society for the Preservation of the Wild Fauna of the Empire (1903–1914)', *Oryx* 37: 251–60.

Pretty, J. N. (1995) *Sustainable Agriculture: policies and practice for sustainability and self-reliance*, Earthscan, London.

Pretty, J. (2002) *Agri-culture: reconnecting people, food and nature*, Earthscan, London.

Pretty, J. (2005) *The Earthscan Reader in Sustainable Agriculture*, Earthscan, London.

Princen, T. (1994a) 'NGOs: creating a niche in environmental diplomacy', pp. 121–59 in T. Princen and M. Finger (eds), *Environmental NGOs in World Politics*, Routledge, London.

Princen, T. (1994b) 'The ivory trade ban: NGOs and international conservation', pp. 121–59 in T. Princen and M. Finger (eds) *Environmental NGOs in World Politics*, Routledge, London.

Princen, T. and Finger, M. (1994) *Environmental NGOs in World Politics: linking the local and the global*, Routledge, London.

Prins, G. and Rayner, S. (2007) 'Time to ditch Kyoto', *Nature* 449: 973–5.

Prior, J. (1993) *Pastoral Development Planning*, Oxfam, Oxford.

Prospero, J. M. and Nees, R. T. (1977) 'Dust concentrations in the atmosphere of the equatorial North Atlantic: possible relationship to the Sahelian drought', *Science* 196: 1196–8.

Purvis, M. and Grainger, A. (2004) (eds) *Exploring Sustainable Development*, Earthscan, London.

Putnam, R. (1993) *Making Democracy Work: civic traditions in modern Italy*, Princeton University Press, Princeton.

Putnam, R. (2000) *Bowling Alone: The Collapse and Revival of American Community*, Simon and Schuster, New York.

Quammen, D. (2003) 'Saving Africa's Eden', *National Geographic* 204 (3): 48–75.

Radcliffe, S. A. (2005a) 'Development and geography II: towards a postcolonial development geography?' *Progress in Human Geography* 29(3): 291–8.

Radcliffe, S. A. (2005b) 'Rethinking development', pp. 200–10 in P. Cloke, P. Crang, and M. Goodwin (eds), *Introducing Human Geographies*, Arnold, London.

Rajan, R. (1998) 'Imperial environmentalism or environmental imperialism? European forestry, colonial foresters and the agendas of forest management in British India, 1800–1900', pp. 3324–71 in R. H. Grove, V. Damodaran and S. Sangwan (eds), *Nature and the Orient: the environmental history of South and South East Asia*, Oxford University Press, Delhi.

Rambo, A. T. (1982) 'Human ecology research on agroecosystems in SE Asia', *Singapore Journal of Tropical Geography* 3: 86–99.

Rangan, H. (2004) 'From Chipko to Uttaranchal: development, environment, and social protest in the Garhwal Himalaya', pp. 205–26 in R. Peet and M. Watts (eds), *Liberation Ecologies: environment, development and social movements*, Routledge, London (2nd edn).

Ranger, T. (1999) *Voices from the Rocks: nature, culture and history in the Matopos Hills of Zimbabwe*, James Currey, Oxford.

Rasid, H. (1979) 'The effects of regime regulation by the Gardiner Dam on downstream geomorphic processes in the South Saskatchewan River', *Canadian Geographer* 23: 140–58.

Ravenel, R. M., and Redford, K. H. (2005) 'Understanding IUCN Protected Area categories', *Natural Areas Journal* 25(4): 381–9.

Ravnborg, H. M. (2003) 'Poverty and environmental degradation in the Nicaraguan hills', *World Development* 31: 1933–46.

Rawcliffe, P. (1998) *Environmental Pressure Groups in Transition*, Manchester University Press, Manchester.

Reardon, T. and Vosti, S. A. (1995) 'Links between rural poverty and the environment in developing countries: asset categories and investment poverty', *World Development* 23: 1495–506.

Redclift, M. R. (1984) *Development and the Environmental Crisis: red or green alternatives?*, Methuen, London.

Redclift, M. R. (1987) *Sustainable Development: exploring the contradictions*, Methuen, London.

Redclift, M. R. (1996) *Wasted: counting the costs of global consumption*, Earthscan, London.

Redclift, M. R. (2000) *Sustainability: life chances and livelihoods*, Routledge, London.

Redclift, M. R. (2005a) 'Sustainable development (1987–2005) – an oxymoron comes of age', *Sustainable Development* 13: 212–27.

Redclift, M. R. (2005b) *Sustainability: critical concepts in the social sciences*, Routledge Major Works (four volumes), Taylor and Francis, London.

Redclift, M. R. and Benton, T. (1994) (eds) *Social Theory and the Global Environment*, Routledge, London.

Redclift, M. R. and Woodgate, G. (1997) (eds) *Developments in Environmental Sociology*, Edward Elgar, Cheltenham.

Redford, K. H. (1990) 'The ecologically noble savage', *Orion*, 9: 25–9.

Redford, K. H. (1992) 'The empty forest', *Bioscience* 42: 412–22.

Redford, K. H. and Sanderson, S. E. (1992) 'The brief, barren marriage of biodiversity and sustainability?', *Bulletin of the Ecological Society of America* 73: 36–9.

Redford, K. H. and Stearman, A. M. (1993) 'Forest-dwelling native Amazonians and the conservation of biodiversity: interests in common or in collision?', *Conservation Biology* 7: 248–55.

Redford, K. H., Brandon, K., and Sanderson, S. E. (1998) 'Holding ground', pp. 455–63 in K. Brandon, K. H. Redford and S. E. Sanderson (eds), *Parks in Peril: people, politics and protected areas*, Island Press, for the Nature Conservancy, Washington.

Redford K. H., Coppolillo, P., Sanderson, E. W., Fonseca, G. A. B. da, Dinerstein, E., Groves, C., Mace, G., Maginnis, S., Mittermeier, R. A., Noss, R., Olson, D., Robinson, J. G., Vedder, A. and Wright, M. (2003) 'Mapping the conservation landscape', *Conservation Biology* 17: 116–31.

Reed, P. and Rothenberg, D. (1993) (eds) *Wisdom in the Open Air: the Norwegian roots of deep ecology*, University of Minnesota Press, Minneapolis.

Reeves, R. R. and Chaudhry, A. A. (1998) 'Status of the Indus River dolphin *Platanista minor*', *Oryx* 32: 35–44.

Reich, C. (1970) *The Greening of America*, Random House, New York.

Repetto, R. (1987) 'Creating incentives for sustainable forest development', *Ambio* 16: 94–9.

Ribot, J. C. (1999) 'A history of fear: imagining deforestation in the West African dryland forests', *Global Ecology and Biogeography* 8: 291–300.

Ribot, J. and Larson, A. M. (2004) (eds), *Decentralization of Natural Resources: experiences in Africa, Asia and Latin America*, Routledge, London.

Ribot, J. and Larson, A. M. (2005) (eds), *Democratic Decentralization through a Natural Resource Lens*, Routledge, London.

Rich, B. (1994) *Mortgaging the Earth: the World Bank, environmental impoverishment and the crisis of development*, Earthscan, London.

Richards, P. (1985) *Indigenous Agricultural Revolution: ecology and food production in West Africa*, Longman, London.

Richards, P. (1986) *Coping with Hunger: hazard and experiment in an African rice-farming system*, Allen and Unwin, London.

Riddell, R. (1981) *Ecodevelopment*, Gower, Aldershot.

Robbins, P. (2004) *Political Ecology: a critical introduction*, Blackwell, Oxford.

Roberts, C. M. (2003) 'Our shifting perspectives on the oceans', *Oryx* 37: 166–77.

Roberts, C. M. (2007) *The Unnatural History of the Sea*, Island Press, Washington.

Roberts, N. ([1989] 1998) *The Holocene: an environmental history*, Blackwell, Oxford (2nd edn).

Robin, L. (1997) 'Ecology: a science of empire?', pp. 63–75 in T. Griffiths and L. Robin (eds), *Ecology and Empire: environmental history of settler societies*, Keele University Press, Keele.

Robinson, J. (2004). 'Squaring the circle? Some thoughts on the idea of sustainable development', *Ecological Economics* 48: 369–84.

Robinson, M. (1993) 'Governance, democracy and conditionality: NGOs and the New Policy Agenda', in A. Clayton (ed.), *Governance, Democracy and Conditionality: what role for NGOs?*, INTRAC, Oxford.

Robinson, N. (1993) (ed.) *Agenda 21: Earth's action plan*, IUCN Environmental Policy and Law Paper 27, Oceana Publications, New York.

Robinson, R. E. (1971) *Developing the Third World: the experience of the 1960s*, Cambridge University Press, Cambridge.

Rocheleau, D., Steinberg, P. E. and Benjamin, P. A. (1995) 'Environment, development, crisis and crusade: Ukambani, Kenya, 1890–1990', *World Development* 23: 1037–51.

Rocheleau, D., Thomas-Slayter, B. and Wangari, E. (1996a) 'Gender and environment: a feminist political ecology perspective', pp. 3–23 in D. Rocheleau, B. Thomas-Slayter and E. Wangari (eds), *Feminist Political Ecology: global issues and local experiences*, Routledge, London.

Rocheleau, D., Thomas-Slayter, B. and Wangari, E. (1996b) (eds) *Feminist Political Ecology: global issues and local experiences*, Routledge, London.

Roddick, J. (1997) 'Earth Summit north and south: building a safe house in the winds of change', *Global Environmental Change* 2: 147–65.

Roder, W. (1994) *Human Adjustment to Kainji Reservoir in Nigeria: an assessment of the economic and environmental consequences of a major man-made lake in Africa*, University Press of America, Lanham, MD.

Rodhe, H. Cowling, E., Galbally, I., Galloway, J. and Herrera, R. (1988) 'Acidification on regional air pollution in the tropics', pp. 3–39 in H. Rodhe and R. Herrera (eds) *Acidification in Tropical Countries*, Wiley, Chichester.

Roe, D. and Elliott, J. (2004) 'Poverty reduction and biodiversity conservation: rebuilding the bridges', *Oryx* 38: 137–9.

Roe, E. (1991) 'Development narratives, or making the best of blueprint development', *World Development* 19: 287–300.

Roe, E. (1994) *Narrative Policy Analysis: Theory and Practice*, Duke University Press, Durham, NC.

Roe, E. (1995) 'Except-Africa: postscript to a special section on development narratives', *World Development* 23: 1065–9.

Roggeri, H. (1985) *African Dams: impacts on the environment*, Environment Liaison Centre, Nairobi.

Rojstaczer, S., Sterling, S. M. and Moore, N. J. (2001) 'Human appropriation of photosynthesis products', *Science* 294(5551): 2549–52.

Rose, C. I. (1993) 'Beyond the struggle for proof: factors changing the environmental movement', *Environmental Values* 2: 285–98.

Rose, C. I. (1998) *The Turning of the Spar*, Greenpeace, London.

Rosenberg, D. M., Bodaly, R. A. and Usher, P. J. (1995) 'Environmental and social impacts of large-scale hydro-electric development: who is listening?', *Global Environmental Change* 5: 127–58.

Rosenfeld Sweeting, A. and Clark, A. P. (2000) *Lightening the Lode: a guide to responsible large-scale mining*, Conservation International, Washington.

Rostow, W. W. (1960) *The Stages of Economic Growth: a non-communist manifesto*, Cambridge University Press, Cambridge.

Rostow, W.W. (1978) *The World Economy: history and prospect*, Macmillan, London.

Roszak, T. (1979) *Person/Planet: the creative disintegration of industrial society*, Victor Gollancz, London.

Rowbotham, E. J. (1996) 'Legal obligations and uncertainties: the climate change convention', pp. 32–50 in T. O'Riordan and J. Jäger (eds), *Politics of Climate Change: a European perspective*, Routledge, London.

Rowell, A. (1996) *Green Backlash: global subversion of the environment movement*, Routledge, London.

Rowley, J. (1999) 'Beyond 6 billion', *People and the Planet* 8(1): 3.

Rubin, N. and Warren, W. M. (1968) (eds) *Dams in Africa: an interdisciplinary study of man-made lakes in Africa*, Frank Cass, London.

Rudé, G. (1980) *Ideology and Popular Protest*, Lawrence and Wishart, London.

Rudel, T. and Roper, J. (1997) 'The paths of rainforest destruction: cross-national patterns of tropical deforestation', *World Development* 25: 53–65.

Rudel, T. K., Perez-Lugo, M. and Zichal, H. (2000) 'When fields revert to forest: development and spontaneous reforestation in post-war Puerto Rico', *Professional Geographer* 52: 386–97.

Runte A. (1987) *National Parks: the American experience*, University of Nebraska Press, Lincoln.

Runte, A. (1990) *Yosemite: the embattled wilderness*, University of Nebraska Press, Lincoln.

Russell, E. J. (1954) *World Population and World Food Supplies,* Allen and Unwin, London.

Rydzewski, J. R. (1990) 'Irrigation: a viable development strategy?', *Geographical Journal* 150: 175–80.

Sachs, I. (1979) 'Ecodevelopment: a definition', *Ambio* 8(2/3): 113.

Sachs, I. (1980) *Stratégies de l'écodéveloppement*, Les Éditions Ouvrières, Paris.

Sachs, J. (2005) *The End of Poverty: how we can make it happen in our lifetime*, Penguin, Harmondsworth.

Sachs, J. D. and McArthur, J. W. (2005) 'The Millennium Project: a plan for meeting the Millennium Development Goals', *The Lancet* 365: 347–53.

Sachs, W. (1992a) 'Introduction', pp. 1–5 in W. Sachs (ed.), *The Development Dictionary: a guide to knowledge as power*, Witwatersrand University Press, Johannesburg, and Zed Books, London.

Sachs, W. (1992b) 'Environment', pp. 26–37 in W. Sachs (ed.), *The Development Dictionary: a guide to knowledge as power*, Witwatersrand University Press, Johannesburg, and Zed Books, London.

Sachs, W. (2002) (ed.) *The Jo'burg Memo. Fairness in a Fragile World: Memorandum for the World Summit on Sustainable Development*. The Heinrich Böll Foundation, Berlin.

Said, E. (1979) *Orientalism*, Pantheon, New York.

Salisbury, E. J. (1964) 'The origin and early years of the British Ecological Society', *Journal of Ecology* 52: 13–18.

Sandbach, F. (1978) 'Ecology and the "limits to growth" debate', *Antipode* 10: 22–32.

Sanderson, S. (2005) 'Poverty and conservation: the new century's "peasant question"?', *World Development* 33: 323–32.

Sanderson, S. and Redford, K. (2003a) 'Contested relationships between biodiversity conservation and poverty alleviation', *Oryx* 37: 389–90.

Sanderson, S. and Redford, K. (2003b) 'The defence of conservation is not an attack on the poor', *Oryx* 38: 146–7.

Sandford, S. (1983) *Management of Pastoral Development in the Third World*, Wiley, Chichester.

Sargisson L. (2001) 'What's wrong with ecofeminism', *Environmental Politics* 10: 52–64.

Saulei, S. M. (1984) 'Natural regeneration following clear-fell logging operations in the Gogol Valley, Papua New Guinea', *Ambio* 13(5–6): 351–4.

Schama, S. (1995) *Landscape and Memory*. HarperCollins, London.

Schech, S. and Haggis, J. (2002) (eds) *Development: a cultural studies reader*, Blackwell, Oxford.

Schmick, M. and Wood, C. H. (1984) (eds) *Frontier Expansion in Amazonia*, University of Florida Press, Gainesville.

Schmidheiny, S. (1992) (ed.) *Changing Course: a global business perspective on development and the environment*, MIT Press, Cambridge, MA.

Schumacher, E. F. (1973) *Small is Beautiful: economics as if people mattered*, Blond and Briggs, London (paperback edn, Abacus, 1974).

Schuurman, F. J. (1993) (ed.) *Beyond the Impasse: new directions in development theory*, Zed Press, London.

Schwela, D., Haq, G., Huizenga, C. Han, W.-J., Fabian, H. and Ajero, M. (2006) *Urban Air Pollution in Asian Cities*, Earthscan, London.

Scoones, I. (1991) 'Wetlands in drylands: key resources for agricultural and pastoral production in Africa', *Ambio* 20: 366–71.

Scoones, I. (1994) *Living with Uncertainty: new directions in pastoral development in Africa*, IT Publications, London.

Scoones, I. (1996) 'Range management science and policy: politics, polemics and pasture in southern Africa', pp. 34–53 in M. Leach and R. Mearns (eds), *The Lie of the Land: challenging received wisdom on the African environment*, James Currey/Heinemann, London.

Scoones, I. (1999) 'New ecology and the social sciences: what prospects for a fruitful engagement?', *Annual Review of Anthropology* 28: 479–507.

Scoones, I. (2007) 'Sustainability', *Development in Practice* 17: 589–96.

Scott, J.C. (1985) *Weapons of the Weak: everyday forms of peasant resistance*, Yale University Press, New Haven.

Scott, J. C. (1998) *Seeing like a State: how certain schemes to improve the human condition have failed*, Yale University Press, New Haven.

Scudder, T. (1975) 'Resettlement', pp. 453–71 in N. F. Stanley and M. P. Alpers (eds), *Man-Made Lakes and Human Health*, Academic Press, London.

Scudder, T. (1980) 'River basin development and local initiative in African savanna environments', pp. 383–405 in D. R. Harris (ed.), *Human Ecology in Savanna Environments*, Academic Press, London.

Scudder, T. (1988) *The African Experience with River Basin Development: achievements to date, the role of institutions and strategies for the future*, Institute of Development Anthropology and Clark University, Binghamton, NY.

Scudder, T. (1991a) 'A sociological framework for the analysis of new land settlements', pp. 148–67 in M. Cernea (ed.), *Putting People First: sociological variables in rural development*, Oxford University Press, Oxford.

Scudder, T. (1991b) 'The need and justification for maintaining transboundary flood regimes: the Africa case', *Natural Resources Journal* 31(1): 75–107.

Scudder, T. (1993) 'Development-induced relocation and refugee studies: 37 years of continuity among Zambia's Gwembe Tonga', *Journal of Refugee Studies* 6: 123–52.

Scudder, T. (2005) *The Future of Large Dams: dealing with the social, environmental and political costs*, Earthscan, London.

Scudder, T. and Acreman, M. C. (1996) 'Water management for the conservation of the Kafue wetlands, Zambia and the practicalities of artificial flood releases', pp. 101–6 in M. C. Acreman and G. E. Hollis (eds), *Water Management and Wetlands in Sub-Saharan Africa*, IUCN, Gland, Switzerland.

Scudder, T. and Colson, E. (1982) 'From welfare to development: a conceptual framework for the analysis of dislocated people', pp. 267–87 in A. Hansen and A. Oliver-Smith (eds), *Involuntary Migration and Resettlement: the problems and responses of dislocated people*, Westview Press, Boulder, CO.

Scudder, T. and Habarad, J. (1991) 'Local responses to involuntary relocation and development in the Zambian portion of the Middle Zambezi Valley', pp. 178–205 in J. A. Mollett (ed.), *Migrants in Agricultural Development*, Macmillan, London.

Secretary of the Convention on Biological Diversity (2006) *Global Biodiversity Outlook 2*, Secretary of the Convention on Biological Diversity, Montreal.

Secrett, C. (1985) *Rainforest: protecting the planet's richest resource*, Friends of the Earth, London.

Seddon, G. (1984) 'Logging in the Gogol Valley, Papua New Guinea', *Ambio* 13: 345–50.

Seeger, A. (1982) 'Native Americans and the conservation of flora and fauna in Brazil', pp. 177–90 in E. G. Hallsworth (ed.), *Socio-Economic Effects and Constraints in Tropical Forest Management*, Wiley, Chichester.

Seers, D. (1980) 'North–South: muddling morality and mutuality', *Third World Quarterly* 2: 681–92.

Sekhar, N. U. (1998) 'Crop and livestock depredation caused by wild animals in protected areas: the case of Sariska Tiger Reserve, Rajasthan, India', *Environmental Conservation* 25: 160–71.

Sen, A. (1981) *Poverty and Famines: an essay on entitlement and deprivation*, Oxford University Press, Oxford.

Sen, A. (1999) *Development as Freedom*, Oxford University Press, Oxford.

Sessions, G. (1995) (ed.) *Deep Ecology for the 21st Century: readings on the philosophy and practice of the new environmentalism*, Shambhala, Boston.

Seyfang, G. and Jordan, A. (2002) 'The Johannesburg Summit and sustainable development: how effective are mega-conferences?', pp. 19–26 in S. Stokke and O. Thommesen (eds), *Yearbook of International Cooperation on Environment and Development*, Earthscan, London.

Shackley, S. (1997) 'The Intergovernmental Panel on Climate Change: consensual knowledge and global politics', *Global Environmental Change* 7: 77–9.

Shankar Raman, T. R., Rawath, G. S. and Johnsingh, A. J. T. (1998) 'Recovery of tropical forest avifauna in relation to vegetation succession following shifting cultivation in Mizoma, north-east India', *Journal of Applied Ecology* 35: 214–31.

Shantz, H. L. and Marbut, C. F. (1923) *The Vegetation and Soils of Africa*, American Geographical Society Research Series 13, American Geographical Society and National Research Council, New York.

Shantz, J. (2003) 'Scarcity and the emergence of fundamentalist ecology', *Critique of Anthropology*, 23: 144–54.

Sharaf el Din, S. H. (1977) 'Effects of the Aswan High Dam on the Nile flood on the estuarine and coastal circulation pattern along the Mediterranean Egyptian coast', *Limnology and Oceanography* 22: 194–207.

Sheail, J. (1976) *Nature in Trust: the history of nature conservation in Great Britain*, Blackie, Glasgow.

Sheail, J. (1984) 'Nature reserves, national parks and post-war reconstruction in Britain', *Environmental Conservation* 11: 29–34.

Sheail, J. (1985) *Pesticides and Nature Conservation: the British experience, 1950–1975*, Clarendon Press, Oxford.

Sheail, J. (1987) *Seventy-Five Years in Ecology: The British Ecological Society*, Blackwell Scientific, Oxford.

Sheail, J. (1996) 'From aspiration to implementation: the establishment of the first National Nature Reserves in Britain', *Landscape Research* 21: 37–54.

Shelton, N. (1985) 'Logging versus the natural habitat in the survival of tropical forests', *Ambio* 14(1): 39–42.

Shiva, V. (1988) *Staying Alive: women, ecology and development*, Zed Books, London.

Shiva, V. (1997) *Biopiracy: the plunder of nature and knowledge*, South End Press, Boston.

Showers, K. B. (2002) 'Water scarcity and urban Africa: an overview of urban–rural water linkages', *World Development* 30: 621–48.

Siedenburg, J. (2006) 'The Machakos case study: solid outcomes, unhelpful hyperbole', *Development Policy Review* 2: 75–85.

Silva, J. P. O. (1997) 'In defence of the Bíobío River', pp. 153–70 in A. D. Usher (ed.), *Dams as Aid: a political anatomy of Nordic development thinking*, Routledge, London.

Simmons, C. S., Walker, R. T., Arima, E. Y., Alkdrichj, S. P. and Caldas, M. M. (2007) 'The Amazon land war in the south of Pará', *Annals of the Association of American Geographers* 97: 567–92.

Simon, J. L. (1981) *The Ultimate Resource*, Princeton University Press, Princeton.

Simonsen, A. H. (1995) 'Where oil kills: the Ogoni story', *Indigenous Affairs* 4: 52–7.

Sinclair, A. R. and Fryxell, J. M. (1985) 'The Sahel of Africa: ecology of a disaster', *Canadian Journal of Zoology* 63: 987–94.

Singh, A. (1986) 'Change detection in the tropical forest environment of northeast India using Landsat', pp. 237–54 in M. J. Eden and J. T. Parry (eds), *Remote Sensing and Tropical Land Management*, Wiley, Chichester.

Singh, J. S., Singh, S. P., Saxena, A. K. and Rawat, Y. S. (1984) 'India's Silent Valley and its threatened rainforest ecosystems', *Environmental Conservation* 11: 223–33.

Singh, S. (1997) *Taming the Waters: the political ecology of large dams in India*, Oxford University Press, Delhi.

Siry, J. P., Cubbage, F. W. and Ahmed, M. R. (2005) 'Sustainable forest management: global trends and opportunities', *Forest Policy and Economics* 7: 551–61.

Skillings, R. F. (1984) 'Economic development of the Brazilian Amazon: opportunities and constraints', *Geographical Journal* 150: 48–54.

Skinner, J. R. (1992) 'Conservation of the Inner Niger Delta in Mali: the interdependence of ecological and socio-economic research', pp. 41–7 in E. Maltby, P. J. Dugan and J. C. Lefeuve (eds), *Conservation and Development: the sustainable use of wetland resources*, IUCN, Gland, Switzerland.

Slater, D. (1993) 'The geopolitical imagination and the enframing of development theory', *Transactions of the Institute of British Geographers* NS 18: 419–37.

Smith, F. M., May, R. M., Pellew, R., Johnson, T. H. and Walter, K. S. (1993) 'Estimating extinction rates', *Nature* 364 (5 August): 494–6.

Smith, N. (1984) *Uneven Development*, Blackwell, Oxford.

Smith, R. (2007) 'Development of the SEEA 2003 and its implementation', *Ecological Economics* 61: 592–99.

Smith, R., Muir, R., Walpole, M., Balmford, A. and Leader-Williams, N. (2003) 'Governance and the loss of biodiversity', *Nature* 426: 67–70.

Smith, W. (2004a) 'Undercutting sustainability: the global problem of illegal logging and trade', *Journal of Sustainable Forestry* 19: 7–30.

Smith, W. (2004b) 'The role of monitoring in cutting crime', *Journal of Sustainable Forestry* 19: 293–317.

Sneddon, C. S. (2000) '"Sustainability" in ecological economics, ecology and livelihoods: a review', *Progress in Human Geography* 24: 521–49.

Sneddon, C. (2003) 'Reconfiguring scale and power: the Khong-Chi-Mun Project in northeast Thailand', *Environment and Planning A* 35: 2229–50.

Sneddon, C. and Binh, T. (2001) 'Politics, ecology and water: the Mekong Delta and development of the Lower Mekong basin', pp. 234–62 in W. N. Adger, P. M. Kelly and N. H. Ninh (eds), *Living with Environmental Change: social vulnerability, adaptation and resilience in Vietnam*, Routledge, London.

Snow, C. P. ([1959] 1998) *The Two Cultures*, Cambridge University Press.

Soerianegara, I. (1982) 'Socio-economic aspects of forest resources management in Indonesia', pp. 73–85 in E. G. Hallsworth (ed.), *Socio-Economic Effects and Constraints in Tropical Forest Management*, Wiley, Chichester.

Spaargaren, G. and Mol, A. P. (1992) 'Sociology, environment and modernity: ecological modernisation as a theory of social change', *Society and Natural Resources* 5: 323–44.

Spash, C. L. (1999) 'The development of environmental thinking in economics', *Environmental Values* 8: 413–35.

Speth, J. G. (2003) 'Perspectives on the Johannesburg Summit,' *Environment*, 45(1): 24–32.

Speth, J. G. and Haas, P. M. (2006) *Global Environmental Governance*, Island Press, Washington.

Spittler, G. (1979) 'Peasants and the state in Niger (West Africa)', *Journal of Peasant Studies* 7: 30–47.

Spivak, G. C. (1990) *The Post-Colonial Critic: interviews, strategies, dialogues*, Routledge, London.

Stamp, L. D. (1938) 'Land utilisation and soil erosion in Nigeria', *Geographical Review* 28: 32–45.

Stamp, L. D. (1940) 'The southern margin of the Sahara: comments on some recent studies on the question of desiccation', *Geograpical Review* 30: 297–300.

Stamp, L. D. (1953) *Our Undeveloped World*, Faber and Faber, London.

Stebbing, E. P. (1935) 'The encroaching Sahara: the threat to the West African Colonies', *Geographical Journal* 85: 506–24.

Stebbing, E. P. (1938) 'The advance of the desert', *Geographical Journal* 91: pp. 356–9.

Stein, R. E. and Johnson, B. (1979) *Banking on the Biosphere? Environmental procedures and practices of nine multilateral aid agencies*, Lexington Press, New York.

Steinhart, E. (1989) 'Hunters, poachers and gamekeepers: towards a social history of hunting in colonial Kenya', *Journal of African History* 30: 247–64.

Stern, D. I. (2004) 'The rise and fall of the Environmental Kuznets Curve', *World Development* 29: 1419–39.

Stern, N. (2007) *The Economics of Climate Change: the Stern Review*, Cambridge University Press, Cambridge.

Steward, T. A., Pickett, V., Parker, T. and Feidler, P. L. (1992) 'The new paradigm in ecology: implications for conservation biology above the species level', pp. 65–88 in P. L. Feidler and S. K. Jain (eds), *Conservation Biology: the theory and practice of nature conservation, preservation and management*, Chapman and Hall, London.

Stewart, F. (1985) *Planning to Meet Basic Needs*, Macmillan, London.

Stier, S. C. and Siebert, S. F. (2002) 'The Kyoto Protocol: an opportunity for biodiversity restoration forestry', *Conservation Biology* 16: 575–6.

Stocking, M. and Perkin, S. (1992) 'Conservation-with-development: an application of the concept in the Usambara Mountains, Tanzania', *Transactions of the Institute of British Geographers* NS 17: 337–49.

Stoddart, D. R. (1970) 'Our environment', *Area* 2(1): 1–3.

Stoddart, D. R. (1986) *On Geography and its History*, Blackwell, Oxford.

Stöhr, W. B. (1981) 'Development from below: the bottom-up and periphery-inward development paradigm', pp. 39–72 in W. B. Stöhr and D. R. F. Taylor (eds), *Development: from above or below?*, Wiley, Chichester.

Stolper, W. (1966) *Planning without Facts*, Harvard University Press, Cambridge, MA.

Stott, P. and Sullivan, S. (2000) (eds) *Political Ecology: science, myth and power*. Arnold, London.

Strange, S. (1986) *Casino Capitalism*, Blackwell, Oxford.

Street, F. A. and Grove, A. T. (1976) 'Environmental and climatic implications of late Quaternary lake-level fluctuations in Africa', *Nature* 261: 385–90.

Struhsaker, T. T. (1999) *Ecology of an African Rain Forest: logging in Kibale and the conflict between conservation and exploitation*, University Press of Florida, Gainesville.

Suckcharoen, S., Nuorteva, P. and Hasanen, E. (1978) 'Alarming signs of mercury pollution in a freshwater area of Thailand', *Ambio* 7: 113–16.

Sugden, R. and Williams, A. (1978) *The Principles of Practical Cost–Benefit Analysis*, Oxford University Press, Oxford.

Sullivan, F. (1993) 'Forest principles', pp. 159–67 in M. Grubb, M. Koch, K. Thompson, A. Munson and F. Sullivan (eds), *The 'Earth Summit' Agreements: a guide and assessment*, Earthscan, London, for the Royal Institute of International Affairs.

Sullivan, S. (1999) 'The impacts of people and livestock on topographically diverse open wood- and shrub-lands in arid north-west Namibia', *Global Ecology and Biogeography* 8: 257–77.

Sullivan, S (2000) 'Getting the science right, or introducing science in the first place? Local "facts", global discourse – "desertification" in north-west Namibia', pp. 15–44 in P. Stott and S. Sullivan (eds), *Political Ecology: science, myth and power*, Arnold, London.

Survival International (1985) *An End to Laughter? Tribal peoples and economic development*, Review 44, Survival International, London.

Sutcliffe, R. B. (1972) *Industry and Underdevelopment*, Addison Wesley, London.

Sutcliffe, R. B. (1984) 'Industry and underdevelopment re-examined', *Journal of Development Studies* 21: 121–33.

Sutton, K. (1977) 'Population resettlement: traumatic upheavals and the Algerian experience', *Journal of Modern African Studies* 15: 279–300.

Svedin, U. (1987) 'The IUCN conference on conservation and development in Ottawa', *Ambio* 16(1): 65.

Swanson, T. and Barbier, E. B. (1992) *Economics for the Wilds: wildlife, wildlands, diversity and development*, Earthscan, London.

Swart, L. and Perry, E. (2007) (eds) *Global Environmental Governance: perspectives on the current debate*, Center for UN Reform, New York.

Swift, J. (1982) 'The future of African hunter-gatherer and pastoral people in Africa', *Development and Change* 13: 159–81.

Swift, J. (1996) 'Desertification narratives; winners and losers', pp. 73–90 in M. Leach and R. Mearns (eds), *The Lie of the Land: challenging received wisdom on the African environment*, James Currey/Heinemann, London.

Swinton, S. M. and Quiroz, R. (2003) 'Is poverty to blame for soil, pasture and forest degradation in Peru's Altiplano?', *World Development* 31: 1903–19.

Swinton, S. M., Escobar, G. and Reardon, T. (2003) 'Poverty and environment in Latin America: concepts, evidence and policy implications', *World Development* 31: 1865–72.

Swyngedouw, E. (1997) 'Power, nature, and the city: the conquest of water and the political ecology of urbanization in Guayaquil, Ecuador, 1880–1990', *Environment and Planning A* 29: 311–32.

Swyngedouw, E. and Heynen, N. (2003) 'Urban political ecology, justice and the politics of scale', *Antipode* 34: 898–918.

Tacconi, L. and Ruchiat, Y. (2006) 'Livelihoods, fire and policy in eastern Indonesia', *Singapore Journal of Tropical Geography* 27: 67–81.

Tacconi, L. and Seymour, F. (2007) *Illegal Logging: law enforcement, livelihoods and the timber trade*, Earthscan, London.

Tansley, A. G. (1911) *Types of British Vegetation*, Cambridge University Press, Cambridge.

Tansley, A. G. (1935) 'The use and abuse of vegetational terms', *Ecology* 14(3): 284–307.

Tansley, A. G. (1939) *The British Islands and their Vegetation*, Cambridge University Press, Cambridge.

Taylor, P. L. (2005) 'A Fair Trade approach to community forest certification? A framework for discussion', *Journal of Rural Studies* 21: 433–47.

Taylor, R. W. D. and Harris, A. H. (1994) 'Control of the larger grain border, *Prostephanus truncatus*, in bagged maize by fumigation under gas-proof sheets', *FAO Plant Protection Bulletin* 42: 129–37.

Terborgh, J. (1999) *Requiem for Nature*. Island Press, Washington.

Thatcher, M. (1988), speech to the Conservative Party Conference, 14 October, at www.margaretthatcher.org/speeches/, accessed 21 November 2007.

Thatcher, M. (2008), speech to the Royal Society, Fishmongers Hall, London, 27 September.

Thatte, C. D. (2001) 'Aftermath, overview and an appraisal of past events leading to some of the imbalances in the Report of the World Commission on Dams', *Water Resources Development* 17(3): 343–51.

Thomas, D. H. L. and Adams, W. M. (1997) 'Space, time and sustainability in the Hadejia-Jama'are wetlands and the Komodugu Yobe basin, Nigeria', *Transactions of the Institute of British Geographers* NS 22: 430–49.

Thomas, D. H. L. and Adams, W. M. (1999) 'Adapting to dams: agrarian change downstream of Tiga Dam, Northern Nigeria', *World Development* 27: 919–35.

Thomas, D. S. G. (1984) 'Ancient ergs of the former arid zones of Zimbabwe, Zambia and Angola', *Transactions of the Institute of British Geographers* NS 9: 75–88.

Thomas, D. S. G. and Middleton, T. (1994) *Desertification: exploding the myth*, Wiley, Chichester.

Thomas, K. (1983) *Man and the Natural World: changing attitudes in England, 1500–1800*, Allen Lane, London (paperback edn, Penguin, Harmondsworth, 1984).

Thomas, S. (1995) 'The next 1000 million people: do we have a choice?', pp. 187–206 in S. Morse and M. Stocking (eds), *People and Environment*, UCL Press, London.

Thomas, U. P. (2004) 'Trade and the environment: stuck in a political impasse at the WTO after the Doha and Cancun Ministerial Conferences', *Global Environmental Politics* 4(3): 9–21.

Thomas, W. L. (1956) (ed.) *Man's Role in Changing the Face of the Earth*, University of Chicago Press, Chicago.

Thompson, M. and Warburton, M. (1988) 'Uncertainty on a Himalayan scale', pp. 1–53 in J. Ives and D. C. Pitt (eds), *Deforestation: social dynamics in watersheds and mountain ecosystems*, Routledge, London.

Throup, D. W. (1987) *Economic and Social Origins of Mau Mau*, James Currey, London.

Tiffen, M. and Mortimore, M. (1994) 'Malthus controverted: the role of capital and technology in growth and environmental recovery in Kenya', *World Development* 22: 997–1010.

Tiffen, M., Mortimore, M. J. and Gichugi, F. (1994) *More People, Less Erosion: environmental recovery in Kenya*, Wiley, Chichester.

Timberlake, L. (1985) *Africa in Crisis: the causes, the cures of environmental bankruptcy*, Earthscan, London.

Timberlake, L. (2006) *Catalyzing Change: a short history of the WBCSD*, World Business Council for Sustainable Development, Geneva, Switzerland.

Timmer, V. and Juma, C. (2005) 'Taking root: biodiversity conservation and poverty reduction come together in the tropics', *Environment* 47(4): 24–44.

Tisdell, C. (1988) 'Sustainable development: differing perspectives of ecologists and economists, and relevance to LDCs', *World Development* 16: 373–84.

Tolba, M. K. (1986) 'Desertification in Africa', *Land Use Policy* 3: 260–8.

Toulmin, C. (1993) *Combating Desertification: setting the agenda for a global convention*, International Institute for Environment and Development, Dryland Networks Programme Paper 42, London.

Townsend, J. T. (1995) (with Arrevillaga, U., Bain, J., Cancino, S., Frenk, S. F., Pacheco, S. and Pérez, E.) *Women's Voices from the Rainforest*, Routledge, London.

Toye, J. ([1987] 1993) *Dilemmas of Development: reflections on the counter-revolution in development theory and practice*, Blackwell, Oxford (2nd edn).

Trapnell, C. G. (1943) *Soils, Vegetation and Agriculture of North-Eastern Rhodesia*, Government Printer, Lusaka.

Trapnell, C. G. and Clothier, J. N. (1937) *The Soils, Vegetation and Agricultural Systems of North-Western Rhodesia*, Government Printer, Lusaka.

Trumper, E. V. and Holt, J. (1998) 'Modeling pest population resurgence due to recolonisation of fields following an insecticide application', *Journal of Applied Ecology* 35: 273–85.

Tuan, N. Q. and Maclaren, V. W. (2005) 'Community concerns about landfills: a case study of Hanoi, Vietnam', *Journal of Environmental Planning and Management* 48: 809–31.

Tucker, C. J., Holben, B. N. and Goff, T. E. (1984) 'Intensive forest clearing in Rondônia, Brazil, as detected by satellite remote sensing', *Remote Sensing of the Environment* 15: 255–61.

Tucker, C. J., Justice, C. O. and Prince, S. D. (1986) 'Monitoring the grasslands of the Sahel, 1984–1985', *International Journal of Remote Sensing* 7: 1571–83.

Tucker, C. J., Townshend, J. E. G. and Goff, T. E. (1985) 'African land cover classification using satellite data', *Science* 277: 369–75.

Tumaneng-Diete, T., Ferguson, I. S. and MacLaren, D. (2005) 'Log export restrictions and trade policies in the Philippines: bane or blessing to sustainable forest management?' *Forest Policy and Economics* 7: 187–98.

Tuntamiroon, N. (1985) 'The environmental impact of industrialisation', *Ecologist* 15(4): 161–4.

Turnbull, C. (1974) *The Mountain People*, Pan Books, London.

Turner, B. L., II, and Meyer, W. B. (1994) 'Global land-use and land-cover change', pp. 3–10 in W. B. Meyer and B. L. Turner, II (eds), *Changes in Land Use and Land Cover: a global perspective*, Cambridge University Press, Cambridge.

Turner, D. J. (1971) 'Dams in ecology', *Civil Engineering* 41: 76–80.

Turner, M. (1993) 'Overstocking the range: a critical analysis of the environmental science of Sahelian pastoralism', *Economic Geography* 69: 402–21.

Turner, R. K. (1988a) 'Sustainability, resource conservation and pollution control: an overview', pp. 17–25 in R. K. Turner (ed.) *Sustainable Environmental Management: principles and practice*, Westview Press, Boulder, CO.

Turner, R. K. (1988b) (ed.) *Sustainable Environmental Management: principles and practice*, Westview Press, Boulder, CO.

Turner, R. K., Bateman, I. and Brooke, J. S. (1992) 'Valuing the benefits of coastal defence: a case study of the Aldeburgh sea-defence scheme', pp. 77–100 in A. Coker and C. Richards (eds), *Valuing the Environment: economic approaches to environmental valuation*, Belhaven, London.

Turton, D. (1987) 'The Mursi and National Park development in the lower Omo Valley', pp. 169–86 in D. M. Anderson and R. H. Grove (eds), *Conservation in Africa: people, policies and practice*, Cambridge University Press, Cambridge.

UNDP (1996) *Human Development Report 1996*, Oxford University Press, Oxford, for the United Nations Development Programme.

UNDP (2006) *Human Development Report 2006: Beyond Scarcity: power, poverty and the global water crisis*, Palgrave Macmillan, London, for the United Nations Development Programme.

UNDP (2007) *Human Development Report 2007/2008: Fighting Climate Change: human solidarity in a divided world*, Palgrave Macmillan, London, for the United Nations Development Programme.

UNEP (1978) *Review of Areas: environment and development and environmental management*, Report No. 3, UNEP Nairobi.

UNEP (1995) *Global Biodiversity Assessment*, Cambridge University Press, Cambridge.

UNEP (2000) *Global Environment Outlook 2000*, Earthscan, London, for the United Nations Environment Programme.

UNESCO (1963) *A Review of the Natural Resources of the African Continent*, UNESCO, Paris.

UNESCO (1973) *Programme on Man and the Biosphere (MAB). Expert Panel on Project 8: Conservation of natural areas and of the genetic material they contain. Final Report*, MAB Report 12, UNESCO, Paris.

UNESCO (1976) *Effects on Man and his Environment of Major Engineering Works*, Man and the Biosphere Report Series 37, UNESCO, Paris.

UNESCO (n.d.) 'Man and the Biosphere Programme', pamphlet, UNESCO, Paris.

UN-HABITAT (2006) *State of the World's Cities 2006/7: the Millennium Development Goals and urban sustainability*, Earthscan, London, for the United Nations Human Settlements Programme.

United Nations (1977) (ed.) *Desertification: its causes and consequences*, Pergamon Press, Oxford.

United Nations (1993) *The Global Partnership for Environment and Development: a guide to Agenda 21*, post-Rio edition, United Nations, New York.

United Nations Millennium Project (2005) *Environment and Human Well-Being: a practical strategy*, summary version of the report of the Task Force on Environmental Sustainability, The Earth Institute at Columbia University, New York.

Upton, C. and Bass, S. (1995) *The Forest Certification Handbook*, Earthscan, London.

Usher, A. D. (1997a) 'About this book, the contributors and what this book is not about', pp. 3–10 in A. D. Usher (ed.), *Dams as Aid: a political anatomy of Nordic development thinking*, Routledge, London.

Usher, A. D. (1997b) 'Kvaerner's game', pp. 133–52 in A. D. Usher (ed.), *Dams as Aid: a political anatomy of Nordic development thinking*, Routledge, London.

Usher, A. D. (1997c) 'The mechanism of pervasive appraisal optimism', pp. 59–75 in A. D. Usher (ed.), *Dams as Aid: a political anatomy of Nordic development thinking*, Routledge, London.

Usher, A. D. (1997d) (ed.) *Dams as Aid: a political anatomy of Nordic development thinking*, Routledge, London.

Usher, A. D. and Ryder, G. (1997) 'Vattenfal abroad: damming the Theun River', pp. 77–104 in A. D. Usher (ed.), *Dams as Aid: a political anatomy of Nordic development thinking*, Routledge, London.

Utting, P. (2002) (ed.) *The Greening of Business in Developing Countries: rhetoric, reality and prospects*, Zed Books, London, for the United Nations Research Institute for Social Development.

van Apeldoorn, G. J. (1980) *Perspectives on Drought and Famine in Nigeria*, Allen and Unwin, Hemel Hempstead.

Van Pelt, M. J. F., Kuyvenhoven, A. and Nijkamp, P. (1990) 'Project appraisal and sustainability: methodological challenges', *Project Appraisal* 5: 139–58.

Vanclay, F. (1999) 'Social impact assessment', pp. 301–26 in J. Petts (ed.) *Handbook of Environmental Impact Assessment*, Blackwell Science, Oxford.

Vanclay, F. and Bronstein, D. A. (1995) (eds) *Environmental and Social Impact Assessment*, Wiley, Chichester.

Vanden, H. E. (2003) 'Globalization in a time of neoliberalism: politicized social movements and the Latin American Response', *Journal of Development Studies* 19: 308–33.

Varma, R. and Varma, D. R. (2005) 'The Bhopal Disaster of 1984', *Bulletin of Science, Technology & Society*, 25: 37–45.

Veldman, M. (1994) *Fantasy, the Bomb and the Greening of Britain: romantic protest, 1945–1980*, Cambridge University Press, Cambridge.

Verstraete, M. M. (1986) 'Defining desertification: a review', *Climatic Change* 9: 5–18.

Vickers, W. T. (1984) 'Indian policy in Amazonian Ecuador', pp. 8–32 in M. Schmick and C. H. Wood (eds), *Frontier Expansion in Amazonia*, University of Florida Press, Gainesville.

Vira, B. (1999) 'Implementing joint forest management in the field: towards an understanding of the community-bureaucracy interface', pp. 254–75 in R. Jeffery and N. Sundar (eds), *A New Moral Economy for India's Forests?* Sage, New Delhi.

Vira, B. (2002) 'Trading with the enemy? Examining North-South perspectives in the climate change debate', pp. 164–80 in D. W. Bromley and J. Paavola (eds), *Economics, Ethics and Environmental Policy: contested choices*, Blackwell, Oxford.

Vitousek, P. M., Ehrlich, P. R., Ehrlich, A. H. and Matson, P. A. (1986) 'Human appropriation of the products of photosynthesis', *BioScience* 36: 368–73.

Vitousek, P. M., Mooney, H. A., Lubchenco, J. and Melillo, J. M. (1997) 'Human domination of earth's ecosystems', *Science* 25 July: 277 (5325): 494–9.

Vlachou, A. (2004) 'Capitalism and ecological sustainability: the shaping of environmental policies', *Review of International Political Economy* 11: 926–52.

Vogt, W. (1949) *Road to Survival*, Victor Gollancz, London.

Wade, R. (1982) 'The system of administrative and political corruption: canal irrigation in south India', *Journal of Development Studies* 18: 287–328.

Wainwright, C. and Wehrmeyer, W. (1998) 'Success in integrating conservation and development? A study from Zambia', *World Development* 26: 933–44.

Walls, J. (1984) 'Summons to action', *Desertification Control Bulletin* 10: 5–14.

Walsh, J. (1985) 'Onchocerciasis: river blindness', pp. 269–94 in A. T. Grove (ed.), *The Niger and its Neighbours: environment, history and hydrobiology, human use and health hazards of the major West African rivers*, Balkema, Rotterdam.

Walter, I. and Ugelow, J. L. (1979) 'Environmental problems in developing countries', *Ambio* 8(2–3): 102–9.

Ward, B. (1966) *Spaceship Earth*, University of Columbia Press, New York.

Ward, B. and Dubos, R. (1972) *Only One Earth: the care and maintenance of a small planet*, André Deutsch, London.

Ward, J. V. and Stanford, J. A. (1979) (eds) *The Ecology of Regulated Streams*, Plenum Press, New York.

Warren, A. (1993) 'Desertification as a global environmental issue', *GeoJournal* 31: 11–14.

Warren, A. (1996) 'Desertification', pp. 342–55 in W. M. Adams, A. S. Goudie and A. Orme (eds), *The Physical Geography of Africa*, Oxford University Press, Oxford.

Warren, A. and Khogali, M. (1992) *Assessment of Desertification and Drought in the Sudano-Sahelian Region, 1985–1991*, United Nations Sudano-Sahelian Office, New York.

Warren, A. and Maizels, J. K. (1977) 'Ecological change and desertification', pp. 169–261 in Secretariat of the United Nations Conference (eds), *Desertification: its causes and consequences*, Pergamon Press, Oxford.

Warren, K. J. (1994) (ed.) *Ecological Feminism*, Routledge, London.

Washbourn, C. (1967) 'Lake levels and Quaternary climates in the eastern Rift Valley of Kenya', *Nature* 216: 672–3.

Waterbury, J. (1979) *Hydropolitics of the Nile Valley*, Syracuse University Press, New York.

Wathern, P. (1988) (ed.) *Environmental Impact Assessment: theory and practice*, Unwin Hyman, London.

Watson, C. (1986) 'Working at the World Bank', pp. 268–75 in T. Hayter and C. Watson (eds) *Aid: rhetoric and reality*, Pluto Press, London.

Watts, M. J. (1983a) *Silent Violence: food, famine and peasantry in northern Nigeria*, University of California Press, Berkeley.

Watts, M. J. (1983b) 'On the poverty of theory: natural hazards research in context', pp. 231–62 in K. Hewitt (ed.), *Interpretations of Calamity from the Viewpoint of Human Ecology*, Allen and Unwin, London.

Watts, M. J. (1984) 'The demise of the moral economy: food and famine in the Sudano-Sahelian region in historical perspective', pp. 128–48 in E. P. Scott (ed.), *Life before the Drought*, Allen and Unwin, London.

Watts, M. J. (1987) 'Drought, environment and food security: some reflections on peasants, pastoralists and commoditisation in dryland East Africa', pp. 171–217 in M. H. Glantz (ed.), *Drought and Hunger in Africa: denying famine a future*, Cambridge University Press, Cambridge.

Watts, M. J. (1989) 'The agrarian question in Africa: debating the crisis', *Progress in Human Geography* 13: 1–14.

Watts, M. J. (1995) 'A new deal in emotions: theory, practice and the crisis of development', pp. 44–62 in J. Crush (ed.), *Power of Development*, Routledge, London.

Watts, M. J. and Bohle, H. G. (1993) 'The space of vulnerability: the causal structure of hunger and famine', *Progress in Human Geography* 17: 43–67.

Watts, M. and Peet. R. (2004) 'Liberating political ecology?', pp. 3–47 in R. Peet and M. Watts (eds), *Liberation Ecologies: environment, development, social movements*, Routledge, London (2nd edn).

WBCSD (2005) *Pathways to 2050: energy and climate change*, World Business Council for Sustainable Development, Geneva, Switzerland.

Weber, M. ([1922] 1978) *Economy and Society: an outline of interpretive sociology*, ed. G. Roth and C. Wittich, University of California Press, Berkeley.

Welcomme, R. L. (1979) *The Fisheries Ecology of Floodplain Rivers*, Longman, London.

Welford, R. and Starkey, R. (1996) *The Earthscan Reader in Business and the Environment*, Earthscan, London.

Wells, M. and Brandon, K. (1992) *People and Parks: linking protected area management with Local Communities*, World Bank, Washington.

West, P. and Brechin, S. R. (1991) *Resident People and National Parks: social dimensions in international conservation*, University of Arizona Press, Tucson.

Western, D. and Wright, R. M. (1994) 'The background to community-based conservation', pp. 1–14 in D. Western, R. M. White and S. C. Strumm (eds), *Natural Connections: perspectives in community-based conservation*, Island Press, Washington.

Western, D., White, R. M. and Strum, S. C. (eds) (1994) *Natural Connections: perspectives in community-based conservation*, Island Press, Washington.

White, S. C. (1996) 'Depoliticising the environment: the uses and abuses of participation', *Development in Practice* 6: 6–15.

Whiteside, K. (2002) *Divided Natures: French contributions to political ecology*, MIT Press, Cambridge, MA.

Wiener, M. J. (1981) *English Culture and the Decline of the Industrial Spirit, 1850–1980*, Cambridge University Press, Cambridge.

Wiggins, S. (1995) 'Change in African farming systems between the mid-1970s and the mid-1980s', *Journal of International Development* 7: 807–48.

Williams, G. (1976) 'Taking the part of the peasants: rural development in Nigeria and Tanzania', pp. 131–54 in P. Gutkind and I. Wallerstein (eds), *The Political Economy of Contemporary Africa*, Sage, Beverly Hills, CA.

Williams, G. (1981) 'The World Bank and the peasant problem', pp. 16–51 in J. Heyer, P. Roberts and G. Williams (eds), *Rural Development in Tropical Africa*, Macmillan, London.

Williams, M. (1989) 'Deforestation: past and present', *Progress in Human Geography* 13: 176–208.

Williams, M. (1994) 'Forests and tree cover', pp. 97–124 in W. B. Meyer and B. L. Turner II (eds), *Changes in Land Use and Land Cover: a global perspective*, Cambridge University Press, Cambridge.

Williams, M. (2003) *Deforesting the Earth*, University of Chicago Press, Chicago.

Williams, M. A. J. (1985) 'Pleistocene aridity in tropical Africa, Australia and Asia', pp. 219–38 in I. Douglas and T. Spencer (eds), *Environmental Change and Tropical Geomorphology*, Allen and Unwin, London.

Williams, M. A. J. and Balling, R. C., Jr (1995) 'Interactions of desertification and climate: an overview', *Desertification Control Bulletin* 26: 8–16.

Willis, K. (2004) *Theories and Practices of Development*, Routledge, London.

Wilshusen, P. R., Brechin, S. R., Fortwangler, C. L. and West, P. C. (2002) 'Reinventing a square wheel: critique of a resurgent "protection paradigm" in international biodiversity conservation', *Society and Natural Resources* 15: 17–40.

Wilson, A. (1992) *The Culture of Nature: North American landscape from Disney to the 'Exxon Valdez'*, Blackwell, Oxford.

Wilson, C. and Tisdell, C. (2001) 'Why farmers continue to use pesticides despite environmental, health and sustainability costs', *Ecological Economics* 39: 449–62.

Wilson, E. O. (1992) *The Diversity of Life*, Penguin, Harmondsworth.

Winid, B. (1981) 'Comments on the development of the Awash Valley, Ethiopia', pp. 147–65 in S. K. Saha and C. J. Barrow (eds), *River Basin Planning: theory and practice*, Wiley, Chichester.

Winter, G. (2003) 'The GATT and Environmental Protection: Problems of Construction', *Journal of Environmental Law* 15: 113–40.

Wissenburg, M. (1993) 'The idea of nature and the nature of distributive justice', pp. 3–20 in A. Dobson and P. Lucardie (eds), *The Politics of Nature: explorations in Green political theory*, Routledge, London.

Withers, C. W. J. (1999) 'Geography, enlightenment and the paradise question', pp. 67–92 in D. N. Livingstone and C. W. J. Withers, *Geography and Enlightenment*, Chicago University Press, Chicago.

Wolfensohn, J. D. and Bourguignon, F. (2004) *Development and Poverty Reduction: looking back, looking ahead*, World Bank, Washington.

Wolmer, W. (2003) 'Transboundary conservation: the politics of ecological integrity in the Great Limpopo Transfrontier Park', *Journal of Southern African Studies* 29: 1: 261–78.

Wolmer, W. (2007) *From Wilderness Vision to Farm Invasions: conservation and development in Zimbabwe's South-East Lowveld*, James Currey, Oxford.

Wood, A. (1950) *The Groundnut Affair*, Bodley Head, London.

Wood, W. B. (1990) 'Tropical deforestation: balancing regional development demands and global environmental concerns', *Global Environmental Change* 1: 23–41.

Woodhouse, P., Bernstein, H. and Hulme, D. (2000) *African Enclosures? The social dynamics of wetlands in drylands*, James Currey, Oxford.

Woodroffe, R., Thirgood, S. and Rabinowitz, A. (2005) (eds) *People and Wildlife: conflict or coexistence?* Cambridge University Press, Cambridge.

World Bank (1981) *Accelerated Development in Sub-Saharan Africa*, World Bank, Washington.

World Bank (1984a) *Tribal Peoples and Economic Development: human ecologic considerations*, World Bank, Washington.

World Bank (1984b) *World Development Report 1984*, Oxford University Press, Oxford and New York.

World Bank (1984c) *Environmental Policies and Procedures of the World Bank*, Office of Environmental and Health Affairs, World Bank, Washington.

World Bank (1992) *Development and the Environment: World Development Report 1992*, Oxford University Press, Oxford, for the World Bank.

World Bank (1996) *From Plan to Market: World Development Report 1996*, Oxford University Press, Oxford, for the World Bank.

World Bank (2000) *Entering the 21st Century: World Development Report 1999/2000*, Oxford University Press, Oxford, for the World Bank.

World Bank (2001) *Attacking Poverty: World Development Report 2000/1*, Oxford University Press for the World Bank, Oxford and New York.

World Commission on Dams (2000) *Dams and Development: a new framework for decision-making*, Earthscan, London.

World Commission on Environment and Development (1987) *Our Common Future*, Oxford University Press, Oxford.

World Conservation Union (IUCN) (2005) *Benefits beyond Boundaries. Proceedings of the Vth World Parks Congress*, World Conservation Union, Cambridge.

Worster, D. (1985) *Nature's Economy: a history of ecological ideas*, Cambridge University Press, Cambridge.

Worster, D. (1993) *The Wealth of Nature: environmental history and the ecological imagination*, Oxford University Press, Oxford.

Worthington, E. B. (1938) *Science in Africa: a review of scientific research relating to tropical and Southern Africa*, Royal Institute of International Affairs, London.

Worthington, E. B. (1958) *Science in the Development of Africa: a review of the contribution of physical and biological knowledge south of the Sahara*, Commission for Technical Cooperation in Africa South of the Sahara and the Scientific Council for Africa South of the Sahara.

Worthington, E. B. (1975) (ed.) *The Evolution of the IBP*, Cambridge University Press, Cambridge.

Worthington, E. B. (1982) 'World campaign for the biosphere', *Environmental Conservation* 9: 93–100.

Worthington, E. B. (1983) *The Ecological Century: a personal appraisal*, Cambridge University Press, Cambridge.

WRI, IUCN, UNEP, in consultation with the FAO and UNESCO (1992) *Global Biodiversity Strategy: guidelines for action to save, study and use Earth's biotic wealth sustainably and equitably*, World Resources Institute, Washington.

WRI, UNEP, UNDP and the World Bank (1996) *World Resources 1996–97: a guide to the global environment*, Oxford University Press, Oxford.

Wright, S. and Nelson, S. (1995) *Power and Participatory Development: theory and practice*, IT Books, London.

Wu, B. and Ci, L. J. (2002) 'Landscape change and desertification development in the Mu Us Sandland, Northern China', *Journal of Arid Environment* 50: 429–44.

WWF (1988) 'Debt-for-nature swap in the Philippines', *WWF News* July/August: 6.

WWF (1991) *Tropical Forest Conservation*, World Wide Fund for Nature Position, Paper 7, Gland, Switzerland.

WWF (1996) *The WWF 1995 Group: the full story*, World Wide Fund for Nature UK, Godalming, Surrey.

Wynne, B. (1992) 'Uncertainty in environmental learning: reconceiving science and policy in the preventative paradigm', *Global Environmental Change* 2: 111–27.

Yapa, L. (1996) 'Improved seeds and constructed scarcity', pp. 69–85 in R. Peet and M. Watts (eds), *Liberation Ecologies: environment, development and social movements*, Routledge, London (1st edn).

York, R. and Rosa, E. A. (2003) 'Key challenges to ecological modernization theory', *Organization & Environment*, 16: 273–88.

Yuefang, D. and Steil, S. (2003) 'China Three Gorges Project Resettlement: policy, planning and implementation', *Journal of Refugee Studies* 16(4): 422–43.

Yuqian, L. and Qishun, Z. (1981) 'Sediment regulation problems in Sanmenxia Reservoir', *Water Supply and Management* 5: 351–60.

Zeidler, J. and Mulongoy, K. J. (2003) 'The Dry and Sub-humid Lands Programme of Work of the Convention on Biological Diversity: connecting the CBD and the UN Convention to Combat Desertification', *Review of European Community and International Environmental Law* 12: 164–75.

Zhao, D. and Sun, B. (1986) 'Air pollution and acid rain in China', *Ambio* 15(1): 2–5.

Zimmerer, K. (2004) 'Soil degradation in Bolivia', pp. 107–24 in R. Peet and M. Watts (eds), *Liberation Ecologies: environment, development, social movements*, Routledge, London (2nd edn).

Zimmerer, K. S. and Bassett, T. J. (2003) (eds) *Political Ecology: an integrative approach to geography and environment-development studies*, Guilford, New York.

Zimmerer, K. S. and Young, K. R. (1998) (eds) *Nature's Geography: new lessons for conservation in developing countries*, University of Wisconsin Press, Madison.

Index